Basic Mathematics

SEVENTH EDITION

Marvin L. Bittinger
Indiana University—Purdue University at Indianapolis

Mervin L. Keedy
Purdue University

ADDISON-WESLEY PUBLISHING COMPANY

Reading, Massachusetts ● *Menlo Park, California* ● *New York*
Don Mills, Ontario ● *Wokingham, England* ● *Amsterdam*
Bonn ● *Sydney* ● *Singapore* ● *Tokyo* ● *Madrid*
San Juan ● *Milan* ● *Paris*

Sponsoring Editor	Jason A. Jordan
Special Projects Editor	Susan Gleason
Managing Editor	Karen Guardino
Production Supervisor	Jack Casteel
Text Design	Bruce Kortebein, The Design Office
Art, Design, Editorial, and Production Services	Geri Davis and Martha Morong, Quadrata, Inc.
Art Consultant	Joseph Vetere
Electronic Illustration	Scientific Illustrators, Precision Graphics, and Monotype, Inc.
Manufacturing Manager	Roy Logan
Cover Design Director	Peter Blaiwas
Cover Design	Marshall Henrichs
Composition	Beacon Graphics Corporation
Printer	Banta Company

Photo Credits

1, © Walter Hodges, Westlight **4,** Pictor Uniphoto, Inc. **79,** Pictor Uniphoto, Inc. **104,** © John Coletti, Tony Stone Images, Inc. **122,** Wide World Photos, Inc. **126,** Professional Bowling Association (PBA) **133,** DPI, Uniphoto, Inc. **140,** NASA **171,** © Scott Markewitz, 1991, FPG International **189,** © Lawrence Manning, Westlight **215,** © Stacy Pic, 1984, Stock Boston **223,** Wide World Photos, Inc. **250,** © Steve Dunwell, The Image Bank **261,** Wide World Photos, Inc. **263,** Rick Haston, Latent Images **271,** Comstock **290,** Wide World Photos, Inc. **291,** © Jim Zuckerman, Westlight **292 (left and right),** Wide World Photos, Inc. **299,** Wide World Photos, Inc. **307,** Professional Bowlers Association (PBA) **309,** © Daemmrich, Stock Boston **352,** Third Coast Stock Source, Inc. **369,** Westlight © Tecmap **377,** Wide World Photos, Inc. **403,** Pictor Uniphoto **417,** Westlight © Nik Wheeler **430,** © Owen Franken, Stock Boston **463,** Dennis O'Clair, © Tony Stone Worldwide **473,** Al Satterwhite, The Image Bank **494,** Tom Walker, Stock Boston **509,** Comstock **547,** Comstock **597,** © J. Pickerell, Westlight

Library of Congress Cataloging-in-Publication Data

Bittinger, Marvin L.
 Basic mathematics/Marvin L. Bittinger, Mervin L. Keedy.—7th ed.
     p. cm.
 Keedy's name appears first on the previous ed.
 ISBN 0-201-59560-5
   1. Arithmetic. I. Keedy, Mervin Laverne. II. Title.
QA107.K43 1994
513' .1—dc20   94-17321
                                                         CIP

1 2 3 4 5 6 7 8 9 10—BAM—97969594

Contents

Preface

This text is appropriate for a one-term course in arithmetic or prealgebra. It is the first in a series of texts that includes the following:

Bittinger/Keedy: *Basic Mathematics,* Seventh Edition,

Bittinger/Keedy: *Introductory Algebra,* Seventh Edition,

Bittinger/Keedy: *Intermediate Algebra,* Seventh Edition.

What's New in the Seventh Edition?

Basic Mathematics, Seventh Edition, is a significant revision of the Sixth Edition, especially with respect to design, an all-new art program, pedagogy, and an enhanced supplements package. Its unique approach, which has been developed and refined over many years, is designed to help students both learn *and* retain mathematical skills. It is our belief that the Seventh Edition will *continue* to help today's students through pedagogical use of full color and updated applications. As part of *MathMax: The Bittinger/Keedy System of Instruction,* it is accompanied by an extremely comprehensive and well-integrated supplements package to provide maximum support for both instructor and student.

The style, format, and approach of the Sixth Edition have been strengthened in this new edition in a number of ways.

USE OF COLOR   The text is now printed in an extremely functional use of full color, evident in striking new design elements and artwork on nearly every page of the text. The use of color has been carried out in a methodical and precise manner so that its use carries a consistent meaning, which enhances the readability of the text for both student and instructor.

For example, when perimeter is considered, figures have a red border to emphasize the perimeter. When area is considered, figures are outlined in black and screened with peach to emphasize the area. Similarly, when volume is considered, figures are three-dimensional, air-brushed blue.

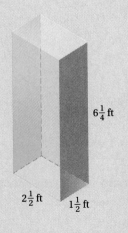

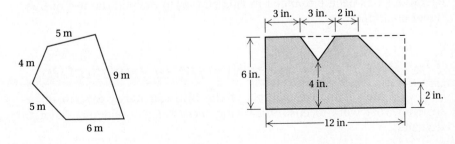

NEW ART PROGRAM   All art in both exposition and the answer section is new. The use of full color in the art program greatly enhances the learning process.

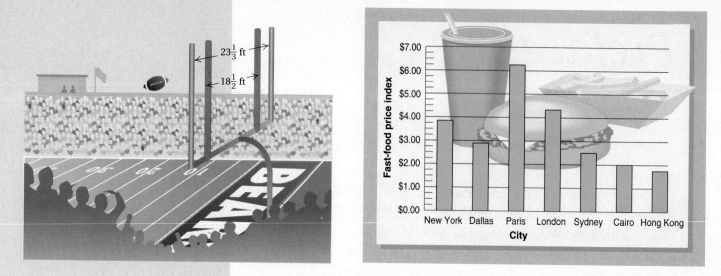

UPDATED APPLICATIONS   Extensive research has been done to make the Seventh Edition's applications even more up-to-date and realistic. Not only are 20% of the exercises new to this edition, but many are drawn from the fields of business and economics, life and physical sciences, social sciences, and areas of general interest such as sports and daily life. To encourage students to "see mathematics" around them every day, many applications use graphs and drawings similar to those found in today's newspapers and magazines.

CRITICAL THINKING   Each chapter now ends with a set of Critical Thinking exercises, which includes Calculator Connections, Extended Synthesis Exercises, and Exercises for Thinking and Writing (all described more fully below).

THE SCIENTIFIC CALCULATOR   Instruction and exercises for the scientific calculator are now covered in many different locations. A calculator icon ▦ highlights exercises that lend themselves to practice with a calculator. Calculator Connection exercises occur in the Critical Thinking sections at the end of each chapter. (These exercises occasionally cover calculator procedures, but for the most part provide critical thinking exercises using a calculator.) Last, there is a special new appendix on calculator keystroke instruction, prepared with the assistance of Rheta Beaver of Valencia Community College.

EXPANDED TREATMENT OF ORDER OF OPERATIONS   The coverage of order of operations has been enhanced in this edition to include the use of fractional notation.

The Bittinger/Keedy System of Instruction

Following are distinctive features of the Bittinger/Keedy System of Instruction that work to ensure learning success for developmental math students.

CAREFUL DEVELOPMENT OF CONCEPTS We have divided each section into discrete and manageable learning objectives. Within the presentation of each objective, there is a careful buildup of understanding through a series of developmental examples. These examples enable students to thoroughly understand the mathematical concepts involved at each step.

FOCUS ON UNDERSTANDING Throughout the text, we present the appropriate mathematical rationale for a topic, rather than simply listing rules and procedures. For example, when simplifying fractional notation, we remove factors of 1 rather than cancel (although cancellation is mentioned with appropriate cautions). This method helps prevent student errors in simplification. The notion of multiplying by 1 is a theme that is carried out throughout the book and the series, providing a rationale for many other procedures, such as the conversion of mixed numerals to fractional notation and the development of the cross-product method for comparing fractions, as well as the understanding behind unit conversions in Chapters 9 and 10.

As another example, finding the least common multiple (LCM) is developed sequentially, through a series of methods, rather than through the presentation of just one rule. Also, the understanding of rounding decimal notation is enhanced by using the number line.

PROBLEM SOLVING We include real-life applications and problem-solving techniques to motivate students and encourage them to think about how mathematics can be used in their everyday life. The basis for problem solving is a five-step process (*Familiarize, Translate, Solve, Check,* and *State*) established early in the text and used henceforth.

Learning Aids

INTERACTIVE WORKTEXT APPROACH The pedagogy of this text is designed as an interaction between the student and the exposition, annotated examples, margin exercises, and exercise sets. This approach provides students with a clear set of learning objectives, involves them with the development of the material, and provides immediate and continual reinforcement.

Section objectives are keyed by letter not only to appropriate objectives of the section, but also to exercises in the exercise sets and answers to review exercises and test questions, so that students can easily find appropriate review material if they are unable to work a particular exercise.

Numerous *margin exercises* throughout the text provide immediate reinforcement of the concepts covered in each section.

FOR EXTRA HELP Many valuable study aids accompany this text. Below each list of section objectives are references to appropriate videotape, audiotape, and tutorial software programs, to make it easy for the student to find the correct support materials. The text exercises that appear on the videotapes are listed in an index at the back of the text as well as in the Instructor's Resource Guide.

EXERCISE SETS The exercises are paired, meaning that each even-numbered exercise is very much like the odd-numbered one that precedes it. Answers to the odd-numbered exercises are given at the back of the book, whereas those for the even-numbered exercises are not. This provides the instructor with many options. If an instructor wants the student to have answers, the odds are assigned. If an instructor wants the student to be able to practice (as on a test) with no answers, the evens are assigned. Thus each exercise set serves as two exercise sets. If an instructor wants the student to have all the answers, a complete answer book is available.

OPPORTUNITIES FOR CRITICAL THINKING In response to the recommendations of both instructors and educational organizations, we provide many opportunities for students to synthesize concepts, verbalize mathematics, and think critically.

Synthesis Exercises at the end of most exercise sets require students to synthesize learning objectives from the section being studied and preceding sections in the book.

Critical Thinking exercise sets occur at the end of each chapter. These exercise sets provide further opportunity for critical thinking by providing three types of exercises:

- *Calculator Connections* review keystrokes and provide exercises for a scientific calculator.

- *Extended Synthesis Exercises* call for students to further synthesize objectives from the chapter being studied and preceding chapters, thereby building critical-thinking skills.

- *Exercises for Thinking and Writing* encourage students to both think and write about key mathematical ideas in the chapter.

SKILL MAINTENANCE A well-received feature of preceding editions, the Skill Maintenance exercises have been enhanced by the inclusion of 50% more exercises in this edition. They occur at the end of most exercise sets. Although these exercises can review any objective of preceding chapters, they tend to focus on four specific objectives, called *Objectives for Review.* These objectives are listed at the beginning of each chapter and are covered in each *Summary and Review* and *Chapter Test* at the end of each chapter. The Objectives for Review are also included in a consistent manner in the Printed Test Bank that the instructor uses for testing.

The *Summary and Review* at the end of each chapter provides an extensive set of review exercises along with a list of important formulas and properties covered in that chapter.

We also include a *Cumulative Review* at the end of each chapter but the first, which reviews material from all preceding chapters.

At the back of the text are answers to all end-of-chapter review exercises, together with section and objective references, so that students know exactly what material to restudy if they miss a review exercise.

TESTING The following assessment opportunities exist in the text.

The *Diagnostic Pretest,* provided at the beginning of the text, can place students in the appropriate chapter for their skill level by identifying familiar material and specific trouble areas.

Chapter Pretests can then be used to place students in a specific section of the chapter, allowing them to concentrate on topics with which they have particular difficulty.

Chapter Tests allow students to review and test comprehension of chapter skills, as well as the four Objectives for Review from earlier chapters. Answers to all Chapter Test questions are found at the back of the book, along with appropriate section and objective references.

Supplements for the Instructor

TEACHER'S EDITION

The Teacher's Edition is a specially bound version of the student text with answers to all exercises in the margins, the exercise sets, and the chapter tests printed in a special color. It also includes answers to all the Critical Thinking exercises at the back of the text.

INSTRUCTOR'S SOLUTIONS MANUAL

The Instructor's Solutions Manual by Judith A. Penna contains brief worked-out solutions to all even-numbered exercises in the exercise sets.

INSTRUCTOR'S RESOURCE GUIDE

The Instructor's Resource Guide contains the following:

- Conversion Guide.
- Extra practice exercises (with answers) for some of the most difficult topics in the text.
- Answers to the Critical Thinking exercises.
- Number lines and grids that can be used as transparency masters for teaching aids and for test preparation.
- Indexes to the videotapes, audiotapes, and tutorial software that accompany the text.
- Instructions for using the Math Hotline.
- Essays on setting up learning labs and testing centers, together with a directory of learning lab coordinators who are available to answer questions.

PRINTED TEST BANK

Prepared by Donna DeSpain, the Printed Test Bank is an extensive collection of alternate chapter test forms, including the following:

- 4 alternate test forms for each chapter, with questions in the same topic order as the objectives presented in the chapter.
- 4 alternate test forms for each chapter, modeled after the Chapter Tests in the text.
- 3 alternate test forms for each chapter, designed for a 50-minute class period.
- 2 multiple-choice test forms for each chapter.
- 2 cumulative review tests for each chapter (with the exception of Chapter 1).
- 8 alternate forms of the final examination, 3 with questions organized by chapter, 3 with questions scrambled, as in the cumulative reviews, and 2 with multiple-choice questions.

ANSWER BOOK

The Answer Book contains answers to all exercises in the exercise sets in the text. Instructors may make quick reference to all answers or have quantities of these booklets made available for sale if they want students to have all the answers.

COMPUTERIZED TESTING: OMNITEST[3]

Addison-Wesley's algorithm-driven computerized testing system for Macintosh and DOS computers features a brand-new graphical user interface for the DOS version and a substantial increase in the number of test items available for each chapter of the text.

The new graphical user interface for DOS is a Windows look-alike. It allows users to choose items by test item number or by reviewing all the test items available for a specific text objective. Users can choose the exact iteration of the test item they wish to have on their test or allow the computer to generate iterations for them. Users can also preview all the items for a test on screen and make changes to them during the preview process. They can control the format of the test, including the appearance of the test header, the spacing between items, and the layout of the test and the answer sheet. In addition, users can now save the exact form of the test they have created so that they can modify it for later use. Users can also enter their own items using Omnitest[3]'s WYSIWYG editor, and have access as well to a library of preloaded graphics.

Both the DOS and Macintosh versions of Omnitest[3] for *Basic Mathematics* contain over 2000 items; 1000 of these are algorithm-driven—capable of generating hundreds of alternative versions. Omnitest[3] for *Basic Mathematics* features at least one algorithm-driven multiple-choice and free-response item for *each* text objective, as well as a selection of static items. Many objectives are covered by *several* multiple-choice and free-response algorithm-based items—the coverage is comparable to the exercise coverage in the text's Summary and Review sections. Each chapter also includes a selection of Thinking and Writing questions.

Omnitest[3] also includes preloaded chapter tests, cumulative tests, and tests designed to parallel state competency examinations to make building your own tests easier than ever before!

COURSE MANAGEMENT AND TESTING SYSTEM

InterAct Math Plus for Windows (available from Addison-Wesley) combines course management and on-line testing with the features of the basic tutorial software (see "Supplements for the Student") to create an invaluable teaching resource. Consult your Addison-Wesley representative for details.

Supplements for the Student

STUDENT'S SOLUTIONS MANUAL

The Student's Solutions Manual by Judith A. Penna contains completely worked-out solutions with step-by-step annotations for all the odd-numbered exercises in the exercise sets in the text. It may be purchased by your students from Addison-Wesley Publishing Company.

"MATH MAKES A DIFFERENCE" VIDEOTAPES

"Math Makes a Difference" is new to this edition of *Basic Mathematics*. It is a complete revision of the existing series of videotapes, based on extensive input from both students and instructors. "Math Makes a Difference" features a team of mathematics teachers who present comprehensive coverage of each section of the text:

Marvin Bittinger, *Indiana University—Purdue University at Indianapolis*

Carilynn Bouie, *Chattanooga State Technical Community College*

Michael Butler, *College of the Redwoods*

Patricia Cleary, *University of Delaware*

Bettyann Daley, *University of Delaware*

Barbara Johnson, *Indiana University—Purdue University at Indianapolis*

Joanne Peeples, *El Paso Community College*

Anita Polk-Conley, *Chattanooga State Technical Community College*

Since the format is a lecture to a group of students, each videotape is interactive and engaging. Lecturers use odd-numbered exercises from the text as examples—these are listed in the videotape indexes in the Instructor's Resource Guide and at the back of the text. Icons ▣ at the beginning of each section reference the appropriate videotape number.

A complete set of "Math Makes a Difference" videotapes is free to qualifying adopters.

AUDIOTAPES

The audiotapes are designed to lead students through the material in each text section. Bill Saler explains solution steps to examples, cautions students about common errors, and instructs them at certain points to stop the tape and do exercises in the margin. He then reviews the margin exercise solutions, pointing out potential errors. Icons ⌒ at the beginning of each section reference the appropriate audiotape number.

The audiotapes are free to qualifying adopters.

THE MATH HOTLINE

Prepared by Larry A. Bittinger, the Math Hotline is open 24 hours a day at 1-800-333-4227 so that students can obtain detailed hints for exercises. Exercises covered include all the odd-numbered exercises in the exercise sets, with the exception of the Skill Maintenance and Synthesis exercises.

INTERACT MATH TUTORIAL SOFTWARE

InterAct Math Tutorial Software, new to this edition of *Basic Mathematics*, has been developed and designed by professional software engineers working closely with a team of experienced developmental math teachers.

InterAct Math Tutorial Software includes exercises that are linked one-to-one with the odd-numbered exercises in the textbook and require the same computational and problem-solving skills as their companion exercises in the text. Each exercise has an example and an interactive guided solution that are designed to involve students in the solution process and to help them identify precisely where they are having trouble. In addition, the software recognizes common student errors and provides students with appropriate customized feedback.

With its sophisticated answer recognition capabilities, InterAct Math Tutorial Software recognizes appropriate forms of the same answer for any kind of input. It also tracks student activity and scores for each section, which can then be printed out. Icons at the beginning of each section 🖫 reference the appropriate disk number.

Available for both DOS-based and Macintosh computers, the software is free to qualifying adopters.

We, your authors, have committed ourselves to writing a usable, understandable, accomplishable, error-free book that will extend the student's knowledge and enjoyment of mathematics. Students and instructors will undoubtedly have many general impressions and attitudes that form during their semester or two in a mathematics course. To help us to continually improve the text and to support the instructor's goals, we invite correspondence from both students and instructors to:

Marv Bittinger and Mike Keedy
c/o Marv Bittinger
3011 Whispering Trail
Carmel, IN 46033

Acknowledgments

Many have helped to mold the Seventh Edition by reviewing, answering surveys, participating in focus groups, filling out questionnaires, and spending time with us on their campuses. We owe a special debt of gratitude to the InterAct Math writers and reviewers for their persistence and pursuit of quality. Our deepest appreciation to all of you and in particular to the following:

Michele Bach, *Kansas City, Kansas Community College*

Rheta Beaver, *Valencia Community College*

Carole Bergen, *Mercy College*

Louise Bernauer, *County College of Morris*

Joan Bookbinder, *Johnson & Wales University*

Jacquelyn P. Briley, *Guilford Technical Community College*

Mary-Jean Brod, *The University of Montana*

Mary Cabral, *Middlesex Community College*

Linda Crabtree, *Longview Community College*

Barbara Davis, *Hillsborough Community College—Dale Mabry Campus*

Gudryn Doherty, *Community College of Denver*

Rita Donnelly, *Indiana Vocational Technical College—South Bend*

Janice Gahan-Rech, *University of Nebraska at Omaha*

Roberta Gansman, *Guilford Technical Community College*

John Garlow, *Tarrant County Junior College*

Becky Harman, *Umpqua Community College*

Judy Holcomb, *Florida Community College at Jacksonville*

Jim Homewood, *Pima County Community College*

Phyllis A. Jore, *Valencia Community College, East Campus*

Robert Kaiden, *Lorain County Community College*

Gerald Krusinski, *College of Du Page*

Bob Malena, *Community College of Allegheny County, South*

Carol Meitler, *Concordia University*

Gail T. Mericle, *Mankato State University*

Jane Pinnow, *University of Wisconsin, Parkside*

Claudinna Rowley, *Community College of Allegheny County*

Randolph J. Taylor, *Las Positas College*

Victor E. Thomas, *Holyoke Community College*

Linda Verceles, *Edison State Community College*

Lorena Wolff, *Westark Community College*

Judith B. Wood, *Central Florida Community College*

We also wish to thank the following people who reviewed the videotapes:

Kathleen Bavelas, *Manchester Community—Technical College*

Julane B. Crabtree, *Johnson County Community College*

Helen Hancock, *Shoreline Community College*

Marti Hidden, *Sacramento City College*

Louis Levy, *Northland Pioneer College*

Aimee Martin, *Amarillo College*

Letty Ann Macdonald, *Piedmont Virginia Community College*

Bill Thieman, *Ventura College*

Cindie Wade, *St. Clair County Community College*

We also wish to thank many people on the Bittinger/Keedy team at Addison-Wesley for the endless hours of hard work and unwavering support. The editorial, design, and production coordination of Quadrata, Inc., was exceptional as always—we especially appreciate their wholehearted dedication and hard work over the many years of our association.

In particular, we thank Judy Beecher for her editorial assistance, without which these books would not exist. Judy Penna has always provided steadfast, quality leadership in the preparation of the solutions manuals and the supervision of all printed supplements. We also gratefully acknowledge a strong supporting cast: Donna DeSpain, for the printed test banks; Bill Saler, for the audiotapes; Larry A. Bittinger, for the Math Hotline; and Larry Bittinger, Linda Collins, Lisa Ford, and Barbara Johnson, for their usual fine quality in the proofreading and pursuit of errors.

M.L.B.
M.L.K.

The Steps to Success

The following six pages show you how to use *Basic Mathematics* to maximize understanding while making studying easier.

Use the chapter opener to begin your work with the chapter.

1

The chapter introduction provides an overall view of the chapter's content.

Read the Objectives for Review, which list four objectives from preceding chapters that will be reinforced for skill maintenance in this chapter and its test.

2

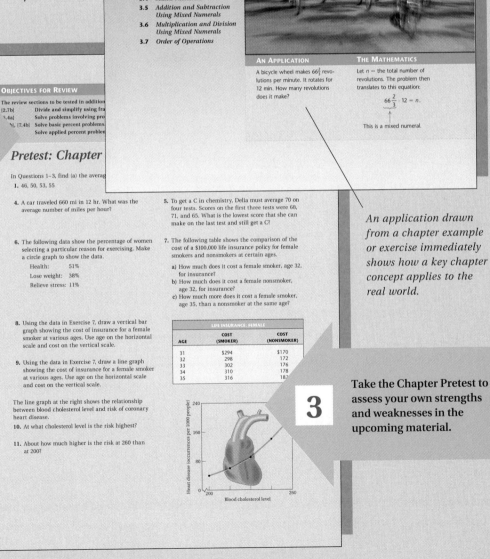

3
Addition and Subtraction: Fractional Notation

INTRODUCTION

In this chapter, we consider addition and subtraction using fractional notation. Also discussed are addition, subtraction, multiplication, and division using mixed numerals. We then apply all these operations to the use of rules for order of operations and to problem solving.

3.1 *Least Common Multiples*
3.2 *Addition*
3.3 *Subtraction and Order*
3.4 *Mixed Numerals*
3.5 *Addition and Subtraction Using Mixed Numerals*
3.6 *Multiplication and Division Using Mixed Numerals*
3.7 *Order of Operations*

AN APPLICATION

A bicycle wheel makes $66\frac{2}{3}$ revolutions per minute. It rotates for 12 min. How many revolutions does it make?

THE MATHEMATICS

Let n = the total number of revolutions. The problem then translates to this equation:

$$66\frac{2}{3} \cdot 12 = n.$$

This is a mixed numeral.

An application drawn from a chapter example or exercise immediately shows how a key chapter concept applies to the real world.

OBJECTIVES FOR REVIEW

The review sections to be tested in addition
[2.7b] Divide and simplify using fra
\.4a] Solve problems involving pro
\], [7.4b] Solve basic percent problems.
Solve applied percent problem

Pretest: Chapter

In Questions 1–3, find (a) the averag
1. 46, 50, 53, 55

4. A car traveled 660 mi in 12 hr. What was the average number of miles per hour?

5. To get a C in chemistry, Delia must average 70 on four tests. Scores on the first three tests were 68, 71, and 65. What is the lowest score that she can make on the last test and still get a C?

6. The following data show the percentage of women selecting a particular reason for exercising. Make a circle graph to show the data.

   Health: 51%
   Lose weight: 38%
   Relieve stress: 11%

7. The following table shows the comparison of the cost of a $100,000 life insurance policy for female smokers and nonsmokers at certain ages.

  a) How much does it cost a female smoker, age 32, for insurance?
  b) How much does it cost a female nonsmoker, age 32, for insurance?
  c) How much more does it cost a female smoker, age 35, than a nonsmoker at the same age?

8. Using the data in Exercise 7, draw a vertical bar graph showing the cost of insurance for a female smoker at various ages. Use age on the horizontal scale and cost on the vertical scale.

9. Using the data in Exercise 7, draw a line graph showing the cost of insurance for a female smoker at various ages. Use age on the horizontal scale and cost on the vertical scale.

LIFE INSURANCE: FEMALE		
AGE	COST (SMOKER)	COST (NONSMOKER)
31	$294	$170
32	298	172
33	302	176
34	310	178
35	316	18?

The line graph at the right shows the relationship between blood cholesterol level and risk of coronary heart disease.

10. At what cholesterol level is the risk highest?

11. About how much higher is the risk at 260 than at 200?

Take the Chapter Pretest to assess your own strengths and weaknesses in the upcoming material.

3

Each section of the chapter is designed as an interaction between you and the written explanations, the annotated examples, the margin exercises, and the exercise set.

Read the objectives listed in the margin and keyed to the text.

Objectives provide an instant outline of the section.

Important definitions, rules, and procedures are highlighted in boxes.

As you study the examples, note the detailed annotations and color highlights that help you on your way.

You are encouraged to do the margin exercises as you work through the material.

The videotape, audiotape, and software references provide extra help for each section.

Answers to the margin exercises are given at the back of the book.

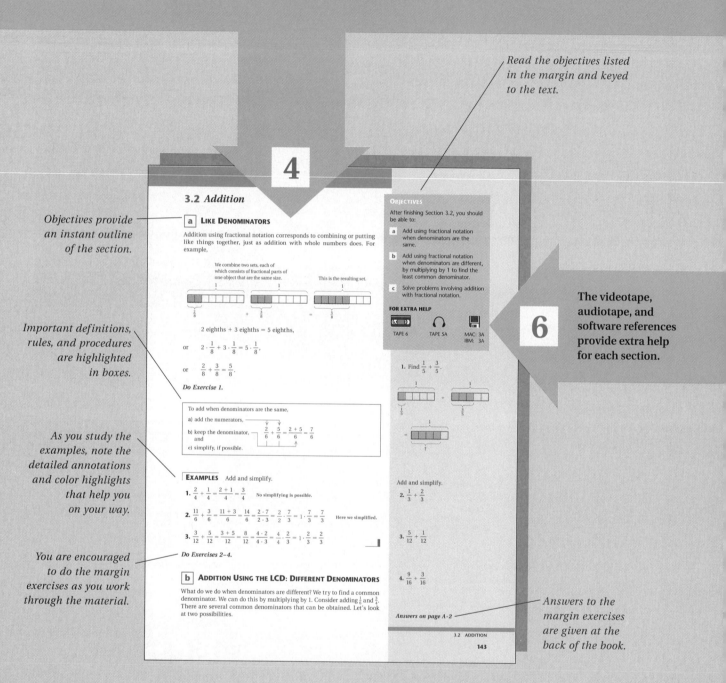

Exercise Sets provide for a wealth
of practice with chapter concepts.

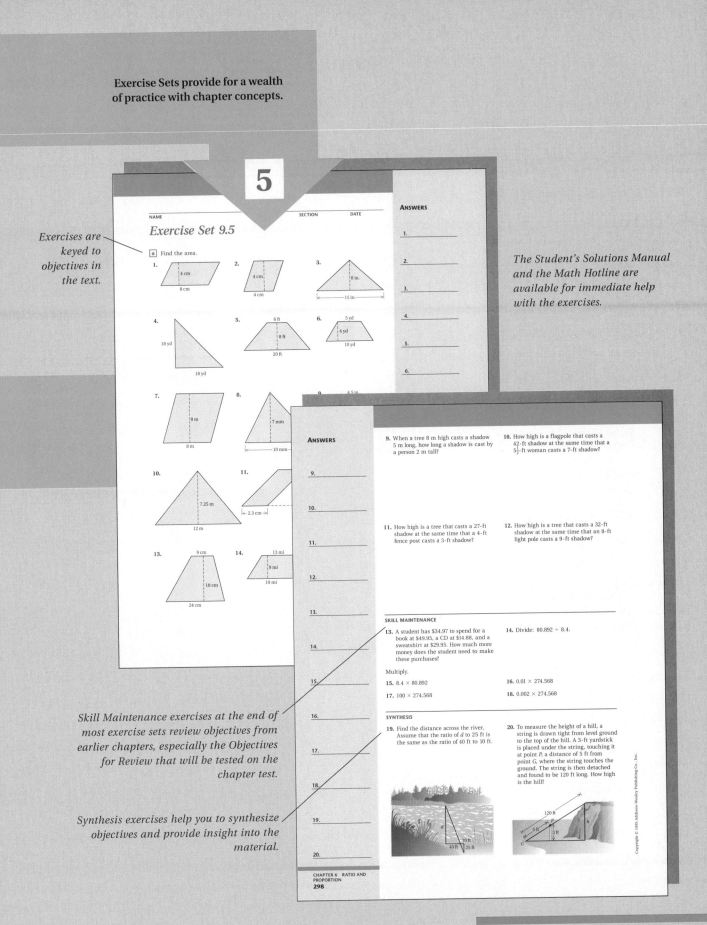

Exercises are keyed to objectives in the text.

The Student's Solutions Manual and the Math Hotline are available for immediate help with the exercises.

Skill Maintenance exercises at the end of most exercise sets review objectives from earlier chapters, especially the Objectives for Review that will be tested on the chapter test.

Synthesis exercises help you to synthesize objectives and provide insight into the material.

5

NAME   SECTION   DATE

Exercise Set 9.5

a Find the area.

1. 4 cm / 8 cm

2. 4 cm / 4 cm

3. 8 in. / 15 in.

4. 18 yd / 18 yd

5. 6 ft / 8 ft / 20 ft

6. 5 yd / 4 yd / 10 yd

7. 8 m / 8 m

8. 7 mm / 10 mm

9. 4.5 in.

10. 7.25 m / 12 m

11. 2.3 cm

13. 9 cm / 18 cm / 24 cm

14. 13 mi / 9 mi / 19 mi

ANSWERS

1. _____
2. _____
3. _____
4. _____
5. _____
6. _____

ANSWERS

9. _____
10. _____
11. _____
12. _____
13. _____
14. _____
15. _____
16. _____
17. _____
18. _____
19. _____
20. _____

9. When a tree 8 m high casts a shadow 5 m long, how long a shadow is cast by a person 2 m tall?

10. How high is a flagpole that casts a 42-ft shadow at the same time that a $5\frac{1}{2}$-ft woman casts a 7-ft shadow?

11. How high is a tree that casts a 27-ft shadow at the same time that a 4-ft fence post casts a 3-ft shadow?

12. How high is a tree that casts a 32-ft shadow at the same time that an 8-ft light pole casts a 9-ft shadow?

SKILL MAINTENANCE

13. A student has $34.97 to spend for a book at $49.95, a CD at $14.88, and a sweatshirt at $29.95. How much more money does the student need to make these purchases?

14. Divide: $80.892 \div 8.4$.

Multiply.

15. 8.4×80.892

16. 0.01×274.568

17. 100×274.568

18. 0.002×274.568

SYNTHESIS

19. Find the distance across the river. Assume that the ratio of *d* to 25 ft is the same as the ratio of 40 ft to 10 ft.

20. To measure the height of a hill, a string is drawn tight from level ground to the top of the hill. A 3-ft yardstick is placed under the string, touching it at point *P*, a distance of 5 ft from point *G*, where the string touches the ground. The string is then detached and found to be 120 ft long. How high is the hill?

THE STEPS TO SUCCESS

Critical Thinking exercises at the end of each chapter (optional) provide further opportunity to synthesize concepts and think critically.

7

Calculator Connections review important keystrokes and provide exercises for the scientific calculator. (See the calculator appendix at the end of the book for basic instruction.)

Extended Synthesis exercises call for you to synthesize many objectives.

CRITICAL THINKING

CALCULATOR CONNECTION

Many calculators have the capability of using fractional notation and mixed numerals for computations, giving answers in such notation as well. The following exercises assume that you are using such a calculator.

1. In the sum below, a and b are digits. Find a and b.

$$\frac{a}{17} + \frac{1b}{23} = \frac{35a}{391}$$

2. Consider only the numbers 2, 3, 4, and 5. Assume each is placed in a blank in the following.

$$\frac{\blacksquare}{\blacksquare} + \frac{\blacksquare}{\blacksquare} = ?$$

What placement of the numbers in the blanks yields the largest sum?

3. Consider only the numbers 3, 4, 5, and 6. Assume each can be placed in a blank in the following.

$$\blacksquare + \frac{\blacksquare}{\blacksquare} \cdot \blacksquare = ?$$

What placement of the numbers in the blanks yields the largest number?

4. Use a standard calculator. Arrange the following in order from largest to smallest.

$$\frac{3}{4}, \frac{17}{21}, \frac{13}{15}, \frac{7}{9}, \frac{15}{17}, \frac{13}{12}, \frac{19}{22}$$

EXTENDED SYNTHESIS EXERCISES

1. a) Simplify each of the following, using fractional notation for your answers.

$$\frac{1}{1 \cdot 2}$$

$$\frac{1}{1 \cdot 2} + \frac{1}{2 \cdot 3}$$

$$\frac{1}{1 \cdot 2} + \frac{1}{2 \cdot 3} + \frac{1}{3 \cdot 4}$$

$$\frac{1}{1 \cdot 2} + \frac{1}{2 \cdot 3} + \frac{1}{3 \cdot 4} + \frac{1}{4 \cdot 5}$$

b) Look for a pattern in your answers to part (a). Then find the following without carrying out the computations.

$$\frac{1}{1 \cdot 2} + \frac{1}{2 \cdot 3} + \frac{1}{3 \cdot 4} + \frac{1}{4 \cdot 5} + \frac{1}{5 \cdot 6}$$
$$+ \frac{1}{6 \cdot 7} + \frac{1}{7 \cdot 8} + \frac{1}{8 \cdot 9} + \frac{1}{9 \cdot 10}$$

2. Each of the following represents a portion of the total area of the square shown below.

$$\frac{1}{2}$$

$$\frac{1}{2} + \frac{1}{4}$$

$$\frac{1}{2} + \frac{1}{4} + \frac{1}{8}$$

$$\frac{1}{2} + \frac{1}{4} + \frac{1}{8} + \frac{1}{16}$$

a) Find each of the areas by simplifying the sums. Use fractional notation for your answers.

b) Look for a pattern in your answers to part (a). Make a conjecture about the total area of the square.

3. Yuri and Olga are orangutans who perform in a circus by riding bicycles around a circular track. It takes Yuri 6 min to make one trip around the track and Olga 4 min. Suppose they start at the same point and then complete their act when they again reach the same point. How long is their act?

4. The students in a math class can be organized into study groups of 8 each such that no students are left out. The same class of students can also be organized into groups of 6 such that no students are left out.

a) Find some class sizes for which this will work.

b) Find the smallest such class size.

5. Find r if

$$\frac{1}{r} = \frac{1}{100} + \frac{1}{150} + \frac{1}{200}.$$

(continued)

CRITICAL THINKING: CHAPTER 3
181

CRITICAL THINKING

6. *Estimation with fractions.* A fraction is very close to 0 when the numerator is very small compared to the denominator. For example, 0 is an estimate for $\frac{2}{17}$. A fraction is very close to $\frac{1}{2}$ when the numerator is about half the denominator. For example, $\frac{1}{2}$ is an estimate for $\frac{11}{23}$. A fraction is very close to 1 when the numerator is very close to the denominator. For example, 1 is an estimate for $\frac{37}{38}$ or $\frac{43}{41}$. Estimate each of the following as either $0, \frac{1}{2}$, or 1.

a) $\frac{2}{47}$ b) $\frac{4}{5}$ c) $\frac{1}{13}$ d) $\frac{7}{8}$

e) $\frac{6}{11}$ f) $\frac{10}{13}$ g) $\frac{7}{15}$ h) $\frac{1}{16}$

i) $\frac{7}{100}$ j) $\frac{5}{9}$ k) $\frac{19}{20}$ l) $\frac{5}{12}$

7. Find a number for the blank so that the fraction is close to but greater than $\frac{1}{2}$. Answers can vary.

a) $\frac{\blacksquare}{11}$ b) $\frac{\blacksquare}{8}$ c) $\frac{\blacksquare}{23}$ d) $\frac{\blacksquare}{35}$

e) $\frac{10}{\blacksquare}$ f) $\frac{7}{\blacksquare}$ g) $\frac{8}{\blacksquare}$ h) $\frac{51}{\blacksquare}$

8. Find a number for the blank so that the fraction is close to but less than 1. Answers can vary.

a) $\frac{7}{\blacksquare}$ b) $\frac{11}{\blacksquare}$ c) $\frac{13}{\blacksquare}$ d) $\frac{27}{\blacksquare}$

e) $\frac{\blacksquare}{15}$ f) $\frac{\blacksquare}{9}$ g) $\frac{\blacksquare}{18}$ h) $\frac{\blacksquare}{100}$

9. Estimate each of the follo... or as a mixed numeral wh... is $\frac{1}{2}$.

a) $2\frac{7}{8}$

c) $12\frac{5}{6}$

10. Estimate the sum.

a) $\frac{4}{5} + \frac{7}{8}$ b) $\frac{1}{12} + \frac{7}{15}$

c) $\frac{2}{3} + \frac{7}{13} + \frac{5}{9}$ d) $\frac{8}{9} + \frac{4}{5} + \frac{11}{12}$

e) $\frac{3}{100} + \frac{1}{10} + \frac{11}{1000}$ f) $\frac{23}{24} + \frac{37}{39} + \frac{51}{50}$

EXERCISES FOR THINKING AND WRITING

1. Explain why $2\frac{1}{4} \cdot 3\frac{2}{5} \neq 6\frac{2}{20}$.

2. Explain why $5 \cdot 3\frac{2}{5} \neq (5 \cdot 3) \cdot \left(5 \cdot \frac{2}{5}\right)$.

3. Discuss the role of least common multiples in adding and subtracting with fractional notation.

4. Find a real-world situation that fits the equation

$$2 \cdot 15\frac{3}{4} + 2 \cdot 28\frac{5}{8} = 88\frac{3}{4}.$$

Exercises for Thinking and Writing encourage you to think and write about key ideas.

CHAPTER 3   ADDITION AND SUBTRACTION: FRACTIONS
182

THE STEPS TO SUCCESS

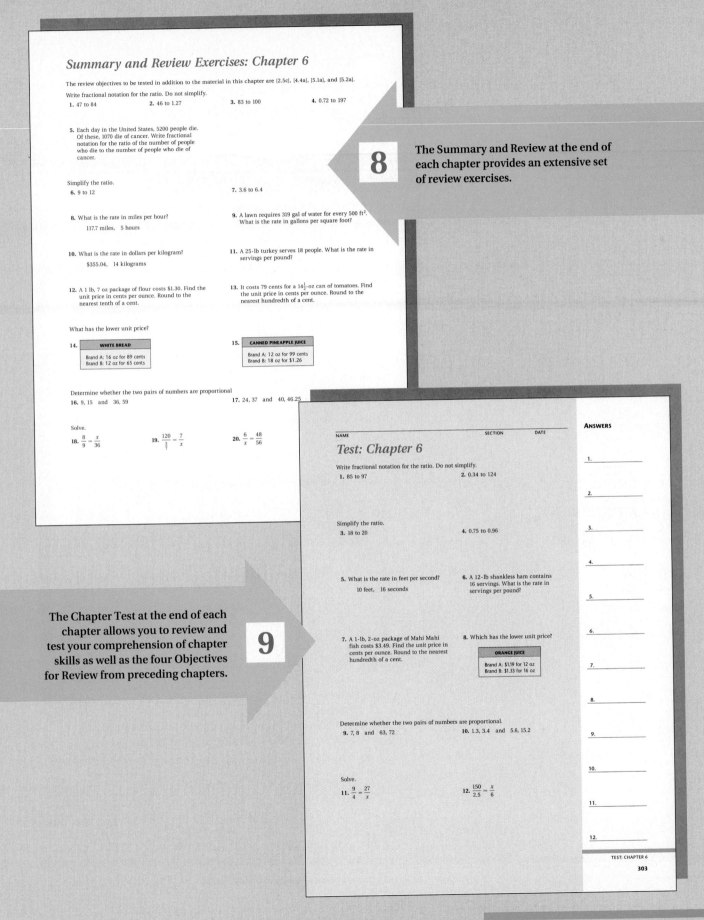

Summary and Review Exercises: Chapter 6

The review objectives to be tested in addition to the material in this chapter are [2.5c], [4.4a], [5.1a], and [5.2a].

Write fractional notation for the ratio. Do not simplify.

1. 47 to 84

2. 46 to 1.27

3. 83 to 100

4. 0.72 to 197

5. Each day in the United States, 5200 people die. Of these, 1070 die of cancer. Write fractional notation for the ratio of the number of people who die to the number of people who die of cancer.

Simplify the ratio.

6. 9 to 12

7. 3.6 to 6.4

8. What is the rate in miles per hour?

117.7 miles, 5 hours

9. A lawn requires 319 gal of water for every 500 ft². What is the rate in gallons per square foot?

10. What is the rate in dollars per kilogram?

$355.04, 14 kilograms

11. A 25-lb turkey serves 18 people. What is the rate in servings per pound?

12. A 1 lb, 7 oz package of flour costs $1.30. Find the unit price in cents per ounce. Round to the nearest tenth of a cent.

13. It costs 79 cents for a $14\frac{1}{2}$-oz can of tomatoes. Find the unit price in cents per ounce. Round to the nearest hundredth of a cent.

What has the lower unit price?

14.
WHITE BREAD
Brand A: 16 oz for 89 cents
Brand B: 12 oz for 65 cents

15.
CANNED PINEAPPLE JUICE
Brand A: 12 oz for 99 cents
Brand B: 18 oz for $1.26

Determine whether the two pairs of numbers are proportional

16. 9, 15 and 36, 59

17. 24, 37 and 40, 46.25

Solve.

18. $\frac{8}{9} = \frac{x}{36}$

19. $\frac{120}{\frac{3}{7}} = \frac{7}{x}$

20. $\frac{6}{x} = \frac{48}{56}$

---

8 The Summary and Review at the end of each chapter provides an extensive set of review exercises.

The Chapter Test at the end of each chapter allows you to review and test your comprehension of chapter skills as well as the four Objectives for Review from preceding chapters. **9**

Test: Chapter 6

Write fractional notation for the ratio. Do not simplify.

1. 85 to 97

2. 0.34 to 124

Simplify the ratio.

3. 18 to 20

4. 0.75 to 0.96

5. What is the rate in feet per second?

10 feet, 16 seconds

6. A 12-lb shankless ham contains 16 servings. What is the rate in servings per pound?

7. A 1-lb, 2-oz package of Mahi Mahi fish costs $3.49. Find the unit price in cents per ounce. Round to the nearest hundredth of a cent.

8. Which has the lower unit price?

ORANGE JUICE
Brand A: $1.19 for 12 oz
Brand B: $1.33 for 16 oz

Determine whether the two pairs of numbers are proportional.

9. 7, 8 and 63, 72

10. 1.3, 3.4 and 5.6, 15.2

Solve.

11. $\frac{9}{4} = \frac{27}{x}$

12. $\frac{150}{2.5} = \frac{x}{6}$

ANSWERS

1. _____
2. _____
3. _____
4. _____
5. _____
6. _____
7. _____
8. _____
9. _____
10. _____
11. _____
12. _____

TEST: CHAPTER 6

303

The Cumulative Review at the end of each chapter provides you with a review of material from all preceding chapters. This is an excellent tool for skill maintenance that provides continual review for the Final Examination.

10

Cumulative Review: Chapters 1–6

Add and simplify.

1. $\begin{array}{r} 2\,7.6\,8 \\ 3.0\,1\,9 \\ +\,4\,8\,3.2\,9\,7 \end{array}$

2. $\begin{array}{r} 2\frac{1}{3} \\ +\,4\frac{5}{12} \end{array}$

3. $\dfrac{6}{35} + \dfrac{5}{28}$

Subtract and simplify.

4. $\begin{array}{r} 4\,0.2 \\ -\,9.7\,0\,9 \end{array}$

5. $73.82 - 0.908$

6. $\dfrac{4}{15} - \dfrac{3}{20}$

Multiply and simplify.

7. $\begin{array}{r} 3\,7.6\,4 \\ \times\ \ \ 5.9 \end{array}$

8. 5.678×100

9. $2\frac{1}{3} \cdot 1\frac{2}{7}$

Divide and simplify.

10. $2.3\,\overline{)\,9\,8.9}$

11. $5\,4\,\overline{)\,4\,8,5\,4\,6}$

12. $\dfrac{7}{11} \div \dfrac{14}{33}$

13. Write expanded notation: 30,074.

14. Write a word name for 120.07.

Which number is larger?

15. 0.7,   0.698

16. 0.799,   0.8

17. Find the prime factorization of 144.

18. Find the LCM of 28 and 35.

19. What part is shaded?

20. Simplify: $\dfrac{90}{144}$.

Calculate.

21. $\dfrac{3}{5} \times 9.53$

22. $\dfrac{1}{3} \times 0.645 - \dfrac{3}{4} \times 0.048$

For more detailed information on how to use MathMax: The Bittinger/Keedy System of Instruction, please see the preface or contact your local Addison-Wesley representative.

Basic Mathematics

SEVENTH EDITION

Diagnostic Pretest

CHAPTER 1

1. Add: $1425 + 382$.

2. Solve: $32 + x = 61$.

3. Multiply: $\begin{array}{r} 321 \\ \times\ 47 \\ \hline \end{array}$

4. A thermos contains 128 oz of juice. How many 6-oz cups can be filled from the thermos? How many ounces will be left over?

CHAPTER 2

5. Solve: $\dfrac{5}{8} \cdot x = \dfrac{3}{16}$.

6. Multiply and simplify: $4 \cdot \dfrac{3}{8}$.

7. Find the prime factorization of 144.

8. A recipe calls for $\frac{3}{4}$ cup of flour. How much is needed to make $\frac{2}{3}$ of a recipe?

CHAPTER 3

9. Find the LCM of 12 and 16.

10. Add: $\dfrac{3}{8} + \dfrac{1}{6}$.

11. Multiply and simplify: $4\dfrac{1}{5} \cdot 3\dfrac{2}{3}$.

12. A 3-m fence post was set $1\frac{2}{5}$ m in the ground. How much was above the ground?

13. A car travels 249 mi on $8\frac{3}{10}$ gal of gas. How many miles per gallon did it get?

CHAPTER 4

14. Which number is larger, 0.00009 or 0.0001?

15. Round to the nearest tenth: 25.562.

16. Add: $\begin{array}{r} 12.035 \\ 0.08 \\ +\ 27.7 \\ \hline \end{array}$

17. A driver bought gasoline when the odometer read 68,123.2. At the next gasoline purchase, the odometer read 68,310.1. How many miles had been driven?

CHAPTER 5

18. Multiply: $\begin{array}{r} 0.012 \\ \times\ 2.5 \\ \hline \end{array}$

19. Find decimal notation: $\dfrac{7}{3}$.

20. Solve: $1.5 \times t = 3.6$.

21. What is the cost of 5 shirts at $23.99 each?

22. Estimate the product 4.68×32.431 by rounding to the nearest one.

CHAPTER 6

23. Solve: $\dfrac{1.2}{x} = \dfrac{0.4}{1.5}$.

24. If 3 cans of green beans cost $1.19, how many cans of green beans can you buy for $4.76?

25. It costs $2.19 for a 22-oz bag of tortilla chips. Find the unit price in cents per ounce. Round to the nearest tenth of a cent.

CHAPTER 7

26. Find percent notation: $\dfrac{1}{8}$.

27. Find decimal notation: 1.35%.

28. The price of a pair of running shoes was reduced from $45 to $27. Find the percent of decrease in price.

29. What is the simple interest on $230 principal at the interest rate of 8.5% for one year?

CHAPTER 8

30. Find the average, the median, and the mode of the following set of numbers:

22, 25, 27, 25, 22, 25.

31. A car traveled 296 mi on 16 gal of gasoline. What was the average number of miles per gallon?

32. In order to get a B in math, a student must average 80 on four tests. Scores on the first three tests were 85, 72, and 78. What is the lowest score the student can get on the last test and still get a B?

CHAPTER 9

Complete.

33. 2 yd = _____ in.

34. 4 cm = _____ km

35. Find the area and the circumference of a circle with a diameter of 12 cm. Leave answers in terms of π.

36. Find the area and the perimeter of a rectangle with length 3 ft and width 2.5 ft.

37. Simplify: $\sqrt{121}$.

CHAPTER 10

Complete.

38. 2 min = _____ sec

39. 5 mg = _____ g

40. 2 yd^2 = _____ ft^2

41. 10 qt = _____ gal

42. The diameter of a ball is 18 cm. Find the volume. Use 3.14 for π.

CHAPTER 11

43. Find the absolute value: $|-4.2|$.

44. Find decimal notation: $-\dfrac{4}{9}$.

Compute and simplify.

45. $-2 - (-1.9)$

46. $\dfrac{5}{6}\left(-\dfrac{1}{10}\right)$

CHAPTER 12

Solve.

47. $2x - 1 = 4x + 5$

48. $\dfrac{1}{3}x + \dfrac{1}{5} = \dfrac{2}{3} - \dfrac{1}{4}x$

49. A student bought a sweater and a pair of jeans. The sweater cost $33.95. This was $8.39 more than the cost of the jeans. How much did the jeans cost?

50. A 22-oz box of cereal costs $3.49. How many boxes of cereal can you buy for $27.92?

1

Operations on the Whole Numbers

INTRODUCTION

In this chapter, we consider addition, subtraction, multiplication, and division of whole numbers. Then we study the solving of simple equations and apply our skills to the solving of problems.

AN APPLICATION

The costs to a student of attending a college for four years are: first year, $11,516; second year, $11,898; third year, $12,270; fourth year, $12,490. Find the total cost of the college education.

THE MATHEMATICS

We let n = the total cost of the college education. Since we are combining costs, addition can be used. We translate the problem to this equation:

$$11{,}516 + 11{,}898 + 12{,}270 + 12{,}490 = n.$$

Here is how addition can occur in problem solving.

Pretest: Chapter 1

1. Write a word name: 3,078,059.

2. Write expanded notation: 6987.

3. Write standard notation: Two billion, forty-seven million, three hundred ninety-eight thousand, five hundred eighty-nine.

4. What does the digit 6 mean in 2,967,342?

5. Round 956,449 to the nearest thousand.

6. Estimate the product 594 · 126 by first rounding the numbers to the nearest hundred.

7. Add.

$$\begin{array}{r} 7\ 3\ 1\ 2 \\ +\ 2\ 9\ 0\ 4 \\ \hline \end{array}$$

8. Subtract.

$$\begin{array}{r} 7\ 0\ 1\ 2 \\ -\ 2\ 9\ 0\ 4 \\ \hline \end{array}$$

9. Multiply: 359 · 64.

10. Divide: 23,149 ÷ 46.

Use either < or > for ▧ to write a true sentence.

11. 346 ▧ 364

12. 54 ▧ 45

Solve.

13. $326 \cdot 17 = m$

14. $y = 924 \div 42$

15. $19 + x = 53$

16. $34 \cdot n = 850$

Solve.

17. Betsy weighs 121 lb and Melissa weighs 109 lb. How much more does Betsy weigh?

18. How many 12-jar cases can be filled with 1512 jars of spaghetti sauce?

19. The population of Illinois is 11,430,602. The population of Ohio is 10,847,115. What is the total population of Illinois and Ohio?

20. A lot measures 48 ft by 54 ft. A pool that is 15 ft by 20 ft is put on the lot. How much area is left over?

Evaluate.

21. 5^2

22. 4^3

Simplify.

23. $8^2 \div 8 \cdot 2 - (2 + 2 \cdot 7)$

24. $108 \div 9 - \{4 \cdot [18 - (5 \cdot 3)]\}$

1.1 *Standard Notation*

We study mathematics in order to be able to solve problems. In this chapter, we learn how to use operations on the whole numbers. We begin by studying how numbers are named.

 a **FROM STANDARD NOTATION TO EXPANDED NOTATION**

To answer questions such as "How many?", "How much?", and "How tall?" we use whole numbers. The set, or collection, of **whole numbers** is

0, 1, 2, 3, 4, 5, 6, 7, 8, 9, 10, 11, 12, ...

The set goes on indefinitely. There is no largest whole number, and the smallest whole number is 0. Each number can be named using various notations. The set 1, 2, 3, 4, 5, ..., without 0, is called the set of **natural numbers**.

As examples, we use data from the bar graph shown here. Note that there were 3014 pairs of bald eagles in 1990.

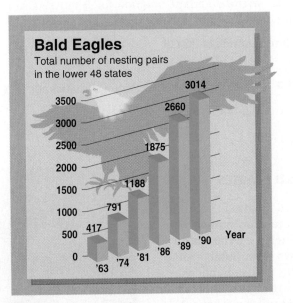

Bald Eagles
Total number of nesting pairs
in the lower 48 states

Standard notation for this number is 3014. We find **expanded notation** for 3014 as follows:

3014 = 3 thousands + 0 hundreds + 1 ten + 4 ones

EXAMPLE 1 Write expanded notation for 2660, the number of pairs of bald eagles in 1989.

2660 = 2 thousands + 6 hundreds + 6 tens + 0 ones

EXAMPLE 2 Write expanded notation for 54,567.

54,567 = 5 ten thousands + 4 thousands
            + 5 hundreds + 6 tens + 7 ones

Do Exercises 1 and 2 (in the margin at the right).

OBJECTIVES

After finishing Section 1.1, you should be able to:

a Convert from standard notation to expanded notation.

b Convert from expanded notation to standard notation.

c Write a word name for a number given standard notation.

d Write standard notation for a number given a word name.

e Given a standard notation like 278,342, tell what 8 means, what 3 means, and so on; identify the hundreds digit, the thousands digit, and so on.

FOR EXTRA HELP

TAPE 1   TAPE 1A   MAC: 1A
                       IBM: 1A

Write expanded notation.

1. 1875

2. 36,223

Answers on page A-1

Write expanded notation.

3. 3021

4. 2009

5. 5700

Write standard notation.

6. 5 thousands + 6 hundreds + 8 tens + 9 ones

7. 8 ten thousands + 7 thousands + 1 hundred + 2 tens + 8 ones

8. 9 thousands + 3 ones

Write a word name.

9. 57

10. 29

11. 88

Answers on page A-1

EXAMPLE 3   Write expanded notation for 7091.

7091 = 7 thousands + 0 hundreds + 9 tens + 1 one,   or
7 thousands + 9 tens + 1 one

EXAMPLE 4   Write expanded notation for 3400.

3400 = 3 thousands + 4 hundreds + 0 tens + 0 ones,   or
3 thousands + 4 hundreds

Do Exercises 3–5.

b   **FROM EXPANDED NOTATION TO STANDARD NOTATION**

EXAMPLE 5   Write standard notation for 2 thousands + 5 hundreds + 7 tens + 5 ones.

Standard notation is 2575

EXAMPLE 6   Write standard notation for 9 ten thousands + 6 thousands + 7 hundreds + 1 ten + 8 ones.

Standard notation is 96,718.

EXAMPLE 7   Write standard notation for 2 thousands + 3 tens.

Standard notation is 2030.

Do Exercises 6–8.

c   **WORD NAMES**

"Three," "two hundred one," and "forty-two" are **word names** for numbers. When we write word names for two-digit numbers like 42, 76, and 91, we use hyphens. For example, the running speed of a giraffe is 32 miles per hour. A word name for 32 is "thirty-two."

EXAMPLES   Write a word name.

8. 42   Forty-two   **9.** 91   Ninety-one

Do Exercises 9–11.

For large numbers, digits are separated into groups of three, called **periods**. Each period has a name like *ones, thousands, millions, billions,* and so on. When we write or read a large number, we start at the left with the largest period. The number named in the period is followed by the name of the period, then a comma is written and the next period is named. Recently, the U.S. national debt was $4,260,517,000,000. We can use a **place-value** chart to illustrate how to use periods to read the number 4,260,517,000,000.

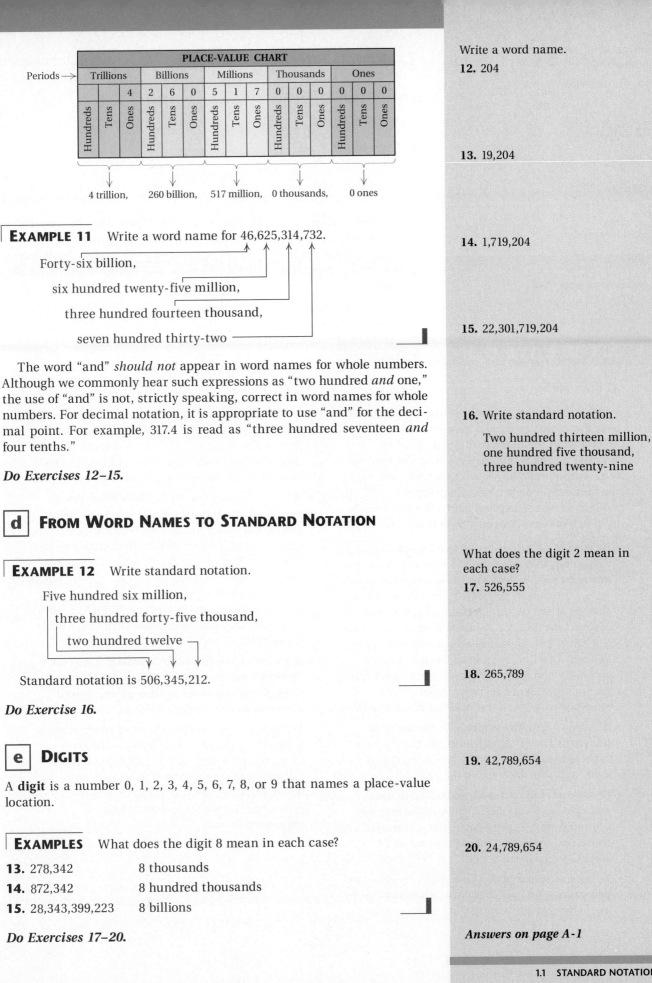

PLACE-VALUE CHART

	Trillions			Billions			Millions			Thousands			Ones		
Periods →			4	2	6	0	5	1	7	0	0	0	0	0	0
	Hundreds	Tens	Ones	Hundreds	Tens	Ones	Hundreds	Tens	Ones	Hundreds	Tens	Ones	Hundreds	Tens	Ones

4 trillion,   260 billion,   517 million,   0 thousands,   0 ones

EXAMPLE 11   Write a word name for 46,625,314,732.

Forty-six billion,

six hundred twenty-five million,

three hundred fourteen thousand,

seven hundred thirty-two

The word "and" *should not* appear in word names for whole numbers. Although we commonly hear such expressions as "two hundred *and* one," the use of "and" is not, strictly speaking, correct in word names for whole numbers. For decimal notation, it is appropriate to use "and" for the decimal point. For example, 317.4 is read as "three hundred seventeen *and* four tenths."

Do Exercises 12–15.

d   **FROM WORD NAMES TO STANDARD NOTATION**

EXAMPLE 12   Write standard notation.

Five hundred six million,

three hundred forty-five thousand,

two hundred twelve

Standard notation is 506,345,212.

Do Exercise 16.

e   **DIGITS**

A **digit** is a number 0, 1, 2, 3, 4, 5, 6, 7, 8, or 9 that names a place-value location.

EXAMPLES   What does the digit 8 mean in each case?

13. 278,342   8 thousands
14. 872,342   8 hundred thousands
15. 28,343,399,223   8 billions

Do Exercises 17–20.

Write a word name.
12. 204

13. 19,204

14. 1,719,204

15. 22,301,719,204

16. Write standard notation.

Two hundred thirteen million, one hundred five thousand, three hundred twenty-nine

What does the digit 2 mean in each case?
17. 526,555

18. 265,789

19. 42,789,654

20. 24,789,654

Answers on page A-1

In 7,890,432, what digit tells the number of:

21. Hundreds?

22. Millions?

23. Ten thousands?

24. Thousands?

Answers on page A-1

| **EXAMPLE 16**   In a recent year, 479,750 athletes participated in sports in the state of Texas. In 479,750, what digit tells the number of:

a) Hundred thousands?   4

b) Thousands?   9

Do Exercises 21–24.

From time to time, you will find some *Sidelights* like the one below. Although they are optional, you may find them helpful and of interest. They will include study tips, career opportunities involving mathematics, applications, computer–calculator exercises, and many other mathematical topics.

S I D E L I G H T S

STUDY TIPS

There are many ways in which you can enhance your use of this book, and they have been illustrated in the diagram entitled "Steps to Success" on the inside front cover. If you have not read that information, do so now, before you begin the exercise set on the following page.

Here we highlight a few points that we consider most helpful when learning from this text.

- **Be sure to note the special symbols** a , b , c , **and so on, that correspond to the objectives you are to be able to perform.** They appear in many places throughout the text. The first time you see them is at the beginning of each section. The second time is in the headings of each subsection. The third time is in the exercise set, as follows. You will also find them in the answers to the summary and review exercises, the chapter tests, and the cumulative reviews. These allow you to reference back when you need to review a topic.

- **Be sure to note the symbols in the margin under the list of objectives at the beginning of each section.** These refer to the many distinctive study aids that accompany the book.

- **Be sure to read and study each step of each example.** The examples include important side comments that explain each step. These carefully chosen examples and notes prepare you for success in the exercise set.

- **Be sure to stop and do the margin exercises as you study a section.** When our students come to us troubled about how they are doing in the course, the first question we ask is "Are you doing the margin exercises when directed to do so?" This is one of the most effective ways to enhance your ability to learn mathematics from this text. Don't deprive yourself of its benefits!

- **When you study the book, don't mark points you think are important, but mark the points you do not understand!** This book includes many design features that highlight important points. Use your efforts to mark where you are having trouble. Then when you go to class, a math lab, or a tutoring session, you will be prepared to ask questions that home in on your difficulties rather than spending time going over what you already understand.

- **If you are having trouble, consider using the Student's Solutions Manual that contains worked-out solutions to the odd-numbered exercises in the exercise sets.**

- **Try to keep one section ahead of your syllabus.** If you study ahead of your lectures, you can concentrate on what is being explained in your lectures, rather than trying to write everything down. You can then take notes only of special points or of questions related to what is happening in class.

Exercise Set 1.1

Always review the objectives before doing an exercise set. See page 3. Note how the objectives are keyed to the exercises.

a Write expanded notation.

1. 5742   **2.** 3897   **3.** 27,342   **4.** 93,986

5. 5609   **6.** 9990   **7.** 2300   **8.** 7020

b Write standard notation.

9. 2 thousands + 4 hundreds + 7 tens + 5 ones

10. 7 thousands + 9 hundreds + 8 tens + 3 ones

11. 6 ten thousands + 8 thousands + 9 hundreds + 3 tens + 9 ones

12. 1 ten thousand + 8 thousands + 4 hundreds + 6 tens + 1 one

13. 7 thousands + 3 hundreds + 0 tens + 4 ones

14. 8 thousands + 0 hundreds + 2 tens + 0 ones

15. 1 thousand + 9 ones

16. 2 thousands + 4 hundreds + 5 tens

c Write a word name.

17. 85   **18.** 48   **19.** 88,000   **20.** 45,987

21. 123,765   **22.** 111,013   **23.** 7,754,211,577   **24.** 43,550,651,808

Write a word name for the number in the sentence.

25. There are 6,469,952 black spots on the dogs in the Walt Disney movie *101 Dalmations.*

26. The diameter of the sun is 865,400 miles.

ANSWERS

1. _____
2. _____
3. _____
4. _____
5. _____
6. _____
7. _____
8. _____
9. _____
10. _____
11. _____
12. _____
13. _____
14. _____
15. _____
16. _____
17. _____
18. _____
19. _____
20. _____
21. _____
22. _____
23. _____
24. _____
25. _____
26. _____

ANSWERS

27. _____

28. _____

29. _____

30. _____

31. _____

32. _____

33. _____

34. _____

35. _____

36. _____

37. _____

38. _____

39. _____

40. _____

41. _____

42. _____

43. _____

44. _____

45. _____

46. _____

47. _____

27. There are 1,954,116 students in junior colleges.

28. In the early part of the next century, there will be 39,568,000 senior citizens in this country.

d Write standard notation.

29. Two million, two hundred thirty-three thousand, eight hundred twelve

30. Three hundred fifty-four thousand, seven hundred two

31. Eight billion

32. Seven hundred million

33. Two hundred seventeen thousand, five hundred three

34. Two hundred thirty billion, forty-three million, nine hundred fifty-one thousand, six hundred seventeen

Write standard notation for the number in the sentence.

35. In a recent year, one billion, one hundred eighty-seven thousand, five hundred forty-two Valentine cards were sent.

36. The population of China is one billion, one hundred fifty-one million, four hundred eighty-six thousand.

37. In one year, Americans use two hundred six million, six hundred fifty-eight thousand pounds of toothpaste.

38. The people of the United States burn seven hundred forty-nine million, five hundred seventy-eight thousand, six hundred fifty-three gallons of fuel annually driving to see motion pictures.

e What does the digit 5 mean in each case?

39. 235,888   **40.** 253,888   **41.** 488,526   **42.** 500,346

In 89,302 what digit tells the number of:

43. Hundreds?

44. Thousands?

45. Tens?

46. Ones?

SYNTHESIS

Any exercises below a line like that above are extra, optional, and usually more challenging, requiring you to put together objectives of this section or preceding sections of the text. They will also help you prepare for the Critical Thinking exercises at the end of each chapter. Any exercises marked with a ▦ are to be worked with a calculator.

47. ▦ What is the largest number that you can name on your calculator? How many digits does that number have? How many periods?

1.2 Addition

a ADDITION AND THE REAL WORLD

Addition of whole numbers corresponds to combining or putting things together. Let us look at various situations in which addition applies.

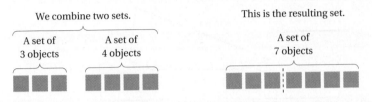

We combine two sets.

A set of 3 objects   A set of 4 objects

This is the resulting set.

A set of 7 objects

The addition that corresponds is

$$3 + 4 = 7.$$

The number of objects in a set can be found by counting. We count and find that the two sets have 3 members and 4 members, respectively. After combining, we count and find that there are 7 objects. We say that the **sum** of 3 and 4 is 7. The numbers added are called **addends**.

$$3 \quad + \quad 4 \quad = \quad 7$$

Addend   Addend   Sum

EXAMPLE 1 Write an addition that corresponds to this situation.

A student has $3 and earns $10 more. How much money does the student have?

The addition that corresponds is $3 + $10 = $13.

Do Exercises 1 and 2.

Addition also corresponds to combining distances or lengths.

EXAMPLE 2 Write an addition that corresponds to this situation.

A car is driven 8 mi (miles) from Topsham to Bath. It is then driven 10 mi from Bath to Wiscasset. How far is it from Topsham to Wiscasset along the same route?

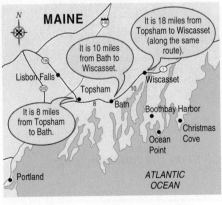

8 miles + 10 miles = 18 miles

Do Exercises 3 and 4.

Write an addition that corresponds to the situation.

1. John has 4 CDs in his backpack. Then he buys 6 more at the bookstore. How many does he have in all?

2. Sue earns $15 in overtime pay on Thursday and $13 on Friday. How much overtime pay does she earn altogether on the two days?

Write an addition that corresponds to the situation.

3. A car is driven 40 mi from Lafayette to Kokomo. Then it is driven 50 mi from Kokomo to Indianapolis. How far is it from Lafayette to Indianapolis along the same route?

4. A rope 5 ft long is tied to a rope 7 ft long. How long is the resulting rope (ignoring the amount of rope it takes to tie the two ropes together)?

Answers on page A-1

Write an addition that corresponds to the situation.

5. Find the distance around (perimeter of) the figure.

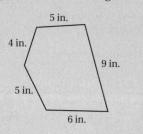

6. Find the distance around (perimeter of) the figure.

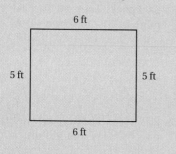

Write an addition that corresponds to the situation.

7. One piece of gift wrapping paper has an area of 80 square inches (in²). Another piece of gift wrapping paper has an area of 90 in². What is the total area of the two pieces of gift wrapping paper?

8. One subdivision contains 100 square miles (mi²). Another subdivision contains 200 mi². What is the total area of the two subdivisions?

Answers on page A-1

When we find the sum of the distances around an object, we are finding its **perimeter**.

EXAMPLE 3   Write an addition that corresponds to this situation.

A sporting goods salesperson travels the following route. How long is the route?

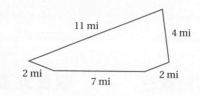

$$2 \text{ mi} + 7 \text{ mi} + 2 \text{ mi} + 4 \text{ mi} + 11 \text{ mi} = 26 \text{ mi}$$

Do Exercises 5 and 6.

Addition also corresponds to combining areas.

EXAMPLE 4   Write an addition that corresponds to this situation.

You have 5 square yards of upholstery fabric. You buy 7 more square yards. How much do you have in all?

You have 5 square yards of upholstery fabric.	You buy 7 more square yards.	You then have 12 square yards of upholstery fabric.

5 square yards   + 7 square yards = 12 square yards

Do Exercises 7 and 8.

Addition corresponds to combining volumes as well.

EXAMPLE 5   Write an addition that corresponds to this situation.

Two trucks haul dirt to a construction site. One hauls 5 cubic yards and the other hauls 7 cubic yards. Altogether, how many cubic yards of dirt have they hauled to the site?

Truck A hauls 5 cubic yards of dirt to a construction site.   Truck B hauls 7 cubic yards of dirt.   Altogether, they haul 12 cubic yards of dirt.

5 cubic yards   +   7 cubic yards   =   12 cubic yards

Do Exercises 9 and 10 on the following page.

b ADDITION OF WHOLE NUMBERS

To add numbers, we add the ones first, then the tens, then the hundreds, and so on.

EXAMPLE 6 Add: 7312 + 2504.

```
    7  3  1  2      Add ones.
 +  2  5  0  4
             6
```

```
    7  3  1  2      Add tens.
 +  2  5  0  4
          1  6
```

We show you this for explanation.

```
    7  3  1  2      Add hundreds.
 +  2  5  0  4
       8  1  6
```

You should write only this.

```
    7  3  1  2      Add thousands.       7  3  1  2
 +  2  5  0  4                        +  2  5  0  4
    9  8  1  6                           9  8  1  6
```

Do Exercise 11.

EXAMPLE 7 Add: 6878 + 4995.

```
          1
    6  8  7  8      Add ones. We get 13 ones, or 1 ten + 3 ones.
 +  4  9  9  5      Write 3 in the ones column and 1 above the tens. This is
             3      called carrying, or regrouping.
```

```
       1  1
    6  8  7  8      Add tens. We get 17 tens, or 1 hundred + 7 tens.
 +  4  9  9  5      Write 7 in the tens column and 1 above the hundreds.
          7  3
```

```
    1  1  1
    6  8  7  8      Add hundreds. We get 18 hundreds, or 1 thousand +
 +  4  9  9  5      8 hundreds.
       8  7  3      Write 8 in the hundreds column and 1 above the thousands.
```

```
    1  1  1
    6  8  7  8      Add thousands. We get 11 thousands.
 +  4  9  9  5
 1  1  8  7  3
```

Do Exercises 12 and 13.

9. A motorist purchases 10 gal (gallons) of gasoline for one family car one day and 18 gal for another car the next day. How many gallons were bought in all?

10. Container A holds 3000 tons of sand and container B holds 7000 tons. How many tons do they hold in all?

11. Add.

```
    6  2  0  3
 +  3  5  4  2
```

Add.

12.
```
    7  9  6  8
 +  5  4  9  7
```

13.
```
    9  8  0  4
 +  6  3  7  8
```

Answers on page A-1

Add from the top.

14.
```
   9
   9
   4
+  5
___
```

15.
```
   8
   6
   9
   7
+  4
___
```

16. Add from the bottom.
```
      9
      9
      4
   +  5
   ___
```

Answers on page A-1

How do we do an addition of three numbers, like $2 + 3 + 6$? We do so by adding 3 and 6, and then 2. We can show this with parentheses:

$$2 + (3 + 6) = 2 + 9 = 11.$$ **Parentheses tell what to do first.**

We could also add 2 and 3, and then 6:

$$(2 + 3) + 6 = 5 + 6 = 11.$$

Either way we get 11. It does not matter how we group the numbers. This illustrates the **associative law of addition,** $a + (b + c) = (a + b) + c$. We can also add whole numbers in any order. That is, $2 + 3 = 3 + 2$. This illustrates the **commutative law of addition,** $a + b = b + a$. Together the commutative and associative laws tell us that to add more than two numbers, we can use any order and grouping we wish.

EXAMPLE 8   Add from the top.

```
   8
   9
   7
+  6
___
```

We first add 8 and 9, getting 17; then 17 and 7, getting 24; then 24 and 6, getting 30.

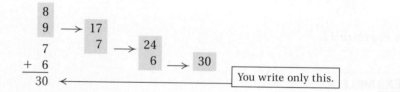

You write only this.

Do Exercises 14 and 15.

EXAMPLE 9   Add from the bottom.

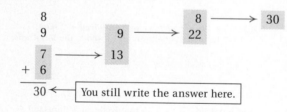

You still write the answer here.

Do Exercise 16.

Sometimes it is easier to look for pairs of numbers whose sums are 10 or 20 or 30, and so on.

EXAMPLES   Add.

10.
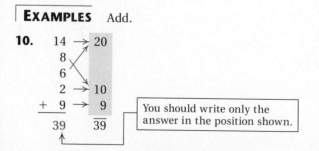

You should write only the answer in the position shown.

11. $23 + 19 + 7 + 21 + 4 = 74$

$$30 + 40 + 4$$
$$74$$

Do Exercises 17–19.

EXAMPLE 12   Add: $2391 + 3276 + 8789 + 1498$.

```
      2
  2 3 9 1
  3 2 7 6
  8 7 8 9
+ 1 4 9 8
        4
```
Add ones: We get 24, so we have 2 tens + 4 ones.
Write 4 in the ones column and 2 above the tens.

```
    3 2
  2 3 9 1
  3 2 7 6
  8 7 8 9
+ 1 4 9 8
      5 4
```
Add tens: We get 35 tens, so we have 30 tens + 5 tens.
This is also 3 hundreds + 5 tens. Write 5 in the tens column
and 3 above the hundreds.

```
  1 3 2
  2 3 9 1
  3 2 7 6
  8 7 8 9
+ 1 4 9 8
    9 5 4
```
Add hundreds: We get 19 hundreds, or 1 thousand +
9 hundreds. Write 9 in the hundreds column and
1 above the thousands.

```
  1 3 2
  2 3 9 1
  3 2 7 6
  8 7 8 9
+ 1 4 9 8
1 5 9 5 4
```
Add thousands: We get 15 thousands.

Do Exercise 20.

Add. Look for pairs of numbers whose sums are 10, 20, 30, and so on.

17.
```
  1 5
    7
    5
    3
+   8
```

18. $6 + 12 + 14 + 8 + 7$

19. $27 + 8 + 13 + 2 + 11$

20. Add.
```
  1 9 3 2
  6 7 2 3
  9 8 7 8
+ 8 9 4 1
```

To the instructor and the student: This section presented a review of addition of whole numbers. Students who are successful should go on to Section 1.3. Those who have trouble should study developmental unit A and then repeat Section 1.2.

Answers on page A-1

NUMBER PATTERNS: MAGIC SQUARES

The following is a *magic square*. The sum along any row, column, or diagonal is the same—in this case, 60. Check this on your calculator.

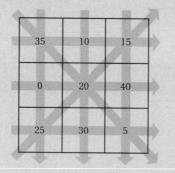

35	10	15
0	20	40
25	30	5

EXERCISES

1. Place the numbers 1 through 9 to form a magic square.

2. Place the numbers 1 through 16 to form a magic square. The sums will all be 34.

1			7
	15		
		8	
16			10

Each of the following is a magic square, but one number is incorrect. Find it.

3.

11	77	62	29
69	22	17	71
27	61	78	12
72	19	21	67

4.

70	25	67	9
18	59	20	75
19	77	15	60
65	10	69	27

5.

3	30	28	8
24	12	14	18
16	20	22	10
26	6	4	32

6.

24	14	33	7	35	16	46
30	27	32	42	19	5	20
37	12	22	9	44	38	13
49	2	47	25	3	48	1
10	39	6	41	28	11	40
21	45	18	8	50	23	29
4	36	17	43	15	34	26

Exercise Set 1.2

a Write an addition that corresponds to the situation.

1. A family rents 3 videotapes one week and 6 videotapes the next week. How many videos did they rent in all?

2. A jogger runs 4 mi one day and 5 mi the next. What total distance was run in the two days?

3. A student earns $23 one day and $31 the next. How much did the student earn in all?

4. You own a 40-acre farm. Then you buy an adjoining 80-acre farm. You now own a farm of how many acres?

Find the distance around (perimeter of) the figure.

5.

325 ft

325 ft   325 ft

325 ft

6.

87 yd

87 yd   87 yd

87 yd   87 yd

87 yd

b Add.

7.
```
  3 6 4
+   2 3
```

8.
```
  1 5 2 1
+   3 4 8
```

9.
```
  1 7 1 6
+ 3 4 8 2
```

10.
```
  7 5 0 3
+ 2 6 8 3
```

1. _____

2. _____

3. _____

4. _____

5. _____

6. _____

7. _____

8. _____

9. _____

10. _____

11. _____

12. _____

13. _____

14. _____

15. _____

16. _____

17. _____

18. _____

19. _____

20. _____

21. _____

22. _____

23. _____

24. _____

25. _____

26. _____

27. _____

28. _____

29. _____

30. _____

31. _____

32. _____

33. _____

34. _____

11.
$$\begin{array}{r} 86 \\ +78 \\ \hline \end{array}$$

12.
$$\begin{array}{r} 73 \\ +69 \\ \hline \end{array}$$

13.
$$\begin{array}{r} 99 \\ +1 \\ \hline \end{array}$$

14.
$$\begin{array}{r} 999 \\ +11 \\ \hline \end{array}$$

15. $789 + 111$

16. $839 + 386$

17. $909 + 101$

18. $707 + 909$

19. $811 + 390$

20. $271 + 333$

21. $356 + 491$

22. $280 + 347$

23.
$$\begin{array}{r} 5093 \\ +3217 \\ \hline \end{array}$$

24.
$$\begin{array}{r} 3654 \\ +2700 \\ \hline \end{array}$$

25.
$$\begin{array}{r} 4825 \\ +1783 \\ \hline \end{array}$$

26.
$$\begin{array}{r} 6775 \\ +1432 \\ \hline \end{array}$$

27.
$$\begin{array}{r} 9999 \\ +6785 \\ \hline \end{array}$$

28.
$$\begin{array}{r} 45,879 \\ +21,786 \\ \hline \end{array}$$

29.
$$\begin{array}{r} 23,443 \\ +10,989 \\ \hline \end{array}$$

30.
$$\begin{array}{r} 67,654 \\ +98,786 \\ \hline \end{array}$$

31.
$$\begin{array}{r} 77,543 \\ +23,767 \\ \hline \end{array}$$

32.
$$\begin{array}{r} 44,654 \\ +4,765 \\ \hline \end{array}$$

33.
$$\begin{array}{r} 99,999 \\ +112 \\ \hline \end{array}$$

34.
$$\begin{array}{r} 127,556 \\ +68,766 \\ \hline \end{array}$$

Add from the top. Then check by adding from the bottom.

35.	36.	37.	38.	39.
7	5	4	8	9
9	6	3	6	4
4	5	9	2	7
+ 8	+ 4	1	3	8
		+ 8	+ 7	+ 7

Add. Look for pairs of numbers whose sums are 10, 20, 30, and so on.

40.	41.	42.	43.	44.
7	2 3	7	4 5	3 8
1 8	1 6	2 4	2 5	2 7
3	1 1	1 5	3 6	3 2
3 7	1 8	6	4 4	1 4
+ 2	+ 1 9	+ 5	+ 8 0	+ 7 6

Add.

45.	46.	47.	48.
2 3	4 3	5 1	3 1
6 2	1 1	3 6	5 3
+ 4 5	+ 3 7	+ 6 2	+ 2 4

49.	50.	51.	52.
2 6	3 2 4	2 0 7	2 4 8
8 2	1 2 6	2 9 5	3 1 4
+ 6 1	+ 4 8 2	+ 3 4 0	+ 6 7 1

53.	54.
3 2 7	9 8 9
4 2 8	5 6 6
5 6 9	8 3 4
7 8 7	9 2 0
+ 2 0 9	+ 7 0 3

35. _____

36. _____

37. _____

38. _____

39. _____

40. _____

41. _____

42. _____

43. _____

44. _____

45. _____

46. _____

47. _____

48. _____

49. _____

50. _____

51. _____

52. _____

53. _____

54. _____

55.
```
  2 0 3 7
  4 9 2 3
  3 4 7 1
+ 1 2 4 8
```

56.
```
  4 5 6 7
  1 0 2 3
  4 8 2 1
+ 3 6 8 3
```

57.
```
  3 4 2 0
  8 7 1 9
  4 3 1 2
+ 6 2 0 3
```

58.
```
  2 0 0 3
    1 4 9
      5 8
+ 3 4 2 6
```

59.
```
  5,6 7 8,9 8 7
  1,4 0 9,3 1 2
    8 9 8,8 8 8
+ 4,7 7 7,9 1 0
```

60.
```
  7 8,8 9 9,3 1 1
    6,7 8 4,1 7 0
  1 1,5 4 1,9 1 3
+     1 0 0,8 1 7
```

55. _____

56. _____

57. _____

58. _____

59. _____

60. _____

61. _____

62. _____

63. _____

64. _____

65. _____

SKILL MAINTENANCE

The exercises that follow are *skill maintenance exercises,* which review any skill previously studied in the text. You can expect such exercises in almost every exercise set.

61. Write standard notation:

$7000 + 900 + 90 + 2.$

62. Write a word name for the number in the following sentence:

Each year Americans eat 5,376,000,000 grams of guacamole dip on Super Bowl Sunday.

63. What does the digit 8 mean in 486,205?

64. Write standard notation:

Twenty-three million.

SYNTHESIS

65. Try to discover a fast way of adding all the numbers from 1 to 100 inclusive.

1.3 *Subtraction*

a | SUBTRACTION AND THE REAL WORLD: TAKE AWAY

Subtraction of whole numbers corresponds to two kinds of situations. The first one is called "take away."

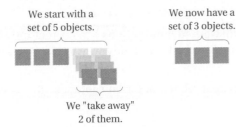

We start with a set of 5 objects.

We now have a set of 3 objects.

We "take away" 2 of them.

The subtraction that corresponds is as follows.

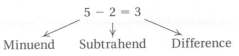

$$5 - 2 = 3$$

Minuend   Subtrahend   Difference

| EXAMPLES | Write a subtraction that corresponds to the situation.

1. A bowler starts with 10 pins and knocks down 8 of them. How many pins are left?

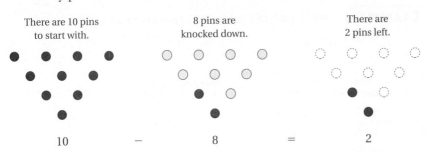

There are 10 pins to start with.

8 pins are knocked down.

There are 2 pins left.

10   −   8   =   2

2. A shopper starts with $100 and spends $65 for groceries. How much money is left?

Amount to begin with

Amount spent for groceries

Amount left

$100   − $65   = $35

Do Exercises 1 and 2.

Write a subtraction that corresponds to the situation.

1. A chef pours 5 oz (ounces) of salsa from a jar that contains 16 oz. How many ounces are left?

2. A ranch covers 400 acres. The owner sells 100 acres of it. How many acres are left?

Answers on page A-1

Write a related addition sentence.

3. $7 - 5 = 2$

4. $17 - 8 = 9$

Write two related subtraction sentences.

5. $5 + 8 = 13$

6. $11 + 3 = 14$

b | RELATED SENTENCES

Subtraction is defined in terms of addition. For example, $5 - 2$ is that number which when added to 2 gives 5. Thus for the subtraction sentence

$$5 - 2 = 3, \quad \text{Taking away 2 from 5 gives 3.}$$

there is a *related addition* sentence

$$5 = 3 + 2. \quad \text{Putting back the 2 gives 5 again.}$$

In fact, we know answers to subtractions are correct only because of the related addition, which provides a handy way to check a subtraction.

EXAMPLE 3   Write a related addition sentence: $8 - 5 = 3$.

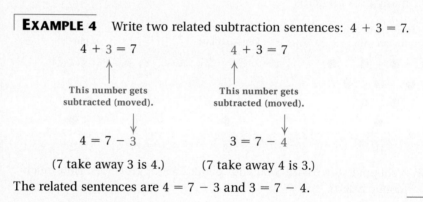

The related sentence is $8 = 3 + 5$.

Do Exercises 3 and 4.

EXAMPLE 4   Write two related subtraction sentences: $4 + 3 = 7$.

$$4 + 3 = 7 \qquad\qquad 4 + 3 = 7$$

This number gets subtracted (moved).   This number gets subtracted (moved).

$$4 = 7 - 3 \qquad\qquad 3 = 7 - 4$$

(7 take away 3 is 4.)   (7 take away 4 is 3.)

The related sentences are $4 = 7 - 3$ and $3 = 7 - 4$.

Do Exercises 5 and 6.

c | HOW MUCH MORE?

The second kind of situation to which subtraction corresponds is called "how much more"? We need the concept of a missing addend for "how-much-more" problems. From the related sentences, we see that finding a *missing addend* is the same as finding a *difference*.

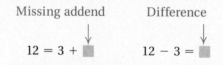

EXAMPLES   Write a subtraction that corresponds to the situation.

5. A student has \$17 and wants to buy a book that costs \$23. How much more is needed to buy the book?

To find the subtraction, we first consider addition.

Amount that the student has	plus	Amount needed	is	Cost of the book
↓	↓	↓	↓	↓
$17	+	■	=	$23

Now we write a related subtraction:

$17 + ■ = 23$

$■ = 23 - 17.$ 17 gets subtracted (moved).

6. It is 134 mi from Los Angeles to San Diego. A driver has gone 90 mi of the trip. How much farther does the driver have to go?

Distance already driven	plus	Distance to drive	is	Distance from L.A. to San Diego
↓	↓	↓	↓	↓
90 mi	+	■	=	134 mi

Now we write a related subtraction:

$90 + ■ = 134$

$■ = 134 - 90.$ 90 gets subtracted (moved).

Do Exercises 7 and 8.

d | SUBTRACTION OF WHOLE NUMBERS

To subtract numbers, we subtract ones first, then tens, then hundreds, and so on.

EXAMPLE 7 Subtract: $9768 - 4320$.

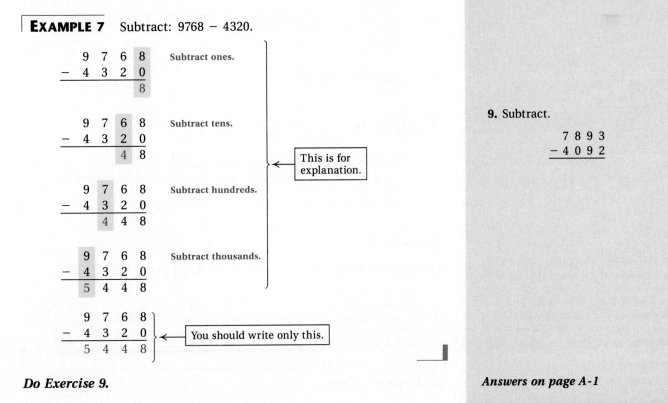

```
  9 7 6 8      Subtract ones.
- 4 3 2 0
        8

  9 7 6 8      Subtract tens.
- 4 3 2 0
      4 8

  9 7 6 8      Subtract hundreds.
- 4 3 2 0
    4 4 8

  9 7 6 8      Subtract thousands.
- 4 3 2 0
  5 4 4 8
```

This is for explanation.

```
  9 7 6 8
- 4 3 2 0
  5 4 4 8
```

You should write only this.

Do Exercise 9.

Answers on page A-1

Write an addition sentence and a related subtraction sentence corresponding to the situation. You need not carry out the subtraction.

7. There are 32 million kangaroos and 15 million people in Australia. How many more kangaroos are there than people?

8. A set of drapes requires 23 yd of material. The decorator has 10 yd of material in stock. How much more must be ordered?

9. Subtract.

```
  7 8 9 3
- 4 0 9 2
```

Subtract. Check by adding.

10. 8 6 8 6   **11.** 7 1 4 5
      − 2 3 5 8   − 2 3 9 8

Subtract.

12. 7 0   **13.** 5 0 3
      − 1 4   − 2 9 8

Subtract.

14. 7 0 0 7   **15.** 6 0 0 0
      − 6 3 4 9   − 3 1 4 9

16. 9 0 3 5
      − 7 4 8 9

To the instructor and the student: This section presented a review of subtraction of whole numbers. Students who are successful should go on to Section 1.4. Those who have trouble should study developmental unit S and then repeat Section 1.3.

Answers on page A-1

Sometimes we need to borrow.

EXAMPLE 8   Subtract: 6246 − 1879.

          3 16
     6 2 4̸ 6̸   We cannot subtract 9 ones from 6 ones, but we can subtract
   − 1 8 7 9   9 ones from 16 ones. We borrow 1 ten to get 16 ones.
           7

           13
       1 3̸ 16
     6 2̸ 4̸ 6̸   We cannot subtract 7 tens from 3 tens, but we can subtract
   − 1 8 7 9   7 tens from 13 tens. We borrow 1 hundred to get 13 tens.
         6 7

       11 13
     5 1̸ 3̸ 16
     6̸ 2̸ 4̸ 6̸   We cannot subtract 8 hundreds from 1 hundred, but we can
   − 1 8 7 9   subtract 8 hundreds from 11 hundreds. We borrow 1 thousand
     4 3 6 7   to get 11 hundreds.

We can always check the answer by adding it to the number being subtracted.

This is what you should write. →

        11 13
      5 1̸ 3̸ 16
      6̸ 2̸ 4̸ 6̸   *Check:*
    − 1 8 7 9
      4 3 6 7

                      1 1 1
                    4 3 6 7
                  + 1 8 7 9
                    6 2 4 6 ←

This answer checks because this is the top number in the subtraction.

Do Exercises 10 and 11.

EXAMPLE 9   Subtract: 902 − 477.

      8 9 12
     9̸ 0̸ 2̸   We cannot subtract 7 ones from 2 ones. We have 9 hundreds, or
   − 4 7 7   90 tens. We borrow 1 ten to get 12 ones. We then have 89 tens.
     4 2 5

Do Exercises 12 and 13.

EXAMPLE 10   Subtract: 8003 − 3667.

      7 9 9 13
     8̸ 0̸ 0̸ 3̸   We have 8 thousands, or 800 tens.
   − 3 6 6 7   We borrow 1 ten to get 13 ones. We then have 799 tens.
     4 3 3 6

EXAMPLES

11. Subtract: 6000 − 3762.

      5 9 9 10
     6̸ 0̸ 0̸ 0̸
   − 3 7 6 2
     2 2 3 8

12. Subtract: 6024 − 2968.

               11
       5 9 1̸ 14
       6̸ 0̸ 2̸ 4̸
     − 2 9 6 8
       3 0 5 6

Do Exercises 14–16.

Exercise Set 1.3

a Write a subtraction that corresponds to the situation. You need not carry out the subtraction.

1. A gasoline station has 2400 gal of lead-free gasoline. One day it sells 800 gal. How many gallons are left in the tank?

2. A consumer has $650 in a checking account and writes a check for $100. How much is left in the account?

b Write a related addition sentence.

3. $7 - 4 = 3$

4. $12 - 5 = 7$

5. $13 - 8 = 5$

6. $9 - 9 = 0$

7. $23 - 9 = 14$

8. $20 - 8 = 12$

9. $43 - 16 = 27$

10. $51 - 18 = 33$

Write two related subtraction sentences.

11. $6 + 9 = 15$

12. $7 + 9 = 16$

13. $8 + 7 = 15$

14. $8 + 0 = 8$

15. $17 + 6 = 23$

16. $11 + 8 = 19$

17. $23 + 9 = 32$

18. $42 + 10 = 52$

c Write an addition sentence and a related subtraction sentence corresponding to the situation. You need not carry out the subtraction.

19. One week before Mother's Day, a florist sets a goal of selling 220 floral bouquets. By Thursday, it has sold 190 bouquets. How many more must it sell by Sunday to meet its goal?

20. Tuition will cost a student $3000. The student has $1250. How much more money is needed?

d Subtract.

21.
$$\begin{array}{r} 1\ 6 \\ -\ \ \ 4 \\ \hline \end{array}$$

22.
$$\begin{array}{r} 8\ 6 \\ -\ 1\ 3 \\ \hline \end{array}$$

23.
$$\begin{array}{r} 6\ 5 \\ -\ 2\ 1 \\ \hline \end{array}$$

24.
$$\begin{array}{r} 8\ 7 \\ -\ 3\ 4 \\ \hline \end{array}$$

25.
$$\begin{array}{r} 8\ 6\ 6 \\ -\ 3\ 3\ 3 \\ \hline \end{array}$$

26.
$$\begin{array}{r} 5\ 2\ 6 \\ -\ 3\ 2\ 3 \\ \hline \end{array}$$

27.
$$\begin{array}{r} 4\ 5\ 4\ 7 \\ -\ 3\ 4\ 2\ 1 \\ \hline \end{array}$$

28.
$$\begin{array}{r} 6\ 8\ 7\ 5 \\ -\ 2\ 1\ 1\ 1 \\ \hline \end{array}$$

29. $86 - 47$

30. $73 - 28$

31. $625 - 327$

32. $726 - 509$

33. $835 - 609$

34. $953 - 246$

35. $981 - 747$

36. $887 - 698$

ANSWERS

1. _____
2. _____
3. _____
4. _____
5. _____
6. _____
7. _____
8. _____
9. _____
10. _____
11. _____
12. _____
13. _____
14. _____
15. _____
16. _____
17. _____
18. _____
19. _____
20. _____
21. _____
22. _____
23. _____
24. _____
25. _____
26. _____
27. _____
28. _____
29. _____
30. _____
31. _____
32. _____
33. _____
34. _____
35. _____
36. _____

37.
$$\begin{array}{r} 7769 \\ -\ 2387 \\ \hline \end{array}$$

38.
$$\begin{array}{r} 6431 \\ -\ 2896 \\ \hline \end{array}$$

39.
$$\begin{array}{r} 3982 \\ -\ 2489 \\ \hline \end{array}$$

40.
$$\begin{array}{r} 7650 \\ -\ 1765 \\ \hline \end{array}$$

41.
$$\begin{array}{r} 5046 \\ -\ 2859 \\ \hline \end{array}$$

42.
$$\begin{array}{r} 6308 \\ -\ 2679 \\ \hline \end{array}$$

43.
$$\begin{array}{r} 7640 \\ -\ 3809 \\ \hline \end{array}$$

44.
$$\begin{array}{r} 8003 \\ -\ \ 599 \\ \hline \end{array}$$

45.
$$\begin{array}{r} 12{,}647 \\ -\ \ 4{,}899 \\ \hline \end{array}$$

46.
$$\begin{array}{r} 16{,}222 \\ -\ \ 5{,}888 \\ \hline \end{array}$$

47.
$$\begin{array}{r} 46{,}771 \\ -\ 12{,}977 \\ \hline \end{array}$$

48.
$$\begin{array}{r} 95{,}654 \\ -\ 48{,}985 \\ \hline \end{array}$$

49.
$$\begin{array}{r} 80 \\ -\ 24 \\ \hline \end{array}$$

50.
$$\begin{array}{r} 40 \\ -\ 37 \\ \hline \end{array}$$

51.
$$\begin{array}{r} 90 \\ -\ 54 \\ \hline \end{array}$$

52.
$$\begin{array}{r} 90 \\ -\ 78 \\ \hline \end{array}$$

53.
$$\begin{array}{r} 140 \\ -\ \ 56 \\ \hline \end{array}$$

54.
$$\begin{array}{r} 470 \\ -\ 188 \\ \hline \end{array}$$

55.
$$\begin{array}{r} 690 \\ -\ 236 \\ \hline \end{array}$$

56.
$$\begin{array}{r} 803 \\ -\ 418 \\ \hline \end{array}$$

57.
$$\begin{array}{r} 903 \\ -\ 132 \\ \hline \end{array}$$

58.
$$\begin{array}{r} 6408 \\ -\ \ 258 \\ \hline \end{array}$$

59.
$$\begin{array}{r} 2300 \\ -\ \ 109 \\ \hline \end{array}$$

60.
$$\begin{array}{r} 3506 \\ -\ 1293 \\ \hline \end{array}$$

61.
$$\begin{array}{r} 6808 \\ -\ 3059 \\ \hline \end{array}$$

62.
$$\begin{array}{r} 7840 \\ -\ 3027 \\ \hline \end{array}$$

63.
$$\begin{array}{r} 8092 \\ -\ 1073 \\ \hline \end{array}$$

64.
$$\begin{array}{r} 6007 \\ -\ 1589 \\ \hline \end{array}$$

65.
$$\begin{array}{r} 7000 \\ -\ 2794 \\ \hline \end{array}$$

66.
$$\begin{array}{r} 8001 \\ -\ 6543 \\ \hline \end{array}$$

67.
$$\begin{array}{r} 48{,}000 \\ -\ 37{,}695 \\ \hline \end{array}$$

68.
$$\begin{array}{r} 17{,}043 \\ -\ 11{,}598 \\ \hline \end{array}$$

SKILL MAINTENANCE

69. What does the digit 7 mean in 6,375,602?

70. Write a word name for 6,375,602.

1.4 *Rounding and Estimating; Order*

a ROUNDING

We round numbers in various situations if we do not need an exact answer. For example, we might round to check if an answer to a problem is reasonable or to check a calculation done by hand or on a calculator. We might also round to see if we are being charged the correct amount in a store.

To understand how to round, we first look at some examples using number lines, even though this is not the way we normally do rounding.

EXAMPLE 1 Round 47 to the nearest ten.

Here is a part of a number line; 47 is between 40 and 50.

Since 47 is closer to 50, we round up to 50.

EXAMPLE 2 Round 42 to the nearest ten.

42 is between 40 and 50.

Since 42 is closer to 40, we round down to 40.

Do Exercises 1–4.

EXAMPLE 3 Round 45 to the nearest ten.

45 is halfway between 40 and 50.

We could round 45 down to 40 or up to 50. We agree to round up to 50.

> When a number is halfway between rounding numbers, round up.

Do Exercises 5–7.

Here is a rule for rounding.

> To round to a certain place:
> **a)** Locate the digit in that place.
> **b)** Then consider the next digit to the right.
> **c)** If the digit to the right is 5 or higher, round up; if the digit to the right is 4 or lower, round down.
> **d)** Change all digits to the right of the rounding location to zeros.

OBJECTIVES

After finishing Section 1.4, you should be able to:

a Round to the nearest ten, hundred, or thousand.

b Estimate sums and differences by rounding.

c Use < or > for ■ to write a true sentence in a situation like 6 ■ 10.

FOR EXTRA HELP

TAPE 1   TAPE 1B   MAC: 1A
                     IBM: 1A

Round to the nearest ten.

1. 37

2. 52

3. 73

4. 98

Round to the nearest ten.

5. 35

6. 75

7. 85

Answers on page A-1

Round to the nearest ten.

8. 137

9. 473

10. 235

11. 285

Round to the nearest hundred.

12. 641

13. 759

14. 750

15. 9325

Round to the nearest thousand.

16. 7896

17. 8459

18. 19,343

19. 68,500

Answers on page A-1

EXAMPLE 4 Round 6485 to the nearest ten.

a) Locate the digit in the tens place.

6 4 8 5
   ↑

b) Then consider the next digit to the right.

6 4 8 5
     ↑

c) Since that digit is 5 or higher, round 8 tens up to 9 tens.

d) Change all digits to the right of the tens digit to zeros.

6 4 9 0 ← This is the answer.

EXAMPLE 5 Round 6485 to the nearest hundred.

a) Locate the digit in the hundreds place.

6 4 8 5
  ↑

b) Then consider the next digit to the right.

6 4 8 5
   ↑

c) Since that digit is 5 or higher, round 4 hundreds up to 5 hundreds.

d) Change all digits to the right of hundreds to zeros.

6 5 0 0 ← This is the answer.

EXAMPLE 6 Round 6485 to the nearest thousand.

a) Locate the digit in the thousands place.

6 4 8 5
↑

b) Then consider the next digit to the right.

6 4 8 5
 ↑

c) Since that digit is 4 or lower, round down, meaning that 6 thousands stays as 6 thousands.

d) Change all digits to the right of thousands to zeros.

6 0 0 0 ← This is the answer.

Caution! 7000 is not a correct answer to Example 6. It is incorrect to round from the ones digit over, as follows:

6485, 6490, 6500, 7000.

Do Exercises 8–19.

There are many methods of rounding. For example, in computer applications, the rounding of 8563 to the nearest hundred might be done using a different rule called **truncating**, meaning that we simply change all digits to the right of the rounding location to zeros. Thus, 8563 would round to 8500, which is not the same answer that we would get using the rule discussed in this section.

b ESTIMATING

Estimating is used to simplify a problem so that it can then be solved easily or mentally. Rounding is used when estimating. There are many ways to estimate.

EXAMPLE 7 Larry Bird and Magic Johnson are retired stars of the National Basketball Association. Larry scored a total of 21,791 points in his career, and Magic scored a total of 17,239. Estimate how many points they scored in all.

There are many ways to get an answer, but there is no one perfect answer based on how the problem is worded. Let's consider a couple of methods.

METHOD 1. Round each number to the nearest thousand and then add.

```
  2 1,7 9 1      2 2,0 0 0
+ 1 7,2 3 9    + 1 7,0 0 0
               3 9,0 0 0  ← Estimated answer
```

METHOD 2. We might use a less formal approach, depending on how specific we want the answer to be. We note that both numbers are close to 20,000, and so the total is close to 40,000. In some contexts, such as a sports commentary, this would be sufficient.

The point to be made is that estimating can be done in many ways and can have many answers, even though in the problems that follow we ask you to round in a specific way.

EXAMPLE 8 Estimate this sum by rounding to the nearest ten:

78 + 49 + 31 + 85.

We round each number to the nearest ten. Then we add.

```
  7 8      8 0
  4 9      5 0
  3 1      3 0
+ 8 5    + 9 0
         2 5 0  ← Estimated answer
```

Do Exercise 20.

EXAMPLE 9 Estimate this sum by rounding to the nearest hundred:

850 + 674 + 986 + 839.

We have

```
  8 5 0      9 0 0
  6 7 4      7 0 0
  9 8 6    1 0 0 0
+ 8 3 9    +  8 0 0
           3 4 0 0
```

Do Exercise 21.

20. Estimate the sum by rounding to the nearest ten. Show your work.

```
  7 4
  2 3
  3 5
+ 6 6
```

21. Estimate the sum by rounding to the nearest hundred. Show your work.

```
  6 5 0
  6 8 5
  2 3 8
+ 1 6 8
```

Answers on page A-1

22. Estimate the difference by rounding to the nearest hundred. Show your work.

$$\begin{array}{r} 9\,2\,8\,5 \\ -\ 6\,7\,3\,9 \end{array}$$

23. Estimate the difference by rounding to the nearest thousand. Show your work.

$$\begin{array}{r} 2\,3{,}2\,7\,8 \\ -\ 1\,1{,}6\,9\,8 \end{array}$$

Use < or > for ▨ to write a true sentence. Draw a number line if necessary.

24. 8 ▨ 12

25. 12 ▨ 8

26. 76 ▨ 64

27. 64 ▨ 76

28. 217 ▨ 345

29. 345 ▨ 217

Answers on page A-1

EXAMPLE 10   Estimate the difference by rounding to the nearest thousand: 9324 − 2849.

We have

$$\begin{array}{r} 9\,3\,2\,4 \\ -\ 2\,8\,4\,9 \end{array} \qquad \begin{array}{r} 9\,0\,0\,0 \\ -\ 3\,0\,0\,0 \\ \hline 6\,0\,0\,0 \end{array}$$

Do Exercises 22 and 23.

The sentence 7 − 5 = 2 says that 7 − 5 is the same as 2. Later we will use the symbol ≈ when rounding. This symbol means **"is approximately equal to."** Thus, when 687 is rounded to the nearest ten, we may write

$$687 \approx 690.$$

C   ORDER

We know that 2 is not the same as 5. We express this by the sentence 2 ≠ 5. We also know that 2 is less than 5. We can see this order on a number line: 2 is to the left of 5.

For any whole numbers a and b:

1. $a < b$ (read "a is less than b") is true when a is to the left of b on a number line.

2. $a > b$ (read "a is greater than b") is true when a is to the right of b on a number line.

We call < and > *inequality* symbols.

EXAMPLE 11   Use < or > for ▨ to write a true sentence: 7 ▨ 11.

Since 7 is to the left of 11, 7 < 11.

EXAMPLE 12   Use < or > for ▨ to write a true sentence: 92 ▨ 87.

Since 92 is to the right of 87, 92 > 87.

A sentence like 8 + 5 = 13 is called an **equation**. A sentence like 7 < 11 is called an **inequality**. The sentence 7 < 11 is a true inequality. The sentence 23 > 69 is a false inequality.

Do Exercises 24–29.

Exercise Set 1.4

ANSWERS

1. _____
2. _____
3. _____
4. _____
5. _____
6. _____
7. _____
8. _____
9. _____
10. _____
11. _____
12. _____
13. _____
14. _____
15. _____
16. _____
17. _____
18. _____
19. _____
20. _____
21. _____
22. _____
23. _____
24. _____
25. _____
26. _____
27. _____
28. _____
29. _____
30. _____
31. _____
32. _____

a Round to the nearest ten.

1. 48   **2.** 17   **3.** 67   **4.** 99

5. 731   **6.** 532   **7.** 895   **8.** 798

Round to the nearest hundred.

9. 146   **10.** 874   **11.** 957   **12.** 650

13. 9079   **14.** 4645   **15.** 2850   **16.** 4402

Round to the nearest thousand.

17. 5876   **18.** 4500   **19.** 7500   **20.** 2001

21. 45,340   **22.** 735,562   **23.** 373,405   **24.** 13,855

b Estimate the sum or difference by rounding to the nearest ten. Show your work.

25.
```
  7 8 2 8
+ 9 7 8 6
```

26.
```
  6 2
  9 7
  4 6
+ 8 8
```

27.
```
  8 0 7 4
- 2 3 4 7
```

28.
```
  6 7 3
-   2 8
```

Estimate the sum by rounding to the nearest ten. Do any of the sums seem to be incorrect? Which ones?

29.
```
  4 5
  7 7
  2 5
+ 5 6
-----
3 4 3
```

30.
```
  4 1
  2 1
  5 5
+ 6 0
-----
1 7 7
```

31.
```
  6 2 2
    7 8
    8 1
+ 1 1 1
-------
  9 3 2
```

32.
```
  8 3 6
  3 7 4
  7 9 4
+ 9 3 8
-------
3 9 4 7
```

33.

34.

35.

36.

37.

38.

39.

40.

41.

42.

43.

44.

45.

46.

47.

48.

49.

50.

51.

52.

53.

54.

55.

56.

57.

58.

59.

60.

61.

62.

63.

64.

Estimate the sum or difference by rounding to the nearest hundred. Show your work.

33.   7 3 4 8
         + 9 2 4 7

34.   5 6 8
           4 7 2
           9 3 8
         + 4 0 2

35.   6 8 5 2
         − 1 7 4 8

36.   9 4 3 8
         − 2 7 8 7

Estimate the sum by rounding to the nearest hundred. Do any of the sums seem to be incorrect? Which ones?

37.   2 1 6
             8 4
           7 4 5
         + 5 9 5
           1 6 4 0

38.   4 8 1
           7 0 2
           6 2 3
         + 1 0 4 3
           1 8 4 9

39.   7 5 0
           4 2 8
             6 3
         + 2 0 5
           1 4 4 6

40.   3 2 6
           2 7 5
           7 5 8
         + 9 4 3
           2 3 0 2

Estimate the sum or difference by rounding to the nearest thousand. Show your work.

41.   9 6 4 3
           4 8 2 1
           8 9 4 3
         + 7 0 0 4

42.   7 6 4 8
           9 3 4 8
           7 8 4 2
         + 2 2 2 2

43.   9 2,1 4 9
         − 2 2,5 5 5

44.   8 4,8 9 0
         − 1 1,1 1 0

$\boxed{\text{c}}$ Use < or > for ▨ to write a true sentence. Draw a number line if necessary.

45. 0 ▨ 17

46. 32 ▨ 0

47. 34 ▨ 12

48. 28 ▨ 18

49. 1000 ▨ 1001

50. 77 ▨ 117

51. 133 ▨ 132

52. 999 ▨ 997

53. 460 ▨ 17

54. 345 ▨ 456

55. 37 ▨ 11

56. 12 ▨ 32

SKILL MAINTENANCE

Add.

57.   6 7,7 8 9
         + 1 8,9 6 5

58.   9 0 0 2
         + 4 5 8 7

Subtract.

59.   6 7,7 8 9
         − 1 8,9 6 5

60.   9 0 0 2
         − 4 5 8 7

SYNTHESIS

61–64. ▦ Use a calculator to find the sums or differences in Exercises 41–44. Since you can still make errors on a calculator—say, by pressing the wrong buttons—you can check your answers by estimating.

1.5 *Multiplication*

a MULTIPLICATION AND THE REAL WORLD

Multiplication of whole numbers corresponds to two kinds of situations.

REPEATED ADDITION

The multiplication 3×5 corresponds to this repeated addition:

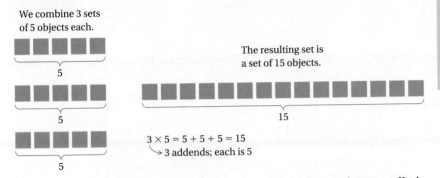

We combine 3 sets of 5 objects each.

The resulting set is a set of 15 objects.

$3 \times 5 = 5 + 5 + 5 = 15$
↘ 3 addends; each is 5

We say that the **product** of 3 and 5 is 15. The numbers 3 and 5 are called **factors**.

$$3 \times 5 \quad = \quad 15$$
↑ ↑   ↑
Factors   Product

RECTANGULAR ARRAYS

The multiplication 3×5 corresponds to this rectangular array:

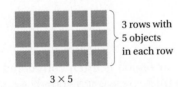

3 rows with 5 objects in each row

3×5

When you write a multiplication corresponding to a real-world situation, you should think of either a rectangular array or repeated addition. In some cases, it may help to think both ways.

We have used an "×" to denote multiplication. A dot "·" is also commonly used. (The dot was invented by the German mathematician Gottfried Wilhelm von Leibniz in 1698.) Parentheses are also used to denote multiplication. For example, $(3)(5) = 15$.

OBJECTIVES

After finishing Section 1.5, you should be able to:

a   Determine what multiplication corresponds to a situation.

b   Multiply whole numbers.

c   Estimate products by rounding.

FOR EXTRA HELP

TAPE 2   TAPE 2A   MAC: 1B
                                       IBM: 1B

Write a multiplication that corresponds to the situation.

1. Rambeau is training for a long-distance running race. He runs 4 mi on each of 8 days. How many miles does he run in all?

2. A lab technician pours 75 mL (milliliters) of acid into each of 10 beakers. How much acid is poured in all?

3. A band is arranged rectangularly in 12 rows with 20 members in each row. How many people are in the band?

Answers on page A-2

EXAMPLES Write a multiplication that corresponds to the situation.

1. It is known that Americans drink 23 million gal of soft drinks per day (*per day* means *each day*). What quantity of soft drinks is consumed every 5 days?

We draw a picture or at least visualize the situation. Repeated addition fits best in this case.

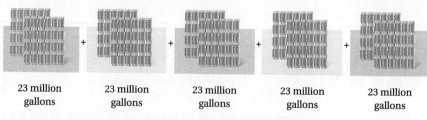

| 23 million gallons | 23 million gallons | 23 million gallons | 23 million gallons | 23 million gallons |

5 · 23 million gallons = 115 million gallons

2. One side of a building has 6 floors with 7 windows on each floor. How many windows are there on that side of the building?

We have a rectangular array and can easily draw a sketch.

6 · 7 = 42

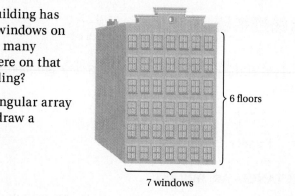

6 floors

7 windows

Do Exercises 1–3.

AREA

The area of a rectangular region is often considered to be the number of square units needed to fill it. Here is a rectangle 4 cm long and 3 cm wide. It takes 12 square centimeters (cm²) to fill it.

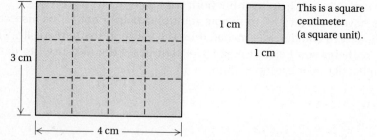

3 cm

4 cm

1 cm
1 cm

This is a square centimeter (a square unit).

In this case, we have a rectangular array. The number of square units is 3 · 4, or 12.

EXAMPLE 3 Write a multiplication that corresponds to this situation.

A rectangular floor is 10 ft long and 8 ft wide. Find its area.

We draw a picture.

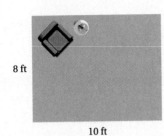

8 ft

10 ft

If we think of filling the rectangle with square feet, we have a rectangular array. The length $l = 10$ ft, and the width $w = 8$ ft. The area A is given by

$$A = l \cdot w = 10 \times 8 = 80 \text{ square feet (ft}^2\text{)}.$$

Do Exercise 4.

b MULTIPLICATION OF WHOLE NUMBERS

Let's find the product

$$\begin{array}{r} 5\ 4 \\ \times\ 3\ 2 \\ \hline \end{array}$$

To do this, we multiply 54 by 2, then 54 by 30, and then add.

$$\begin{array}{r} 5\ 4 \\ \times\ 2 \\ \hline 1\ 0\ 8 \end{array} \qquad \begin{array}{r} 1 \\ 5\ 4 \\ \times\ 3\ 0 \\ \hline 1\ 6\ 2\ 0 \end{array}$$

Since we are going to add the results, let's write the work this way.

$$\begin{array}{r} 5\ 4 \\ \times\ 3\ 2 \\ \hline 1\ 0\ 8 \qquad \text{Multiplying by 2} \\ 1\ 6\ 2\ 0 \qquad \text{Multiplying by 30} \\ \hline 1\ 7\ 2\ 8 \qquad \text{Adding} \end{array}$$

The fact that we can do this is based on a property called the **distributive law.** It says that to multiply a number by a sum, $a \cdot (b + c)$, we can multiply the parts by a and then add like this: $(a \cdot b) + (a \cdot c)$. Thus, $a \cdot (b + c) = (a \cdot b) + (a \cdot c)$. Applied to the above example, the distributive law gives us

$$54 \cdot 32 = 54 \cdot (30 + 2) = (54 \cdot 30) + (54 \cdot 2).$$

4. What is the area of this pool table?

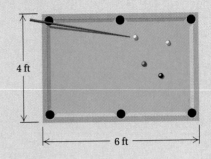

4 ft

6 ft

Answer on page A-1

Multiply.

5.
$$\begin{array}{r} 4\ 5 \\ \times\ 2\ 3 \\ \hline \end{array}$$

6. 48×63

Multiply.

7.
$$\begin{array}{r} 7\ 4\ 6 \\ \times\ \ \ 6\ 2 \\ \hline \end{array}$$

8. 245×837

EXAMPLE 4   Multiply: 43×57.

$$\begin{array}{r} \overset{2}{} \\ 5\ 7 \\ \times\ 4\ 3 \\ \hline 1\ 7\ 1 \end{array}$$ Multiplying by 3

$$\begin{array}{r} \overset{2}{\overset{2}{}} \\ 5\ 7 \\ \times\ 4\ 3 \\ \hline 1\ 7\ 1 \\ 2\ 2\ 8\ 0 \end{array}$$ Multiplying by 40. (We write a 0 and then multiply 57 by 4.)

You may have learned that such a 0 does not have to be written. You may omit it if you wish. If you do omit it, remember, when multiplying by tens, to put the answer in the tens place.

$$\begin{array}{r} \overset{2}{\overset{2}{}} \\ 5\ 7 \\ \times\ 4\ 3 \\ \hline 1\ 7\ 1 \\ 2\ 2\ 8\ 0 \\ \hline 2\ 4\ 5\ 1 \end{array}$$ Adding

Do Exercises 5 and 6.

EXAMPLE 5   Multiply: 457×683.

$$\begin{array}{r} \overset{5}{}\ \overset{2}{} \\ 6\ 8\ 3 \\ \times\ 4\ 5\ 7 \\ \hline 4\ 7\ 8\ 1 \end{array}$$ Multiplying 683 by 7

$$\begin{array}{r} \overset{4}{}\ \overset{1}{} \\ \overset{5}{}\ \overset{2}{} \\ 6\ 8\ 3 \\ \times\ 4\ 5\ 7 \\ \hline 4\ 7\ 8\ 1 \\ 3\ 4\ 1\ 5\ 0 \end{array}$$ Multiplying 683 by 50

$$\begin{array}{r} \overset{3}{}\ \overset{1}{} \\ \overset{4}{}\ \overset{1}{} \\ \overset{5}{}\ \overset{2}{} \\ 6\ 8\ 3 \\ \times\ 4\ 5\ 7 \\ \hline 4\ 7\ 8\ 1 \\ 3\ 4\ 1\ 5\ 0 \\ 2\ 7\ 3\ 2\ 0\ 0 \\ \hline 3\ 1\ 2,1\ 3\ 1 \end{array}$$ Multiplying 683 by 400

Adding

Do Exercises 7 and 8.

ZEROS IN MULTIPLICATION

EXAMPLE 6   Multiply: 306×274.

Note that $306 = 3$ hundreds $+ 6$ ones.

```
        2 7 4
    ×   3 0 6
        1 6 4 4      Multiplying by 6
      8 2 2 0 0      Multiplying by 3 hundreds. (We write 00 and then
                     multiply 274 by 3.)
      8 3,8 4 4      Adding
```

Do Exercises 9–11.

EXAMPLE 7   Multiply: 360×274.

Note that $360 = 3$ hundreds $+ 6$ tens.

```
        2 7 4       Multiplying by 6 tens. (We write 0
    ×   3 6 0       and then multiply 274 by 6.)
      1 6 4 4 0     Multiplying by 3 hundreds. (We write 00
      8 2 2 0 0     and then multiply 274 by 3.)
      9 8,6 4 0     Adding
```

Do Exercises 12–15.

Note the following.

$$3 \cdot 5 = 15 \qquad 5 \cdot 3 = 15$$

If we rotate the array on the left, we get the array on the right. The answers are the same. This illustrates the **commutative law of multiplication.** It says that we can multiply numbers in any order: $a \cdot b = b \cdot a$.

Do Exercise 16.

Multiply.

9.
```
      4 7 2
  ×   3 0 6
```

10. 408×704

11.
```
    2 3 4 4
  × 6 0 0 5
```

Multiply.

12.
```
      4 7 2
  ×   8 3 0
```

13.
```
    2 3 4 4
  × 7 4 0 0
```

14. 100×562

15. 1000×562

16. a) Find $23 \cdot 47$.

b) Find $47 \cdot 23$.

c) Compare your answers to parts (a) and (b).

Answers on page A-1

Multiply.

17. $5 \cdot 2 \cdot 4$

18. $5 \cdot 1 \cdot 3$

19. Estimate the product by rounding to the nearest ten and the nearest hundred. Show your work.

$$\begin{array}{r} 8\ 3\ 7 \\ \times\ 2\ 4\ 5 \\ \hline \end{array}$$

To multiply three or more numbers, we usually group them so that we multiply two at a time. Consider $2 \cdot (3 \cdot 4)$ and $(2 \cdot 3) \cdot 4$. The parentheses tell what to do first:

$$2 \cdot (3 \cdot 4) = 2 \cdot (12) = 24. \qquad \text{We multiply 3 and 4, then 2.}$$

We can also multiply 2 and 3, then 4:

$$(2 \cdot 3) \cdot 4 = (6) \cdot 4 = 24.$$

Either way we get 24. It does not matter how we group the numbers. This illustrates that **multiplication is associative**: $a \cdot (b \cdot c) = (a \cdot b) \cdot c$. Together the commutative and associative laws tell us that to multiply more than two numbers, we can use any order and grouping we wish.

Do Exercises 17 and 18.

c ROUNDING AND ESTIMATING

EXAMPLE 8 Estimate the following product by rounding to the nearest ten and to the nearest hundred: 683×457.

Exact	*Nearest ten*	*Nearest hundred*
$\begin{array}{r} 6\ 8\ 3 \\ \times\ 4\ 5\ 7 \\ \hline 4\ 7\ 8\ 1 \\ 3\ 4\ 1\ 5\ 0 \\ 2\ 7\ 3\ 2\ 0\ 0 \\ \hline 3\ 1\ 2\ 1\ 3\ 1 \end{array}$	$\begin{array}{r} 6\ 8\ 0 \\ \times\ 4\ 6\ 0 \\ \hline 4\ 0\ 8\ 0\ 0 \\ 2\ 7\ 2\ 0\ 0\ 0 \\ \hline 3\ 1\ 2\ 8\ 0\ 0 \end{array}$	$\begin{array}{r} 7\ 0\ 0 \\ \times\ 5\ 0\ 0 \\ \hline 3\ 5\ 0\ 0\ 0\ 0 \end{array}$

Why does the rounding in Example 8 give a larger answer than the exact one?

Do Exercise 19.

To the instructor and the student: This section presented a review of multiplication of whole numbers. Students who are successful should go on to Section 1.6. Those who have trouble should study developmental unit M and then repeat Section 1.5.

Answers on page A-1

Exercise Set 1.5

a Write a multiplication that corresponds to the situation.

1. A store sold 32 calculators at $10 each. How much money did the store receive for the calculators?

2. There are 7 days in a week. How many days are there in 16 weeks?

3. A checkerboard contains 8 rows with 8 squares in each row. How many squares in all are there on a checkerboard?

4. A beverage carton contains 8 cans, each of which holds 12 oz. How many ounces are there in the carton?

What is the area of the region?

5.

```
        3 ft
  6 ft
```

6.

```
     7 mi
  7 mi
```

b Multiply.

7.
```
   8 7
 × 1 0
```

8.
```
   1 0 0
 ×    9 6
```

9.
```
   2 3 4 0
 × 1 0 0 0
```

10.
```
   8 0 0
 ×    7 0
```

11.
```
   6 5
 ×   8
```

12.
```
   8 7
 ×   4
```

13.
```
   9 4
 ×   6
```

14.
```
   7 6
 ×   9
```

15. 3 · 509

16. 7 · 806

17. 7 · 9229

18. 4 · 7867

19. 90 · 53

20. 60 · 78

21. 47 · 85

22. 34 · 87

23.
```
   6 4 0
 ×    7 2
```

24.
```
   6 6 6
 ×    6 6
```

25.
```
   4 4 4
 ×    3 3
```

26.
```
   5 0 9
 ×    8 8
```

27.
```
   5 0 9
 × 4 0 8
```

28.
```
   4 3 2
 × 3 7 5
```

29.
```
   8 5 3
 × 9 3 6
```

30.
```
   3 4 6
 × 6 5 0
```

ANSWERS

1. _____
2. _____
3. _____
4. _____
5. _____
6. _____
7. _____
8. _____
9. _____
10. _____
11. _____
12. _____
13. _____
14. _____
15. _____
16. _____
17. _____
18. _____
19. _____
20. _____
21. _____
22. _____
23. _____
24. _____
25. _____
26. _____
27. _____
28. _____
29. _____
30. _____

31. $\begin{array}{r} 489 \\ \times\ 340 \\ \hline \end{array}$
32. $\begin{array}{r} 7080 \\ \times\ 160 \\ \hline \end{array}$
33. $\begin{array}{r} 4378 \\ \times 2694 \\ \hline \end{array}$
34. $\begin{array}{r} 8007 \\ \times\ 480 \\ \hline \end{array}$

35. $\begin{array}{r} 6428 \\ \times 3224 \\ \hline \end{array}$
36. $\begin{array}{r} 8928 \\ \times 3172 \\ \hline \end{array}$
37. $\begin{array}{r} 3482 \\ \times\ 104 \\ \hline \end{array}$
38. $\begin{array}{r} 6408 \\ \times 6064 \\ \hline \end{array}$

39. $\begin{array}{r} 5006 \\ \times 4008 \\ \hline \end{array}$
40. $\begin{array}{r} 6789 \\ \times 2330 \\ \hline \end{array}$
41. $\begin{array}{r} 5608 \\ \times 4500 \\ \hline \end{array}$
42. $\begin{array}{r} 4560 \\ \times 7890 \\ \hline \end{array}$

43. $\begin{array}{r} 876 \\ \times 345 \\ \hline \end{array}$
44. $\begin{array}{r} 355 \\ \times 299 \\ \hline \end{array}$
45. $\begin{array}{r} 7889 \\ \times 6224 \\ \hline \end{array}$
46. $\begin{array}{r} 6501 \\ \times 3449 \\ \hline \end{array}$

47. $\begin{array}{r} 555 \\ \times\ 55 \\ \hline \end{array}$
48. $\begin{array}{r} 888 \\ \times\ 88 \\ \hline \end{array}$
49. $\begin{array}{r} 734 \\ \times 407 \\ \hline \end{array}$
50. $\begin{array}{r} 5080 \\ \times\ 302 \\ \hline \end{array}$

c Estimate the product by rounding to the nearest ten. Show your work.

51. $\begin{array}{r} 45 \\ \times 67 \\ \hline \end{array}$
52. $\begin{array}{r} 51 \\ \times 78 \\ \hline \end{array}$
53. $\begin{array}{r} 34 \\ \times 29 \\ \hline \end{array}$
54. $\begin{array}{r} 63 \\ \times 54 \\ \hline \end{array}$

Estimate the product by rounding to the nearest hundred. Show your work.

55. $\begin{array}{r} 876 \\ \times 345 \\ \hline \end{array}$
56. $\begin{array}{r} 355 \\ \times 299 \\ \hline \end{array}$
57. $\begin{array}{r} 432 \\ \times 199 \\ \hline \end{array}$
58. $\begin{array}{r} 789 \\ \times 434 \\ \hline \end{array}$

Estimate the product by rounding to the nearest thousand. Show your work.

59. $\begin{array}{r} 5608 \\ \times 4576 \\ \hline \end{array}$
60. $\begin{array}{r} 2344 \\ \times 6123 \\ \hline \end{array}$
61. $\begin{array}{r} 7888 \\ \times 6224 \\ \hline \end{array}$
62. $\begin{array}{r} 6501 \\ \times 3449 \\ \hline \end{array}$

SKILL MAINTENANCE

63. Add.

$$\begin{array}{r} 20 \\ 850 \\ +3500 \\ \hline \end{array}$$

64. Subtract.

$$\begin{array}{r} 6003 \\ -2894 \\ \hline \end{array}$$

65. Round 2345 to the nearest ten, then to the nearest hundred, and then to the nearest thousand.

1.6 *Division*

a DIVISION AND THE REAL WORLD

Division of whole numbers corresponds to two kinds of situations. Consider the division $20 \div 5$, read "20 divided by 5." We can think of 20 objects arranged in a rectangular array. We ask "How many rows, each with 5 objects, are there?"

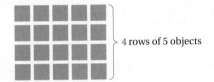

4 rows of 5 objects

Since there are 4 rows of 5 objects each, we have

$20 \div 5 = 4$.

We can also ask, "If we make 5 rows, how many objects will there be in each row?"

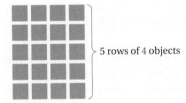

5 rows of 4 objects

Since there are 4 objects in each of the 5 rows, we have

$20 \div 5 = 4$.

We also write a division such as $20 \div 5$ as

$20/5 \quad \text{or} \quad \dfrac{20}{5}$.

EXAMPLE 1 Write a division that corresponds to this situation.

A parent gives \$24 to 3 children, with each child getting the same amount. How much does each child get?

We think of an array with 3 rows. Each row will go to a child. How many dollars will be in each row?

3 rows with 8 in each row

$24 \div 3 = 8$

OBJECTIVES

After finishing Section 1.6, you should be able to:

a Write a division that corresponds to a situation.

b Given a division sentence, write a related multiplication sentence; and given a multiplication sentence, write two related division sentences.

c Divide whole numbers.

FOR EXTRA HELP

TAPE 2   TAPE 2A   MAC: 1B
                                 IBM: 1B

Write a division that corresponds to the situation. You need not carry out the division.

1. There are 112 students in a college band, and they are marching with 14 in each row. How many rows are there?

EXAMPLE 2 Write a division that corresponds to this situation. You need not carry out the division.

How many cassette players that cost $45 each can be purchased for $495?

We think of an array with 45 one-dollar bills in each row. The money in each row will buy a cassette player. How many rows will there be?

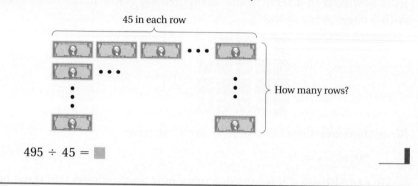

45 in each row

How many rows?

$$495 \div 45 = \blacksquare$$

Whenever we have a rectangular array, we know the following:

(The total number) ÷ (The number of rows) =
(The number in each row).

Also:

(The total number) ÷ (The number in each row) =
(The number of rows).

Do Exercises 1 and 2.

2. A college band is in a rectangular array. There are 112 students in the band, and they are marching in 8 rows. How many students are there in each row?

b RELATED SENTENCES

By looking at rectangular arrays, we can see how multiplication and division are related. The following array shows that $4 \cdot 5 = 20$.

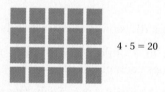

$4 \cdot 5 = 20$

The array also shows the following:

$$20 \div 5 = 4 \quad \text{and} \quad 20 \div 4 = 5.$$

Division is actually defined in terms of multiplication. For example, $20 \div 5$ is defined to be the number that when multiplied by 5 gives 20. Thus, for every division sentence, there is a related multiplication sentence.

$20 \div 5 = 4$ **Division sentence**

$20 = 4 \cdot 5$ **Related multiplication sentence**

To get the related multiplication sentence, we move the 5 to the other side and then write a multiplication.

EXAMPLE 3 Write a related multiplication sentence: $12 \div 6 = 2$.

We have

$12 \div 6 = 2$ **This number moves to the right.**

The related multiplication sentence is $12 = 2 \cdot 6$.

By the commutative law of multiplication, there is also another multiplication sentence: $12 = 6 \cdot 2$.

Do Exercises 3 and 4.

For every multiplication sentence, we can write related divisions, as we can see from the preceding array. We move one of the factors to the opposite side and then write a division.

EXAMPLE 4 Write two related division sentences: $7 \cdot 8 = 56$.

We move a factor to the other side and then write a division:

$7 \cdot 8 = 56$ $7 \cdot 8 = 56$

$7 = 56 \div 8$ $8 = 56 \div 7$

Do Exercises 5 and 6.

With multiplication and division, we use the following words.

$$14 \quad = \quad 7 \quad \cdot \quad 2$$
Product   Factor   Factor

$$14 \quad \div \quad 7 \quad = \quad 2$$
Dividend   Divisor   Quotient

c | DIVISION OF WHOLE NUMBERS

Multiplication can be thought of as repeated addition. Division can be thought of as repeated subtraction. Compare.

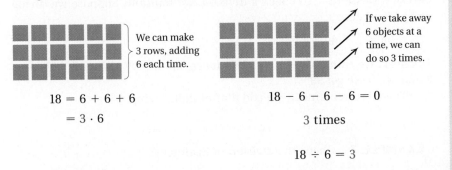

We can make 3 rows, adding 6 each time.

$18 = 6 + 6 + 6$
$\quad = 3 \cdot 6$

If we take away 6 objects at a time, we can do so 3 times.

$18 - 6 - 6 - 6 = 0$

3 times

$18 \div 6 = 3$

Write a related multiplication sentence.

3. $15 \div 3 = 5$

4. $72 \div 8 = 9$

Write two related division sentences.

5. $6 \cdot 2 = 12$

6. $7 \cdot 6 = 42$

Answers on page A-1

Divide by repeated subtraction. Then check.

7. 54 ÷ 9

8. 61 ÷ 9

9. 53 ÷ 12

10. 157 ÷ 24

To divide by repeated subtraction, we keep track of the number of times we subtract.

EXAMPLE 5 Divide by repeated subtraction: 20 ÷ 4.

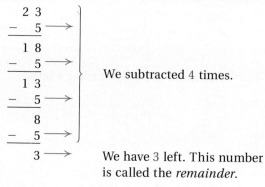

We subtracted 5 times,
so 20 ÷ 4 = 5.

EXAMPLE 6 Divide by repeated subtraction: 23 ÷ 5.

$$
\begin{array}{r}
2\;3 \\
-\quad 5 \longrightarrow \\
\hline
1\;8 \\
-\quad 5 \longrightarrow \\
\hline
1\;3 \\
-\quad 5 \longrightarrow \\
\hline
8 \\
-\quad 5 \longrightarrow \\
\hline
3 \longrightarrow
\end{array}
$$

We subtracted 4 times.

We have 3 left. This number is called the *remainder*.

We write

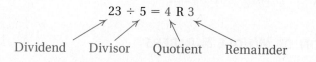

23 ÷ 5 = 4 R 3

Dividend   Divisor   Quotient   Remainder

CHECKING DIVISIONS. To check a division, we multiply. Suppose we divide 98 by 2 and get 49:

98 ÷ 2 = 49.

To check, we think of the related sentence 49 · 2 = . We multiply 49 by 2 and see if we get 98.

If there is a remainder, we add it after multiplying.

EXAMPLE 7 Check the division in Example 6.

We found that 23 ÷ 5 = 4 R 3. To check, we multiply 5 by 4. This gives us 20. Then we add 3 to get 23. The dividend is 23, so the answer checks.

Do Exercises 7–10.

When we use the process of long division, we are doing repeated subtraction, even though we are going about it in a different way.

To divide, we start from the digit of highest place value in the dividend and work down to the lowest through the remainders. At each step we ask if there are multiples of the divisor in the quotient.

EXAMPLE 8   Divide and check: $3642 \div 5$.

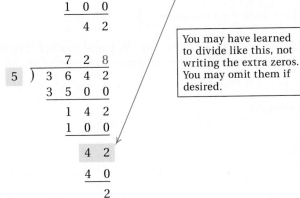

1. We start with the thousands digit in the dividend. Are there any thousands in the thousands place of the quotient? No; $5 \cdot 1000 = 5000$, and 5000 is larger than 3000.

2. Now we go to the hundreds place in the dividend. Are there any hundreds in the hundreds place of the quotient? Think of the dividend as 36 hundreds. Estimate 7 hundreds. Write 7 in the hundreds place, multiply 700 by 5, write the answer below 3642, and subtract.

3. a) We go to the tens place of the first remainder. Are there any tens in the tens place of the quotient? To answer the question, think of the first remainder as 14 tens. Estimate 3 tens. When we multiply, we get 150, which is too large.

   b) We lower our estimate to 2 tens. Write 2 in the tens place, multiply 20 by 5, and subtract.

4. We go to the ones place of the second remainder. Are there any ones in the ones place of the quotient? To answer the question, think of the second remainder as 42 ones. Estimate 8 ones. Write 8 in the ones place, multiply 8 by 5, and subtract.

You may have learned to divide like this, not writing the extra zeros. You may omit them if desired.

```
     7 2 8
5 ) 3 6 4 2
    3 5
      1 4
      1 0
        4 2
        4 0
          2
```

The answer is 728 R 2. To check, we multiply 728 by 5. This gives us 3640. Then we add 2 to get 3642. The dividend is 3642, so the answer checks.

Do Exercises 11–13.

Divide and check.

11. $4 \overline{) 2\ 3\ 9}$

12. $6 \overline{) 8\ 8\ 5\ 5}$

13. $5 \overline{) 5\ 0\ 7\ 5}$

Answers on page A-1

Divide.

14. $4\,5\,\overline{)\,6\,0\,3\,0\,}$

Sometimes rounding the divisor helps us find estimates.

EXAMPLE 9   Divide: $8904 \div 42$.

We mentally round 42 to 40.

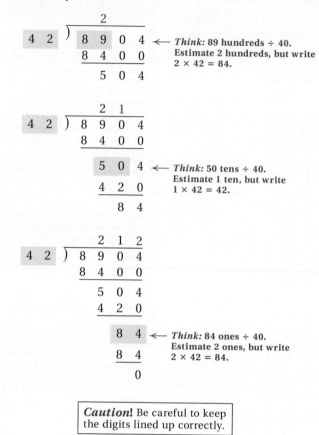

$\begin{array}{r} 2 \\ 4\,2\,\overline{)\,8\,9\,0\,4\,} \\ 8\,4\,0\,0 \\ \hline 5\,0\,4 \end{array}$ ← *Think:* 89 hundreds ÷ 40.
   Estimate 2 hundreds, but write
   $2 \times 42 = 84$.

← *Think:* 50 tens ÷ 40.
   Estimate 1 ten, but write
   $1 \times 42 = 42$.

← *Think:* 84 ones ÷ 40.
   Estimate 2 ones, but write
   $2 \times 42 = 84$.

> **Caution!** Be careful to keep the digits lined up correctly.

The answer is 212. *Remember:* If after estimating and multiplying you get a number that is larger than the divisor, you cannot subtract, so lower your estimate.

Do Exercises 14 and 15.

15. $5\,2\,\overline{)\,3\,2\,8\,8\,}$

Answers on page A-1

ZEROS IN QUOTIENTS

EXAMPLE 10 Divide: 6341 ÷ 7.

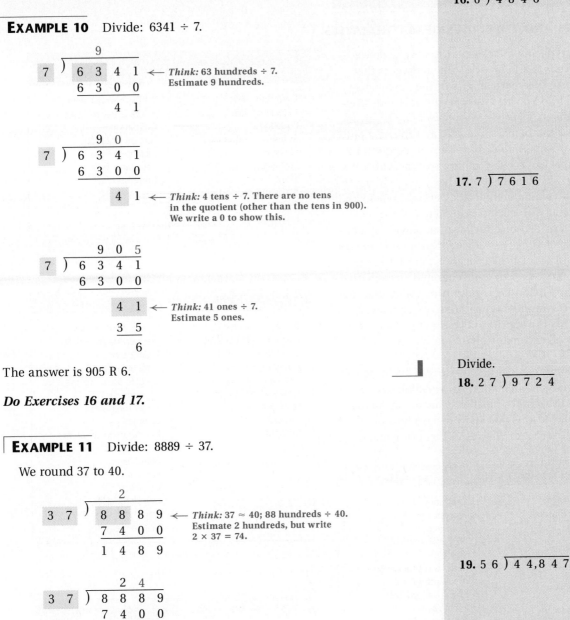

Think: 63 hundreds ÷ 7.
Estimate 9 hundreds.

Think: 4 tens ÷ 7. There are no tens
in the quotient (other than the tens in 900).
We write a 0 to show this.

Think: 41 ones ÷ 7.
Estimate 5 ones.

The answer is 905 R 6.

Do Exercises 16 and 17.

EXAMPLE 11 Divide: 8889 ÷ 37.

We round 37 to 40.

Think: 37 ≈ 40; 88 hundreds ÷ 40.
Estimate 2 hundreds, but write
2 × 37 = 74.

Think: 148 tens ÷ 40.
Estimate 4 tens, but write
4 × 37 = 148.

Think: 9 ones ÷ 40.
There are no ones in the quotient.

The answer is 240 R 9.

Do Exercises 18 and 19.

Divide.

16. 6) 4 8 4 6

17. 7) 7 6 1 6

Divide.

18. 2 7) 9 7 2 4

19. 5 6) 4 4,8 4 7

*To the instructor and the student: This
section presented a review of division
of whole numbers. Students who are
successful should go on to Section 1.7.
Those who have trouble should study
developmental unit D and then repeat
Section 1.6.*

Answers on page A-1

CAREERS AND THEIR USES OF MATHEMATICS

Students typically ask the question "Why do we have to study mathematics?" One answer is that you will use this mathematics in the next course. Although it is a correct answer, it sometimes frustrates students, because this answer can be given in the next mathematics course, and the next one, and so on. Sometimes an answer can be given by applications like those you have seen or will see in this book. Another answer is that you are living in a society in which mathematics becomes more and more critical with each passing day. Evidence of this was provided recently by a nationwide symposium sponsored by the National Research Council's Mathematical Sciences Education Board. Results showed that "Other than demographic factors, the *strongest* predictor of earnings nine years after high school is the number of mathematics courses taken." This is a significant testimony to the need for you to take as many mathematics courses as possible.

We try to provide other answers to "Why do we have to study mathematics?" in what follows. We have listed several occupations that are attractive and popular to students and various kinds of mathematics that are useful in that occupation.

ACCOUNTANT OR BUSINESS MANAGER	TRAVEL AGENT
Computer skills	Whole-number skills
Calculator skills	Fraction/decimal skills
Equations	Estimation
Systems of equations	Percent notation
Formulas	Equations
Probability	Calculator skills
Statistics	Computer skills
Ratio and proportion	
Percent notation	
Estimation	

LIBRARIAN	MACHINIST
Whole-number skills	Whole-number skills
Fraction/decimal skills	Fraction/decimal skills
Estimation	Estimation
Percent notation	Percent notation
Ratio and proportion	Length, area, volume, and perimeter
Area and perimeter	Angle measures
Formulas	Geometry
Calculator skills	Pythagorean theorem
Computer skills	Square roots
	Equations
	Formulas
	Graphing
	Calculator skills
	Computer skills
	Metric measures

DOCTOR	LAWYER
Equations	Equations
Percent notation	Percent notation
Graphing	Graphing
Statistics	Probability
Geometry	Statistics
Measurement	Ratio and proportion
Estimation	Area and volume
Exponents	Negative numbers
Logic	Formulas
	Calculator skills

PILOT	FIREFIGHTER
Equations	Percent notation
Percent notation	Graphing
Graphing	Estimation
Trigonometry	Formulas
Angles and geometry	Angles and geometry
Calculator skills	Probability
Computer skills	Statistics
Ratio and proportion	Area and geometry
Vectors	Square roots
	Exponents
	Pythagorean theorem

NURSE	POLICE OFFICER
Whole-number skills	Whole-number skills
Fraction/decimal skills	Fraction/decimal skills
Estimation	Estimation
Percent notation	Percent notation
Ratio and proportion	Ratio and proportion
Estimation	Geometry
Equations	Negative numbers
English/Metric measurement	Probability
Probability	Statistics
Statistics	Calculator skills
Formulas	
Exponents and scientific notation	
Calculator skills	
Computer skills	

Exercise Set 1.6

a Write a division that corresponds to the situation. You need not carry out the division.

1. A cheese factory made 176 lb (pounds) of Monterey Jack cheese. The cheese was placed in 4-lb boxes. How many boxes were filled?

2. A beverage company put 222 cans of soda into 6-can cartons. How many cartons did they fill?

3. A family divides an inheritance of $184,000 among its children, giving each of them $23,000. How many children are there?

4. A lab technician pours 455 mL of acid into 5 beakers, putting the same amount in each. How much acid is in each beaker?

b Write a related multiplication sentence.

5. $18 \div 3 = 6$

6. $72 \div 9 = 8$

7. $22 \div 22 = 1$

8. $32 \div 1 = 32$

9. $54 \div 6 = 9$

10. $72 \div 8 = 9$

11. $37 \div 1 = 37$

12. $28 \div 28 = 1$

Write two related division sentences.

13. $9 \times 5 = 45$

14. $2 \cdot 7 = 14$

15. $37 \cdot 1 = 37$

16. $4 \cdot 12 = 48$

17. $8 \times 8 = 64$

18. $9 \cdot 7 = 63$

19. $11 \cdot 6 = 66$

20. $1 \cdot 43 = 43$

c Divide.

21. $277 \div 5$

22. $699 \div 3$

23. $864 \div 8$

24. $869 \div 8$

25. $4\,\overline{)\,1\,2\,2\,8}$

26. $3\,\overline{)\,2\,1\,2\,4}$

27. $6\,\overline{)\,4\,5\,2\,1}$

28. $9\,\overline{)\,9\,1\,1\,0}$

29. $297 \div 4$

30. $389 \div 2$

31. $738 \div 8$

32. $881 \div 6$

33. $5\,\overline{)\,8\,5\,1\,5}$

34. $3\,\overline{)\,6\,0\,2\,7}$

35. $9\,\overline{)\,8\,8\,8\,8}$

36. $8\,\overline{)\,4\,1\,3\,9}$

37. $7\,0\,\overline{)\,3\,6\,9\,2}$

38. $2\,0\,\overline{)\,5\,7\,9\,8}$

39. $3\,0\,\overline{)\,8\,7\,5}$

40. $4\,0\,\overline{)\,9\,8\,7}$

ANSWERS

1.
2.
3.
4.
5.
6.
7.
8.
9.
10.
11.
12.
13.
14.
15.
16.
17.
18.
19.
20.
21.
22.
23.
24.
25.
26.
27.
28.
29.
30.
31.
32.
33.
34.
35.
36.
37.
38.
39.
40.

41.

42.

43.

44.

45.

46.

47.

48.

49.

50.

51.

52.

53.

54.

55.

56.

57.

58.

59.

60.

61.

62.

63.

64.

65.

66.

67.

68.

69.

41. $852 \div 21$

42. $942 \div 23$

43. $85 \overline{)7672}$

44. $54 \overline{)2729}$

45. $111 \overline{)3219}$

46. $102 \overline{)5612}$

47. $8 \overline{)843}$

48. $7 \overline{)749}$

49. $5 \overline{)8047}$

50. $9 \overline{)7273}$

51. $5 \overline{)5036}$

52. $7 \overline{)7074}$

53. $1058 \div 46$

54. $7242 \div 24$

55. $3425 \div 32$

56. $48 \overline{)4899}$

57. $24 \overline{)8880}$

58. $36 \overline{)7563}$

59. $28 \overline{)17{,}067}$

60. $36 \overline{)28{,}929}$

61. $80 \overline{)24{,}320}$

62. $90 \overline{)88{,}560}$

63. $285 \overline{)999{,}999}$

64. $306 \overline{)888{,}888}$

65. $456 \overline{)3{,}679{,}920}$

66. $803 \overline{)5{,}622{,}606}$

SKILL MAINTENANCE

67. Write expanded notation for 7882.

68. Use < or > for ▪ to write a true sentence.

888 ▪ 788.

SYNTHESIS

69. A group of 1231 college students is going to take a bus on a field trip. The bus company informs the planning committee that each bus it provides can hold 42 students. How many buses are needed?

1.7 *Solving Equations*

a SOLUTIONS BY TRIAL

Let's find a number that we can put in the blank to make this sentence true:

$$9 = 3 + \blacksquare.$$

We are asking "9 is 3 plus what number?" The answer is 6.

$$9 = 3 + \boxed{6}$$

Do Exercises 1 and 2.

A sentence with = is called an **equation**. A **solution** of an equation is a number that makes the sentence true. Thus, 6 is a solution of

$$9 = 3 + \blacksquare \quad \text{because} \quad 9 = 3 + \boxed{6} \quad \text{is true.}$$

But 7 is not a solution of

$$9 = 3 + \blacksquare \quad \text{because} \quad 9 = 3 + \boxed{7} \quad \text{is false.}$$

Do Exercises 3 and 4.

We can use a letter instead of a blank. For example,

$$9 = 3 + x.$$

We call x a **variable** because it can represent any number.

> A *solution* is a replacement for the letter that makes the equation true. When we find the solutions, we say that we have *solved* the equation.

EXAMPLE 1 Solve $x + 12 = 27$ by trial.

We replace x by several numbers.

If we replace x by 13, we get a false equation: $13 + 12 = 27$.
If we replace x by 14, we get a false equation: $14 + 12 = 27$.
If we replace x by 15, we get a true equation: $15 + 12 = 27$.

No other replacement makes the equation true, so the solution is 15.

EXAMPLES Solve.

2. $7 + n = 22$
(7 plus what number is 22?)
The solution is 15.

3. $8 \cdot 23 = y$
(8 times 23 is what?)
The solution is 184.

Note, as in Example 3, that when the letter is alone on one side of the equation, the other side shows us what calculations to do in order to find the solution.

Do Exercises 5–8.

OBJECTIVES

After finishing Section 1.7, you should be able to:

a Solve simple equations by trial.

b Solve equations like $t + 28 = 54$, $28 \cdot x = 168$, and $98 \div 2 = y$.

FOR EXTRA HELP

TAPE 2   TAPE 2B   MAC: 1B
IBM: 1B

Find a number that makes the sentence true.

1. $8 = 1 + \blacksquare$

2. $\blacksquare + 2 = 7$

3. Determine whether 7 is a solution of $\blacksquare + 5 = 9$.

4. Determine whether 4 is a solution of $\blacksquare + 5 = 9$.

Solve by trial.

5. $n + 3 = 8$

6. $x - 2 = 8$

7. $45 \div 9 = y$

8. $10 + t = 32$

Answers on page A-1

Solve.

9. $346 \times 65 = y$

10. $x = 2347 + 6675$

11. $4560 \div 8 = t$

12. $x = 6007 - 2346$

Solve.

13. $x + 9 = 17$

14. $77 = m + 32$

Answers on page A-1

b | SOLVING EQUATIONS

We now begin to develop more efficient ways to solve certain equations. When an equation has a variable alone on one side, it is easy to see the solution or to compute it. For example, the solution of

$$x = 12$$

is 12.

When a calculation is on one side and the variable is alone on the other, we can find the solution by carrying out the calculation.

EXAMPLE 4   Solve: $x = 245 \times 34$.

To solve the equation, we carry out the calculation.

$$
\begin{array}{r}
2\ 4\ 5 \\
\times\quad 3\ 4 \\
\hline
9\ 8\ 0 \\
7\ 3\ 5\ 0 \\
\hline
8\ 3\ 3\ 0
\end{array}
$$

The solution is 8330.

Do Exercises 9–12.

Look at the equation

$$x + 12 = 27.$$

We can get x alone on one side of the equation by writing a related subtraction sentence:

$$x = 27 - 12 \qquad \text{12 gets subtracted to find the related subtraction.}$$
$$x = 15. \qquad \text{Doing the subtraction}$$

It is useful to think of this as "subtracting 12 *on both sides.*" Thus,

$$x + 12 - 12 = 27 - 12 \qquad \text{Subtracting 12 on both sides}$$
$$x + 0 = 15 \qquad \text{Carrying out the subtraction}$$
$$x = 15.$$

To solve $x + a = b$, subtract a on both sides.

If we can get an equation in a form with the letter alone on one side, we can "see" the solution.

EXAMPLE 5   Solve: $t + 28 = 54$.

We have

$$t + 28 = 54$$
$$t + 28 - 28 = 54 - 28 \qquad \text{Subtracting 28 on both sides}$$
$$t + 0 = 26$$
$$t = 26.$$

The solution is 26.

Do Exercises 13 and 14.

EXAMPLE 6 Solve: $182 = 65 + n$.

We have

$$182 = 65 + n$$
$$182 - 65 = 65 + n - 65 \qquad \text{Subtracting 65 on both sides}$$
$$117 = 0 + n \qquad \text{65 plus } n \text{ minus 65 is } 0 + n$$
$$117 = n$$

The solution is 117.

Do Exercise 15.

EXAMPLE 7 Solve: $7381 + x = 8067$.

We have

$$7381 + x = 8067$$
$$7381 + x - 7381 = 8067 - 7381 \qquad \text{Subtracting 7381 on both sides}$$
$$x = 686.$$

The solution is 686.

Do Exercises 16 and 17.

We now learn to solve equations like $8 \cdot n = 96$. Look at

$$8 \cdot n = 96.$$

We can get n alone by writing a related division sentence:

$$n = 96 \div 8 = \frac{96}{8} \qquad \text{We move 8 to the other side and write a division.}$$

$$n = 12. \qquad \text{Doing the division}$$

Note that $n = 12$ is easier to solve than $8 \cdot n = 96$. This is because we see easily that if we replace n on the left side by 12, we get a true sentence: $12 = 12$. The solution of $n = 12$ is 12, which is also the solution of $8 \cdot n = 96$.

It is useful to think of the preceding as "dividing by 8 *on both sides.*" Thus,

$$\frac{8 \cdot n}{8} = \frac{96}{8} \qquad \text{Dividing by 8 on both sides}$$

$$n = 12. \qquad \text{8 times } n \text{ divided by 8 is } n.$$

> To solve $a \cdot x = b$, divide by a on both sides.

15. Solve: $155 = t + 78$.

Solve.

16. $4566 + x = 7877$

17. $8172 = h + 2058$

Answers on page A-1

Solve.

18. $8 \cdot x = 64$

19. $144 = 9 \cdot n$

20. Solve: $5152 = 8 \cdot t$.

21. Solve: $18 \cdot y = 1728$.

22. Solve: $n \cdot 48 = 4512$.

Answers on page A-1

EXAMPLE 8   Solve: $10 \cdot x = 240$.

We have

$$10 \cdot x = 240$$

$$\frac{10 \cdot x}{10} = \frac{240}{10} \qquad \text{Dividing by 10 on both sides}$$

$$x = 24.$$

The solution is 24.

Do Exercises 18 and 19.

EXAMPLE 9   Solve: $5202 = 9 \cdot t$.

We have

$$5202 = 9 \cdot t$$

$$\frac{5202}{9} = \frac{9 \cdot t}{9} \qquad \text{Dividing by 9 on both sides}$$

$$578 = t.$$

The solution is 578.

Do Exercise 20.

EXAMPLE 10   Solve: $14 \cdot y = 1092$.

We have

$$14 \cdot y = 1092$$

$$\frac{14 \cdot y}{14} = \frac{1092}{14} \qquad \text{Dividing by 14 on both sides}$$

$$y = 78.$$

The solution is 78.

Do Exercise 21.

EXAMPLE 11   Solve: $n \cdot 56 = 4648$.

We have

$$n \cdot 56 = 4648$$

$$\frac{n \cdot 56}{56} = \frac{4648}{56} \qquad \text{Dividing by 56 on both sides}$$

$$n = 83.$$

The solution is 83.

Do Exercise 22.

Exercise Set 1.7

a Solve by trial.

1. $x + 0 = 14$ **2.** $x - 7 = 18$ **3.** $y \cdot 17 = 0$ **4.** $56 \div m = 7$

b Solve.

5. $13 + x = 42$ **6.** $15 + t = 22$ **7.** $12 = 12 + m$

8. $16 = t + 16$ **9.** $3 \cdot x = 24$ **10.** $6 \cdot x = 42$

11. $112 = n \cdot 8$ **12.** $162 = 9 \cdot m$ **13.** $45 \times 23 = x$

14. $23 \times 78 = y$ **15.** $t = 125 \div 5$ **16.** $w = 256 \div 16$

17. $p = 908 - 458$ **18.** $9007 - 5667 = m$ **19.** $x = 12{,}345 + 78{,}555$

20. $5678 + 9034 = t$ **21.** $3 \cdot m = 96$ **22.** $4 \cdot y = 96$

23. $715 = 5 \cdot z$ **24.** $741 = 3 \cdot t$ **25.** $10 + x = 89$

26. $20 + x = 57$ **27.** $61 = 16 + y$ **28.** $53 = 17 + w$

29. $6 \cdot p = 1944$ **30.** $4 \cdot w = 3404$ **31.** $5 \cdot x = 3715$

32. $9 \cdot x = 1269$ **33.** $47 + n = 84$ **34.** $56 + p = 92$

ANSWERS

1.
2.
3.
4.
5.
6.
7.
8.
9.
10.
11.
12.
13.
14.
15.
16.
17.
18.
19.
20.
21.
22.
23.
24.
25.
26.
27.
28.
29.
30.
31.
32.
33.
34.

35. _____

36. _____

37. _____

38. _____

39. _____

40. _____

41. _____

42. _____

43. _____

44. _____

45. _____

46. _____

47. _____

48. _____

49. _____

50. _____

51. _____

52. _____

53. _____

54. _____

55. _____

56. _____

57. _____

58. _____

59. _____

60. _____

61. _____

35. $x + 78 = 144$

36. $z + 67 = 133$

37. $165 = 11 \cdot n$

38. $660 = 12 \cdot n$

39. $624 = t \cdot 13$

40. $784 = y \cdot 16$

41. $x + 214 = 389$

42. $x + 221 = 333$

43. $567 + x = 902$

44. $438 + x = 807$

45. $18 \cdot x = 1872$

46. $19 \cdot x = 6080$

47. $40 \cdot x = 1800$

48. $20 \cdot x = 1500$

49. $2344 + y = 6400$

50. $9281 = 8322 + t$

51. $8322 + 9281 = x$

52. $9281 - 8322 = y$

53. $234 \times 78 = y$

54. $10{,}534 \div 458 = q$

55. $58 \cdot m = 11{,}890$

56. $233 \cdot x = 22{,}135$

57. $x \cdot 198 = 10{,}890$

SKILL MAINTENANCE

58. Write two related subtraction sentences: $7 + 8 = 15$.

59. Write two related division sentences: $6 \cdot 8 = 48$.

Use $>$ or $<$ for ▨ to write a true sentence.

60. 123 ▨ 789

61. 342 ▨ 339

1.8 Solving Problems

a To solve a problem using the operations on the whole numbers, we first look at the situation. We try to translate the problem to an equation. Then we solve the equation. We check to see if the solution of the equation is a solution of the original problem. Thus we are using the following five-step strategy.

PROBLEM-SOLVING TIPS

1. *Familiarize* yourself with the situation. If it is described in words, as in a textbook, *read carefully*. In any case, think about the situation. Draw a picture whenever it makes sense to do so. Choose a letter, or *variable*, to represent the unknown quantity to be solved for.

2. *Translate* the problem to an equation.

3. *Solve* the equation.

4. *Check* the answer in the original wording of the problem.

5. *State* the answer to the problem clearly with appropriate units.

EXAMPLE 1 The bar graph at the right shows the costs to a student of attending a college for four years. Find the total cost of the college education.

1. **Familiarize.** We can make a drawing or at least visualize the situation.

$$\underbrace{\$11{,}516}_{\substack{\text{First}\\\text{year}}} + \underbrace{\$11{,}898}_{\substack{\text{Second}\\\text{year}}} + \underbrace{\$12{,}270}_{\substack{\text{Third}\\\text{year}}} + \underbrace{\$12{,}490}_{\substack{\text{Fourth}\\\text{year}}} = n$$

Since we are combining objects, addition can be used. First we define the unknown. We let $n =$ the total cost of the college education.

2. **Translate.** We translate to an equation by writing a number sentence that corresponds to the situation:

$$11{,}516 + 11{,}898 + 12{,}270 + 12{,}490 = n.$$

3. **Solve.** We solve the equation by carrying out the addition.

$$
\begin{array}{r}
{\scriptstyle 2\ 2\ 1} \\
1\,2{,}4\,9\,0 \\
1\,2{,}2\,7\,0 \\
1\,1{,}8\,9\,8 \\
+\ 1\,1{,}5\,1\,6 \\
\hline
4\,8{,}1\,7\,4
\end{array}
$$

Thus, $48{,}174 = n$ or $n = 48{,}174$.

4. **Check.** We check $48,174 in the original problem. There are many ways to check. For example, we can repeat the calculation. We leave this to the student. Another way is to check the reasonableness of the answer. We would expect our answer to be larger than any of the separate costs, which it is. We can also estimate by rounding:

$$11{,}516 + 11{,}898 + 12{,}270 + 12{,}490 \approx 12{,}000 + 12{,}000 + 12{,}000 + 12{,}000$$

$$= 48{,}000 \approx 48{,}174.$$

College Costs

Bar graph titled "College Costs" with vertical axis "College costs (in dollars)" showing values 11,000 / 12,000 / 13,000, and horizontal axis "Year" with bars for First ($11,516), Second ($11,898), Third ($12,270), and Fourth ($12,490).

1. A family must replace some appliances in their home. They buy a large-capacity washer for $339, an electric dryer for $289, a four-cycle built-in dishwasher for $399, and a gas range for $297. How much do they spend in all?

2. On a long four-day trip, a family bought the following amounts of gasoline for their motor home:

   23 gallons, 24 gallons,

   26 gallons, 25 gallons.

   How much gasoline did they buy in all?

3. In an average year, 123,000 hip replacements and 95,000 knee replacements are performed. How many replacements of both kinds are performed?

4. In a recent year, there were 325,848 hotel rooms in Holiday Inns and 266,123 rooms in Best Western Hotels. How many rooms are there altogether in these two hotel chains?

Answers on page A-1

If we had gotten an estimate like 36,000 or 58,000, we might be suspicious that our calculated answer is incorrect. Since our estimated answer is close to our calculation, we are further convinced that our answer checks.

5. **State.**   The total cost of the college education is $48,174.

Do Exercises 1 and 2.

In the real world, problems are not usually given in words. You must still become familiar with the situation before you can solve the problem.

EXAMPLE 2   The John Hancock Building in Chicago is 1127 ft tall. It has two 342-ft antennas on top. How far are the tops of the antennas from the ground?

1. ***Familiarize.***   We first make a drawing.

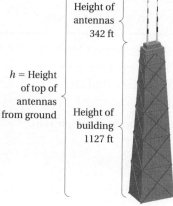

Height of antennas 342 ft

h = Height of top of antennas from ground

Height of building 1127 ft

Since we are combining lengths, addition can be used. To define the unknown, we let h = the height of the top of the antennas from the ground.

2. ***Translate.***   We translate the problem to the following addition sentence:

$$1127 + 342 = h.$$

3. ***Solve.***   To solve the equation, we carry out the addition.

$$\begin{array}{r} 1\ 1\ 2\ 7 \\ +\ \ 3\ 4\ 2 \\ \hline 1\ 4\ 6\ 9 \end{array}$$

Thus, $1469 = h$, or $h = 1469$.

4. ***Check.***   We check the height of 1469 ft in the original problem. We can repeat the calculation. We can also check the reasonableness of the answer. We would expect our answer to be larger than either of the heights, which it is. We can also find an estimated answer by rounding:

$$1127 + 342 \approx 1100 + 300 = 1400 \approx 1469.$$

The answer checks.

5. ***State.***   The height of the top of the antennas from the ground is 1469 ft.

Do Exercises 3 and 4.

EXAMPLE 3   A farm contains 2679 acres. If the owner sells 1884 acres, how many acres are left?

1. Familiarize.   We first draw a picture or at least visualize the situation. We let A = the number of acres left.

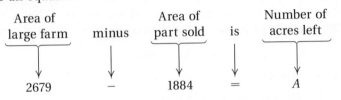

2. Translate.   We see that this is a "take-away" situation. We translate to an equation.

Area of large farm	minus	Area of part sold	is	Number of acres left
↓	↓	↓	↓	↓
2679	−	1884	=	A

3. Solve.   This sentence tells us what to do. We subtract.

$$
\begin{array}{r}
\overset{\scriptstyle 15}{}\\[-4pt]
\overset{\scriptstyle 1\;\,\cancel{5}\;17}{\cancel{2}\,\cancel{6}\,\cancel{7}\,9}\\
-\;1\,8\,8\,4\\
\hline
7\,9\,5
\end{array}
$$

Thus, $795 = A$, or $A = 795$.

4. Check.   We check 795 acres. We can repeat the calculation. We note that the answer should be less than the original acreage, 2679 acres, which it is. We can add the answer, 795, to the number being subtracted, 1884: $1884 + 795 = 2679$. We can also estimate:

$$2679 - 1884 \approx 2700 - 1900 = 800 \approx 795.$$

The answer checks.

5. State.   There are 795 acres left.

Do Exercises 5 and 6.

EXAMPLE 4   It is 1154 mi from Indianapolis to Denver. A driver has traveled 685 mi of that distance. How much farther is it to Denver?

1. Familiarize.   We first make a drawing or at least visualize the situation. We let x = the remaining distance to Denver.

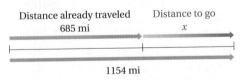

2. Translate.   We see that this is a "how-much-more" situation. We translate to an equation.

Distance already traveled	plus	Distance to go	is	Total distance of trip
↓	↓	↓	↓	↓
685	+	x	=	1154

5. A person has $948 in a checking account. A check is written for $427 to pay property taxes. How much is left in the checking account?

6. As shown in the bar graph below, 479,750 athletes participated in sports in the state of Texas in a recent year. In California, 461,794 participated; in New York, 291,591; and in Ohio, 274,224. How many more participated in Texas than in California? How many participated in all?

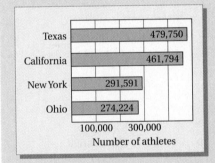

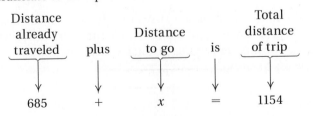

7. In a recent year, there were 107,588 female league bowlers in Detroit and 97,461 in Los Angeles. How many more were there in Detroit?

8. The average income in New Jersey is $25,372 per person. In Utah, it is $14,529. How much greater is the income in New Jersey?

9. A certain type of shelf can hold 40 books. How many books can 70 shelves hold?

Answers on page A-1

3. Solve.   We solve the equation.

$$685 + x = 1154$$
$$685 + x - 685 = 1154 - 685 \quad \text{Subtracting 685 on both sides}$$
$$x = 469$$

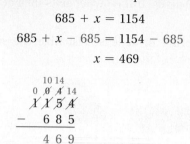

4. Check.   We check 469 mi in the original problem. This number should be less than the total distance, 1154 mi, which it is. We can add the answer, 469, to the number being subtracted, 685: $685 + 469 = 1154$. We can also estimate,

$$1154 - 685 \approx 1200 - 700 = 500 \approx 469.$$

The answer checks.

5. State.   It is 469 mi farther to Denver.

Do Exercises 7 and 8.

EXAMPLE 5   A ream of paper contains 500 sheets. How many sheets are in 9 reams?

1. Familiarize.   We first draw a picture or at least visualize the situation. We can think of this situation as a stack of reams. We let $n =$ the total number of sheets in 9 reams.

2. Translate.   Then we translate and solve as follows.

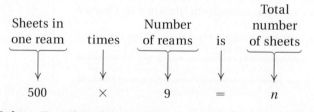

3. Solve.   To solve the equation, we multiply: $500 \times 9 = 4500$. Thus, $4500 = n$, or $n = 4500$.

4. Check.   In this case, an estimated answer is almost a repeated calculation. Certainly our answer should be larger than 500, since we are multiplying, so the answer seems reasonable. The answer checks.

5. State.   There are 4500 sheets in 9 reams.

Do Exercise 9.

EXAMPLE 6   What is the total cost of 5 compact 8-mm remote camcorders if each one costs $597?

1. **Familiarize.** We first draw a picture or at least visualize the situation. We let n = the cost of 5 camcorders. Repeated addition works well here.

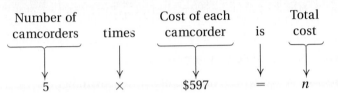

2. **Translate.**   We translate to an equation and solve.

Number of camcorders	times	Cost of each camcorder	is	Total cost
5	×	$597	=	n

3. **Solve.**   This sentence tells us what to do. We multiply.

$$\begin{array}{r} {\scriptstyle 4\ 3} \\ 5\ 9\ 7 \\ \times \qquad 5 \\ \hline 2\ 9\ 8\ 5 \end{array}$$ Thus, $n = 2985$.

4. **Check.**   We have an answer that is much larger than the cost of any individual camcorder, which is reasonable. We can repeat our calculation. We can also check by estimating:

   $5 \times 597 \approx 5 \times 600 = 3000 \approx 2985.$

   The answer checks.

5. **State.**   The cost of 5 camcorders is $2985.

Do Exercise 10.

EXAMPLE 7   The state of Colorado is roughly the shape of a rectangle that is 270 mi by 380 mi. What is its area?

1. **Familiarize.**   We first make a drawing. We let A = the area.

2. **Translate.**   Using a formula for area, we have

   $A = l \cdot w = 380 \cdot 270.$

270 mi

380 mi

3. **Solve.**   We carry out the multiplication.

$$\begin{array}{r} 3\ 8\ 0 \\ \times \qquad 2\ 7\ 0 \\ \hline 2\ 6\ 6\ 0\ 0 \\ 7\ 6\ 0\ 0\ 0 \\ \hline 1\ 0\ 2\ 6\ 0\ 0 \end{array}$$ Thus, $A = 102{,}600.$

10. An electronics firm sells 38 VCRs one month, each at a price of $249. How much money did it receive from the VCRs?

Answer on page A-1

11. The state of Wyoming is roughly the shape of a rectangle that is 275 mi by 365 mi. What is its area?

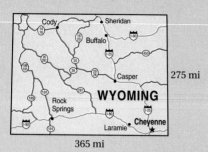

275 mi

365 mi

12. There are 60 minutes in an hour. How many minutes are there in 72 hours?

Answers on page A-1

4. Check. We repeat our calculation. We also note that the answer is larger than either the length or the width, which it should be. (This might not be the case if we were using decimals.) The answer checks.

5. State. The area is 102,600 square miles (mi²).

Do Exercise 11.

EXAMPLE 8   There are 24 hours in a day and 7 days in a week. How many hours are there in a week?

1. Familiarize. We first make a drawing. We let $y =$ the number of hours in a week. Repeated addition works well here.

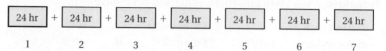

1   2   3   4   5   6   7

2. Translate. We translate to a number sentence.

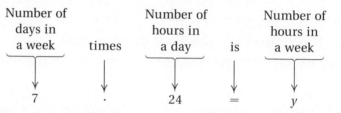

Number of days in a week	times	Number of hours in a day	is	Number of hours in a week
7	·	24	=	y

3. Solve. To solve the equation, we carry out the multiplication.

$$\begin{array}{r} \overset{2}{2}\,4 \\ \times\ \ 7 \\ \hline 1\,6\,8 \end{array}$$ Thus, $y = 168$.

4. Check. We check our answer by estimating:

$$7 \times 24 \approx 7 \times 20 = 140 \approx 168.$$

We can also check by repeating the calculation. We note that there are more hours in a week than in a day, which we would expect.

5. State. There are 168 hours in a week.

Do Exercise 12.

EXAMPLE 9   A beverage company produces 2269 cans of soda. How many 6-can packages can be filled? How many cans will be left over?

1. Familiarize. We first draw a picture. We let $n =$ the number of 6-can packages to be filled.

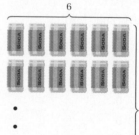

6 cans in each row. How many rows?

2., 3. Translate and **Solve.** We translate to an equation and solve as follows:

$2269 \div 6 = n.$

$$
\begin{array}{r}
3\ 7\ 8 \\
6\ )\overline{2\ 2\ 6\ 9} \\
\underline{1\ 8\ 0\ 0} \\
4\ 6\ 9 \\
\underline{4\ 2\ 0} \\
4\ 9 \\
\underline{4\ 8} \\
1
\end{array}
$$

4. Check. We can check by multiplying the number of packages by 6 and adding the remainder of 1:

$6 \cdot 378 = 2268, \qquad 2268 + 1 = 2269.$

5. State. Thus, 378 six-can packages can be filled. There will be 1 can left over.

Do Exercises 13 and 14.

| **EXAMPLE 10** An automobile with a 5-speed transmission gets 27 miles to the gallon in city driving. How many gallons will it use to travel 7020 mi of city driving?

1. Familiarize. We first draw a picture. It is often helpful to be descriptive about how you define a variable. In this example, we let $g =$ the number of gallons (g comes from "gallons").

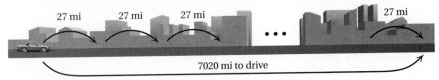

2. Translate. Repeated addition applies here. Thus the following multiplication corresponds to the situation.

Miles per gallon times Number of gallons needed is Number of miles to drive

$27 \qquad \cdot \qquad g \qquad = \qquad 7020$

3. Solve. To solve the equation, we divide on both sides by 27.

$27 \cdot g = 7020$

$\dfrac{27 \cdot g}{27} = \dfrac{7020}{27}$

$g = 260$

$$
\begin{array}{r}
2\ 6\ 0 \\
27\ )\overline{7\ 0\ 2\ 0} \\
\underline{5\ 4\ 0\ 0} \\
1\ 6\ 2\ 0 \\
\underline{1\ 6\ 2\ 0} \\
0
\end{array}
$$

4. Check. To check, we multiply 260 by 27: $27 \cdot 260 = 7020.$

5. State. The automobile will use 260 gal.

Do Exercise 15.

---

13. A beverage company produces 2205 cans of soda. How many 8-can packages can be filled? How many cans will be left over?

14. "Cheers" is the longest-running comedy in the history of television, with 271 episodes. A local station picks up the syndicated reruns. At 5 episodes per week, how many full weeks will pass before it must start over with past episodes? How many episodes will be left for the next week?

15. An automobile with a 5-speed transmission gets 33 miles to the gallon in city driving. How many gallons will it use to travel 1485 mi?

Answers on page A-1

16. Use the information in the figure below. How long must you swim in order to lose one pound?

To burn off 100 calories, you must:
- Run for 8 min at a brisk pace, or
- Swim for 2 min at a brisk pace, or
- Bicycle for 15 min at 9 mph, or
- Do aerobic exercises for 15 min.

17. There are 27 bones in each human hand and 26 bones in each human foot. How many bones are there in all in the hands and feet?

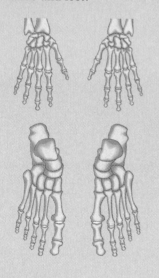

MULTISTEP PROBLEMS

Sometimes we must use more than one operation to solve a problem. We do so in the following example.

EXAMPLE 11   Referring to the figure in the margin at left, we see that to lose one pound, you must burn off about 3500 calories. How long must you run in order to lose one pound?

1. Familiarize.   We first draw a picture.

ONE POUND				
3500 calories				
100 cal	100 cal	. . .		100 cal
8 min	8 min	. . .		8 min

2. Translate.   Repeated addition applies here. Thus the following multiplication corresponds to the situation. We must find out how many 100's there are in 3500. We let x = the number of 100's in 3500.

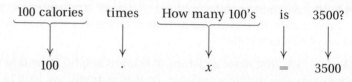

100 calories   times   How many 100's   is   3500?

$$100 \cdot x = 3500$$

3. Solve.   To solve the equation, we divide on both sides by 100.

$$100 \cdot x = 3500$$
$$\frac{100 \cdot x}{100} = \frac{3500}{100}$$
$$x = 35$$

$$
\begin{array}{r}
35 \\
100 \overline{)3500} \\
3000 \\
\hline
500 \\
500 \\
\hline
0
\end{array}
$$

We know that running for 8 min will burn off 100 calories. To do this 35 times will burn off one pound, so you must run for 35 times 8 minutes in order to burn off one pound. We let t = the time it takes to run off one pound.

$$35 \times 8 = t$$

$$
\begin{array}{r}
35 \\
\times \ 8 \\
\hline
280
\end{array}
$$

4. Check.   Suppose you run for 280 minutes. If we divide 280 by 8, we get 35, and 35 times 100 is 3500, the number of calories it takes to lose one pound.

5. State.   It will take 280 min, or 4 hr 40 min, of running to lose one pound.

Do Exercises 16 and 17.

Answers on page A-1

Exercise Set 1.8

a Solve.

1. The bar graph below shows the numbers of career hits for the top four all-time hitters in baseball.

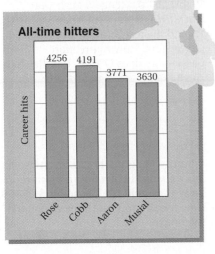

a) Ty Cobb got 4191 hits in his career. Pete Rose got 4256. How many did they hit altogether? How many more did Rose get than Cobb?

b) Hank Aaron got 3771 hits in his career. Stan Musial got 3630. How many did they hit altogether? How many more did Aaron get than Musial?

3. The Empire State building is 381 m (meters) tall. It has a 68-m antenna on top. How far is the top of the antenna from the ground?

5. A medical researcher poured first 2340 cubic centimeters of water and then 655 cubic centimeters of alcohol into a beaker. How much liquid was poured altogether?

2. The bar graph below shows the areas of the world's largest islands.

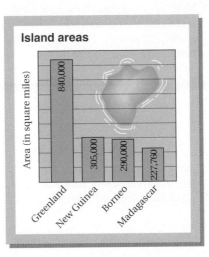

a) The largest island is Greenland, which has an area of 840,000 square miles (mi^2). The second largest is New Guinea, with an area of 305,000 mi^2. What is the combined area of the two islands? How much larger is Greenland than New Guinea?

b) The island of Borneo has an area of 290,000 mi^2, and the island of Madagascar has an area of 227,760 mi^2. What is the combined area of the two islands? How much larger is Borneo than Madagascar?

4. In a recent year, 1,580,000 people visited the United States from Japan. In addition, 1,200,000 came from England, 800,000 from Mexico, and 510,000 from West Germany. How many, in all, came from these countries?

6. A person's water loss each day includes 400 cubic centimeters from the lungs and 500 cubic centimeters from the skin. How much is lost from both?

1. a) _____

b) _____

2. a) _____

b) _____

3. _____

4. _____

5. _____

6. _____

7. _____

8. _____

9. _____

10. _____

11. _____

12. _____

13. _____

14. _____

15. _____

16. _____

17. _____

18. _____

7. Pam has a 170-megabyte hard disk on her computer. She installs word processing software that uses 15 megabytes of memory from the hard disk, a database program that uses 6 megabytes, and a spreadsheet program that uses 12 megabytes. How much memory is left on her computer?

8. The computer reading on a copy machine is 11,543 before Adam makes copies. After he is finished, the reading is 11,715. How many copies did he make?

9. In the Tokyo–Yokohama area, there are 29,971,000 people. In the New York–northeastern New Jersey and Long Island area, there are 18,087,521 people. How many more people are there in the Tokyo–Yokohama area?

10. O'Hare International Airport is the busiest in the world, handling 59,900,000 passengers each year. Hartsfield Atlanta International is the second busiest, handling 48,000,000 passengers. How many more passengers does O'Hare handle than Hartsfield?

11. You wrote checks of $45, $78, and $32. Your balance before that was $246. What is your new balance?

12. A family bought a house for $108,900. They spent $16,500 for an extra room. They then sold the house for $137,780. How much profit did they make on the house?

13. Playing golf for an hour burns 133 calories. How many calories are burned in 5 hours?

14. Each day 200 people in the United States become millionaires. How many people become millionaires in one year? (Use 365 days for one year.)

15. There are 60 seconds in a minute and 60 minutes in an hour. How many seconds are there in an hour?

16. The average heartbeat is 73 beats per minute. How many beats are there in one hour? one day? one year?

17. A standard-sized tennis court is a rectangle that is 36 ft by 78 ft. What is the area of a tennis court?

18. A rectangular field measures 48 m by 85 m. What is the area of the field?

19. The diameter of a circle is the length of a line through its center. The diameter of Jupiter is about 85,965 mi. The diameter of Jupiter is about 11 times the diameter of the earth. What is the diameter of the earth?

20. Sound travels at a speed of about 1087 ft per second (*per* means *for each*). How long does it take the sound of an airplane engine to reach your ear when the plane is 9783 ft overhead?

21. A customer buys 8 suits at $495 each and 3 shirts at $46 each. How much is spent?

22. What is the cost of 11 CD players at $340 each and 6 television sets at $736 each?

23. A college has a vacant rectangular lot that is 324 yd by 25 yd. On the lot students dug a garden that was 165 yd by 18 yd. How much area was left over?

24. A college student pays $257 a month for rent and $188 a month for food during 9 months at college. What is the total cost?

25. How many 16-oz bottles can be filled with 608 oz of catsup?

26. How many 24-can cases can be filled with 768 cans of mandarin oranges?

27. There are 225 members in a band, and they are marching with 15 in each row. How many rows are there?

28. There are 225 members in a band, and they are marching in 5 rows. How many band members are there in each row?

29. A loan of $324 will be paid off in 12 monthly payments. How much is each payment?

30. A loan of $1404 will be paid off in 36 monthly payments. How much is each payment?

ANSWERS

19. _____

20. _____

21. _____

22. _____

23. _____

24. _____

25. _____

26. _____

27. _____

28. _____

29. _____

30. _____

31.

32.

33.

34.

35.

36.

37.

38.

39.

40.

41.

42.

43.

44.

31. How many 23-kg (kilogram) bags can be filled by 885 kg of sand? How many kilograms of sand will be left over?

32. A vial contains 50 cubic centimeters of penicillin. How many 3-cubic-centimeter injections can be filled from the vial? How much will be left over?

33. A wholesaler bought 5 coats at $64 each and paid for them with $20 bills. How many $20 bills did it take?

34. How many $10 bills would it take to buy the 5 coats mentioned in Exercise 33?

35. A map has a scale of 55 miles to the inch. How far apart *on the map* are two cities that, in reality, are 605 mi apart? How far apart *in reality* are two cities that are 14 in. apart on the map?

36. A map has a scale of 46 miles to the inch. How far apart *in reality* are two cities that are 15 in. apart on the map? How far apart *on the map* are two cities that, in reality, are 552 mi apart?

37. A beverage company fills 640 12-oz cans with soda. How many 16-oz cans can be filled with the same amount of soda?

38. A rectangular piece of cardboard measures 64 in. by 15 in. Suppose the width were changed to 16 in. What would the length have to be in order to have the same area?

39. Use the information from the table in Example 11. How long must you bicycle at 9 mph in order to lose one pound?

40. Use the information from the table in Example 11. How long must you do aerobic exercises in order to lose one pound?

SKILL MAINTENANCE

41. Round to the nearest thousand: 234,562.

42. Round to the nearest hundred: 234,562.

SYNTHESIS

43. 🔲 Light travels at a speed of 8,370,000 mi in 45 sec. How far does it travel in 1 sec?

44. The thickness of paper is often measured by weights. For example, 20-lb paper is that weight of paper for which 1000 letter-size sheets weigh 20 lb. A ream of paper is 500 sheets. How much does a ream of 70-lb paper weigh?

1.9 Exponential Notation and Order of Operations

OBJECTIVES

After finishing Section 1.9, you should be able to:

a Write exponential notation for products such as 4 · 4 · 4.

b Evaluate exponential notation.

c Simplify expressions using the rules for order of operations.

d Remove parentheses within parentheses.

FOR EXTRA HELP

TAPE 3   TAPE 3A   MAC: 1B
                    IBM: 1B

a EXPONENTIAL NOTATION

Consider the product $3 \cdot 3 \cdot 3 \cdot 3$. Such products occur often enough that mathematicians have found it convenient to invent a shorter notation, called **exponential notation,** explained as follows.

$$\underbrace{3 \cdot 3 \cdot 3 \cdot 3}_{\text{4 factors}} \text{ is shortened to } 3^4 \leftarrow \text{exponent}$$

We read 3^4 as "three to the fourth power," 5^3 as "five cubed," and 5^2 as "five squared." The latter comes from the fact that a square of side s has area A given by $A = s^2$.

EXAMPLE 1 Write exponential notation for $10 \cdot 10 \cdot 10 \cdot 10 \cdot 10$.

Exponential notation is 10^5.   5 is the *exponent*.
10 is the *base*.

EXAMPLE 2 Write exponential notation for $2 \cdot 2 \cdot 2$.

Exponential notation is 2^3.

Do Exercises 1–4.

b EVALUATING EXPONENTIAL NOTATION

We evaluate exponential notation by rewriting it as a product and computing the product.

EXAMPLE 3 Evaluate: 10^3.

$10^3 = 10 \cdot 10 \cdot 10 = 1000$

EXAMPLE 4 Evaluate: 5^4.

$5^4 = 5 \cdot 5 \cdot 5 \cdot 5 = 625$

Do Exercises 5–8.

Write exponential notation.

1. $5 \cdot 5 \cdot 5 \cdot 5$

2. $5 \cdot 5 \cdot 5 \cdot 5 \cdot 5$

3. $10 \cdot 10$

4. $10 \cdot 10 \cdot 10 \cdot 10$

Evaluate.

5. 10^4

6. 10^2

7. 8^3

8. 2^5

Answers on page A-1

Simplify.

9. $93 - 14 \cdot 3$

10. $104 \div 4 + 4$

11. $25 \cdot 26 - (56 + 10)$

12. $75 \div 5 + (83 - 14)$

Simplify and compare.

13. $64 \div (32 \div 2)$ and $(64 \div 32) \div 2$

14. $(28 + 13) + 11$ and $28 + (13 + 11)$

Answers on page A-1

c   SIMPLIFYING EXPRESSIONS

Suppose we have a calculation like the following:

$$8 \cdot 6 - 1.$$

How do we find the answer? Do we subtract 1 from 6 and then multiply by 8, or do we multiply 8 by 6 and then subtract 1? In the first case, the answer is 40. In the second case, the answer is 47.

Consider the calculation

$$7 \cdot 14 - (12 + 18).$$

What do the parentheses mean? To deal with these questions, we must make some agreement regarding the order in which we perform operations. The rules are as follows.

RULES FOR ORDER OF OPERATIONS

1. Do all calculations within parentheses before operations outside.

2. Evaluate all exponential expressions.

3. Do all multiplications and divisions in order from left to right.

4. Do all additions and subtractions in order from left to right.

It is worth noting that these are the rules that a computer uses to do computations. In order to program a computer, you must know these rules.

EXAMPLE 5   Simplify: $8 \cdot 6 - 1$.

There are no parentheses or exponents, so we start with the third step.

$$8 \cdot 6 - 1 = 48 - 1 \qquad \text{Doing all multiplications and divisions in order from left to right}$$

$$= 47 \qquad \text{Doing all additions and subtractions in order from left to right}$$

EXAMPLE 6   Simplify: $7 \cdot 14 - (12 + 18)$.

$$7 \cdot 14 - (12 + 18) = 7 \cdot 14 - 30 \qquad \text{Carrying out operations inside parentheses}$$

$$= 98 - 30 \qquad \text{Doing all multiplications and divisions}$$

$$= 68 \qquad \text{Doing all additions and subtractions}$$

Do Exercises 9–12.

EXAMPLE 7   Simplify and compare: $23 - (10 - 9)$ and $(23 - 10) - 9$.

We have

$$23 - (10 - 9) = 23 - 1 = 22;$$

$$(23 - 10) - 9 = 13 - 9 = 4.$$

We can see that $23 - (10 - 9)$ and $(23 - 10) - 9$ represent different numbers.

Do Exercises 13 and 14.

EXAMPLE 8 Simplify: $7 \cdot 2 - (12 + 0) \div 3 - (5 - 2)$.

$7 \cdot 2 - (12 + 0) \div 3 - (5 - 2) = 7 \cdot 2 - 12 \div 3 - 3$ Carrying out operations inside parentheses

$\qquad\qquad\qquad\qquad = 14 - 4 - 3$ Doing all multiplications and divisions in order from left to right

$\qquad\qquad\qquad\qquad = 7$ Doing all additions and subtractions in order from left to right

Do Exercise 15.

EXAMPLE 9 Simplify: $15 \div 3 \cdot 2 \div (10 - 8)$.

$15 \div 3 \cdot 2 \div (10 - 8) = 15 \div 3 \cdot 2 \div 2$ Carrying out operations inside parentheses

$\qquad\qquad\qquad\qquad = 5 \cdot 2 \div 2$ Doing all multiplications and divisions in order from left to right

$\qquad\qquad\qquad\qquad = 10 \div 2$

$\qquad\qquad\qquad\qquad = 5$

Do Exercises 16–18.

EXAMPLE 10 Simplify and compare: $(3 + 5)^2$ and $3^2 + 5^2$.

$(3 + 5)^2 = 8^2 = 64$;

$3^2 + 5^2 = 9 + 25 = 34$.

We see that $(3 + 5)^2$ and $3^2 + 5^2$ do not represent the same numbers.

Do Exercise 19.

EXAMPLE 11 Simplify: $6^3 \div (10 - 8)^2$.

$6^3 \div (10 - 8)^2 = 6^3 \div 2^2$ Carrying out operations inside parentheses first

$\qquad\qquad\qquad\quad = 216 \div 4$ Evaluating exponential expressions second

$\qquad\qquad\qquad\quad = 54$ Dividing

EXAMPLE 12 Simplify: $2^4 + 51 \cdot 4 - 2 \cdot (37 + 23 \cdot 2)$.

$2^4 + 51 \cdot 4 - 2 \cdot (37 + 23 \cdot 2)$

$\quad = 2^4 + 51 \cdot 4 - 2 \cdot (37 + 46)$ Carrying out operations inside parentheses. To do this, we first multiply 23 by 2.

$\quad = 2^4 + 51 \cdot 4 - 2 \cdot 83$ Completing the addition inside parentheses

$\quad = 16 + 51 \cdot 4 - 2 \cdot 83$ Evaluating exponential expressions

$\quad = 16 + 204 - 166$ Doing the multiplications

$\quad = 220 - 166$

$\quad = 54$ Doing the additions and subtractions in order from left to right

Do Exercises 20–22.

15. Simplify:

$\quad 9 \times 4 - (20 + 4) \div 8 - (6 - 2)$.

Simplify.

16. $5 \cdot 5 \cdot 5 + 26 \cdot 71$
$\quad - (16 + 25 \cdot 3)$

17. $4 \cdot 4 \cdot 4 + 10 \cdot 20 + 8 \cdot 8 - 23$

18. $95 - 2 \cdot 2 \cdot 2 \cdot 5 \div (24 - 4)$

19. Simplify and compare:

$\quad (4 + 6)^2 \quad \text{and} \quad 4^2 + 6^2$.

Simplify.

20. $5^3 + 26 \cdot 71 - (16 + 25 \cdot 3)$

21. $(1 + 3)^3 + 10 \cdot 20 + 8^2 - 23$

22. $95 - 2^3 \cdot 5 \div (24 - 4)$

Answers on page A-2

23. The calories per ounce of certain meats is given in the bar graph below. Find the average number of calories per ounce.

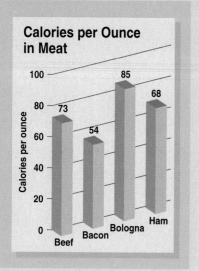

Simplify.

24. $9 \times 5 + \{6 \div [14 - (5 + 3)]\}$

25. $[18 - (2 + 7) \div 3]$
$- (31 - 10 \times 2)$

Answers on page A-2

EXAMPLE 13 The expenditure on education per pupil in each of three countries is given in the bar graph at the right. Find the average. To find the **average** of a set of numbers, we first add the numbers and then divide by the number of addends.

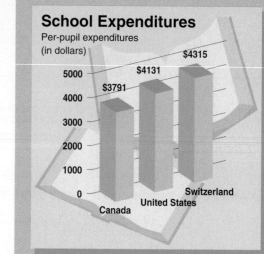

School Expenditures
Per-pupil expenditures (in dollars)

The average is given by

$(3791 + 4131 + 4315) \div 3$.

To find the average, we carry out the computation using the rules for order of operations:

$(3791 + 4131 + 4315) \div 3 = 12{,}237 \div 3$
$= 4079.$

Thus the average expenditure for education in the three countries is \$4079.

Do Exercise 23.

d | PARENTHESES WITHIN PARENTHESES

When parentheses occur within parentheses, we can make them different shapes, such as [ ] (also called "brackets") and { } (also called "braces"). All of these have the same meaning. When parentheses occur within parentheses, computations in the innermost ones are to be done first.

EXAMPLE 14 Simplify: $16 \div 2 + \{40 - [13 - (4 + 2)]\}$.

$16 \div 2 + \{40 - [13 - (4 + 2)]\}$

$= 16 \div 2 + \{40 - [13 - 6]\}$ Doing the calculations in the innermost parentheses first

$= 16 \div 2 + \{40 - 7\}$ Again, doing the calculations in the innermost parentheses

$= 16 \div 2 + 33$

$= 8 + 33$ Doing all multiplications and divisions in order from left to right

$= 41$ Doing all additions and subtractions in order from left to right

EXAMPLE 15 Simplify: $[25 - (4 + 3) \times 3] \div (11 - 7)$.

$[25 - (4 + 3) \times 3] \div (11 - 7) = [25 - 7 \times 3] \div (11 - 7)$
$= [25 - 21] \div (11 - 7)$
$= 4 \div 4$
$= 1$

Do Exercises 24 and 25.

Exercise Set 1.9

a Write exponential notation.

1. $3 \cdot 3 \cdot 3 \cdot 3$

2. $2 \cdot 2 \cdot 2 \cdot 2 \cdot 2$

3. $5 \cdot 5$

4. $13 \cdot 13 \cdot 13$

5. $7 \cdot 7 \cdot 7 \cdot 7 \cdot 7$

6. $10 \cdot 10$

7. $10 \cdot 10 \cdot 10$

8. $1 \cdot 1 \cdot 1 \cdot 1$

b Evaluate.

9. 7^2

10. 5^3

11. 9^3

12. 10^2

13. 12^4

14. 10^5

15. 11^2

16. 6^3

c Simplify.

17. $12 + (6 + 4)$

18. $(12 + 6) + 18$

19. $52 - (40 - 8)$

20. $(52 - 40) - 8$

21. $1000 \div (100 \div 10)$

22. $(1000 \div 100) \div 10$

23. $(256 \div 64) \div 4$

24. $256 \div (64 \div 4)$

25. $(2 + 5)^2$

26. $2^2 + 5^2$

27. $2 + 5^2$

28. $2^2 + 5$

29. $16 \cdot 24 + 50$

30. $23 + 18 \cdot 20$

31. $83 - 7 \cdot 6$

32. $10 \cdot 7 - 4$

33. $10 \cdot 10 - 3 \cdot 4$

34. $90 - 5 \cdot 5 \cdot 2$

35. $4^3 \div 8 - 4$

36. $8^2 - 8 \cdot 2$

37. $17 \cdot 20 - (17 + 20)$

38. $1000 \div 25 - (15 + 5)$

39. $6 \cdot 10 - 4 \cdot 10$

40. $3 \cdot 8 + 5 \cdot 8$

ANSWERS

1.
2.
3.
4.
5.
6.
7.
8.
9.
10.
11.
12.
13.
14.
15.
16.
17.
18.
19.
20.
21.
22.
23.
24.
25.
26.
27.
28.
29.
30.
31.
32.
33.
34.
35.
36.
37.
38.
39.
40.

ANSWERS

41.

42.

43.

44.

45.

46.

47.

48.

49.

50.

51.

52.

53.

54.

55.

56.

57.

58.

59.

60.

61.

62.

63.

64.

65.

66.

41. $300 \div 5 + 10$

42. $144 \div 4 - 2$

43. $3 \cdot (2 + 8)^2 - 5 \cdot (4 - 3)^2$

44. $7 \cdot (10 - 3)^2 - 2 \cdot (3 + 1)^2$

45. $4^2 + 8^2 \div 2^2$

46. $6^2 - 3^4 \div 3^3$

47. $10^3 - 10 \cdot 6 - (4 + 5 \cdot 6)$

48. $7^2 + 20 \cdot 4 - (28 + 9 \cdot 2)$

49. $6 \cdot 11 - (7 + 3) \div 5 - (6 - 4)$

50. $8 \times 9 - (12 - 8) \div 4 - (10 - 7)$

51. $120 - 3^3 \cdot 4 \div (30 - 24)$

52. $80 - 2^4 \cdot 15 \div (35 - 15)$

53. Find the average of $64, $97, and $121.

54. Find the average of four test grades of 86, 92, 80, and 78.

d Simplify.

55. $8 \times 13 + \{42 \div [18 - (6 + 5)]\}$

56. $72 \div 6 - \{2 \times [9 - (4 \times 2)]\}$

57. $[14 - (3 + 5) \div 2] - [18 \div (8 - 2)]$

58. $[92 \times (6 - 4) \div 8] + [7 \times (8 - 3)]$

59. $(82 - 14) \times [(10 + 45 \div 5) - (6 \cdot 6 - 5 \cdot 5)]$

60. $(18 \div 2) \cdot \{[(9 \cdot 9 - 1) \div 2] - [5 \cdot 20 - (7 \cdot 9 - 2)]\}$

61. $4 \times \{(200 - 50 \div 5) - [(35 \div 7) \cdot (35 \div 7) - 4 \times 3]\}$

62. $\{[18 - 2 \cdot 6] - [40 \div (17 - 9)]\} + \{48 - 13 \times 3 + [(50 - 7 \cdot 5) + 2]\}$

SYNTHESIS

Each of the expressions in Exercises 63–65 is incorrect. First find the correct answer. Then place as many parentheses as needed in the expression in order to make the incorrect answer correct.

63. $1 + 5 \cdot 4 + 3 = 36$

64. $12 \div 4 + 2 \cdot 3 - 2 = 2$

65. $12 \div 4 + 2 \cdot 3 - 2 = 4$

66. Use any grouping symbols and one occurrence each of 1, 2, 3, 4, 5, 6, 7, 8, and 9 to represent 100.

These Critical Thinking exercises extend the concept of the Synthesis exercises, which have been introduced at the ends of exercise sets. They require you to synthesize objectives from several sections, thereby building your critical thinking skills. The exercises are of three types: Calculator Connections, which call for you to use your calculator; Extended Synthesis exercises, which are similar to the Synthesis exercises in the exercise sets, but may require more time; and Thinking and Writing exercises, which encourage you to both think and write about the ideas you have studied in this chapter.

CALCULATOR CONNECTION

In the following, use a calculator in any way you can to ease your work. Discussion of the general use of a calculator is presented in the section titled "Using a Scientific Calculator" at the back of the book. If you are unfamiliar with the use of your particular calculator, consult its manual or your instructor.

Complete the following exercises, using your calculator for trial-and-error computations.

1. Calculators do not always follow the rules for order of operations. To find out how your calculator is programmed, consider the computation

$$3 \boxed{+} 4 \boxed{\cdot} 2 \boxed{=} \rightarrow ?$$

Carry out this computation on your calculator.

If you get 11, the correct answer, then your calculator is programmed to follow the order of operations. If you get 14, then you know that you must always enter the operations in the correct order:

$$4 \boxed{\cdot} 2 \boxed{+} 3 \boxed{=} \rightarrow 11$$

to get the correct answer.

   a) Does your calculator follow the rules for order of operations?
   b) Find $84 - 5 \cdot 7$.

2. In the product below, d is a digit, which means that d is 0, 1, 2, 3, 4, 5, 6, 7, 8, or 9. Find d.

$$\begin{array}{r} 9d \\ \times\, d2 \\ \hline 8036 \end{array}$$

3. a) Look for a pattern in the following list, and find the missing numbers.

   2, 6, 18, 54, ___, ___, ___, ___, ___, ___, ___.

   b) If the list in part (a) were to continue, at what position in the list would the number 774,840,978 be located?

4. The list of six numbers below is written by starting with two given numbers, and multiplying them to get the next number in the list. Find the missing numbers.

   ___; ___; ___; 567; 35721; 20,253,807.

5. In the division below, a and b are digits. Find a and b.

$$2b1)\overline{236,421} \quad 9a1$$

6. Consider only the numbers 3, 5, 7, and 9. Assume that each is placed in a blank in the following. What placement of the numbers in the blanks yields the largest number? Explain why there are two answers.

$$\blacksquare + \blacksquare \cdot \blacksquare - \blacksquare = ?$$

7. Consider only the numbers 2, 4, 6, and 8. Assume that each is placed in a blank in the following. What placement of the numbers in the blanks yields the largest number? Explain why there is just one answer.

$$\blacksquare + \blacksquare^2 \cdot \blacksquare - \blacksquare = ?$$

In Exercises 8 and 9, place one of $+$, $-$, $\times$, and $\div$ in each blank to make a true sentence.

8. $31 \,\blacksquare\, (87 \,\blacksquare\, 19) = 2108$
9. $2^9 \,\blacksquare\, 8^2 \,\blacksquare\, 3^2 = 72$

EXTENDED SYNTHESIS EXERCISES

1. A mining company estimates that a crew needs to tunnel 2100 ft into a mountain to reach the desired deposit of copper ore. Each day the crew is able to tunnel about 500 ft. Each night about 200 ft of loose rocks roll back into the excavated tunnel. How many days will it take the mining company to reach the copper deposits?

2. A crew that works for a large shopping center is asked to rope off a rectangular area in the parking lot for an automobile show. The area roped off is 350 ft by 400 ft. Posts are to be placed every 25 ft around the area. How many posts are needed?

3. An **even number** is a number like 0, 2, 4, 6, 8, and so on, that can be expressed as $2 \cdot k$, where k is any whole number. For example, we know that 6 is an even number because we can express 6 as $6 = 2 \cdot 3$. An **odd number** is a number like 1, 3, 5, 7, 9, 11, and so on, that can be expressed as $2 \cdot k + 1$, where k is any whole number. For

(continued)

example, we know that 11 is odd because we can express it as $11 = 2 \cdot 5 + 1$.

a) Show that 38 is even.

b) Show that 0 is even.

c) Show that 47 is odd.

d) What can be said about the sum of two even numbers?

e) What can be said about the sum of two odd numbers?

f) What can be said about the sum of an even number and an odd number?

g) What can be said about the product of an even number and an odd number?

h) What can be said about the expression $a^2 + a \cdot b$, where a is odd and b is even? Explain.

4. Words like *radar* and *toot* read the same backward and forward. A number that reads the same backward and forward is called a **palindrome.** For example,

   11, 121, 202, and 34543

are palindrome numbers. Many numbers can be transformed to palindromes by reversing the digits and adding, and so on. For example,

257 ←	Not palindrome
752 ←	Reverse digits
1009 ←	Add
9001 ←	Reverse digits
10010 ←	Add
01001 ←	Reverse digits
11011 ←	Palindrome

a) To what palindrome can 356 be transformed?

b) To what palindrome can 471 be transformed?

5. Ahmad and Felicia attend Ag-State Community College. After visiting one of the college farms, Ahmad noted that a particular building contained 22 animals consisting of pigs and chickens. Felicia noted that the animals had a total of 74 legs. How many of each type of animal were there?

6. Explain the reasoning behind arranging these numbers as follows:

   8, 5, 4, 9, 1, 7, 6, 10, 3, 2.

7. Consider the squares of whole numbers: 1^2, 2^2, 3^2, 4^2, and so on. We get 1, 4, 9, 16, 25, and so on, which are called **perfect squares.** List all the perfect squares that have five digits or fewer and are also palindromes. See Exercise 4. Is the square of a palindrome a palindrome?

8. Consider the cubes of whole numbers: 1^3, 2^3, 3^3, 4^3, and so on. We get 1, 8, 27, 64, and so on, which are called **perfect cubes.** List all the perfect cubes that have seven digits or fewer and are also palindromes. See Exercise 4.

Square Numbers   Consider the numbers and arrays below. We call 1 the first *square number*, 4 the second square number, 9 the third, 16 the fourth, and so on, because these numbers can be arranged in square arrays as shown. Look for patterns in the following.

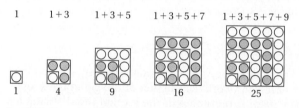

9. Find the eighth square number.

10. Find the thirteenth square number.

11. How are square numbers related to odd numbers?

12. Which square number is 841?

13. Explain a fast way to find the following sum without adding.

$$1 + 3 + 5 + 7 + 9 + 11 + 13 + 15 + 17 + 19 + 21 + 23 + 25 + 27 + 29 + 31$$

14. Find a 3-digit number such that:

a) the number is a palindrome (see Exercise 4);

b) the sum of the digits is a square;

c) the product of the digits is a cube.

EXERCISES FOR THINKING AND WRITING

1. Discuss the difference between a "take away" problem situation and a "how much more" problem situation in subtraction.

2. Discuss at least three reasons for rounding and estimating.

3. Describe two ways in which multiplication can occur in the real world.

4. Describe two ways in which division can occur in the real world.

5. Find a real-world situation that fits the equation $7 \cdot 29 + 3 = 206$.

Summary and Review Exercises: Chapter 1

The review exercises that follow are for practice. Answers are at the back of the book. If you miss an exercise, restudy the section and objective indicated alongside the answer.

1. Write expanded notation: 2793.

2. Write a word name: 2,781,427.

3. What does the digit 7 mean in 4,678,952?

4. Write standard notation for the number in this sentence: The gross national product is five trillion, six hundred eighty-five billion, eight hundred million dollars.

Add.

5.
$$\begin{array}{r} 3\,8\,4\,7 \\ +\,2\,1\,3\,2 \\ \hline \end{array}$$

6.
$$\begin{array}{r} 2\,7,6\,0\,9 \\ +\,3\,8,4\,1\,5 \\ \hline \end{array}$$

7.
$$\begin{array}{r} 2\,7\,4\,3 \\ 4\,1\,2\,5 \\ 6\,2\,7\,4 \\ +\,8\,9\,5\,6 \\ \hline \end{array}$$

8.
$$\begin{array}{r} 9\,1,4\,2\,6 \\ +\ \ 7,4\,9\,5 \\ \hline \end{array}$$

Subtract.

9.
$$\begin{array}{r} 8\,4\,6\,5 \\ -\,7\,3\,1\,2 \\ \hline \end{array}$$

10.
$$\begin{array}{r} 3\,7\,4\,3 \\ -\,2\,5\,9\,6 \\ \hline \end{array}$$

11.
$$\begin{array}{r} 6\,0\,0\,3 \\ -\,3\,7\,2\,9 \\ \hline \end{array}$$

12.
$$\begin{array}{r} 3\,7,4\,0\,5 \\ -\,1\,9,6\,4\,8 \\ \hline \end{array}$$

13. $678 - 234$

14. $6000 - 1234$

Multiply.

15.
$$\begin{array}{r} 7\,0\,0 \\ \times\,6\,0\,0 \\ \hline \end{array}$$

16.
$$\begin{array}{r} 7\,8\,4\,6 \\ \times\ \ 8\,0\,0 \\ \hline \end{array}$$

17. 9×76

18. 7×6394

19.
$$\begin{array}{r} 7\,4 \\ \times\,4\,6 \\ \hline \end{array}$$

20.
$$\begin{array}{r} 7\,2\,6 \\ \times\,6\,9\,8 \\ \hline \end{array}$$

21.
$$\begin{array}{r} 5\,8\,7 \\ \times\ \ 4\,7 \\ \hline \end{array}$$

22.
$$\begin{array}{r} 3\,4\,5\,6 \\ \times\,1\,0\,0\,0 \\ \hline \end{array}$$

Divide.

23. $80 \div 16$

24. $63 \div 5$

25. $7 \overline{)5\,6\,0}$

26. $4 \overline{)8\,3\,0}$

27. $8 \overline{)3\,0\,7\,3}$

28. $6\,0 \overline{)2\,8\,6}$

29. $7\,9 \overline{)4\,2\,6\,6}$

30. $3\,8 \overline{)1\,7,1\,7\,6}$

31. $1\,4 \overline{)7\,0,1\,1\,2}$

32. $1\,2 \overline{)5\,2,6\,6\,8}$

Solve.

33. $47 + x = 92$

34. $x = 782 - 236$

35. $46 \cdot n = 368$

Solve.

36. In 1909, the first "Lincoln-head" pennies were minted. Seventy-three years later, these pennies were first minted with a decreased copper content. In what year was the copper content reduced?

37. A farmer harvested 625 bu (bushels) of corn, 865 bu of wheat, 698 bu of soybeans, and 597 bu of potatoes. What was the farmer's total harvest?

38. A family budgets $4950 yearly for food and clothing and $3585 for entertainment. The yearly income of the family was $28,283. How much of this income remained after these two allotments?

39. A certain cottage cheese contains 113 calories per ounce. A bulk container of this cheese contains 25 oz. What is the caloric content of this container?

40. A sweater costs $28 and a skirt costs $37. Find the total cost of a wardrobe of 6 sweaters and 9 skirts.

41. A chemist has 2753 mL (milliliters) of acid. How many 20-mL beakers can be filled? How much will be left over?

42. A student buys 8 books at $25 each and pays for them with $20 bills. How many $20 bills does it take?

Round 345,759 to the nearest:

43. Hundred.   **44.** Ten.   **45.** Thousand.

Estimate the sum, difference, or product by rounding to the nearest hundred. Show your work.

46.
$$\begin{array}{r} 4\,1,3\,4\,8 \\ +\,1\,9,7\,4\,9 \\ \hline \end{array}$$

47.
$$\begin{array}{r} 3\,8,6\,5\,2 \\ -\,2\,4,5\,4\,9 \\ \hline \end{array}$$

48.
$$\begin{array}{r} 3\,9\,6 \\ \times\,7\,4\,8 \\ \hline \end{array}$$

Use < or > for ▓ to write a true sentence.

49. 67 ▓ 56

50. 1 ▓ 23

51. Write exponential notation: $8 \cdot 8 \cdot 8$.

Evaluate.

52. 2^4

53. 6^2

Simplify.

54. $8 \times 6 + 17$

55. $10 \times 24 - (18 + 2) \div 4 - (9 - 7)$

56. $7 + (4 + 3)^2$

57. $7 + 4^2 + 3^2$

58. $(80 \div 16) \times [(20 - 56 \div 8) + (8 \cdot 8 - 5 \cdot 5)]$

59. Find the average of 157, 170, and 168.

Find the distance around (perimeter of) and the area of the figure.

60.

61.

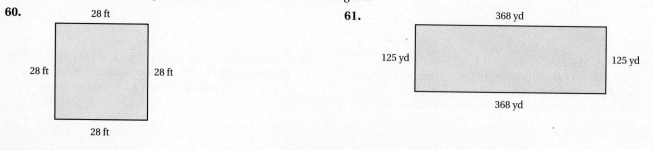

Test: Chapter 1

1. Write expanded notation: 8843.

2. Write a word name: 38,403,277.

3. In the number 546,789, which digit tells the number of hundred thousands?

Add.

4.
```
   6 8 1 1
 + 3 1 7 8
```

5.
```
   4 5,8 8 9
 + 1 7,9 0 2
```

6.
```
   1 2
      8
      3
      7
 +    4
```

7.
```
   6 2 0 3
 + 4 3 1 2
```

Subtract.

8.
```
   7 9 8 3
 - 4 3 5 3
```

9.
```
   2 9 7 4
 - 1 9 3 5
```

10.
```
   8 9 0 7
 - 2 0 5 9
```

11.
```
   2 3,0 6 7
 - 1 7,8 9 2
```

Multiply.

12.
```
   4 5 6 8
 ×       9
```

13.
```
   8 8 7 6
 ×     6 0 0
```

14.
```
     6 5
 ×   3 7
```

15.
```
     6 7 8
 ×   7 8 8
```

Divide.

16. $15 \div 4$

17. $420 \div 6$

18. $89 \overline{)8633}$

19. $44 \overline{)35,428}$

Solve.

20. James Dean was 24 years old when he died. He was born in 1931. In what year did he die?

21. A beverage company produces 739 cans of soda. How many 8-can packages can be filled? How many cans will be left over?

22. A customer buys 15 pieces of lumber at $12 each and pays for them with $10 bills. How many $10 bills does it take?

23. A rectangular lot measures 200 m by 600 m. What is the area of the lot?

24. A sack of oranges weighs 27 lb. A sack of apples weighs 32 lb. Find the total weight of 16 bags of oranges and 43 bags of apples.

25. A box contains 5000 staples. How many staplers can be filled from the box if each stapler holds 250 staples?

ANSWERS

1. _____

2. _____

3. _____

4. _____

5. _____

6. _____

7. _____

8. _____

9. _____

10. _____

11. _____

12. _____

13. _____

14. _____

15. _____

16. _____

17. _____

18. _____

19. _____

20. _____

21. _____

22. _____

23. _____

24. _____

25. _____

26. _____

27. _____

28. _____

29. _____

30. _____

31. _____

32. _____

33. _____

34. _____

35. _____

36. _____

37. _____

38. _____

39. _____

40. _____

41. _____

42. _____

43. _____

44. _____

45. _____

46. _____

47. _____

48. _____

49. _____

50. _____

26. The area of Vermont is 9609 sq mi. The area of New Hampshire is 9304 sq mi. How much larger is Vermont?

27. A professional bowler rolled a game of 245. Then the bowler rolled a game of 189. How much higher was the first game?

28. Listed below are the areas, in square miles, of the New England states. What is the total area of New England?

Maine	33,215
Massachusetts	8,093
New Hampshire	9,304
Vermont	9,609
Connecticut	5,009
Rhode Island	1,214

29. You have $345 in a checking account. You write checks for $45 and $29. How much money is left in the checking account?

Solve.

30. $28 + x = 74$

31. $169 \div 13 = n$

32. $38 \cdot y = 532$

Round 34,578 to the nearest:

33. Thousand.

34. Ten.

35. Hundred.

Estimate the sum, difference, or product by rounding to the nearest hundred. Show your work.

36.
$$\begin{array}{r} 2\,3,6\,4\,9 \\ +\,5\,4,7\,4\,6 \end{array}$$

37.
$$\begin{array}{r} 5\,4,7\,5\,1 \\ -\,2\,3,6\,4\,9 \end{array}$$

38.
$$\begin{array}{r} 8\,2\,4 \\ \times\,4\,8\,9 \end{array}$$

Use < or > for ▆ to write a true sentence.

39. 34 ▆ 17

40. 117 ▆ 157

41. Write exponential notation: $12 \cdot 12 \cdot 12 \cdot 12$.

Evaluate.

42. 7^3

43. 2^3

Simplify.

44. $(10 - 2)^2$

45. $10^2 - 2^2$

46. $(25 - 15) \div 5$

47. $8 \times \{(20 - 11) \cdot [(12 + 48) \div 6 - (9 - 2)]\}$

48. $2^4 + 24 \div 12$

49. Find the average of 97, 98, 87, and 86.

50. Find the distance around (perimeter of) and the area of the figure.

86 ft

27 ft

2

Multiplication and Division: Fractional Notation

INTRODUCTION

We consider multiplication and division using fractional notation in this chapter. To aid our study, the chapter begins with factorizations and rules for divisibility. After multiplication and division are discussed, those skills are used to solve equations and problems.

AN APPLICATION

How many test tubes, each containing $\frac{3}{5}$ mL, can a nursing student fill from a container of 60 mL?

THE MATHEMATICS

Let n = the number of test tubes that can be filled from a container of 60 mL. The problem then translates to this equation:

$$n = \underbrace{60 \div \frac{3}{5}}.$$

This division using fractional notation occurs in problem solving.

OBJECTIVES FOR REVIEW

The review objectives to be tested in addition to the material in this chapter are as follows.

[1.3d] Subtract whole numbers.

[1.6c] Divide whole numbers.

[1.7b] Solve equations like $t + 28 = 54$, $28 \cdot x = 168$, and $98 \div 2 = y$.

[1.8a] Solve problems involving addition, subtraction, multiplication, or division of whole numbers.

Pretest: Chapter 2

1. Determine whether 59 is prime, composite, or neither.

2. Find the prime factorization of 140.

3. Determine whether 1503 is divisible by 9.

4. Determine whether 788 is divisible by 8.

Simplify.

5. $\dfrac{57}{57}$

6. $\dfrac{68}{1}$

7. $\dfrac{0}{50}$

8. $\dfrac{8}{32}$

Multiply and simplify.

9. $\dfrac{1}{3} \cdot \dfrac{18}{5}$

10. $\dfrac{5}{6} \cdot 24$

11. $\dfrac{2}{5} \cdot \dfrac{25}{8}$

Find the reciprocal.

12. $\dfrac{7}{8}$

13. 11

Divide and simplify.

14. $15 \div \dfrac{5}{8}$

15. $\dfrac{2}{3} \div \dfrac{8}{9}$

16. Solve:

$\dfrac{7}{10} \cdot x = 21.$

17. Use $=$ or $\neq$ for ▨ to write a true sentence:

$\dfrac{5}{11}$ ▨ $\dfrac{1}{2}$.

Solve.

18. A person earns \$48 for working a full day. How much is earned for working $\dfrac{3}{4}$ of a day?

19. A piece of tubing $\dfrac{5}{8}$ m long is to be cut into 15 pieces of the same length. What is the length of each piece?

2.1 Factorizations

In this chapter, we begin our work with fractions. Certain skills make such work easier. For example, in order to simplify

$$\frac{12}{32},$$

it is important that we be able to *factor* the 12 and the 32, as follows:

$$\frac{12}{32} = \frac{4 \cdot 3}{4 \cdot 8}.$$

Then we "remove" a factor of 1:

$$\frac{4 \cdot 3}{4 \cdot 8} = \frac{4}{4} \cdot \frac{3}{8} = 1 \cdot \frac{3}{8} = \frac{3}{8}.$$

Thus factoring is an important skill in working with fractions.

a   FACTORS AND FACTORIZATION

In Sections 2.1 and 2.2, we consider only the **natural numbers** 1, 2, 3, and so on.

Let's look at the product $3 \cdot 4 = 12$. We say that 3 and 4 are **factors** of 12. Since $12 = 12 \cdot 1$, we also know that 12 and 1 are factors of 12.

> A *factor* of a given number is a number multiplied in a product.
>
> A *factorization* of a number is an expression for the number that shows it as a product of natural numbers.

For example, each of the following gives a factorization of 12.

$12 = 4 \cdot 3$	This factorization shows that 4 and 3 are factors of 12.
$12 = 12 \cdot 1$	This factorization shows that 12 and 1 are factors of 12.
$12 = 6 \cdot 2$	This factorization shows that 6 and 2 are factors of 12.
$12 = 2 \cdot 3 \cdot 2$	This factorization shows that 2 and 3 are factors of 12.

Since $n = n \cdot 1$, every number has a factorization, and every number has factors even if its only factors are itself and 1.

EXAMPLE 1   Find all the factors of 24.

We first find some factorizations.

$$24 = 1 \cdot 24 \qquad 24 = 3 \cdot 8$$
$$24 = 2 \cdot 12 \qquad 24 = 4 \cdot 6$$

Note that all but one of the factors of a natural number are *less* than the number.

Factors: 1, 2, 3, 4, 6, 8, 12, 24.

Do Exercises 1–4.

OBJECTIVES

After finishing Section 2.1, you should be able to:

a   Find the factors of a number.

b   Find some multiples of a number, and determine whether a number is divisible by another.

c   Given a number from 1 to 100, tell whether it is prime, composite, or neither.

d   Find the prime factorization of a composite number.

FOR EXTRA HELP

TAPE 3   TAPE 3A   MAC: 2
                             IBM: 2

Find all the factors of the number. (*Hint:* Find some factorizations of the number.)

1. 6

2. 8

3. 10

4. 32

Answers on page A-2

The following is a table of the prime numbers from 2 to 157. There are more extensive tables, but these prime numbers will be the most helpful to you in this text.

A TABLE OF PRIMES

2, 3, 5, 7, 11, 13, 17, 19, 23, 29, 31, 37, 41, 43, 47, 53, 59, 61, 67, 71, 73, 79, 83, 89, 97, 101, 103, 107, 109, 113, 127, 131, 137, 139, 149, 151, 157

---

d | PRIME FACTORIZATIONS

To factor a composite number into a product of primes is to find a **prime factorization** of the number. To do this, we consider the primes

2, 3, 5, 7, 11, 13, 17, 19, 23, and so on,

and determine whether a given number is divisible by the primes.

EXAMPLE 8 Find the prime factorization of 39.

a) We divide by the first prime, 2.

$$\frac{19}{2)\overline{39}} \quad R = 1$$

Because the remainder is not 0, 2 is not a factor of 39.

b) We divide by the next prime, 3.

$$\frac{13}{3)\overline{39}} \quad R = 0$$

Because 13 is prime, we are finished. The prime factorization is

$$39 = 3 \cdot 13.$$

EXAMPLE 9 Find the prime factorization of 76.

a) We divide by the first prime, 2.

$$\frac{38}{2)\overline{76}} \quad R = 0$$

b) Because 38 is composite, we start with 2 again:

$$\frac{19}{2)\overline{38}} \quad R = 0$$

Because 19 is a prime, we are finished. The prime factorization is

$$76 = 2 \cdot 2 \cdot 19.$$

We abbreviate our procedure as follows.

$$\frac{19}{\frac{2)\overline{38}}{2)\overline{76}}}$$

$$76 = 2 \cdot 2 \cdot 19$$

Multiplication is commutative so a factorization such as $2 \cdot 2 \cdot 19$ could also be expressed as $2 \cdot 19 \cdot 2$ or $19 \cdot 2 \cdot 2$, but the prime factors are still the same. For this reason, we agree that any of these is "the" prime factorization of 76.

Every number has just one (unique) prime factorization.

EXAMPLE 10   Find the prime factorization of 72.

$$
\begin{array}{r}
3 \\
3\overline{)9} \\
2\overline{)18} \\
2\overline{)36} \\
2\overline{)72}
\end{array}
$$

$$72 = 2 \cdot 2 \cdot 2 \cdot 3 \cdot 3$$

Another way to find a prime factorization is by using a **factor tree** as follows:

```
        72
       /  \
      8    9
     /\   /\
    2  4 3  3
   / /\  \  \
  2 2  2  3  3
```

EXAMPLE 11   Find the prime factorization of 189.

We can use a string of successive divisions:

$$
\begin{array}{r}
7 \\
3\overline{)21} \\
3\overline{)63} \\
3\overline{)189}
\end{array}
$$

189 is not divisible by 2. We move to 3.
63 is not divisible by 2. We move to 3.
21 is not divisible by 2. We move to 3.

$$189 = 3 \cdot 3 \cdot 3 \cdot 7$$

We can also use a factor tree.

```
      189
     /   \
    3    63
   /    /  \
  3    7    9
 /    /    /\
3    7    3  3
```

Find the prime factorization of the number.

12. 6

13. 12

14. 45

15. 98

16. 126

17. 144

To the student and the instructor: Recall that the Skill Maintenance exercises, which occur at the end of most exercise sets, review any skill that has been studied before in the text. Beginning with this chapter, however, certain objectives from four particular sections, along with the material of this chapter, will be tested on the chapter test. For this chapter, the review sections and objectives to be tested are Sections [1.3d], [1.6c], [1.7b], and [1.8a].

Answers on page A-2

EXAMPLE 12 Find the prime factorization of 65.

We can use a string of successive divisions.

$$\begin{array}{r} 13 \\ 5\overline{)65} \end{array}$$ 65 is not divisible by 2 or 3. We move to 5.

$$65 = 5 \cdot 13$$

We can also use a factor tree.

$$\begin{array}{c} 65 \\ \diagup \diagdown \\ 5 \quad 13 \end{array}$$

Do Exercises 12–17 on the preceding page.

S I D E L I G H T S

FACTORS AND SUMS

To *factor* a number is to express it as a product. Since $15 = 5 \cdot 3$, we say that 15 is *factored* and that 5 and 3 are *factors* of 15. In the table below, the top number in each column has been factored in such a way that the sum of the factors is the bottom number in the column. For example, in the first column, 56 has been factored as $7 \cdot 8$, and $7 + 8 = 15$, the bottom number. Such thinking will be important in understanding the meaning of a factor and in algebra.

PRODUCT	56	63	36	72	140	96		168	110			
FACTOR	7									9	24	3
FACTOR	8					8	8			10	18	
SUM	15	16	20	38	24	20	14		21			24

EXERCISE
Find the missing numbers in the table.

Exercise Set 2.1

a Find all the factors of the number.

1. 18   **2.** 16   **3.** 54   **4.** 48

5. 4   **6.** 9   **7.** 7   **8.** 11

9. 1   **10.** 3   **11.** 98   **12.** 100

b Multiply by 1, 2, 3, and so on, to find ten multiples of the number.

13. 4   **14.** 11   **15.** 20   **16.** 50

17. 3   **18.** 5   **19.** 12   **20.** 13

21. 10   **22.** 6   **23.** 9   **24.** 14

25. Determine whether 26 is divisible by 6.

26. Determine whether 29 is divisible by 9.

27. Determine whether 1880 is divisible by 8.

28. Determine whether 4227 is divisible by 3.

29. Determine whether 256 is divisible by 16.

30. Determine whether 102 is divisible by 4.

31. Determine whether 4227 is divisible by 9.

32. Determine whether 200 is divisible by 25.

33. Determine whether 8650 is divisible by 16.

34. Determine whether 4143 is divisible by 7.

ANSWERS

1. _____
2. _____
3. _____
4. _____
5. _____
6. _____
7. _____
8. _____
9. _____
10. _____
11. _____
12. _____
13. _____
14. _____
15. _____
16. _____
17. _____
18. _____
19. _____
20. _____
21. _____
22. _____
23. _____
24. _____
25. _____
26. _____
27. _____
28. _____
29. _____
30. _____
31. _____
32. _____
33. _____
34. _____

ANSWERS

35.

36.

37.

38.

39.

40.

41.

42.

43.

44.

45.

46.

47.

48.

49.

50.

51.

52.

53.

54.

55.

56.

57.

58.

59.

60.

61.

62.

63.

64.

65.

66.

67.

68.

69.

70.

71.

72.

73.

74.

75.

76.

c Determine whether the number is prime, composite, or neither.

35. 1   **36.** 2   **37.** 9   **38.** 19

39. 11   **40.** 27   **41.** 29   **42.** 49

d Find the prime factorization of the number.

43. 8   **44.** 16   **45.** 14   **46.** 15

47. 42   **48.** 32   **49.** 25   **50.** 40

51. 50   **52.** 62   **53.** 169   **54.** 140

55. 100   **56.** 110   **57.** 35   **58.** 70

59. 72   **60.** 86   **61.** 77   **62.** 99

63. 156   **64.** 142   **65.** 300   **66.** 175

SKILL MAINTENANCE

Multiply.

67. $2 \cdot 13$ **68.** $8 \cdot 32$ **69.** $17 \cdot 25$ **70.** $25 \cdot 168$

Divide.

71. $0 \div 22$ **72.** $22 \div 1$ **73.** $22 \div 22$ **74.** $66 \div 22$

SYNTHESIS

75. Describe an arrangement of 54 objects that corresponds to the factorization $54 = 6 \times 9$.

76. Describe an arrangement of 24 objects that corresponds to the factorization $24 = 2 \cdot 3 \cdot 4$.

2.2 Divisibility

Suppose you are asked to find the simplest fractional notation for

$$\frac{117}{225}.$$

Since the numbers are quite large, you might feel that the task is difficult. However, both the numerator and the denominator have 9 as a factor. If you knew this, you could factor and simplify quickly as follows:

$$\frac{117}{225} = \frac{9 \cdot 13}{9 \cdot 25} = \frac{9}{9} \cdot \frac{13}{25} = 1 \cdot \frac{13}{25} = \frac{13}{25}.$$

How did we know that both numbers have 9 as a factor? There are fast tests for such determinations. If the sum of the digits of a number is divisible by 9, then the number is divisible by 9; that is, it has 9 as a factor. Since $1 + 1 + 7 = 9$ and $2 + 2 + 5 = 9$, both numbers have 9 as a factor.

a RULES FOR DIVISIBILITY

In this section, we learn fast ways of determining whether numbers are divisible by 2, 3, 4, 5, 6, 8, 9, and 10. This will make simplifying fractional notation much easier.

DIVISIBILITY BY 2

You may already know the test for divisibility by 2.

> A number is divisible by 2 (is *even*) if it has a ones digit of 0, 2, 4, 6, or 8 (that is, it has an even ones digit).

Let us see why. Consider 354, which is

3 hundreds + 5 tens + 4.

Hundreds and tens are both multiples of 2. If the last digit is a multiple of 2, then the entire number is a multiple of 2.

EXAMPLES   Determine whether the number is divisible by 2.

1. 355   is not a multiple of 2; 5 is *not* even.
2. 4786   is a multiple of 2; 6 is even.
3. 8990   is a multiple of 2; 0 is even.
4. 4261   is not a multiple of 2; 1 is *not* even.

Do Exercises 1–4.

OBJECTIVE

After finishing Section 2.2, you should be able to:

a   Determine whether a number is divisible by 2, 3, 4, 5, 6, 8, 9, or 10.

FOR EXTRA HELP

TAPE 4   TAPE 3B   MAC: 2
                                 IBM: 2

Determine whether the number is divisible by 2.

1. 84

2. 59

3. 998

4. 2225

Answers on page A-2

Determine whether the number is divisible by 3.

5. 111

6. 1111

7. 309

8. 17,216

Determine whether the number is divisible by 6.

9. 420

10. 106

11. 321

12. 444

Answers on page A-2

DIVISIBILITY BY 3

> A number is divisible by 3 if the sum of its digits is divisible by 3.

EXAMPLES Determine whether the number is divisible by 3.

5. 18 $\quad 1 + 8 = 9$

6. 93 $\quad 9 + 3 = 12$ $\qquad$ All divisible by 3 because the sums of their digits are divisible by 3.

7. 201 $\quad 2 + 0 + 1 = 3$

8. 256 $\quad 2 + 5 + 6 = 13$ $\qquad$ The sum is not divisible by 3, so 256 is not divisible by 3.

Do Exercises 5–8.

DIVISIBILITY BY 6

A number divisible by 6 is a multiple of 6. But $6 = 2 \cdot 3$, so the number is also a multiple of 2 and 3. Thus we have the following.

> A number is divisible by 6 if its ones digit is 0, 2, 4, 6, or 8 (is even) and the sum of its digits is divisible by 3.

EXAMPLES Determine whether the number is divisible by 6.

9. 720

Because 720 is even, it is divisible by 2. Also, $7 + 2 + 0 = 9$, so 720 is divisible by 3. Thus, 720 is divisible by 6.

$$720 \qquad 7 + 2 + 0 = 9$$

↑ Even $\qquad$ ↑ Divisible by 3

10. 73

73 is *not* divisible by 6 because it is *not* even.

73

↑ Not even

11. 256

256 is *not* divisible by 6 because the sum of its digits is *not* divisible by 3.

$$2 + 5 + 6 = 13$$

↑ Not divisible by 3

Do Exercises 9–12.

DIVISIBILITY BY 9

The test for divisibility by 9 is similar to the test for divisibility by 3.

A number is divisible by 9 if the sum of its digits is divisible by 9.

EXAMPLE 12 The number 6984 is divisible by 9 because

$$6 + 9 + 8 + 4 = 27$$

and 27 is divisible by 9.

EXAMPLE 13 The number 322 is *not* divisible by 9 because

$$3 + 2 + 2 = 7$$

and 7 is not divisible by 9.

Do Exercises 13–16.

DIVISIBILITY BY 10

A number is divisible by 10 if its ones digit is 0.

We know that this test works because the product of 10 and *any* number has a ones digit of 0.

EXAMPLES Determine whether the number is divisible by 10.

14. 3440 is divisible by 10 because the ones digit is 0.
15. 3447 is *not* divisible by 10 because the ones digit is not 0.

Do Exercises 17–20.

DIVISIBILITY BY 5

A number is divisible by 5 if its ones digit is 0 or 5.

EXAMPLES Determine whether the number is divisible by 5.

16. 220 is divisible by 5 because the ones digit is 0.
17. 475 is divisible by 5 because the ones digit is 5.
18. 6514 is *not* divisible by 5 because the ones digit is neither a 0 nor a 5.

Do Exercises 21–24.

Let us see why the test for 5 works. Consider 7830:

$$7830 = 10 \cdot 783 = 5 \cdot 2 \cdot 783.$$

Since 7830 is divisible by 10 and 5 is a factor of 10, 7830 is divisible by 5.

Determine whether the number is divisible by 9.

13. 16

14. 117

15. 930

16. 29,223

Determine whether the number is divisible by 10.

17. 305

18. 300

19. 847

20. 8760

Determine whether the number is divisible by 5.

21. 5780

22. 3427

23. 34,678

24. 7775

Answers on page A-2

Determine whether the number is divisible by 4.

25. 216

26. 217

27. 5865

28. 23,524

Determine whether the number is divisible by 8.

29. 7564

30. 7864

31. 17,560

32. 25,716

Answers on page A-2

Consider 6734:

$$6734 = 673 \text{ tens} + 4.$$

Tens are multiples of 5, so the only number that must be checked is the ones digit. If the last digit is a multiple of 5, the entire number is. In this case, 4 is not a multiple of 5, so 6734 is not divisible by 5.

DIVISIBILITY BY 4

The test for divisibility by 4 is similar to the test for divisibility by 2.

> A number is divisible by 4 if the number named by its last *two* digits is divisible by 4.

EXAMPLES   Determine whether the number is divisible by 4.

19. 8212   is divisible by 4 because 12 is divisible by 4.
20. 5216   is divisible by 4 because 16 is divisible by 4.
21. 8211   is *not* divisible by 4 because 11 is *not* divisible by 4.
22. 7515   is *not* divisible by 4 because 15 is *not* divisible by 4.

Do Exercises 25–28.

To see why the test for divisibility by 4 works, consider 516:

$$516 = 5 \text{ hundreds} + 16.$$

Hundreds are multiples of 4. If the number named by the last two digits is a multiple of 4, then the entire number is a multiple of 4.

DIVISIBILITY BY 8

The test for divisibility by 8 is an extension of the tests for divisibility by 2 and 4.

> A number is divisible by 8 if the number named by its last *three* digits is divisible by 8.

EXAMPLES   Determine whether the number is divisible by 8.

23.   5648   is divisible by 8 because 648 is divisible by 8.
24. 96,088   is divisible by 8 because 88 is divisible by 8.
25.   7324   is *not* divisible by 8 because 324 is *not* divisible by 8.
26. 13,420   is *not* divisible by 8 because 420 is *not* divisible by 8.

Do Exercises 29–32.

A NOTE ABOUT DIVISIBILITY BY 7

There are several tests for divisibility by 7, but all of them are more complicated than simply dividing by 7. So if you want to test for divisibility by 7, divide by 7.

Exercise Set 2.2

a To answer Exercises 1–8, consider the following numbers.

46	300	85
224	36	711
19	45,270	13,251
555	4444	254,765

1. Which of the above are divisible by 2?   **2.** Which of the above are divisible by 3?

3. Which of the above are divisible by 4?   **4.** Which of the above are divisible by 5?

5. Which of the above are divisible by 6?   **6.** Which of the above are divisible by 8?

7. Which of the above are divisible by 9?   **8.** Which of the above are divisible by 10?

To answer Exercises 9–16, consider the following numbers.

56	200	75
324	42	812
784	501	2345
55,555	3009	2001

9. Which of the above are divisible by 3?   **10.** Which of the above are divisible by 2?

1. _____

2. _____

3. _____

4. _____

5. _____

6. _____

7. _____

8. _____

9. _____

10. _____

11. _____

12. _____

13. _____

14. _____

15. _____

16. _____

17. _____

18. _____

19. _____

20. _____

21. _____

22. _____

23. _____

24. _____

25. _____

26. _____

11. Which of the above are divisible by 5? **12.** Which of the above are divisible by 4?

13. Which of the above are divisible by 9? **14.** Which of the above are divisible by 6?

15. Which of the above are divisible by 10? **16.** Which of the above are divisible by 8?

SKILL MAINTENANCE

Solve.

17. $56 + x = 194$

18. $y + 124 = 263$

19. $18 \cdot t = 1008$

20. $24 \cdot m = 624$

21. Find the total cost of 12 shirts at $37 each and 4 pairs of pants at $59 each.

22. Divide:

$$4\,5\,\overline{)\,1\,8\,0{,}1\,3\,5}$$

SYNTHESIS

Use the tests of divisibility to find the prime factorization of the number.

23. 7800   **24.** 2520   **25.** 2772   **26.** 1998

2.3 Fractions

The study of arithmetic begins with the set of whole numbers

0, 1, 2, 3, 4, 5, 6, 7, 8, 9, 10, 11, and so on.

The need soon arises for fractional parts of numbers such as halves, thirds, fourths, and so on. Here are some examples:

$\frac{1}{25}$ of the parking spaces in a commercial area in the state of Indiana are to be marked for the handicapped.

For $\frac{1}{10}$ of the people in the United States, English is not the primary language.

$\frac{1}{4}$ of the minimum daily requirement of calcium is provided by a cup of frozen yogurt.

$\frac{43}{100}$ of all corporate travel money is spent on airfares.

a | IDENTIFYING NUMERATORS AND DENOMINATORS

The following are some additional examples of fractions:

$$\frac{1}{2}, \quad \frac{3}{4}, \quad \frac{8}{5}, \quad \frac{11}{23}.$$

This way of writing number names is called **fractional notation.** The top number is called the **numerator** and the bottom number is called the **denominator**.

EXAMPLE 1 Identify the numerator and the denominator.

$$\frac{7}{8} \quad \begin{array}{l} \leftarrow \text{Numerator} \\ \leftarrow \text{Denominator} \end{array}$$

Do Exercises 1–3.

b | FRACTIONS AND THE REAL WORLD

EXAMPLE 2 What part is shaded?

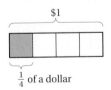

$1

$\frac{1}{4}$ of a dollar

When an object is divided into 4 parts of the same size, each of these parts is $\frac{1}{4}$ of the object. Thus, $\frac{1}{4}$ (*one-fourth*) is shaded.

Do Exercises 4–7.

Identify the numerator and the denominator.

1. $\frac{1}{6}$ **2.** $\frac{5}{7}$ **3.** $\frac{22}{3}$

What part is shaded?

4.

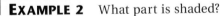

$1

5.

1 mile

6.

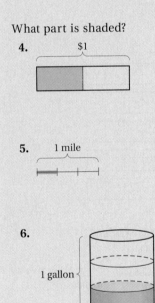

1 gallon

7.

Answers on page A-2

What part is shaded?

8.

$1

9.

1 mile

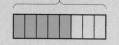

10.

1 gallon

11.

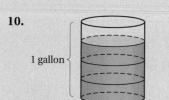

What part is shaded?

12.

2 miles

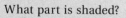

13.

$1

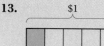

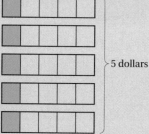

5 dollars

What part is shaded?

14.

1 mile

2 miles

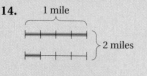

15.

1 gallon

2 gallons

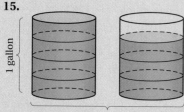

Answers on page A-2

EXAMPLE 3   What part is shaded?

$1

$\frac{3}{4}$ of a dollar

The object is divided into 4 parts of the same size, and 3 of them are shaded. This is $3 \cdot \frac{1}{4}$, or $\frac{3}{4}$. Thus, $\frac{3}{4}$ (*three-fourths*) of the object is shaded.

Do Exercises 8–11.

The fraction $\frac{3}{4}$ corresponds to another situation. We take 3 objects, divide them into fourths, and take $\frac{1}{4}$ of the entire amount. This is $\frac{1}{4} \cdot 3$, or $\frac{3}{4}$, or $3 \div 4$.

EXAMPLE 4   What part is shaded?

$1

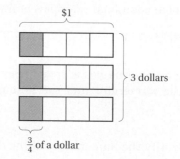

3 dollars

$\frac{3}{4}$ of a dollar

Thus, $\$\frac{3}{4}$ is shaded.

Do Exercises 12 and 13.

Fractions greater than 1 correspond to situations like the following.

EXAMPLE 5   What part is shaded?

$1 $1

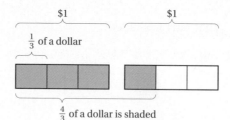

$\frac{1}{3}$ of a dollar

$\frac{4}{3}$ of a dollar is shaded

We divide the two objects into 3 parts each and take 4 of those parts. We have more than one whole object. In this case, it is $4 \cdot \frac{1}{3}$, or $\frac{4}{3}$.

Do Exercises 14 and 15.

Circle graphs, or pie charts, are often used to show the relationships of fractional parts of a whole. The following graph shows time spent at shopping malls.

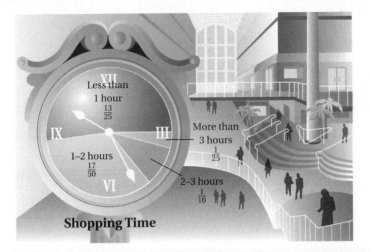

Shopping Time

Fractional notation also corresponds to situations involving part of a set.

EXAMPLE 6 What part of this set, or collection, of tools are wrenches?

There are 5 tools, and 3 are wrenches. We say that three-fifths of the tools are wrenches; that is, $\frac{3}{5}$ of the set consists of wrenches.

Do Exercises 16–18.

c | SOME FRACTIONAL NOTATION FOR WHOLE NUMBERS

FRACTIONAL NOTATION FOR 1

The number 1 corresponds to situations like the following.

If we divide an object into n parts and take n of them, we get all of the object (1 whole object).

$$\frac{n}{n} = 1, \quad \text{for any whole number } n \text{ that is not 0.}$$

16. What part of the set of tools in Example 6 are hammers?

17. What part of this set is shaded?

18. What part of this set are or were United States presidents? are recording stars?

Abraham Lincoln
Whitney Houston
Garth Brooks
Bill Clinton
Linda Ronstadt
Gloria Estefan

Simplify.

19. $\frac{1}{1}$ **20.** $\frac{4}{4}$

21. $\frac{34}{34}$ **22.** $\frac{100}{100}$

23. $\frac{2347}{2347}$ **24.** $\frac{103}{103}$

Answers on page A-2

Simplify, if possible.

25. $\dfrac{0}{1}$ **26.** $\dfrac{0}{8}$

27. $\dfrac{0}{107}$ **28.** $\dfrac{4-4}{567}$

29. $\dfrac{15}{0}$ **30.** $\dfrac{0}{3-3}$

Simplify.

31. $\dfrac{8}{1}$ **32.** $\dfrac{10}{1}$

33. $\dfrac{346}{1}$ **34.** $\dfrac{24-1}{23}$

Answers on page A-2

EXAMPLES   Simplify.

7. $\dfrac{5}{5} = 1$ **8.** $\dfrac{9}{9} = 1$ **9.** $\dfrac{23}{23} = 1$

Do Exercises 19–24 on the preceding page.

FRACTIONAL NOTATION FOR 0

Consider $\frac{0}{4}$. This corresponds to dividing an object into 4 parts and taking none of them. We get 0.

> $\dfrac{0}{n} = 0$,   for any whole number n that is not 0.

EXAMPLES   Simplify.

10. $\dfrac{0}{1} = 0$ **11.** $\dfrac{0}{9} = 0$ **12.** $\dfrac{0}{23} = 0$

Fractional notation with a denominator of 0, such as $n/0$, is meaningless because we cannot speak of an object divided into *zero* parts. (If it is not divided at all, then we say that it is undivided and remains in one part.)

> $\dfrac{n}{0}$ is not defined for any whole number n.

Do Exercises 25–30.

OTHER WHOLE NUMBERS

Consider $\frac{4}{1}$. This corresponds to taking 4 objects and dividing each into 1 part. (We do not divide them.) We have 4 objects.

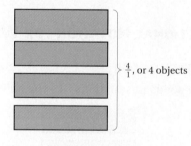

$\frac{4}{1}$, or 4 objects

> Any whole number divided by 1 is the whole number. That is,
>
> $\dfrac{n}{1} = n$,   for any whole number n.

EXAMPLES   Simplify.

13. $\dfrac{2}{1} = 2$ **14.** $\dfrac{9}{1} = 9$ **15.** $\dfrac{34}{1} = 34$

Do Exercises 31–34.

Exercise Set 2.3

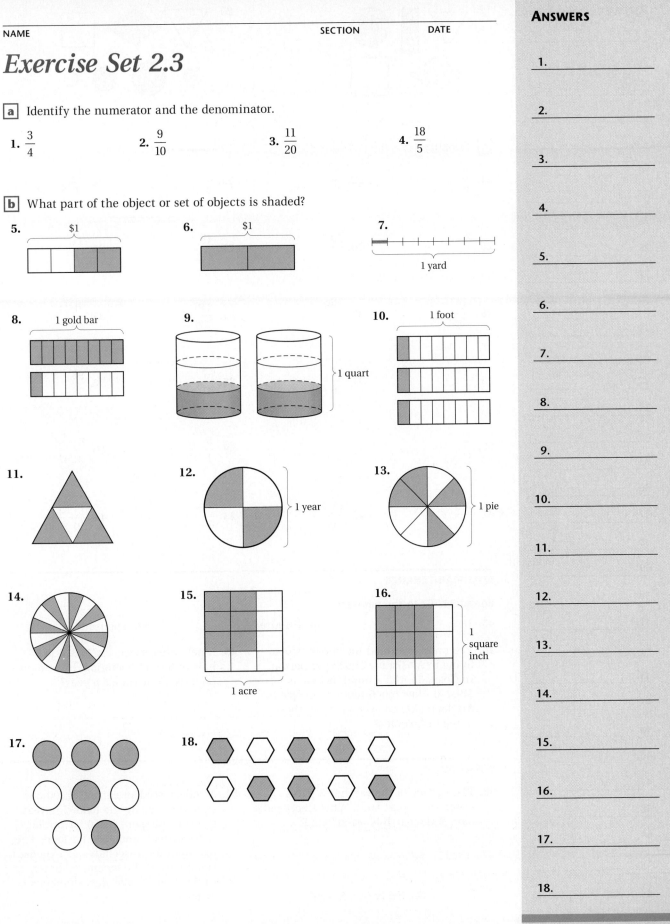

a Identify the numerator and the denominator.

1. $\dfrac{3}{4}$ 2. $\dfrac{9}{10}$ 3. $\dfrac{11}{20}$ 4. $\dfrac{18}{5}$

b What part of the object or set of objects is shaded?

5. $1

6. $1

7. 1 yard

8. 1 gold bar

9. 1 quart

10. 1 foot

11.

12. 1 year

13. 1 pie

14.

15. 1 acre

16. 1 square inch

17.

18.

ANSWERS

1. _____

2. _____

3. _____

4. _____

5. _____

6. _____

7. _____

8. _____

9. _____

10. _____

11. _____

12. _____

13. _____

14. _____

15. _____

16. _____

17. _____

18. _____

ANSWERS

19.

20.

21.

22.

23.

24.

25.

26.

27.

28.

29.

30.

31.

32.

33.

34.

35.

36.

37.

38.

39.

40.

41.

42.

43.

44.

45.

46.

47.

48.

49.

50.

51.

19.

20.

[c] Simplify.

21. $\dfrac{0}{8}$ **22.** $\dfrac{8}{8}$ **23.** $\dfrac{8}{1}$ **24.** $\dfrac{16}{1}$

25. $\dfrac{20}{20}$ **26.** $\dfrac{20}{1}$ **27.** $\dfrac{45}{45}$ **28.** $\dfrac{11}{1}$

29. $\dfrac{0}{238}$ **30.** $\dfrac{238}{1}$ **31.** $\dfrac{238}{238}$ **32.** $\dfrac{0}{16}$

33. $\dfrac{3}{3}$ **34.** $\dfrac{56}{56}$ **35.** $\dfrac{87}{87}$ **36.** $\dfrac{98}{98}$

37. $\dfrac{8}{8}$ **38.** $\dfrac{0}{8}$ **39.** $\dfrac{8}{1}$ **40.** $\dfrac{8-8}{1247}$

41. $\dfrac{729}{0}$ **42.** $\dfrac{1317}{0}$ **43.** $\dfrac{5}{6-6}$ **44.** $\dfrac{13}{10-10}$

SKILL MAINTENANCE

Round 34,562 to the nearest:

45. Ten.   **46.** Hundred.   **47.** Thousand.

48. The average annual income of people living in Alaska is $21,932 per person. In Colorado, the annual income is $19,440. How much more do people in Alaska make, on average, than those living in Colorado?

49. A pet care service cut 29,824 lb of hair in 8 yr of operation. How many pounds did it cut each year?

SYNTHESIS

50. The surface of the earth is 3 parts water and 1 part land. What fractional part of the earth is water? land?

51. A college student earned $2700 one summer. During the following year, the student spent $1200 for tuition, $540 for rent, and $360 for food. The rest went for miscellaneous expenses. What part of the income went for tuition? rent? food? miscellaneous expenses?

2.4 Multiplication

a | MULTIPLICATION BY A WHOLE NUMBER

We can find $3 \cdot \frac{1}{4}$ by thinking of repeated addition. We add three $\frac{1}{4}$'s.

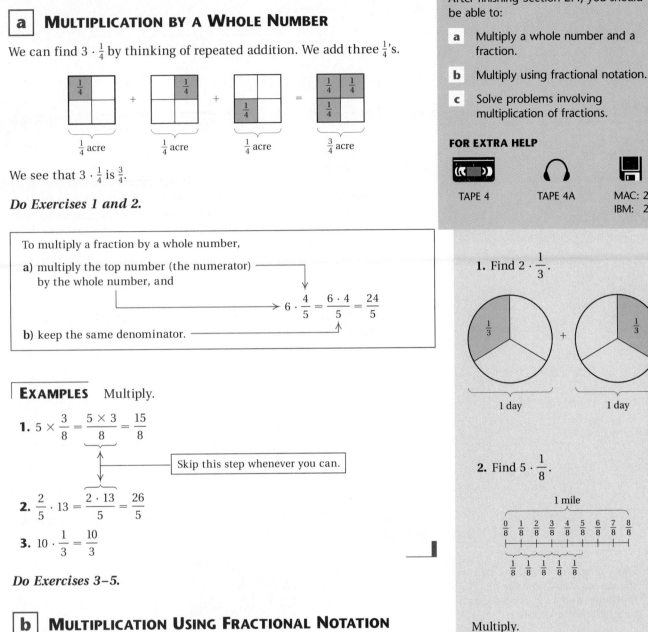

We see that $3 \cdot \frac{1}{4}$ is $\frac{3}{4}$.

Do Exercises 1 and 2.

To multiply a fraction by a whole number,

a) multiply the top number (the numerator) by the whole number, and

$$6 \cdot \frac{4}{5} = \frac{6 \cdot 4}{5} = \frac{24}{5}$$

b) keep the same denominator.

EXAMPLES   Multiply.

1. $5 \times \frac{3}{8} = \frac{5 \times 3}{8} = \frac{15}{8}$

> Skip this step whenever you can.

2. $\frac{2}{5} \cdot 13 = \frac{2 \cdot 13}{5} = \frac{26}{5}$

3. $10 \cdot \frac{1}{3} = \frac{10}{3}$

Do Exercises 3–5.

b | MULTIPLICATION USING FRACTIONAL NOTATION

We find a product such as $\frac{9}{7} \cdot \frac{3}{4}$ as follows.

To multiply a fraction by a fraction,

a) multiply the numerators, and

$$\frac{9}{7} \cdot \frac{3}{4} = \frac{9 \cdot 3}{7 \cdot 4} = \frac{27}{28}$$

b) multiply the denominators.

1. Find $2 \cdot \frac{1}{3}$.

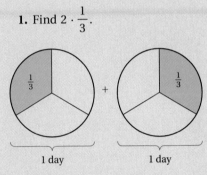

2. Find $5 \cdot \frac{1}{8}$.

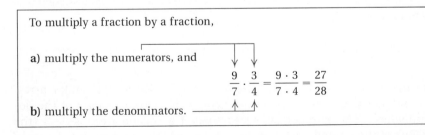

Multiply.

3. $5 \times \frac{2}{3}$

4. $11 \times \frac{3}{8}$

5. $23 \cdot \frac{2}{5}$

Answers on page A-2

Multiply.

6. $\dfrac{3}{8} \cdot \dfrac{5}{7}$

7. $\dfrac{4}{3} \times \dfrac{8}{5}$

8. $\dfrac{3}{10} \cdot \dfrac{1}{10}$

9. $7 \cdot \dfrac{2}{3}$

10. Draw diagrams like those in the text to show how the multiplication $\dfrac{1}{3} \cdot \dfrac{4}{5}$ corresponds to a real-world situation.

Answers on page A-2

EXAMPLES Multiply.

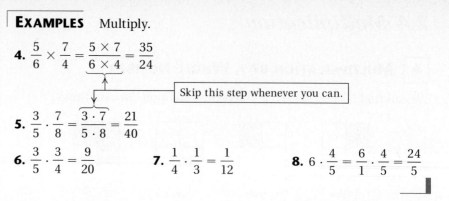

4. $\dfrac{5}{6} \times \dfrac{7}{4} = \dfrac{5 \times 7}{6 \times 4} = \dfrac{35}{24}$

Skip this step whenever you can.

5. $\dfrac{3}{5} \cdot \dfrac{7}{8} = \dfrac{3 \cdot 7}{5 \cdot 8} = \dfrac{21}{40}$

6. $\dfrac{3}{5} \cdot \dfrac{3}{4} = \dfrac{9}{20}$

7. $\dfrac{1}{4} \cdot \dfrac{1}{3} = \dfrac{1}{12}$

8. $6 \cdot \dfrac{4}{5} = \dfrac{6}{1} \cdot \dfrac{4}{5} = \dfrac{24}{5}$

Do Exercises 6–9.

Unless one of the factors is a whole number, multiplication using fractional notation does not correspond to repeated addition. Let us see how multiplication of fractions corresponds to situations in the real world. We consider the multiplication

$$\dfrac{3}{5} \cdot \dfrac{3}{4}.$$

We first consider some object and take $\frac{3}{4}$ of it. We divide it into 4 parts and take 3 of them. That is shown in the shading below.

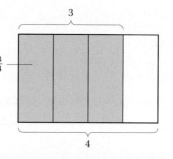

Next, we take $\frac{3}{5}$ of the result. We divide the shaded part into 5 parts and take 3 of them. That is shown below.

The entire object has been divided into 20 parts, and we have shaded 9 of them for a second time:

$$\dfrac{3}{5} \cdot \dfrac{3}{4} = \dfrac{3 \cdot 3}{5 \cdot 4} = \dfrac{9}{20}.$$

The figure above shows a rectangular array inside a rectangular array. The number of pieces in the entire array is $5 \cdot 4$ (the product of the denominators). The number of pieces shaded a second time is $3 \cdot 3$ (the product of the numerators). For the answer, we take 9 pieces out of a set of 20 to get $\frac{9}{20}$.

Do Exercise 10.

c SOLVING PROBLEMS

Most problems that can be solved by multiplying fractions can be thought of in terms of rectangular arrays.

EXAMPLE 9 A rancher owns a square mile of land. He gives $\frac{4}{5}$ of it to his daughter and she gives $\frac{2}{3}$ of her share to her son. How much land goes to the son?

1. **Familiarize.** We draw a picture to help solve the problem. The land may not be square. It could be in a shape like A or B below, or it could even be in more than one piece. But to think out the problem, we can think of it as a square, as shown by shape C.

1 square mile   1 square mile   1 square mile

The daughter gets $\frac{4}{5}$ of the land. We shade $\frac{4}{5}$.

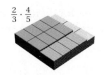

Her son gets $\frac{2}{3}$ of her part. We shade that.

2. **Translate.** We let n = the part of the land that goes to the son. We are taking "two-thirds of four-fifths." The word "of" corresponds to multiplication. Thus the following multiplication sentence corresponds to the situation:

$$\frac{2}{3} \cdot \frac{4}{5} = n.$$

3. **Solve.** The number sentence tells us what to do. We multiply:

$$\frac{2}{3} \cdot \frac{4}{5} = \frac{8}{15}.$$

4. **Check.** We can check partially by noting that the answer is smaller than the original area, 1, which we expect since the rancher is giving parts of the land away. Thus, $\frac{8}{15}$ is a reasonable answer. We can also check this in the figure above, where we see that 8 of 15 parts have been shaded a second time.

5. **State.** The son gets $\frac{8}{15}$ of a square mile of land.

Do Exercise 11.

11. A resort hotel uses $\frac{3}{4}$ of its extra land for recreational purposes. Of that, $\frac{1}{2}$ is used for swimming pools. What part of the land is used for swimming pools?

Answer on page A-2

12. The length of a button on a fax machine is $\frac{9}{10}$ cm. The width is $\frac{7}{10}$ cm. What is the area?

Example 9 and the preceding discussion indicate that the area of a rectangular region can be found by multiplying length by width. That is true whether length and width are whole numbers or not. Remember, the area of a rectangular region is given by the formula

$$A = l \cdot w.$$

EXAMPLE 10 The length of a rectangular button on a calculator is $\frac{7}{10}$ cm (centimeter). The width is $\frac{3}{10}$ cm. What is the area?

1. Familiarize. Recall that area is length times width. We draw a picture, letting A = the area of the calculator button.

2. Translate. Then we translate.

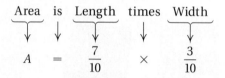

Area is Length times Width

$$A = \frac{7}{10} \times \frac{3}{10}$$

3. Solve. The sentence tells us what to do. We multiply:

$$\frac{7}{10} \cdot \frac{3}{10} = \frac{7 \cdot 3}{10 \cdot 10} = \frac{21}{100}.$$

4. Check. We check by repeating the calculation. This is left to the student.

5. State. The area is $\frac{21}{100}$ cm^2.

Do Exercise 12.

13. Of the students at Overton Junior College, $\frac{1}{8}$ participate in sports and $\frac{3}{5}$ of these play football. What fractional part of the students play football?

EXAMPLE 11 A recipe calls for $\frac{3}{4}$ cup of cornmeal. A chef is making $\frac{1}{2}$ of the recipe. How much cornmeal should the chef use?

1. Familiarize. We draw a picture or at least visualize the situation. We let $n =$ the amount of cornmeal the chef should use.

$\frac{3}{4}$ cup in recipe

$\frac{1}{2}$ of $\frac{3}{4}$ cup

2. Translate. The multiplication sentence $\frac{1}{2} \cdot \frac{3}{4} = n$ corresponds to the situation.

3. Solve. We carry out the multiplication:

$$\frac{1}{2} \cdot \frac{3}{4} = \frac{1 \cdot 3}{2 \cdot 4} = \frac{3}{8}.$$

4. Check. We check by repeating the calculation. This is left to the student.

5. State. The chef should use $\frac{3}{8}$ cup of cornmeal.

Do Exercise 13.

Answers on page A-2

Exercise Set 2.4

a Multiply.

1. $3 \cdot \dfrac{1}{5}$

2. $2 \cdot \dfrac{1}{3}$

3. $5 \times \dfrac{1}{8}$

4. $4 \times \dfrac{1}{5}$

5. $\dfrac{2}{11} \cdot 4$

6. $\dfrac{2}{5} \cdot 3$

7. $10 \cdot \dfrac{7}{9}$

8. $9 \cdot \dfrac{5}{8}$

9. $\dfrac{2}{5} \cdot 1$

10. $\dfrac{3}{8} \cdot 1$

11. $\dfrac{2}{5} \cdot 3$

12. $\dfrac{3}{5} \cdot 4$

13. $7 \cdot \dfrac{3}{4}$

14. $7 \cdot \dfrac{2}{5}$

15. $17 \times \dfrac{5}{6}$

16. $\dfrac{3}{7} \cdot 40$

b Multiply.

17. $\dfrac{1}{2} \cdot \dfrac{1}{3}$

18. $\dfrac{1}{6} \cdot \dfrac{1}{4}$

19. $\dfrac{1}{4} \times \dfrac{1}{10}$

20. $\dfrac{1}{3} \times \dfrac{1}{10}$

21. $\dfrac{2}{3} \times \dfrac{1}{5}$

22. $\dfrac{3}{5} \times \dfrac{1}{5}$

23. $\dfrac{2}{5} \cdot \dfrac{2}{3}$

24. $\dfrac{3}{4} \cdot \dfrac{3}{5}$

25. $\dfrac{3}{4} \cdot \dfrac{3}{4}$

26. $\dfrac{3}{7} \cdot \dfrac{4}{5}$

27. $\dfrac{2}{3} \cdot \dfrac{7}{13}$

28. $\dfrac{3}{11} \cdot \dfrac{4}{5}$

ANSWERS

1. _____
2. _____
3. _____
4. _____
5. _____
6. _____
7. _____
8. _____
9. _____
10. _____
11. _____
12. _____
13. _____
14. _____
15. _____
16. _____
17. _____
18. _____
19. _____
20. _____
21. _____
22. _____
23. _____
24. _____
25. _____
26. _____
27. _____
28. _____

29. $\dfrac{1}{10} \cdot \dfrac{7}{10}$ **30.** $\dfrac{3}{10} \cdot \dfrac{3}{10}$ **31.** $\dfrac{7}{8} \cdot \dfrac{7}{8}$ **32.** $\dfrac{4}{5} \cdot \dfrac{4}{5}$

33. $\dfrac{1}{10} \cdot \dfrac{1}{100}$ **34.** $\dfrac{3}{10} \cdot \dfrac{7}{100}$ **35.** $\dfrac{14}{15} \cdot \dfrac{13}{19}$ **36.** $\dfrac{12}{13} \cdot \dfrac{12}{13}$

c Solve.

37. A rectangular table top measures $\frac{4}{5}$ m long by $\frac{3}{5}$ m wide. What is its area?

38. If each piece of pie is $\frac{1}{6}$ of a pie, how much of the pie is $\frac{1}{2}$ of a piece?

39. One of 39 high school football players plays college football. One of 39 college players plays professional football. What fractional part of high school players plays professional football?

40. A gasoline can holds $\frac{7}{8}$ L. How much will it hold when it is $\frac{1}{2}$ full?

41. A cereal recipe calls for $\frac{3}{4}$ cup of granola. How much is needed to make $\frac{1}{2}$ of a recipe?

42. It takes $\frac{2}{3}$ yd of ribbon to make a bow. How much ribbon is needed to make 5 bows?

43. Of every 100 containers of juice bought in grocery stores, 56 are orange juice. What fractional part of juice purchased is orange juice?

44. Of every 1000 people who attend movies, 230 are in the 16–20 age group. What fractional part of all moviegoers are in the 16–20 age group?

SKILL MAINTENANCE

Divide.

45. $35\,)\,\overline{7\,1\,4\,0}$ **46.** $46\,)\,\overline{3\,2{,}2\,0\,0}$ **47.** $9\,)\,\overline{2\,7{,}0\,0\,9}$ **48.** $35\,)\,\overline{7\,1\,4\,8}$

49. What does the digit 6 mean in 4,678,952?

50. What does the digit 8 mean in 4,678,952?

2.5 *Simplifying*

a | MULTIPLYING BY 1

Recall the following:

$$1 = \frac{1}{1} = \frac{2}{2} = \frac{3}{3} = \frac{4}{4} = \frac{10}{10} = \frac{45}{45} = \frac{100}{100} = \frac{n}{n}.$$

Any nonzero number divided by itself is 1.

When we multiply a number by 1, we get the same number.

$$\frac{3}{5} = \frac{3}{5} \cdot 1 = \frac{3}{5} \cdot \frac{4}{4} = \frac{12}{20}$$

Since $\frac{3}{5} \cdot 1 = \frac{12}{20}$, we know that $\frac{3}{5}$ and $\frac{12}{20}$ are two names for the same number. We also say that $\frac{3}{5}$ and $\frac{12}{20}$ are **equivalent**.

Do Exercises 1–4.

Suppose we want to find a name for $\frac{2}{3}$, but one that has a denominator of 9. We can multiply by 1 to find equivalent fractions:

$$\frac{2}{3} = \frac{2}{3} \cdot \frac{3}{3} = \frac{2 \cdot 3}{3 \cdot 3} = \frac{6}{9}.$$

We chose $\frac{3}{3}$ for 1 in order to get a denominator of 9.

EXAMPLE 1   Find a name for $\frac{1}{4}$ with a denominator of 24.

Since $4 \cdot 6 = 24$, we multiply by $\frac{6}{6}$.

$$\frac{1}{4} = \frac{1}{4} \cdot \frac{6}{6} = \frac{1 \cdot 6}{4 \cdot 6} = \frac{6}{24}.$$

EXAMPLE 2   Find a name for $\frac{2}{5}$ with a denominator of 35.

Since $5 \cdot 7 = 35$, we multiply by $\frac{7}{7}$:

$$\frac{2}{5} = \frac{2}{5} \cdot \frac{7}{7} = \frac{2 \cdot 7}{5 \cdot 7} = \frac{14}{35}.$$

Do Exercises 5–9.

b | SIMPLIFYING

All of the following are names for three-fourths:

$$\frac{3}{4}, \quad \frac{6}{8}, \quad \frac{9}{12}, \quad \frac{12}{16}, \quad \frac{15}{20}.$$

We say that $\frac{3}{4}$ is **simplest** because it has the smallest numerator and the smallest denominator. That is, the numerator and the denominator have no common factor other than 1.

Multiply.

1. $\dfrac{1}{2} \cdot \dfrac{8}{8}$ 2. $\dfrac{3}{5} \cdot \dfrac{10}{10}$

3. $\dfrac{13}{25} \cdot \dfrac{4}{4}$ 4. $\dfrac{8}{3} \cdot \dfrac{25}{25}$

Find another name for the number, but with the denominator indicated. Use multiplying by 1.

5. $\dfrac{4}{3} = \dfrac{?}{9}$ 6. $\dfrac{3}{4} = \dfrac{?}{24}$

7. $\dfrac{9}{10} = \dfrac{?}{100}$ 8. $\dfrac{3}{15} = \dfrac{?}{45}$

9. $\dfrac{8}{7} = \dfrac{?}{49}$

Answers on page A-2

Simplify.

10. $\dfrac{2}{8}$

11. $\dfrac{10}{12}$

12. $\dfrac{40}{8}$

13. $\dfrac{24}{18}$

To simplify, we reverse the process of multiplying by 1:

$$\dfrac{12}{18} = \dfrac{2 \cdot 6}{3 \cdot 6} \begin{matrix} \leftarrow \text{Factoring the numerator} \\ \leftarrow \text{Factoring the denominator} \end{matrix}$$

$$= \dfrac{2}{3} \cdot \dfrac{6}{6} \qquad \text{Factoring the fraction}$$

$$= \dfrac{2}{3} \cdot 1 \qquad \dfrac{6}{6} = 1$$

$$= \dfrac{2}{3}. \qquad \text{Removing a factor of 1: } \dfrac{2}{3} \cdot 1 = \dfrac{2}{3}$$

EXAMPLES Simplify.

3. $\dfrac{8}{20} = \dfrac{2 \cdot 4}{5 \cdot 4} = \dfrac{2}{5} \cdot \dfrac{4}{4} = \dfrac{2}{5}$

4. $\dfrac{2}{6} = \dfrac{1 \cdot 2}{3 \cdot 2} = \dfrac{1}{3} \cdot \dfrac{2}{2} = \dfrac{1}{3}$ The number 1 allows for pairing of factors in the numerator and the denominator.

5. $\dfrac{30}{6} = \dfrac{5 \cdot 6}{1 \cdot 6} = \dfrac{5}{1} \cdot \dfrac{6}{6} = \dfrac{5}{1} = 5 \leftarrow$ We could also simplify $\dfrac{30}{6}$ by doing the division $30 \div 6$. That is, $\dfrac{30}{6} = 30 \div 6 = 5$.

Do Exercises 10–13.

The use of prime factorizations can be helpful for larger numbers.

EXAMPLE 6   Simplify: $\dfrac{90}{84}$.

$$\dfrac{90}{84} = \dfrac{2 \cdot 3 \cdot 3 \cdot 5}{2 \cdot 2 \cdot 3 \cdot 7} \qquad \begin{matrix} \text{Factoring the numerator and} \\ \text{the denominator into primes} \end{matrix}$$

$$= \dfrac{2 \cdot 3 \cdot 3 \cdot 5}{2 \cdot 3 \cdot 2 \cdot 7} \qquad \begin{matrix} \text{Changing the order so that like primes} \\ \text{are above and below each other} \end{matrix}$$

$$= \dfrac{2}{2} \cdot \dfrac{3}{3} \cdot \dfrac{3 \cdot 5}{2 \cdot 7} \qquad \text{Factoring the fraction}$$

$$= 1 \cdot 1 \cdot \dfrac{3 \cdot 5}{2 \cdot 7}$$

$$= \dfrac{3 \cdot 5}{2 \cdot 7} \qquad \text{Removing factors of 1}$$

$$= \dfrac{15}{14}$$

We could have shortened the preceding example had we recalled our tests for divisibility (Section 2.2) and noted that 6 is a factor of both the numerator and the denominator. Then

$$\dfrac{90}{84} = \dfrac{6 \cdot 15}{6 \cdot 14} = \dfrac{6}{6} \cdot \dfrac{15}{14} = \dfrac{15}{14}.$$

The tests for divisibility are very helpful in simplifying.

Answers on page A-2

EXAMPLE 7 Simplify: $\dfrac{603}{207}$.

At first glance this looks difficult. But note, using the test for divisibility by 9 (sum of digits divisible by 9), that both the numerator and the denominator are divisible by 9. Thus we can factor 9 from both numbers:

$$\frac{603}{207} = \frac{9 \cdot 67}{9 \cdot 23} = \frac{9}{9} \cdot \frac{67}{23} = \frac{67}{23}.$$

Do Exercises 14–18.

CANCELING Canceling is a shortcut that you may have used for removing a factor of 1 when working with fractional notation. With *great* concern, we mention it as a possibility for speeding up your work. Canceling may be done only when removing common factors in numerators and denominators. Each common factor allows us to remove a factor of 1 in a product. Canceling may not be done in sums. Our concern is that canceling be done with care and understanding. In effect, slashes are used to indicate factors of 1 that have been removed. For instance, Example 6 might have been done faster as follows:

$$\frac{90}{84} = \frac{2 \cdot 3 \cdot 3 \cdot 5}{2 \cdot 2 \cdot 3 \cdot 7} \quad \text{Factoring the numerator and the denominator}$$

$$= \frac{\cancel{2} \cdot \cancel{3} \cdot 3 \cdot 5}{2 \cdot \cancel{2} \cdot \cancel{3} \cdot 7} \quad \begin{array}{l}\text{When a factor of 1 is noted}\\ \text{it is "canceled" as shown: } \frac{2 \cdot 3}{2 \cdot 3} = 1.\end{array}$$

$$= \frac{3 \cdot 5}{2 \cdot 7} = \frac{15}{14}.$$

Caution! The difficulty with canceling is that it is often applied incorrectly in situations like the following:

$$\frac{\cancel{2} + 3}{\cancel{2}} = 3; \qquad \frac{\cancel{4} + 1}{\cancel{4} + 2} = \frac{1}{2}; \qquad \frac{1\cancel{5}}{\cancel{5}4} = \frac{1}{4}.$$

Wrong!   Wrong!   Wrong!

The correct answers are

$$\frac{2 + 3}{2} = \frac{5}{2}; \qquad \frac{4 + 1}{4 + 2} = \frac{5}{6}; \qquad \frac{15}{54} = \frac{5}{18}.$$

In each situation, the number canceled was *not* a factor of 1. Factors are parts of products. For example, in $2 \cdot 3$, 2 and 3 are factors, but in $2 + 3$, 2 and 3 are *not* factors.

> If you cannot factor, do not cancel! If in doubt, do not cancel!

c | A TEST FOR EQUALITY

When denominators are the same, we say that fractions have a **common denominator.** Suppose we want to compare $\frac{2}{4}$ and $\frac{3}{6}$. We find a common denominator and compare numerators. To do this, we multiply by 1 using symbols for 1 formed by looking at opposite denominators.

$$\left.\begin{array}{l}\dfrac{3}{6} = \dfrac{3}{6} \cdot \dfrac{4}{4} = \dfrac{3 \cdot 4}{6 \cdot 4} = \dfrac{12}{24} \\[2mm] \dfrac{2}{4} = \dfrac{2}{4} \cdot \dfrac{6}{6} = \dfrac{2 \cdot 6}{4 \cdot 6} = \dfrac{12}{24}\end{array}\right\} \quad \text{We see that } \frac{3}{6} = \frac{2}{4}.$$

Simplify.

14. $\dfrac{35}{40}$

15. $\dfrac{801}{702}$

16. $\dfrac{24}{21}$

17. $\dfrac{75}{300}$

18. Simplify each fraction in this circle graph.

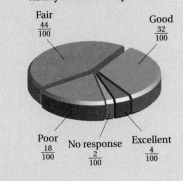

High School Students Rate the Schools' Ability to Teach Study Habits

Fair $\dfrac{44}{100}$

Good $\dfrac{32}{100}$

Poor $\dfrac{18}{100}$ No response $\dfrac{2}{100}$ Excellent $\dfrac{4}{100}$

Answers on page A-2

Use = or ≠ for ■ to write a true sentence.

19. $\frac{2}{6} \blacksquare \frac{3}{9}$

20. $\frac{2}{3} \blacksquare \frac{14}{20}$

Note in the preceding that if

$$\frac{3}{6} = \frac{2}{4}, \quad \text{then} \quad 3 \cdot 4 = 6 \cdot 2.$$

We need only check the products $3 \cdot 4$ and $6 \cdot 2$ to compare the fractions.

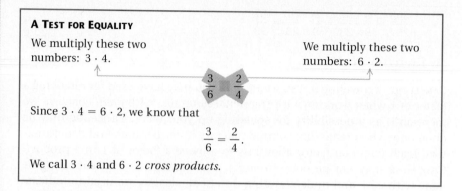

A Test for Equality

We multiply these two numbers: $3 \cdot 4$.   We multiply these two numbers: $6 \cdot 2$.

Since $3 \cdot 4 = 6 \cdot 2$, we know that

$$\frac{3}{6} = \frac{2}{4}.$$

We call $3 \cdot 4$ and $6 \cdot 2$ *cross products*.

If a sentence $a = b$ is true, it means that a and b name the same number. If a sentence $a \neq b$ is true, it means that a and b do *not* name the same number.

EXAMPLE 8   Use = or ≠ for ■ to write a true sentence:

$$\frac{6}{7} \blacksquare \frac{7}{8}.$$

We multiply these two numbers: $6 \cdot 8 = 48$.   We multiply these two numbers: $7 \cdot 7 = 49$.

Because $48 \neq 49$ (read "48 is not the same as 49"), $\frac{6}{7} = \frac{7}{8}$ is not a true sentence. Thus,

$$\frac{6}{7} \neq \frac{7}{8}.$$

EXAMPLE 9   Use = or ≠ for ■ to write a true sentence:

$$\frac{6}{10} \blacksquare \frac{3}{5}.$$

We multiply these two numbers: $6 \cdot 5 = 30$.   We multiply these two numbers: $10 \cdot 3 = 30$.

Because the cross products are the same, we have

$$\frac{6}{10} = \frac{3}{5}.$$

Do Exercises 19 and 20.

Exercise Set 2.5

ANSWERS
1. _____
2. _____
3. _____
4. _____
5. _____
6. _____
7. _____
8. _____
9. _____
10. _____
11. _____
12. _____
13. _____
14. _____
15. _____
16. _____
17. _____
18. _____
19. _____
20. _____
21. _____
22. _____
23. _____
24. _____
25. _____
26. _____
27. _____
28. _____
29. _____
30. _____
31. _____
32. _____
33. _____
34. _____
35. _____
36. _____

a Find another name for the given number, but with the denominator indicated. Use multiplying by 1.

1. $\dfrac{1}{2} = \dfrac{?}{10}$

2. $\dfrac{1}{6} = \dfrac{?}{18}$

3. $\dfrac{5}{8} = \dfrac{?}{32}$

4. $\dfrac{2}{9} = \dfrac{?}{18}$

5. $\dfrac{9}{10} = \dfrac{?}{30}$

6. $\dfrac{5}{6} = \dfrac{?}{48}$

7. $\dfrac{7}{8} = \dfrac{?}{32}$

8. $\dfrac{2}{5} = \dfrac{?}{25}$

9. $\dfrac{5}{12} = \dfrac{?}{48}$

10. $\dfrac{3}{8} = \dfrac{?}{56}$

11. $\dfrac{17}{18} = \dfrac{?}{54}$

12. $\dfrac{11}{16} = \dfrac{?}{256}$

13. $\dfrac{5}{3} = \dfrac{?}{45}$

14. $\dfrac{11}{5} = \dfrac{?}{30}$

15. $\dfrac{7}{22} = \dfrac{?}{132}$

16. $\dfrac{10}{21} = \dfrac{?}{126}$

b Simplify.

17. $\dfrac{2}{4}$

18. $\dfrac{4}{8}$

19. $\dfrac{6}{8}$

20. $\dfrac{8}{12}$

21. $\dfrac{3}{15}$

22. $\dfrac{8}{10}$

23. $\dfrac{24}{8}$

24. $\dfrac{36}{9}$

25. $\dfrac{18}{24}$

26. $\dfrac{42}{48}$

27. $\dfrac{14}{16}$

28. $\dfrac{15}{25}$

29. $\dfrac{12}{10}$

30. $\dfrac{16}{14}$

31. $\dfrac{16}{48}$

32. $\dfrac{100}{20}$

33. $\dfrac{150}{25}$

34. $\dfrac{19}{76}$

35. $\dfrac{17}{51}$

36. $\dfrac{425}{525}$

ANSWERS

37. _____

38. _____

39. _____

40. _____

41. _____

42. _____

43. _____

44. _____

45. _____

46. _____

47. _____

48. _____

49. _____

50. _____

51. _____

52. _____

53. _____

54. _____

55. _____

56. _____

57. _____

58. _____

59. _____

60. _____

61. _____

62. _____

c Use = or ≠ for ▨ to write a true sentence.

37. $\dfrac{3}{4}$ ▨ $\dfrac{9}{12}$ **38.** $\dfrac{4}{8}$ ▨ $\dfrac{3}{6}$ **39.** $\dfrac{1}{5}$ ▨ $\dfrac{2}{9}$ **40.** $\dfrac{1}{4}$ ▨ $\dfrac{2}{9}$

41. $\dfrac{3}{8}$ ▨ $\dfrac{6}{16}$ **42.** $\dfrac{2}{6}$ ▨ $\dfrac{6}{18}$ **43.** $\dfrac{2}{5}$ ▨ $\dfrac{3}{7}$ **44.** $\dfrac{1}{3}$ ▨ $\dfrac{1}{4}$

45. $\dfrac{12}{9}$ ▨ $\dfrac{8}{6}$ **46.** $\dfrac{16}{14}$ ▨ $\dfrac{8}{7}$ **47.** $\dfrac{5}{2}$ ▨ $\dfrac{17}{7}$ **48.** $\dfrac{3}{10}$ ▨ $\dfrac{7}{24}$

49. $\dfrac{3}{10}$ ▨ $\dfrac{30}{100}$ **50.** $\dfrac{700}{1000}$ ▨ $\dfrac{70}{100}$ **51.** $\dfrac{5}{10}$ ▨ $\dfrac{520}{1000}$ **52.** $\dfrac{49}{100}$ ▨ $\dfrac{50}{1000}$

SKILL MAINTENANCE

Solve.

53. A playing field is 78 ft long and 64 ft wide. What is its area?

54. A landscaper buys 13 small maple trees and 17 small oak trees for a project. A maple costs $23 and an oak costs $37. How much is spent altogether for the trees?

Subtract.

55. $34 - 23$ **56.** $50 - 18$ **57.** $803 - 617$ **58.** $8344 - 5607$

SYNTHESIS

59. Sociologists have found that 4 of 10 people are shy. Write fractional notation for the part of the population that is shy; that is not shy. Simplify.

60. Is $\dfrac{5}{6}$ mi the same as $\dfrac{7}{8}$ mi? Why or why not?

61. ▦ In a recent year, Bernard Gilkey of the St. Louis Cardinals got 116 hits in 384 times at bat. Bip Roberts of the Cincinnati Reds got 172 hits in 532 times at bat. Did they have the same fraction of hits (batting average)? Why or why not?

62. ▦ On a test of 82 questions, a student got 63 correct. On another test of 100 questions, the student got 77 correct. Did the student get the same portion of each test correct? Why or why not?

2.6 *Multiplying and Simplifying*

a SIMPLIFYING AFTER MULTIPLYING

We usually simplify after we multiply. To make such simplifying easier, it is generally best not to carry out the products in the numerator and the denominator, but to factor and simplify before multiplying. Consider the product

$$\frac{3}{8} \cdot \frac{4}{9}.$$

We proceed as follows:

$$\frac{3}{8} \cdot \frac{4}{9} = \frac{3 \cdot 4}{8 \cdot 9}$$ We write the products in the numerator and the denominator, but we do not carry them out.

$$= \frac{3 \cdot 2 \cdot 2}{2 \cdot 2 \cdot 2 \cdot 3 \cdot 3}$$ Factoring the numerator and the denominator

$$= \frac{3 \cdot 2 \cdot 2}{3 \cdot 2 \cdot 2} \cdot \frac{1}{2 \cdot 3}$$ Factoring the fraction

$$= 1 \cdot \frac{1}{2 \cdot 3}$$

$$= \frac{1}{2 \cdot 3}$$ Removing a factor of 1

$$= \frac{1}{6}.$$

The procedure could have been shortened had we noticed that 4 is a factor of the 8 in the denominator:

$$\frac{3}{8} \cdot \frac{4}{9} = \frac{3 \cdot 4}{8 \cdot 9} = \frac{3 \cdot 4}{4 \cdot 2 \cdot 3 \cdot 3} = \frac{3 \cdot 4}{3 \cdot 4} \cdot \frac{1}{2 \cdot 3} = 1 \cdot \frac{1}{2 \cdot 3} = \frac{1}{2 \cdot 3} = \frac{1}{6}.$$

To multiply and simplify:

a) Write the products in the numerator and the denominator, but do not carry out the products.

b) Factor the numerator and the denominator.

c) Factor the fraction to remove factors of 1.

d) Carry out the remaining products.

EXAMPLES   Multiply and simplify.

1. $\dfrac{2}{3} \cdot \dfrac{9}{4} = \dfrac{2 \cdot 9}{3 \cdot 4} = \dfrac{2 \cdot 3 \cdot 3}{3 \cdot 2 \cdot 2} = \dfrac{2 \cdot 3}{2 \cdot 3} \cdot \dfrac{3}{2} = 1 \cdot \dfrac{3}{2} = \dfrac{3}{2}$

2. $\dfrac{6}{7} \cdot \dfrac{5}{3} = \dfrac{6 \cdot 5}{7 \cdot 3} = \dfrac{3 \cdot 2 \cdot 5}{7 \cdot 3} = \dfrac{3}{3} \cdot \dfrac{2 \cdot 5}{7} = 1 \cdot \dfrac{2 \cdot 5}{7} = \dfrac{2 \cdot 5}{7} = \dfrac{10}{7}$

3. $40 \cdot \dfrac{7}{8} = \dfrac{40 \cdot 7}{8} = \dfrac{8 \cdot 5 \cdot 7}{8 \cdot 1} = \dfrac{8}{8} \cdot \dfrac{5 \cdot 7}{1} = 1 \cdot \dfrac{5 \cdot 7}{1} = \dfrac{5 \cdot 7}{1} = 35$

OBJECTIVES

After finishing Section 2.6, you should be able to:

a Multiply and simplify using fractional notation.

b Solve problems involving multiplication.

FOR EXTRA HELP

TAPE 5   TAPE 4B   MAC: 2
                               IBM: 2

Multiply and simplify.

1. $\dfrac{2}{3} \cdot \dfrac{7}{8}$

2. $\dfrac{4}{5} \cdot \dfrac{5}{12}$

3. $16 \cdot \dfrac{3}{8}$

4. $\dfrac{5}{8} \cdot 4$

5. A landscaper uses $\frac{2}{5}$ lb of peat moss for a rosebush. How much will be needed for 25 rosebushes?

Answers on page A-2

Caution!   Canceling can be used as follows for these examples.

1. $\dfrac{2}{3} \cdot \dfrac{9}{4} = \dfrac{2 \cdot 9}{3 \cdot 4} = \dfrac{2 \cdot 3 \cdot 3}{3 \cdot 2 \cdot 2} = \dfrac{3}{2}$ Removing a factor of 1: $\dfrac{2 \cdot 3}{2 \cdot 3} = 1$

2. $\dfrac{6}{7} \cdot \dfrac{5}{3} = \dfrac{6 \cdot 5}{7 \cdot 3} = \dfrac{3 \cdot 2 \cdot 5}{7 \cdot 3} = \dfrac{2 \cdot 5}{7} = \dfrac{10}{7}$ Removing a factor of 1: $\dfrac{3}{3} = 1$

3. $40 \cdot \dfrac{7}{8} = \dfrac{40 \cdot 7}{8} = \dfrac{8 \cdot 5 \cdot 7}{8 \cdot 1} = \dfrac{5 \cdot 7}{1} = 35$ Removing a factor of 1: $\dfrac{8}{8} = 1$

Remember, if you can't factor, you can't cancel!

Do Exercises 1–4.

b   SOLVING PROBLEMS

EXAMPLE 4   How much steak will be needed to serve 30 people if each person gets $\frac{2}{3}$ lb?

1. Familiarize. We first draw a picture or at least visualize the situation. Repeated addition will work here.

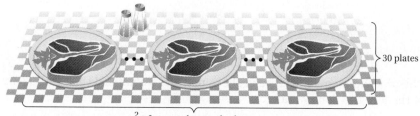

$\frac{2}{3}$ of a pound on each plate

We let n = the number of pounds of steak needed.

2. Translate.   The problem translates to the following equation:

$$n = 30 \cdot \dfrac{2}{3}.$$

3. Solve.   To solve the equation, we carry out the multiplication:

$$n = 30 \cdot \dfrac{2}{3} = \dfrac{30 \cdot 2}{3} \qquad \text{Multiplying}$$

$$= \dfrac{3 \cdot 10 \cdot 2}{3 \cdot 1}$$

$$= \dfrac{3}{3} \cdot \dfrac{10 \cdot 2}{1}$$

$$= 20. \qquad \text{Simplifying}$$

4. Check.   We check by repeating the calculation. (We leave the check to the student.) We can also think about the reasonableness of the answer. We are multiplying 30 by a number less than 1, so the product will be less than 30. Since 20 is less than 30, we have a partial check of the reasonableness of the answer. The number 20 checks.

5. State.   Thus, 20 lb of steak will be needed.

Do Exercise 5.

Exercise Set 2.6

a   Multiply and simplify. | Don't forget to simplify! |

1. $\dfrac{2}{3} \cdot \dfrac{1}{2}$ **2.** $\dfrac{3}{8} \cdot \dfrac{1}{3}$ **3.** $\dfrac{7}{8} \cdot \dfrac{1}{7}$ **4.** $\dfrac{4}{9} \cdot \dfrac{1}{4}$

5. $\dfrac{1}{8} \cdot \dfrac{4}{5}$ **6.** $\dfrac{2}{5} \cdot \dfrac{1}{6}$ **7.** $\dfrac{1}{4} \cdot \dfrac{2}{3}$ **8.** $\dfrac{4}{6} \cdot \dfrac{1}{6}$

9. $\dfrac{12}{5} \cdot \dfrac{9}{8}$ **10.** $\dfrac{16}{15} \cdot \dfrac{5}{4}$ **11.** $\dfrac{10}{9} \cdot \dfrac{7}{5}$ **12.** $\dfrac{25}{12} \cdot \dfrac{4}{3}$

13. $9 \cdot \dfrac{1}{9}$ **14.** $4 \cdot \dfrac{1}{4}$ **15.** $\dfrac{1}{3} \cdot 3$ **16.** $\dfrac{1}{6} \cdot 6$

17. $\dfrac{7}{10} \cdot \dfrac{10}{7}$ **18.** $\dfrac{8}{9} \cdot \dfrac{9}{8}$ **19.** $\dfrac{7}{5} \cdot \dfrac{5}{7}$ **20.** $\dfrac{2}{11} \cdot \dfrac{11}{2}$

1. _____

2. _____

3. _____

4. _____

5. _____

6. _____

7. _____

8. _____

9. _____

10. _____

11. _____

12. _____

13. _____

14. _____

15. _____

16. _____

17. _____

18. _____

19. _____

20. _____

21. _____

22. _____

23. _____

24. _____

25. _____

26. _____

27. _____

28. _____

29. _____

30. _____

31. _____

32. _____

33. _____

34. _____

35. _____

36. _____

37. _____

38. _____

39. _____

40. _____

21. $\frac{1}{4} \cdot 8$ **22.** $\frac{1}{3} \cdot 18$ **23.** $24 \cdot \frac{1}{6}$ **24.** $16 \cdot \frac{1}{2}$

25. $12 \cdot \frac{3}{4}$ **26.** $18 \cdot \frac{5}{6}$ **27.** $\frac{3}{8} \cdot 24$ **28.** $\frac{2}{9} \cdot 36$

29. $13 \cdot \frac{2}{5}$ **30.** $15 \cdot \frac{1}{6}$ **31.** $\frac{7}{10} \cdot 28$ **32.** $\frac{5}{8} \cdot 34$

33. $\frac{1}{6} \cdot 360$ **34.** $\frac{1}{3} \cdot 120$ **35.** $240 \cdot \frac{1}{8}$ **36.** $150 \cdot \frac{1}{5}$

37. $\frac{4}{10} \cdot \frac{5}{10}$ **38.** $\frac{7}{10} \cdot \frac{34}{150}$ **39.** $\frac{8}{10} \cdot \frac{45}{100}$ **40.** $\frac{3}{10} \cdot \frac{8}{10}$

41. $\dfrac{11}{24} \cdot \dfrac{3}{5}$

42. $\dfrac{15}{22} \cdot \dfrac{4}{7}$

43. $\dfrac{10}{21} \cdot \dfrac{3}{4}$

44. $\dfrac{17}{18} \cdot \dfrac{3}{5}$

b Solve.

45. Anna receives $36 for working a full day doing inventory at a hardware store. How much will she receive for working $\frac{3}{4}$ of the day?

46. After Jack completes 60 hr of teacher training in college, he can earn $45 for working a full day as a substitute teacher. How much will he receive for working $\frac{1}{5}$ of a day?

47. Business people have determined that $\frac{1}{4}$ of the addresses on a mailing list will change in one year. A business has a mailing list of 2500 people. After one year, how many addresses on that list will be incorrect?

48. Sociologists have determined that $\frac{2}{5}$ of the people in the world are shy. A sales manager is interviewing 650 people for an aggressive sales position. How many of these people might be shy?

49. A recipe calls for $\frac{2}{3}$ cup of flour. A chef is making $\frac{1}{2}$ of the recipe. How much flour should the chef use?

50. Of the students in the freshman class, $\frac{2}{5}$ have cameras; $\frac{1}{4}$ of these students also join the college photography club. What fraction of the students in the freshman class join the photography club?

ANSWERS

41. _____

42. _____

43. _____

44. _____

45. _____

46. _____

47. _____

48. _____

49. _____

50. _____

51. _____

52. _____

53. _____

54. _____

55. _____

56. _____

57. _____

58. _____

59. _____

60. _____

61. _____

62. _____

51. A student's tuition was $2400. A loan was obtained for $\frac{2}{3}$ of the tuition. How much was the loan?

52. A student's tuition was $2800. A loan was obtained for $\frac{3}{4}$ of the tuition. How much was the loan?

53. On a map, 1 in. represents 240 mi. How much does $\frac{2}{3}$ in. represent?

54. On a map, 1 in. represents 120 mi. How much does $\frac{3}{4}$ in. represent?

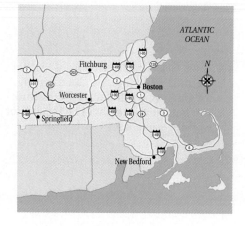

55. A family has an annual income of $27,000. Of this, $\frac{1}{4}$ is spent for food, $\frac{1}{5}$ for housing, $\frac{1}{10}$ for clothing, $\frac{1}{9}$ for savings, $\frac{1}{4}$ for taxes, and the rest for other expenses. How much is spent for each?

56. A family has an annual income of $25,200. Of this, $\frac{1}{4}$ is spent for food, $\frac{1}{5}$ for housing, $\frac{1}{10}$ for clothing, $\frac{1}{9}$ for savings, $\frac{1}{4}$ for taxes, and the rest for other expenses. How much is spent for each?

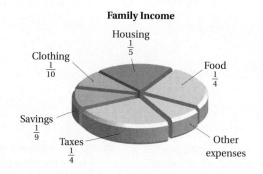

Family Income

SKILL MAINTENANCE

Solve.

57. $48 \cdot t = 1680$

58. $74 \cdot x = 6290$

59. $t + 28 = 5017$

60. $456 + x = 9002$

Subtract.

61.
$$\begin{array}{r} 9\ 0\ 6\ 0 \\ -4\ 3\ 8\ 7 \end{array}$$

62.
$$\begin{array}{r} 7\ 8\ 0\ 0 \\ -2\ 4\ 6\ 2 \end{array}$$

2.7 Reciprocals and Division

a | RECIPROCALS

Look at these products:

$$8 \cdot \frac{1}{8} = \frac{8 \cdot 1}{8} = \frac{8}{8} = 1; \qquad \frac{2}{3} \cdot \frac{3}{2} = \frac{2 \cdot 3}{3 \cdot 2} = \frac{6}{6} = 1.$$

> If the product of two numbers is 1, we say that they are *reciprocals* of each other. To find a reciprocal, interchange the numerator and the denominator.
>
> $$\text{Number} \longrightarrow \frac{3}{4} \qquad \frac{4}{3} \longleftarrow \text{Reciprocal}$$

EXAMPLES Find the reciprocal.

1. The reciprocal of $\frac{4}{5}$ is $\frac{5}{4}$. $\frac{4}{5} \cdot \frac{5}{4} = \frac{20}{20} = 1$

2. The reciprocal of $\frac{8}{7}$ is $\frac{7}{8}$. $\frac{8}{7} \cdot \frac{7}{8} = \frac{56}{56} = 1$

3. The reciprocal of 8 is $\frac{1}{8}$.   Think of 8 as $\frac{8}{1}$: $\frac{8}{1} \cdot \frac{1}{8} = \frac{8}{8} = 1.$

4. The reciprocal of $\frac{1}{3}$ is 3. $\frac{1}{3} \cdot 3 = \frac{3}{3} = 1$

Do Exercises 1–4.

Does 0 have a reciprocal? If it did, it would have to be a number x such that

$$0 \cdot x = 1.$$

But 0 times any number is 0. Thus,

> The number 0, or $\frac{0}{n}$, has no reciprocal. $\left(\text{Recall that } \frac{n}{0} \text{ is not defined.}\right)$

b | DIVISION

Recall that $a \div b$ is the number that when multiplied by b gives a. Consider the division $\frac{3}{4} \div \frac{1}{8}$. We are asking how many $\frac{1}{8}$'s are in $\frac{3}{4}$. We can answer this by looking at the figure below.

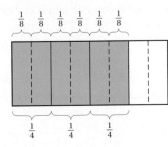

OBJECTIVES

After finishing Section 2.7, you should be able to:

a Find the reciprocal of a number.

b Divide and simplify using fractional notation.

c Solve equations of the type $a \cdot x = b$ and $x \cdot a = b$, where a and b may be fractions.

d Solve problems involving division.

FOR EXTRA HELP

| TAPE 5 | TAPE 4B | MAC: 2 |
| | | IBM: 2 |

Find the reciprocal.

1. $\frac{2}{5}$

2. $\frac{10}{7}$

3. 9

4. $\frac{1}{5}$

Answers on page A-2

Divide and simplify.

5. $\dfrac{6}{7} \div \dfrac{3}{4}$

6. $\dfrac{2}{3} \div \dfrac{1}{4}$

7. $\dfrac{4}{5} \div 8$

8. $60 \div \dfrac{3}{5}$

9. $\dfrac{3}{5} \div \dfrac{3}{5}$

Answers on page A-2

We see that there are six $\frac{1}{8}$'s in $\frac{3}{4}$. Thus,

$$\frac{3}{4} \div \frac{1}{8} = 6.$$

We can check this by multiplying:

$$6 \cdot \frac{1}{8} = \frac{6}{8} = \frac{3}{4}.$$

Here is a faster way to divide.

To divide fractions, multiply the dividend by the reciprocal of the divisor:

$$\frac{2}{5} \div \frac{3}{4} = \frac{2}{5} \cdot \frac{4}{3} = \frac{2 \cdot 4}{5 \cdot 3} = \frac{8}{15}.$$

Multiply by the reciprocal of the divisor.

EXAMPLES   Divide and simplify.

5. $\dfrac{5}{6} \div \dfrac{2}{3} = \dfrac{5}{6} \cdot \dfrac{3}{2} = \dfrac{5 \cdot 3}{6 \cdot 2} = \dfrac{5 \cdot 3}{3 \cdot 2 \cdot 2} = \dfrac{3}{3} \cdot \dfrac{5}{2 \cdot 2} = \dfrac{5}{2 \cdot 2} = \dfrac{5}{4}$

6. $\dfrac{3}{4} \div \dfrac{1}{8} = \dfrac{3}{4} \cdot 8 = \dfrac{3 \cdot 8}{4} = \dfrac{3 \cdot 4 \cdot 2}{4 \cdot 1} = \dfrac{4}{4} \cdot \dfrac{3 \cdot 2}{1} = \dfrac{3 \cdot 2}{1} = 6$

7. $\dfrac{2}{5} \div 6 = \dfrac{2}{5} \cdot \dfrac{1}{6} = \dfrac{2 \cdot 1}{5 \cdot 6} = \dfrac{2 \cdot 1}{5 \cdot 2 \cdot 3} = \dfrac{2}{2} \cdot \dfrac{1}{5 \cdot 3} = \dfrac{1}{5 \cdot 3} = \dfrac{1}{15}$

8. $\dfrac{3}{5} \div \dfrac{1}{2} = \dfrac{3}{5} \cdot 2 = \dfrac{3 \cdot 2}{5} = \dfrac{6}{5}$

Caution!   Canceling can be used as follows for Examples 5–7.

5. $\dfrac{5}{6} \div \dfrac{2}{3} = \dfrac{5}{6} \cdot \dfrac{3}{2} = \dfrac{5 \cdot 3}{6 \cdot 2} = \dfrac{5 \cdot \cancel{3}}{\cancel{3} \cdot 2 \cdot 2} = \dfrac{5}{2 \cdot 2} = \dfrac{5}{4}$ Removing a factor of 1: $\dfrac{3}{3} = 1$

6. $\dfrac{3}{4} \div \dfrac{1}{8} = \dfrac{3}{4} \cdot 8 = \dfrac{3 \cdot 8}{4} = \dfrac{3 \cdot \cancel{4} \cdot 2}{\cancel{4} \cdot 1} = \dfrac{3 \cdot 2}{1} = 6$ Removing a factor of 1: $\dfrac{4}{4} = 1$

7. $\dfrac{2}{5} \div 6 = \dfrac{2}{5} \cdot \dfrac{1}{6} = \dfrac{2 \cdot 1}{5 \cdot 6} = \dfrac{\cancel{2} \cdot 1}{5 \cdot \cancel{2} \cdot 3} = \dfrac{1}{5 \cdot 3} = \dfrac{1}{15}$ Removing a factor of 1: $\dfrac{2}{2} = 1$

Remember, if you can't factor, you can't cancel!

Do Exercises 5–9.

Why do we multiply by a reciprocal when dividing? To see this, let us consider $\frac{2}{3} \div \frac{7}{5}$. We will multiply by 1. The name for 1 that we will use is $(5/7)/(5/7)$:

$$\frac{2}{3} \div \frac{7}{5} = \frac{\dfrac{2}{3}}{\dfrac{7}{5}}.$$ Writing fractional notation for the division

Then

$$= \frac{\frac{2}{3}}{\frac{7}{5}} \cdot 1 \qquad \text{Multiplying by 1}$$

$$= \frac{\frac{2}{3}}{\frac{7}{5}} \cdot \frac{\frac{5}{7}}{\frac{5}{7}} \qquad \text{Multiplying by 1; } \frac{5}{7} \text{ is the reciprocal of } \frac{7}{5} \text{ and } \frac{\frac{5}{7}}{\frac{5}{7}} = 1$$

$$= \frac{\frac{2}{3} \cdot \frac{5}{7}}{\frac{7}{5} \cdot \frac{5}{7}} \qquad \text{Multiplying the numerators and the denominators}$$

$$= \frac{\frac{2}{3} \cdot \frac{5}{7}}{1} = \frac{2}{3} \cdot \frac{5}{7} = \frac{10}{21}.$$

After we multiplied, we got 1 for the denominator. The numerator (in color) shows the multiplication by the reciprocal.

Do Exercise 10.

10. Divide by multiplying by 1:

$$\frac{\frac{4}{5}}{\frac{6}{7}}.$$

c | SOLVING EQUATIONS

Now let us solve equations $a \cdot x = b$ and $x \cdot a = b$, where a and b may be fractions. Proceeding as we have before, we divide on both sides by a.

EXAMPLE 9 Solve: $\frac{4}{3} \cdot x = \frac{6}{7}$.

$$\frac{4}{3} \cdot x = \frac{6}{7}$$

$$x = \frac{6}{7} \div \frac{4}{3} \qquad \text{Dividing on both sides by } \frac{4}{3}$$

$$= \frac{6}{7} \cdot \frac{3}{4} \qquad \text{Multiplying by the reciprocal}$$

$$= \frac{2 \cdot 3 \cdot 3}{7 \cdot 2 \cdot 2} = \frac{2}{2} \cdot \frac{3 \cdot 3}{7 \cdot 2} = \frac{3 \cdot 3}{7 \cdot 2} = \frac{9}{14}$$

The solution is $\frac{9}{14}$.

EXAMPLE 10 Solve: $t \cdot \frac{4}{5} = 80$.

Dividing on both sides by $\frac{4}{5}$, we get

$$t = 80 \div \frac{4}{5} = 80 \cdot \frac{5}{4} = \frac{80 \cdot 5}{4} = \frac{4 \cdot 20 \cdot 5}{4 \cdot 1} = \frac{4}{4} \cdot \frac{20 \cdot 5}{1} = \frac{20 \cdot 5}{1} = 100.$$

The solution is 100.

Do Exercises 11 and 12.

Solve.

11. $\frac{5}{6} \cdot y = \frac{2}{3}$

12. $\frac{3}{4} \cdot n = 24$

d | SOLVING PROBLEMS

EXAMPLE 11 How many test tubes, each containing $\frac{3}{5}$ mL, can a nursing student fill from a container of 60 mL?

Answers on page A-2

1. Familiarize. Repeated addition will apply here. We let $n =$ the number of test tubes in all. We draw a picture.

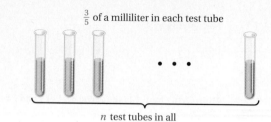

$\frac{3}{5}$ of a milliliter in each test tube

n test tubes in all

2. Translate. The equation that corresponds to the situation is

$$n \cdot \frac{3}{5} = 60.$$

3. Solve. We solve the equation by dividing on both sides by $\frac{3}{5}$ and carrying out the division:

$$n = 60 \div \frac{3}{5} = 60 \cdot \frac{5}{3} = \frac{60 \cdot 5}{3} = \frac{3 \cdot 20 \cdot 5}{3 \cdot 1} = \frac{3}{3} \cdot \frac{20 \cdot 5}{1} = 100.$$

4. Check. We check by repeating the calculation.

5. State. Thus, 100 test tubes can be filled.

Do Exercise 13.

13. Each loop in a spring uses $\frac{3}{8}$ in. of wire. How many loops can be made from 120 in. of wire?

EXAMPLE 12 After driving 210 mi, $\frac{5}{6}$ of a sales trip was completed. How long was the total trip?

1. Familiarize. We first draw a picture or at least visualize the situation. We let $n =$ the length of the trip.

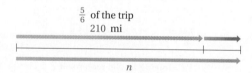

$\frac{5}{6}$ of the trip
210 mi

n

2. Translate. We translate to an equation.

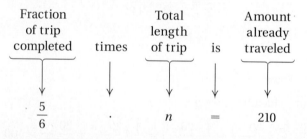

Fraction of trip completed	times	Total length of trip	is	Amount already traveled
$\frac{5}{6}$	$\cdot$	n	$=$	210

3. Solve. The equation that corresponds to the situation is $\frac{5}{6} \cdot n = 210$. We solve the equation by dividing on both sides by $\frac{5}{6}$ and carrying out the division:

$$n = 210 \div \frac{5}{6} = 210 \cdot \frac{6}{5} = \frac{210 \cdot 6}{5} = \frac{5 \cdot 42 \cdot 6}{5 \cdot 1} = \frac{5}{5} \cdot \frac{42 \cdot 6}{1} = 252.$$

4. Check. We check by repeating the calculation.

5. State. The total trip was 252 mi.

Do Exercise 14.

14. A service station tank had 175 gal of oil when it was $\frac{7}{8}$ full. How much could it hold altogether?

Answers on page A-2

Exercise Set 2.7

a Find the reciprocal.

1. $\dfrac{5}{6}$

2. $\dfrac{7}{8}$

3. 6

4. 4

5. $\dfrac{1}{6}$

6. $\dfrac{1}{4}$

7. $\dfrac{10}{3}$

8. $\dfrac{17}{4}$

b Divide and simplify.   Don't forget to simplify!

9. $\dfrac{3}{5} \div \dfrac{3}{4}$

10. $\dfrac{2}{3} \div \dfrac{3}{4}$

11. $\dfrac{3}{5} \div \dfrac{9}{4}$

12. $\dfrac{6}{7} \div \dfrac{3}{5}$

13. $\dfrac{4}{3} \div \dfrac{1}{3}$

14. $\dfrac{10}{9} \div \dfrac{1}{3}$

15. $\dfrac{1}{3} \div \dfrac{1}{6}$

16. $\dfrac{1}{4} \div \dfrac{1}{5}$

17. $\dfrac{3}{8} \div 3$

18. $\dfrac{5}{6} \div 5$

19. $\dfrac{12}{7} \div 4$

20. $\dfrac{18}{5} \div 2$

21. $12 \div \dfrac{3}{2}$

22. $24 \div \dfrac{3}{8}$

23. $28 \div \dfrac{4}{5}$

24. $40 \div \dfrac{2}{3}$

25. $\dfrac{5}{8} \div \dfrac{5}{8}$

26. $\dfrac{2}{5} \div \dfrac{2}{5}$

27. $\dfrac{8}{15} \div \dfrac{4}{5}$

28. $\dfrac{6}{13} \div \dfrac{3}{26}$

29. $\dfrac{9}{5} \div \dfrac{4}{5}$

30. $\dfrac{5}{12} \div \dfrac{25}{36}$

31. $120 \div \dfrac{5}{6}$

32. $360 \div \dfrac{8}{7}$

c Solve.

33. $\dfrac{4}{5} \cdot x = 60$

34. $\dfrac{3}{2} \cdot t = 90$

35. $\dfrac{5}{3} \cdot y = \dfrac{10}{3}$

36. $\dfrac{4}{9} \cdot m = \dfrac{8}{3}$

ANSWERS

1. _____
2. _____
3. _____
4. _____
5. _____
6. _____
7. _____
8. _____
9. _____
10. _____
11. _____
12. _____
13. _____
14. _____
15. _____
16. _____
17. _____
18. _____
19. _____
20. _____
21. _____
22. _____
23. _____
24. _____
25. _____
26. _____
27. _____
28. _____
29. _____
30. _____
31. _____
32. _____
33. _____
34. _____
35. _____
36. _____

37. $x \cdot \dfrac{25}{36} = \dfrac{5}{12}$ **38.** $p \cdot \dfrac{4}{5} = \dfrac{8}{15}$ **39.** $n \cdot \dfrac{8}{7} = 360$ **40.** $y \cdot \dfrac{5}{6} = 120$

37. _____

38. _____

39. _____

40. _____

41. _____

42. _____

43. _____

44. _____

45. _____

46. _____

47. _____

48. _____

49. _____

50. _____

51. _____

52. _____

53. _____

54. _____

55. _____

d Solve.

41. A piece of wire $\frac{3}{5}$ m long is to be cut into 6 pieces of the same length. What is the length of each piece?

42. A piece of coaxial cable $\frac{4}{5}$ m long is to be cut into 8 pieces of the same length. What is the length of each piece?

43. A sporting goods manufacturer requires $\frac{3}{4}$ yd of a particular fabric to make a pair of soccer shorts. How many pairs of shorts can be made from 24 yd of the fabric?

44. A child's baseball shirt requires $\frac{5}{6}$ yd of a certain fabric. How many shirts can be made from 25 yd of the fabric?

45. How many $\frac{2}{3}$-cup sugar bowls can be filled from 16 cups of sugar?

46. How many $\frac{2}{3}$-cup cereal bowls can be filled from 10 cups of cereal?

47. A bucket had 12 L of water in it when it was $\frac{3}{4}$ full. How much could it hold altogether?

48. A tank had 20 L of gasoline in it when it was $\frac{4}{5}$ full. How much could it hold altogether?

49. After driving 180 km, $\frac{5}{8}$ of a trip is completed. How long is the total trip? How many kilometers are left to drive?

50. After driving 240 km, $\frac{3}{5}$ of a trip is completed. How long is the total trip? How many kilometers are left to drive?

SKILL MAINTENANCE

Divide.

51. $268 \div 4$ **52.** $268 \div 8$ **53.** $6842 \div 24$ **54.** $8765 \div 85$

SYNTHESIS

55. If $\frac{1}{3}$ of a number is $\frac{1}{4}$, what is $\frac{1}{2}$ of the number?

CRITICAL THINKING

CALCULATOR CONNECTION

Use your calculator for trial-and-error computations in the following exercises.

1. In the simplification below, find a.

$$\frac{111,486}{183,624} = \frac{17}{28} \cdot \frac{a}{a} = \frac{17}{28}$$

2. In the multiplication below, find a and b.

$$\frac{332,644}{808,776} = \frac{13}{24} \cdot \frac{a}{b}$$

3. In the division below, find a and b.

$$\frac{19}{34} \div \frac{a}{b} = \frac{187,853}{268,226}$$

4. Use a calculator to help you determine whether each of the following is a prime number. Try dividing the number by smaller prime numbers such as 3, 5, 7, 11, and so on. Consult the list on p. 84.

159, 161, 286, 289, 315, 421, 587,

611, 685, 687

EXTENDED SYNTHESIS EXERCISES

1. Show that each of the following is true by simplifying each side of the equation. Then look for a pattern.

$$1 = \frac{1 \cdot 2}{2}$$

$$1 + 2 = \frac{2 \cdot 3}{2}$$

$$1 + 2 + 3 = \frac{3 \cdot 4}{2}$$

$$1 + 2 + 3 + 4 = \frac{4 \cdot 5}{2}$$

$$1 + 2 + 3 + 4 + 5 = \frac{5 \cdot 6}{2}$$

Use the pattern above to find the following sums without adding.

2. $1 + 2 + 3 + 4 + 5 + 6$

3. $1 + 2 + 3 + 4 + 5 + 6 + 7 + 8$

4. $1 + 2 + 3 + 4 + 5 + 6 + 7 + 8 + 9 + 10$

5. $1 + 2 + 3 + \cdots + 100$
   (The dots stand for the symbols we did not write.)

Triangular Numbers   Consider the numbers and arrays below. We call 1 the first *triangular number*, 3 the second triangular number, 6 the third, 10 the fourth, and so on, because these numbers can be arranged in triangular arrays as shown. Look for patterns in the following exercises and in the preceding Exercises 1–5.

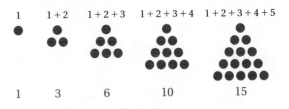

6. Find the eighth triangular number.

7. Find the thirteenth triangular number.

8. Find the 200th triangular number.

9. Which triangular number is 136?

10. Is 406 a triangular number? Explain.

11. Is 423 a triangular number? Explain.

Consecutive Numbers   The numbers 13 and 14 are said to be *consecutive* because one is found from the other by adding 1. The page numbers in this book are examples of consecutive numbers.

12. List several pairs of consecutive numbers.

13. How can you relate the idea of consecutive numbers to even and odd numbers?

14. How can you relate the idea of consecutive numbers to triangular numbers?

15. A prime number that becomes another prime number when we reverse its digits is called a **palindrome prime.** For example, 17 is a palindrome prime because both 17 and 71 are primes. Which of the following numbers are palindrome primes?

13, 91, 16, 11, 15, 24, 29, 101, 201

16. A pair of numbers such as 11, 13 are called **twin primes**—that is, primes that differ by 2. Another pair of twin primes is 17, 19. Find as many pairs of twin primes as you can that are less than 200.

17. In 1742, mathematician Christian Goldbach stated his famous *Goldbach conjecture*, which asserted that every even number greater than 4 is the sum of two odd prime numbers. For example, $6 = 3 + 3$, $18 = 7 + 11$, and so on. To date, this conjecture has never been proven. Show that the Goldbach conjecture holds for every even number less than 50.

(continued)

18. Look for a pattern in the following list and find the missing numbers.

$$1, \frac{2}{5}, \frac{4}{25}, \frac{8}{125}, \frac{16}{625}, \underline{\quad}, \underline{\quad}, \underline{\quad}, \underline{\quad}, \underline{\quad}.$$

19. Below is a pie chart that shows the breakdown of different types of bowling leagues in the United States.

Number of Leagues

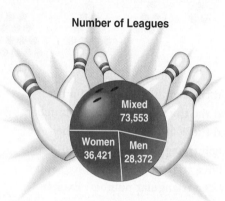

a) What fractional part of the number of bowling leagues are mixed leagues?

b) What fractional part of the number of bowling leagues are men's leagues?

c) What fractional part of the number of bowling leagues are women's leagues?

EXERCISES FOR THINKING AND WRITING

1. Describe the process of simplifying when using fractional notation.

2. Describe the advantage of prime factorization and the tests for divisibility when working with fractions.

3. Find a real-world situation that fits the equation

$$\frac{7}{8} \cdot \frac{9}{10} = \frac{63}{80}.$$

4. A student claims, "taking $\frac{1}{2}$ of a number is the same as dividing by $\frac{1}{2}$." Explain the error in this reasoning.

5. Which of the years from 1990 to 2010, if any, also happen to be prime numbers? Explain at least two ways you might go about solving this problem.

6. It has been proven that there is an infinite number of prime numbers and that there is no largest prime number. But finding large prime numbers is no easy matter. Recently, Cray Research announced the discovery that

$$2^{756,839} - 1 \quad \text{is a prime number.}$$

It contains 227,832 digits and took a Cray-2 supercomputer nineteen hours to make the determination. Do some research and make a report on the search for large prime numbers through the use of computers.

Summary and Review Exercises: Chapter 2

Beginning with this chapter, material from certain sections of preceding chapters will be covered on the chapter tests. Accordingly, the review exercises and the chapter tests will contain Skill Maintenance exercises. The review objectives to be tested in addition to the material in this chapter are [1.3d], [1.6c], [1.7b], and [1.8a].

Find the prime factorization of the number.

1. 70

2. 30

3. 45

4. 150

5. Determine whether 2432 is divisible by 6.

6. Determine whether 182 is divisible by 4.

7. Determine whether 4344 is divisible by 8.

8. Determine whether 4344 is divisible by 9.

9. Determine whether 37 is prime, composite, or neither.

10. Identify the numerator and the denominator of $\frac{2}{7}$.

11. What fractional part is shaded?

12. Simplify, if possible, the fractions on this circle graph.

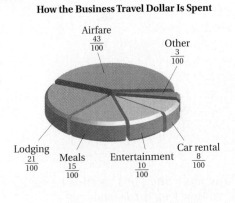

How the Business Travel Dollar Is Spent

Airfare $\frac{43}{100}$

Other $\frac{3}{100}$

Lodging $\frac{21}{100}$

Meals $\frac{15}{100}$

Entertainment $\frac{10}{100}$

Car rental $\frac{8}{100}$

Simplify.

13. $\frac{0}{4}$

14. $\frac{23}{23}$

15. $\frac{48}{1}$

16. $\frac{48}{8}$

17. $\frac{10}{15}$

18. $\frac{7}{28}$

19. $\frac{21}{21}$

20. $\frac{0}{25}$

21. $\frac{12}{30}$

22. $\frac{18}{1}$

23. $\frac{32}{8}$

24. $\frac{9}{27}$

25. $\frac{18}{0}$

26. $\frac{5}{8-8}$

Use = or ≠ for ▨ to write a true sentence.

27. $\dfrac{3}{5}$ ▨ $\dfrac{4}{6}$

28. $\dfrac{4}{7}$ ▨ $\dfrac{8}{14}$

29. $\dfrac{4}{5}$ ▨ $\dfrac{5}{6}$

30. $\dfrac{4}{3}$ ▨ $\dfrac{28}{21}$

Multiply and simplify.

31. $4 \cdot \dfrac{3}{8}$

32. $\dfrac{7}{3} \cdot 24$

33. $9 \cdot \dfrac{5}{18}$

34. $\dfrac{6}{5} \cdot 20$

35. $\dfrac{3}{4} \cdot \dfrac{8}{9}$

36. $\dfrac{5}{7} \cdot \dfrac{1}{10}$

37. $\dfrac{3}{7} \cdot \dfrac{14}{9}$

38. $\dfrac{1}{4} \cdot \dfrac{2}{11}$

Find the reciprocal.

39. $\dfrac{4}{5}$

40. 3

41. $\dfrac{1}{9}$

42. $\dfrac{47}{36}$

Divide and simplify.

43. $\dfrac{5}{9} \div \dfrac{5}{18}$

44. $6 \div \dfrac{4}{3}$

45. $\dfrac{1}{6} \div \dfrac{1}{11}$

46. $\dfrac{3}{14} \div \dfrac{6}{7}$

47. $180 \div \dfrac{3}{5}$

48. $\dfrac{1}{4} \div \dfrac{1}{9}$

49. $\dfrac{23}{25} \div \dfrac{23}{25}$

50. $\dfrac{2}{3} \div \dfrac{3}{2}$

Solve.

51. $\dfrac{5}{4} \cdot t = \dfrac{3}{8}$

52. $x \cdot \dfrac{2}{3} = 160$

Solve.

53. After driving 60 km, $\frac{3}{8}$ of a vacation trip is complete. How long is the total trip?

54. A recipe calls for $\frac{2}{3}$ cup of diced bell peppers. In making $\frac{1}{2}$ of this recipe, how much diced pepper should be used?

55. A person usually earns $42 for working a full day. How much is received for working $\frac{1}{7}$ of a day?

56. How many $\frac{2}{3}$-cup sugar bowls can be filled from 12 cups of sugar?

SKILL MAINTENANCE

Solve.

57. $17 \cdot x = 408$

58. $765 + t = 1234$

59. An economy car gets 43 miles per gallon on the highway. How far can it go on a full tank of 26 gal?

60. You wrote checks for $78, $97, and $102. Your balance before that was $789. What is your new balance?

61. Divide: $36 \overline{)14{,}697}$

62. Subtract: $\begin{array}{r} 5\,6\,0\,4 \\ -\,1\,9\,9\,7 \\ \hline \end{array}$

Test: Chapter 2

Find the prime factorization of the number.

1. 18

2. 60

3. Determine whether 1784 is divisible by 8.

4. Determine whether 784 is divisible by 9.

5. Identify the numerator and the denominator of $\frac{4}{9}$.

6. What part is shaded?

Simplify.

7. $\dfrac{26}{1}$

8. $\dfrac{12}{12}$

9. $\dfrac{0}{16}$

10. $\dfrac{12}{24}$

11. $\dfrac{42}{7}$

12. $\dfrac{2}{28}$

13. $\dfrac{9}{0}$

14. $\dfrac{7}{2-2}$

Use $=$ or $\neq$ for ▩ to write a true sentence.

15. $\dfrac{3}{4}$ ▩ $\dfrac{6}{8}$

16. $\dfrac{5}{4}$ ▩ $\dfrac{9}{7}$

Multiply and simplify.

17. $\dfrac{4}{3} \cdot 24$

18. $5 \cdot \dfrac{3}{10}$

19. $\dfrac{2}{3} \cdot \dfrac{15}{4}$

20. $\dfrac{3}{5} \cdot \dfrac{1}{6}$

ANSWERS

1. _____

2. _____

3. _____

4. _____

5. _____

6. _____

7. _____

8. _____

9. _____

10. _____

11. _____

12. _____

13. _____

14. _____

15. _____

16. _____

17. _____

18. _____

19. _____

20. _____

Find the reciprocal.

21. $\dfrac{5}{8}$ **22.** $\dfrac{1}{4}$ **23.** 18

21. _____

22. _____

23. _____

Divide and simplify.

24. $\dfrac{3}{8} \div \dfrac{5}{4}$ **25.** $\dfrac{1}{5} \div \dfrac{1}{8}$ **26.** $12 \div \dfrac{2}{3}$

24. _____

25. _____

26. _____

Solve.

27. $\dfrac{7}{8} \cdot x = 56$ **28.** $\dfrac{2}{5} \cdot t = \dfrac{7}{10}$

27. _____

28. _____

29. _____

Solve.

29. It takes $\frac{7}{8}$ lb of salt to use in the ice of one batch of homemade ice cream. How much salt is required for 32 batches?

30. A strip of taffy $\frac{9}{10}$ m long is cut into 12 equal pieces. What is the length of each piece?

30. _____

31. _____

32. _____

Solve.

31. $x + 198 = 2003$ **32.** $47 \cdot t = 4747$

33. _____

33. It is 2060 mi from San Francisco to Winnipeg, Canada. It is 1575 mi from Winnipeg to Atlanta. What is the total length of a route from San Francisco to Winnipeg to Atlanta?

34. _____

34. Divide: $2\,4\,\overline{)\,9\,1\,2\,7}$

35. Subtract: $\begin{array}{r} 8\,0\,0\,1 \\ -\,3\,5\,6\,7 \\ \hline \end{array}$

35. _____

Cumulative Review: Chapters 1–2

1. Write standard notation for the number in the following sentence: The earth travels five hundred eighty-four million, seventeen thousand, eight hundred miles around the sun.

2. Write a word name: 5,380,621.

3. In the number 2,751,043, which digit tells the number of hundreds?

Add.

4.
$$
\begin{array}{r}
1\,4{,}8\,6\,2 \\
+\ \ \ 2{,}9\,3\,5 \\
\hline
\end{array}
$$

5.
$$
\begin{array}{r}
7\,9\,8\,9 \\
7\,9\,8 \\
+\ \ \ \ \ 7\,9 \\
\hline
\end{array}
$$

Subtract.

6.
$$
\begin{array}{r}
5\,3\,7\,6 \\
-\ \ \ 4\,3\,0 \\
\hline
\end{array}
$$

7.
$$
\begin{array}{r}
2\,0\,0\,4 \\
-\ \ \ 5\,7\,9 \\
\hline
\end{array}
$$

Multiply and simplify.

8.
$$
\begin{array}{r}
6\,2\,1 \\
\times\ \ \ 2\,7 \\
\hline
\end{array}
$$

9.
$$
\begin{array}{r}
2\,5\,0\,5 \\
\times\,3\,3\,0\,0 \\
\hline
\end{array}
$$

10. $5 \times \dfrac{3}{100}$

11. $\dfrac{4}{9} \cdot \dfrac{3}{8}$

Divide and simplify.

12. $1\,9\,\overline{)\,4\,5\,8\,0}$

13. $6\,2\,\overline{)\,3\,8\,4\,4}$

14. $\dfrac{3}{10} \div 5$

15. $\dfrac{8}{9} \div \dfrac{15}{6}$

16. Round 427,931 to the nearest thousand.

17. Round 5309 to the nearest hundred.

Estimate the sum or product by rounding to the nearest hundred. Show your work.

18.
$$
\begin{array}{r}
7\,4\,9{,}5\,5\,9 \\
+\,3\,0\,1{,}3\,6\,2 \\
\hline
\end{array}
$$

19.
$$
\begin{array}{r}
7\,4\,9 \\
\times\,5\,3\,1 \\
\hline
\end{array}
$$

20. Use < or > for ▓ to write a true sentence:

26 ▓ 17.

21. Use = or ≠ for ▓ to write a true sentence:

$\dfrac{7}{10}$ ▓ $\dfrac{5}{7}$.

22. Evaluate: 3^4.

Simplify.

23. $35 - 25 \div 5 + 2 \times 3$

24. $\{17 - [8 - (5 - 2 \times 2)]\} \div (3 + 12 \div 6)$

25. Find all the factors of 28.

26. Find the prime factorization of 28.

27. Determine whether 39 is prime, composite, or neither.

28. Determine whether 32,712 is divisible by 3.

29. Determine whether 32,712 is divisible by 5.

Simplify.

30. $\dfrac{35}{1}$

31. $\dfrac{77}{11}$

32. $\dfrac{28}{98}$

33. $\dfrac{0}{47}$

Solve.

34. $x + 13 = 50$

35. $\dfrac{1}{5} \cdot t = \dfrac{3}{10}$

36. $13 \cdot y = 39$

37. $384 \div 16 = n$

Solve.

38. Emerson Fittipaldi won a recent Indianapolis 500 race. His earnings were $1,155,304. Arie Luyendyk won second place and received $681,303. How much more did Fittipaldi receive?

39. Four of the largest hotels in the United States are in Las Vegas. One has 3174 rooms, the second has 2920 rooms, the third 2832 rooms, and the fourth 2793 rooms. What is the total number of rooms in these four hotels?

40. A student is offered a part-time job paying $3900 a year. How much is each weekly paycheck?

41. One $\frac{3}{4}$-cup serving of macaroni and cheese contains 290 calories. A box makes 4 servings. How many cups of macaroni and cheese does the box make?

42. It takes 6 hr to paint the trim on a certain house. If the painter can work only $\frac{3}{4}$ hr per day, how many days will it take to finish the job?

43. Eastside Appliance sells a refrigerator for $600 and $30 tax with no delivery charge. Westside Appliance sells the same model for $560 and $28 tax plus a $25 delivery charge. Which is the better buy?

SYNTHESIS

44. A student works 35 hr a week and earns $6 an hour. The employer withholds $\frac{1}{4}$ of the total salary for taxes. Room and board expenses are $65 a week and tuition is $800 for a 16-week semester. Books cost $150 for the semester. Will the student make enough to cover the listed expenses? If so, how much will be left over at the end of the semester?

45. A can of mixed nuts is 1 part cashews, 1 part almonds, 1 part pecans, and 3 parts peanuts. What part of the mixture is peanuts?

3

Addition and Subtraction: Fractional Notation

INTRODUCTION

In this chapter, we consider addition and subtraction using fractional notation. Also discussed are addition, subtraction, multiplication, and division using mixed numerals. We then apply all these operations to the use of rules for order of operations and to problem solving.

AN APPLICATION

A bicycle wheel makes $66\frac{2}{3}$ revolutions per minute. It rotates for 12 min. How many revolutions does it make?

THE MATHEMATICS

Let n = the total number of revolutions. The problem then translates to this equation:

$$66\underbrace{\frac{2}{3}} \cdot 12 = n.$$

This is a mixed numeral.

The review objectives to be tested in addition to the material in this chapter are as follows.

[1.5b] Multiply whole numbers.
[1.8a] Solve problems involving addition, subtraction, multiplication, or division of whole numbers.
[2.6a] Multiply and simplify using fractional notation.
[2.7b] Divide and simplify using fractional notation.

Pretest: Chapter 3

1. Find the LCM of 15 and 24.

2. Use < or > for ▇ to write a true sentence:

$$\frac{7}{9} \; ▇ \; \frac{4}{5}.$$

3. Convert to fractional notation: $7\frac{5}{8}$.

4. Convert to a mixed numeral: $\frac{11}{2}$.

5. Divide. Write a mixed numeral for the answer.

$$1\,2\;)\overline{\,4\,7\,8\,9\,}$$

6. Add. Write a mixed numeral for the answer.

$$8\frac{11}{12}$$
$$+\,2\frac{3}{5}$$

7. Subtract. Write a mixed numeral for the answer.

$$14$$
$$-\,7\frac{5}{6}$$

8. Multiply. Write a mixed numeral for the answer.

$$3 \cdot 4\frac{8}{15}$$

9. Multiply. Write a mixed numeral for the answer.

$$6\frac{2}{3} \cdot 3\frac{1}{4}$$

10. Divide. Write a mixed numeral for the answer.

$$35 \div 5\frac{5}{6}$$

11. Divide. Write a mixed numeral for the answer.

$$5\frac{5}{12} \div 3\frac{1}{4}$$

12. Solve:

$$\frac{2}{3} + x = \frac{8}{9}.$$

13. At a summer camp, the cook bought 100 lb of potatoes and used $78\frac{3}{4}$ lb. How many pounds were left?

14. The weight of water is $62\frac{1}{2}$ lb per cubic foot. How many cubic feet would be occupied by $265\frac{5}{8}$ lb of water?

15. A courier drove $214\frac{3}{10}$ mi one day and $136\frac{9}{10}$ mi the next. How far did she travel in all?

16. A cake recipe calls for $3\frac{3}{4}$ cups of flour. How much flour would be used to make 6 cakes?

3.1 *Least Common Multiples*

In this chapter, we study addition and subtraction using fractional notation. Suppose we want to add $\frac{2}{3}$ and $\frac{1}{2}$. To do so, we find the least common multiple of the denominators: $\frac{2}{3} + \frac{1}{2} = \frac{4}{6} + \frac{3}{6}$. Then we add the numerators and keep the common denominator, 6. Before we do this, though, we study finding the **least common denominator,** or **least common multiple,** of the denominators.

FINDING LEAST COMMON MULTIPLES

> The *least common multiple,* or LCM, of two natural numbers is the smallest number that is a multiple of both.

EXAMPLE 1 Find the LCM of 20 and 30.

a) First list some multiples of 20 by multiplying 20 by 1, 2, 3, and so on:

  20, 40, 60, 80, 100, 120, 140, 160, 180, 200, 220, 240, ...

b) Then list some multiples of 30 by multiplying 30 by 1, 2, 3, and so on:

  30, 60, 90, 120, 150, 180, 210, 240, ...

c) Now list the numbers *common* to both lists, the common multiples:

  60, 120, 180, 240, ...

d) These are the common multiples of 20 and 30. Which is the smallest? The LCM of 20 and 30 is 60.

Do Exercise 1.

Next we develop two efficient methods for finding LCMs. You may choose to learn either method (consult with your instructor), or both, but if you are going on to a study of algebra, you should definitely learn method 2.

METHOD 1: FINDING LCMs USING ONE LIST OF MULTIPLES

> *Method 1.* To find the LCM of a set of numbers (9, 12):
>
> a) Determine whether the largest number is a multiple of the others. If it is, it is the LCM. That is, if the largest number has the others as factors, the LCM is that number.
>
>   (12 is not a multiple of 9)
>
> b) If not, check multiples of the largest number until you get one that is a multiple of the others.
>
>   (2 · 12 = 24, not a multiple of 9)
>
>   (3 · 12 = 36, a multiple of 9)
>
> c) That number is the LCM.
>
>   LCM = 36

OBJECTIVE

After finishing Section 3.1, you should be able to:

a Find the LCM of two or more numbers using a list of multiples or factorizations.

FOR EXTRA HELP

TAPE 5   TAPE 5A   MAC: 3A
                     IBM: 3A

1. By examining lists of multiples, find the LCM of 9 and 15.

Answer on page A-2

2. By examining lists of multiples, find the LCM of 8 and 10.

Find the LCM.
3. 10, 15

4. 6, 8

Find the LCM.
5. 5, 10

6. 20, 40, 80

EXAMPLE 2   Find the LCM of 12 and 15.

a) 15 is not a multiple of 12.

b) Check multiples:

$$2 \cdot 15 = 30, \quad \text{Not a multiple of 12}$$
$$3 \cdot 15 = 45, \quad \text{Not a multiple of 12}$$
$$4 \cdot 15 = 60. \quad \text{A multiple of 12}$$

c) The LCM = 60.

Do Exercise 2.

EXAMPLE 3   Find the LCM of 4 and 6.

a) 6 is not a multiple of 4.

b) Check multiples:

$$2 \cdot 6 = 12. \quad \text{A multiple of 4}$$

c) The LCM = 12.

Do Exercises 3 and 4.

EXAMPLE 4   Find the LCM of 4 and 8.

a) 8 is a multiple of 4, so it is the LCM.

c) The LCM = 8.

EXAMPLE 5   Find the LCM of 10, 100, and 1000.

a) 1000 is a multiple of 10 and 100, so it is the LCM.

c) The LCM = 1000.

Do Exercises 5 and 6.

METHOD 2: FINDING LCMs USING FACTORIZATIONS

A second method for finding LCMs uses prime factorizations. Consider again 20 and 30. Their prime factorizations are

$$20 = 2 \cdot 2 \cdot 5 \quad \text{and} \quad 30 = 2 \cdot 3 \cdot 5.$$

Let's look at these prime factorizations in order to find the LCM. Any multiple of 20 will have to have *two* 2's as factors and *one* 5 as a factor. Any multiple of 30 will have to have *one* 2, *one* 3, and *one* 5 as factors. The smallest number satisfying these conditions is

Two 2's, one 5; makes 20 a factor
$$2 \cdot 2 \cdot 3 \cdot 5.$$
One 2, one 3, one 5; makes 30 a factor

The LCM must have all the factors of 20 and all the factors of 30, but the factors need not be repeated when they are common to both numbers.

The greatest number of times a 2 occurs as a factor of either 20 or 30 is two, and the LCM has 2 as a factor twice. The greatest number of times a 3 occurs as a factor of either 20 or 30 is one, and the LCM has 3 as a factor once. The greatest number of times that 5 occurs as a factor of either 20 or 30 is one, and the LCM has 5 as a factor once.

Use prime factorizations to find the LCM.

7. 8, 10

Method 2. To find the LCM of a set of numbers using prime factorizations:

a) Find the prime factorization of each number.

b) Create a product of factors, using each factor the greatest number of times it occurs in any one factorization.

EXAMPLE 6 Find the LCM of 6 and 8.

a) Find the prime factorization of each number.

$$6 = 2 \cdot 3, \qquad 8 = 2 \cdot 2 \cdot 2$$

b) Create a product by writing factors, using each the greatest number of times it occurs in any one factorization.

Consider the factor 2. The greatest number of times that 2 occurs in any one factorization is three. We write 2 as a factor three times.

$$2 \cdot 2 \cdot 2 \cdot ?$$

Consider the factor 3. The greatest number of times that 3 occurs in any one factorization is one. We write 3 as a factor one time.

$$2 \cdot 2 \cdot 2 \cdot 3 \cdot ?$$

Since there are no other prime factors in either factorization, the

LCM is $2 \cdot 2 \cdot 2 \cdot 3$, or 24.

8. 18, 40

EXAMPLE 7 Find the LCM of 24 and 36.

a) Find the prime factorization of each number.

$$24 = 2 \cdot 2 \cdot 2 \cdot 3, \qquad 36 = 2 \cdot 2 \cdot 3 \cdot 3$$

b) Create a product by writing factors, using each the greatest number of times it occurs in any one factorization.

Consider the factor 2. The greatest number of times that 2 occurs in any one factorization is three. We write 2 as a factor three times:

$$2 \cdot 2 \cdot 2 \cdot ?$$

9. 32, 54

Consider the factor 3. The greatest number of times that 3 occurs in any one factorization is two. We write 3 as a factor two times:

$$2 \cdot 2 \cdot 2 \cdot 3 \cdot 3 \cdot ?$$

Since there are no other prime factors in either factorization, the

LCM is $2 \cdot 2 \cdot 2 \cdot 3 \cdot 3$, or 72.

Do Exercises 7–9.

Answers on page A-2

10. Find the LCM of 24, 35, and 45.

Find the LCM.

11. 3, 18

12. 12, 24

Answers on page A-2

EXAMPLE 8   Find the LCM of 27, 90, and 84.

a) Find the prime factorization of each number.

$$27 = 3 \cdot 3 \cdot 3, \qquad 90 = 2 \cdot 3 \cdot 3 \cdot 5, \qquad 84 = 2 \cdot 2 \cdot 3 \cdot 7$$

b) Create a product by writing factors, using each the greatest number of times it occurs in any one factorization.

Consider the factor 2. The greatest number of times that 2 occurs in any one factorization is two. We write 2 as a factor two times:

$$2 \cdot 2 \cdot ?$$

Consider the factor 3. The greatest number of times that 3 occurs in any one factorization is three. We write 3 as a factor three times:

$$2 \cdot 2 \cdot 3 \cdot 3 \cdot 3 \cdot ?$$

Consider the factor 5. The greatest number of times that 5 occurs in any one factorization is one. We write 5 as a factor one time:

$$2 \cdot 2 \cdot 3 \cdot 3 \cdot 3 \cdot 5 \cdot ?$$

Consider the factor 7. The greatest number of times that 7 occurs in any one factorization is one. We write 7 as a factor one time:

$$2 \cdot 2 \cdot 3 \cdot 3 \cdot 3 \cdot 5 \cdot 7 \cdot ?$$

Since no other prime factors are possible in any of the factorizations, the

$$\text{LCM is } 2 \cdot 2 \cdot 3 \cdot 3 \cdot 3 \cdot 5 \cdot 7, \text{ or } 3780.$$

Do Exercise 10.

EXAMPLE 9   Find the LCM of 7 and 21.

a) Find the prime factorization of each number. Because 7 is prime, it has no prime factorization. We think of $7 = 7$ as a "factorization" in order to carry out our procedure.

$$7 = 7, \qquad 21 = 3 \cdot 7$$

b) Create a product by writing factors, using each the greatest number of times it occurs in any one factorization.

Consider the factor 7. The greatest number of times that 7 occurs in any one factorization is one. We write 7 as a factor one time:

$$7 \cdot ?$$

Consider the factor 3. The greatest number of times that 3 occurs in any one factorization is one. We write 3 as a factor one time:

$$7 \cdot 3 \cdot ?$$

Since no other prime factors are possible in any of the factorizations, the

$$\text{LCM is } 7 \cdot 3, \text{ or } 21.$$

Note, in Example 9, that 7 is a factor of 21. We stated earlier that if one number is a factor of another, the LCM is the larger of the numbers. Thus, if you note this at the outset, you can find the LCM quickly without using factorizations.

Do Exercises 11 and 12.

EXAMPLE 10   Find the LCM of 8 and 9.

a) Find the prime factorization of each number.

$$8 = 2 \cdot 2 \cdot 2, \qquad 9 = 3 \cdot 3$$

b) Create a product by writing factors, using each the greatest number of times it occurs in any one factorization.

Consider the factor 2. The greatest number of times that 2 occurs in any one factorization is three. We write 2 as a factor three times.

$$2 \cdot 2 \cdot 2 \cdot ?$$

Consider the factor 3. The greatest number of times that 3 occurs in any one factorization is two. We write 3 as a factor two times.

$$2 \cdot 2 \cdot 2 \cdot 3 \cdot 3 \cdot ?$$

Since no other prime factors are possible in any of the factorizations, the

LCM is $2 \cdot 2 \cdot 2 \cdot 3 \cdot 3$, or 72.

Note in Example 10, that the two numbers, 8 and 9, have no common prime factor. When this happens, the LCM is just the product of the two numbers. Thus, when you note this at the outset, you can find the LCM quickly by multiplying the two numbers.

Do Exercises 13 and 14.

Let's compare the two methods considered for finding LCMs: the multiples method and the factorization method.

Method 1, the **multiples method,** can be longer than the factorization method when the LCM is large or when there are more than two numbers. But this method is faster and easier to use mentally for two numbers.

Method 2, the **factorization method,** works well for several numbers. It is just like a method used in algebra. If you are going to study algebra, you should definitely learn the factorization method.

METHOD 3: A THIRD METHOD FOR FINDING LCMs (OPTIONAL)

Here is another method for finding LCMs that may work well for you. Suppose you want to find the LCM of 48, 72, and 80. If possible, find a prime number that divides any two of these numbers with no remainder. Do the division and bring the third number down, unless the third number is divisible by the prime also. Repeat the process until you can divide no more. Multiply, as shown at the right, all the numbers at the side by all the numbers at the bottom. The LCM is

2	48	72	80
3	24	36	40
2	8	12	40
2	4	6	20
2	2	3	10
	1	3	5

$$2 \cdot 3 \cdot 2 \cdot 2 \cdot 2 \cdot 1 \cdot 3 \cdot 5, \text{ or } 720.$$

Do Exercises 15 and 16.

Find the LCM.

13. 4, 9

14. 5, 6, 7

Find the LCM using the optional method.

15. 24, 35, 45

16. 27, 90, 84

Answers on page A-2

APPLICATION OF LCMs: PLANET ORBITS

The earth, Jupiter, Saturn, and Uranus all revolve around the sun. The earth takes 1 year, Jupiter 12 years, Saturn 30 years, and Uranus 84 years to make a complete revolution. On a certain night, you look at those three distant planets and wonder how many years it will take before they have the same position again. (*Hint:* To find out, you find the LCM of 12, 30, and 84. It will be that number of years.)

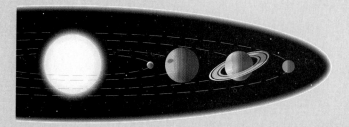

EXERCISES

1. How often will Jupiter and Saturn appear in the same direction in the night sky as seen from the earth?
2. How often will Saturn and Uranus appear in the same direction in the night sky as seen from the earth?
3. How often will Jupiter, Saturn, and Uranus appear in the same direction in the night sky as seen from the earth?

Hubble space telescope image of Saturn.

Exercise Set 3.1

a Find the LCM of the set of numbers. Do so mentally, if possible.

1. 2, 4

2. 3, 15

3. 10, 25

4. 10, 15

5. 20, 40

6. 8, 12

7. 18, 27

8. 9, 11

9. 30, 50

10. 24, 36

11. 30, 40

12. 21, 27

13. 18, 24

14. 12, 18

15. 60, 70

16. 35, 45

17. 16, 36

18. 18, 20

19. 32, 36

20. 36, 48

21. 2, 3, 5

22. 5, 18, 3

23. 3, 5, 7

24. 6, 12, 18

25. 24, 36, 12

26. 8, 16, 22

27. 5, 12, 15

28. 12, 18, 40

29. 9, 12, 6

30. 8, 16, 12

31. 3, 6, 8

32. 12, 8, 4

ANSWERS

1.
2.
3.
4.
5.
6.
7.
8.
9.
10.
11.
12.
13.
14.
15.
16.
17.
18.
19.
20.
21.
22.
23.
24.
25.
26.
27.
28.
29.
30.
31.
32.

33. 8, 48   **34.** 16, 32   **35.** 5, 50   **36.** 12, 72

37. 11, 13   **38.** 13, 14   **39.** 12, 35   **40.** 23, 25

41. 54, 63   **42.** 56, 72   **43.** 81, 90   **44.** 75, 100

SKILL MAINTENANCE

45. A performing arts center was sold out for a musical. Its seats sell for $13 each. Total receipts were $3250. How many seats does this auditorium contain?

46. Multiply: $23 \cdot 345$.

47. Multiply and simplify: $\dfrac{4}{5} \cdot \dfrac{10}{12}$.

48. Divide and simplify: $\dfrac{4}{5} \div \dfrac{7}{10}$.

SYNTHESIS

49. A pencil company uses two sizes of boxes, 6 in. and 8 in. long. These are packed in bigger cartons to be shipped. What is the shortest length carton that will accommodate boxes of either size without any room left over? (Each carton can contain only boxes of one size; no mixing is allowed.)

50. Consider 8 and 12. Determine whether each of the following is the LCM of 8 and 12. Tell why or why not.
   a) $2 \cdot 2 \cdot 3 \cdot 3$
   b) $2 \cdot 2 \cdot 3$
   c) $2 \cdot 3 \cdot 3$
   d) $2 \cdot 2 \cdot 2 \cdot 3$

▦ Use a calculator and the multiples method to find the LCM of the pair of numbers.

51. 288, 324

52. 2700, 7800

3.2 Addition

a | LIKE DENOMINATORS

Addition using fractional notation corresponds to combining or putting like things together, just as addition with whole numbers does. For example,

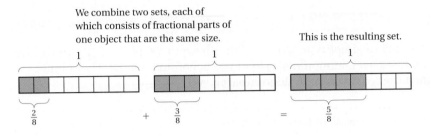

We combine two sets, each of which consists of fractional parts of one object that are the same size.

This is the resulting set.

2 eighths + 3 eighths = 5 eighths,

or
$$2 \cdot \frac{1}{8} + 3 \cdot \frac{1}{8} = 5 \cdot \frac{1}{8},$$

or
$$\frac{2}{8} + \frac{3}{8} = \frac{5}{8}.$$

Do Exercise 1.

To add when denominators are the same,

a) add the numerators,

b) keep the denominator, and

c) simplify, if possible.

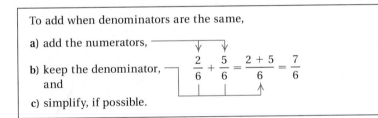

$$\frac{2}{6} + \frac{5}{6} = \frac{2+5}{6} = \frac{7}{6}$$

EXAMPLES   Add and simplify.

1. $\dfrac{2}{4} + \dfrac{1}{4} = \dfrac{2+1}{4} = \dfrac{3}{4}$ **No simplifying is possible.**

2. $\dfrac{11}{6} + \dfrac{3}{6} = \dfrac{11+3}{6} = \dfrac{14}{6} = \dfrac{2 \cdot 7}{2 \cdot 3} = \dfrac{2}{2} \cdot \dfrac{7}{3} = 1 \cdot \dfrac{7}{3} = \dfrac{7}{3}$ **Here we simplified.**

3. $\dfrac{3}{12} + \dfrac{5}{12} = \dfrac{3+5}{12} = \dfrac{8}{12} = \dfrac{4 \cdot 2}{4 \cdot 3} = \dfrac{4}{4} \cdot \dfrac{2}{3} = 1 \cdot \dfrac{2}{3} = \dfrac{2}{3}$

Do Exercises 2–4.

b | ADDITION USING THE LCD: DIFFERENT DENOMINATORS

What do we do when denominators are different? We try to find a common denominator. We can do this by multiplying by 1. Consider adding $\frac{1}{6}$ and $\frac{3}{4}$. There are several common denominators that can be obtained. Let's look at two possibilities.

1. Find $\dfrac{1}{5} + \dfrac{3}{5}$.

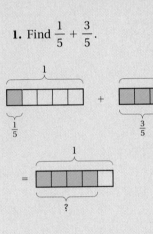

Add and simplify.

2. $\dfrac{1}{3} + \dfrac{2}{3}$

3. $\dfrac{5}{12} + \dfrac{1}{12}$

4. $\dfrac{9}{16} + \dfrac{3}{16}$

Answers on page A-2

5. Add. Find the least common denominator.

$$\frac{2}{3} + \frac{1}{6}$$

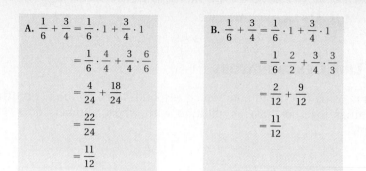

A. $\dfrac{1}{6} + \dfrac{3}{4} = \dfrac{1}{6} \cdot 1 + \dfrac{3}{4} \cdot 1$

$= \dfrac{1}{6} \cdot \dfrac{4}{4} + \dfrac{3}{4} \cdot \dfrac{6}{6}$

$= \dfrac{4}{24} + \dfrac{18}{24}$

$= \dfrac{22}{24}$

$= \dfrac{11}{12}$

B. $\dfrac{1}{6} + \dfrac{3}{4} = \dfrac{1}{6} \cdot 1 + \dfrac{3}{4} \cdot 1$

$= \dfrac{1}{6} \cdot \dfrac{2}{2} + \dfrac{3}{4} \cdot \dfrac{3}{3}$

$= \dfrac{2}{12} + \dfrac{9}{12}$

$= \dfrac{11}{12}$

We had to simplify in (A). We didn't have to simplify in (B). In (B), we used the least common multiple of the denominators, 12. That number is called the **least common denominator,** or LCD.

To add when denominators are different:

a) Find the least common multiple of the denominators. That number is the least common denominator, LCD.

b) Multiply by 1, using an appropriate notation, *n/n,* to express each number in terms of the LCD.

c) Add and simplify, if appropriate.

---

6. Add: $\dfrac{3}{8} + \dfrac{5}{6}$.

EXAMPLE 4 Add: $\dfrac{3}{4} + \dfrac{1}{8}$.

The LCD is 8.   4 is a factor of 8 so the LCM of 4 and 8 is 8.

$\dfrac{3}{4} + \dfrac{1}{8} = \dfrac{3}{4} \cdot 1 + \dfrac{1}{8}$ ← This fraction already has the LCD as its denominator.

$= \dfrac{3}{4} \cdot \dfrac{2}{2} + \dfrac{1}{8}$ ⎡ *Think:* $4 \times \blacksquare = 8$. The answer is 2, so we multiply by 1, using $\frac{2}{2}$.

$= \dfrac{6}{8} + \dfrac{1}{8}$

$= \dfrac{7}{8}$

Do Exercise 5.

EXAMPLE 5 Add: $\dfrac{1}{9} + \dfrac{5}{6}$.

The LCD is 18. $9 = 3 \cdot 3$ and $6 = 2 \cdot 3$, so the LCM of 9 and 6 is $2 \cdot 3 \cdot 3$, or 18.

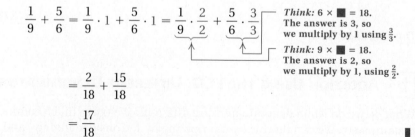

$\dfrac{1}{9} + \dfrac{5}{6} = \dfrac{1}{9} \cdot 1 + \dfrac{5}{6} \cdot 1 = \dfrac{1}{9} \cdot \dfrac{2}{2} + \dfrac{5}{6} \cdot \dfrac{3}{3}$ ⎡ *Think:* $6 \times \blacksquare = 18$. The answer is 3, so we multiply by 1 using $\frac{3}{3}$.
⎣ *Think:* $9 \times \blacksquare = 18$. The answer is 2, so we multiply by 1, using $\frac{2}{2}$.

$= \dfrac{2}{18} + \dfrac{15}{18}$

$= \dfrac{17}{18}$

Do Exercise 6.

Answers on page A-2

EXAMPLE 6 Add: $\frac{5}{9} + \frac{11}{18}$.

The LCD is 18.

$$\frac{5}{9} + \frac{11}{18} = \frac{5}{9} \cdot \frac{2}{2} + \frac{11}{18}$$

$$= \frac{10}{18} + \frac{11}{18}$$

$$= \frac{21}{18}$$

$$= \frac{7}{6}$$

> We may still have to simplify, but it is usually easier if we have used the LCD.

Do Exercise 7.

EXAMPLE 7 Add: $\frac{1}{10} + \frac{3}{100} + \frac{7}{1000}$.

Since 10 and 100 are factors of 1000, the LCD is 1000. Then

$$\frac{1}{10} + \frac{3}{100} + \frac{7}{1000} = \frac{1}{10} \cdot \frac{100}{100} + \frac{3}{100} \cdot \frac{10}{10} + \frac{7}{1000}$$

$$= \frac{100}{1000} + \frac{30}{1000} + \frac{7}{1000} = \frac{137}{1000}.$$

Look back over this example. Try to think it out so that you can do it mentally.

EXAMPLE 8 Add: $\frac{13}{70} + \frac{11}{21} + \frac{6}{15}$.

We have

$$\frac{13}{70} + \frac{11}{21} + \frac{6}{15} = \frac{13}{2 \cdot 5 \cdot 7} + \frac{11}{3 \cdot 7} + \frac{6}{3 \cdot 5}. \qquad \text{Factoring denominators}$$

The LCD is $2 \cdot 3 \cdot 5 \cdot 7$, or 210. Then

$$\frac{13}{70} + \frac{11}{21} + \frac{6}{15} = \frac{13}{2 \cdot 5 \cdot 7} \cdot \frac{3}{3} + \frac{11}{3 \cdot 7} \cdot \frac{2 \cdot 5}{2 \cdot 5} + \frac{6}{3 \cdot 5} \cdot \frac{7 \cdot 2}{7 \cdot 2}$$

$$= \frac{13 \cdot 3}{2 \cdot 5 \cdot 7 \cdot 3} + \frac{11 \cdot 2 \cdot 5}{3 \cdot 7 \cdot 2 \cdot 5} + \frac{6 \cdot 7 \cdot 2}{3 \cdot 5 \cdot 7 \cdot 2}$$

$$= \frac{39}{3 \cdot 5 \cdot 7 \cdot 2} + \frac{110}{3 \cdot 5 \cdot 7 \cdot 2} + \frac{84}{3 \cdot 5 \cdot 7 \cdot 2}$$

$$= \frac{233}{3 \cdot 5 \cdot 7 \cdot 2}$$

$$= \frac{233}{210} \qquad \text{We left 210 factored until we knew we could not simplify.}$$

> In each case, we multiply by 1 to obtain the LCD. In other words, look at the prime factorization of the LCD. Multiply each number by 1 to obtain what is missing in the LCD.

Do Exercises 8–10.

7. Add: $\frac{1}{6} + \frac{7}{18}$.

Add.

8. $\frac{4}{10} + \frac{1}{100} + \frac{3}{1000}$

9. $\frac{7}{10} + \frac{5}{100} + \frac{9}{1000}$
(Try to do this one mentally.)

10. $\frac{7}{10} + \frac{2}{21} + \frac{1}{7}$

Answers on page A-2

11. A consumer bought $\frac{1}{2}$ lb of peanuts and $\frac{3}{5}$ lb of cashews. How many pounds of nuts were bought altogether?

Answer on page A-2

---

c | SOLVING PROBLEMS

EXAMPLE 9 A jogger ran $\frac{4}{5}$ mi, rested, and then ran another $\frac{1}{10}$ mi. How far did she run in all?

1. Familiarize. We first draw a picture. We let $D =$ the distance run in all.

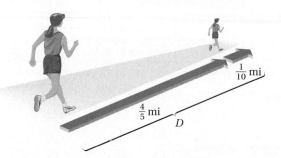

2. Translate. The problem can be translated to an equation as follows.

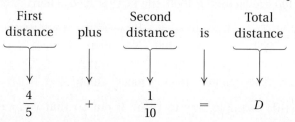

$$\frac{4}{5} + \frac{1}{10} = D$$

3. Solve. To solve the equation, we carry out the addition. The LCM of the denominators is 10 because 5 is a factor of 10. We multiply by 1 in order to obtain the LCD:

$$\frac{4}{5} \cdot \frac{2}{2} + \frac{1}{10} = D$$

$$\frac{8}{10} + \frac{1}{10} = D$$

$$\frac{9}{10} = D.$$

4. Check. We check by repeating the calculation. We also note that the sum should be larger than either of the individual distances, which it is. This gives us a partial check on the reasonableness of the answer.

5. State. In all, the jogger ran $\frac{9}{10}$ mi.

Do Exercise 11.

Exercise Set 3.2

a , b Add and simplify.

1. $\dfrac{7}{8} + \dfrac{1}{8}$ 2. $\dfrac{2}{5} + \dfrac{3}{5}$ 3. $\dfrac{1}{8} + \dfrac{5}{8}$ 4. $\dfrac{3}{10} + \dfrac{3}{10}$

5. $\dfrac{2}{3} + \dfrac{5}{6}$ 6. $\dfrac{5}{6} + \dfrac{1}{9}$ 7. $\dfrac{1}{8} + \dfrac{1}{6}$ 8. $\dfrac{1}{6} + \dfrac{3}{4}$

9. $\dfrac{4}{5} + \dfrac{7}{10}$ 10. $\dfrac{3}{4} + \dfrac{1}{12}$ 11. $\dfrac{5}{12} + \dfrac{3}{8}$ 12. $\dfrac{7}{8} + \dfrac{1}{16}$

13. $\dfrac{3}{20} + \dfrac{3}{4}$ 14. $\dfrac{2}{15} + \dfrac{2}{5}$ 15. $\dfrac{5}{6} + \dfrac{7}{9}$ 16. $\dfrac{5}{8} + \dfrac{5}{6}$

17. $\dfrac{3}{10} + \dfrac{1}{100}$ 18. $\dfrac{9}{10} + \dfrac{3}{100}$ 19. $\dfrac{5}{12} + \dfrac{4}{15}$ 20. $\dfrac{3}{16} + \dfrac{1}{12}$

21. $\dfrac{9}{10} + \dfrac{99}{100}$ 22. $\dfrac{3}{10} + \dfrac{27}{100}$ 23. $\dfrac{7}{8} + \dfrac{0}{1}$ 24. $\dfrac{0}{1} + \dfrac{5}{6}$

25. $\dfrac{3}{8} + \dfrac{1}{6}$ 26. $\dfrac{5}{8} + \dfrac{1}{6}$ 27. $\dfrac{5}{12} + \dfrac{7}{24}$ 28. $\dfrac{1}{18} + \dfrac{7}{12}$

29. $\dfrac{3}{16} + \dfrac{5}{16} + \dfrac{4}{16}$ 30. $\dfrac{3}{8} + \dfrac{1}{8} + \dfrac{2}{8}$ 31. $\dfrac{8}{10} + \dfrac{7}{100} + \dfrac{4}{1000}$

ANSWERS

1. _____
2. _____
3. _____
4. _____
5. _____
6. _____
7. _____
8. _____
9. _____
10. _____
11. _____
12. _____
13. _____
14. _____
15. _____
16. _____
17. _____
18. _____
19. _____
20. _____
21. _____
22. _____
23. _____
24. _____
25. _____
26. _____
27. _____
28. _____
29. _____
30. _____
31. _____

ANSWERS

32. _____

33. _____

34. _____

35. _____

36. _____

37. _____

38. _____

39. _____

40. _____

41. _____

42. _____

43. _____

44. _____

45. _____

46. _____

47. _____

48. _____

49. _____

50. _____

32. $\dfrac{1}{10} + \dfrac{2}{100} + \dfrac{3}{1000}$ 33. $\dfrac{3}{8} + \dfrac{5}{12} + \dfrac{8}{15}$ 34. $\dfrac{1}{2} + \dfrac{3}{8} + \dfrac{1}{4}$

35. $\dfrac{15}{24} + \dfrac{7}{36} + \dfrac{91}{48}$ 36. $\dfrac{5}{7} + \dfrac{25}{52} + \dfrac{7}{4}$

c Solve.

37. A shopper bought $\frac{1}{3}$ lb of orange pekoe tea and $\frac{1}{2}$ lb of English cinnamon tea. How many pounds of tea were bought?

38. A shopper bought $\frac{1}{4}$ lb of bonbons and $\frac{1}{2}$ lb of caramels. How many pounds of candy were bought?

39. A student walked $\frac{7}{6}$ mi to a friend's dormitory, and then $\frac{3}{4}$ mi to class. How far did the student walk?

40. A student walked $\frac{7}{8}$ mi to the student union, and then $\frac{2}{5}$ mi to class. How far did the student walk?

41. A cubic meter of concrete mix contains 420 kg of cement, 150 kg of stone, and 120 kg of sand. What is the total weight of the cubic meter of concrete mix? What part is cement? stone? sand? Add these amounts. What is the result?

42. A recipe for strawberry punch calls for $\frac{1}{5}$ qt of ginger ale and $\frac{3}{5}$ qt of strawberry soda. How much liquid is needed? If the recipe is doubled, how much liquid is needed? If the recipe is halved, how much liquid is needed?

43. A board $\frac{9}{10}$ in. thick is glued to a board $\frac{8}{10}$ in. thick. The glue is $\frac{3}{100}$ in. thick. How thick is the result?

44. A baker used $\frac{1}{2}$ lb of flour for rolls, $\frac{1}{4}$ lb for donuts, and $\frac{1}{3}$ lb for cookies. How much flour was used?

SKILL MAINTENANCE

Multiply.

45. $408 \cdot 516$ 46. $1125 \cdot 3728$ 47. $423 \cdot 8009$

Solve.

48. A shopper has $3458 in a checking account and writes checks for $329 and $52. How much is left in the account?

49. What is the total cost of 5 sweaters at $89 each and 6 shirts at $49 each?

SYNTHESIS

50. A student finds a part-time job with some friends on a weekend, cleaning and waxing rebuilt cars. As a result, the student does one-half of four-ninths of the job on Saturday and two-fifths of five-ninths of the job on Sunday. What part of the total job does the student do on the two days?

3.3 *Subtraction and Order*

a | SUBTRACTION

LIKE DENOMINATORS

We can consider the difference $\frac{4}{8} - \frac{3}{8}$ as we did before, as either "take away" or "how much more." Let us consider "take away."

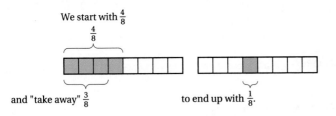

We start with $\frac{4}{8}$

and "take away" $\frac{3}{8}$ to end up with $\frac{1}{8}$.

We start with 4 eighths and take away 3 eighths:

$$4 \text{ eighths} - 3 \text{ eighths} = 1 \text{ eighth},$$

or $\quad 4 \cdot \frac{1}{8} - 3 \cdot \frac{1}{8} = \frac{1}{8},\quad$ or $\quad \frac{4}{8} - \frac{3}{8} = \frac{1}{8}.$

To subtract when denominators are the same,

a) subtract the numerators,

b) keep the denominator, and

c) simplify, if possible.

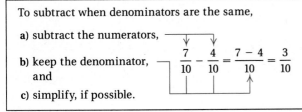

$$\frac{7}{10} - \frac{4}{10} = \frac{7-4}{10} = \frac{3}{10}$$

Answers should be simplified, if possible.

EXAMPLES   Subtract and simplify.

1. $\dfrac{7}{10} - \dfrac{3}{10} = \dfrac{7-3}{10} = \dfrac{4}{10} = \dfrac{2 \cdot 2}{5 \cdot 2} = \dfrac{2}{5} \cdot \dfrac{2}{2} = \dfrac{2}{5} \cdot 1 = \dfrac{2}{5}$

2. $\dfrac{8}{9} - \dfrac{2}{9} = \dfrac{8-2}{9} = \dfrac{6}{9} = \dfrac{2 \cdot 3}{3 \cdot 3} = \dfrac{2}{3} \cdot \dfrac{3}{3} = \dfrac{2}{3} \cdot 1 = \dfrac{2}{3}$

3. $\dfrac{32}{12} - \dfrac{25}{12} = \dfrac{32-25}{12} = \dfrac{7}{12}$

Do Exercises 1–3.

DIFFERENT DENOMINATORS

To subtract when denominators are different:

a) Find the least common multiple of the denominators. That number is the least common denominator, LCD.

b) Multiply by 1, using an appropriate notation, n/n, to express each number in terms of the LCD.

c) Subtract and simplify, if appropriate.

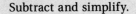
Subtract and simplify.

1. $\dfrac{7}{8} - \dfrac{3}{8}$

2. $\dfrac{10}{16} - \dfrac{4}{16}$

3. $\dfrac{8}{10} - \dfrac{3}{10}$

Answers on page A-2

4. Subtract: $\dfrac{3}{4} - \dfrac{2}{3}$.

Subtract.

5. $\dfrac{5}{6} - \dfrac{1}{9}$

6. $\dfrac{4}{5} - \dfrac{3}{10}$

7. Subtract: $\dfrac{11}{28} - \dfrac{5}{16}$.

EXAMPLE 4   Subtract: $\dfrac{2}{5} - \dfrac{3}{8}$.

The LCM of 5 and 8 is 40. The LCD is 40.

$$\frac{2}{5} - \frac{3}{8} = \frac{2}{5} \cdot \frac{8}{8} - \frac{3}{8} \cdot \frac{5}{5}$$

Think: $8 \times \blacksquare = 40$. The answer is 5, so we multiply by 1, using $\frac{5}{5}$.

Think: $5 \times \blacksquare = 40$. The answer is 8, so we multiply by 1, using $\frac{8}{8}$.

$$= \frac{16}{40} - \frac{15}{40}$$

$$= \frac{16 - 15}{40} = \frac{1}{40}$$

Do Exercise 4.

EXAMPLE 5   Subtract: $\dfrac{5}{6} - \dfrac{7}{12}$.

Since 6 is a factor of 12, the LCM of 6 and 12 is 12. The LCD is 12.

$$\frac{5}{6} - \frac{7}{12} = \frac{5}{6} \cdot \frac{2}{2} - \frac{7}{12}$$

$$= \frac{10}{12} - \frac{7}{12}$$

$$= \frac{10 - 7}{12} = \frac{3}{12}$$

$$= \frac{3 \cdot 1}{3 \cdot 4} = \frac{3}{3} \cdot \frac{1}{4}$$

$$= \frac{1}{4}$$

Do Exercises 5 and 6.

EXAMPLE 6   Subtract: $\dfrac{17}{24} - \dfrac{4}{15}$.

We have

$$\frac{17}{24} - \frac{4}{15} = \frac{17}{3 \cdot 2 \cdot 2 \cdot 2} - \frac{4}{5 \cdot 3}.$$

The LCD is $3 \cdot 2 \cdot 2 \cdot 2 \cdot 5$, or 120. Then

$$\frac{17}{24} - \frac{4}{15} = \frac{17}{3 \cdot 2 \cdot 2 \cdot 2} \cdot \frac{5}{5} - \frac{4}{5 \cdot 3} \cdot \frac{2 \cdot 2 \cdot 2}{2 \cdot 2 \cdot 2}$$

$$= \frac{17 \cdot 5}{3 \cdot 2 \cdot 2 \cdot 2 \cdot 5} - \frac{4 \cdot 2 \cdot 2 \cdot 2}{5 \cdot 3 \cdot 2 \cdot 2 \cdot 2}$$

$$= \frac{85}{120} - \frac{32}{120} = \frac{53}{120}.$$

In each case, we multiply by 1 to obtain the LCD. In other words, look at the prime factorization of the LCD. Multiply each number by 1 to obtain what is missing in the LCD.

Do Exercise 7.

b **ORDER**

We see from this figure that $\frac{4}{5} > \frac{3}{5}$. That is, $\frac{4}{5}$ is greater than $\frac{3}{5}$.

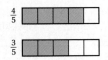

To determine which of two numbers is greater when there is a common denominator, compare the numerators:

$$\frac{4}{5}, \frac{3}{5}, \qquad 4 > 3 \qquad \frac{4}{5} > \frac{3}{5}.$$

Do Exercises 8 and 9.

When denominators are different, we multiply by 1 to make the denominators the same.

EXAMPLE 7 Use < or > for ▮ to write a true sentence:

$$\frac{2}{5} \ \blacksquare \ \frac{3}{4}.$$

We have

$$\frac{2}{5} \cdot \frac{4}{4} = \frac{8}{20}; \qquad \text{We multiply by 1 using } \tfrac{4}{4} \text{ to get the LCD.}$$

$$\frac{3}{4} \cdot \frac{5}{5} = \frac{15}{20}. \qquad \text{We multiply by 1 using } \tfrac{5}{5} \text{ to get the LCD.}$$

Since $8 < 15$, it follows that $\frac{8}{20} < \frac{15}{20}$, so

$$\frac{2}{5} < \frac{3}{4}.$$

EXAMPLE 8 Use < or > for ▮ to write a true sentence:

$$\frac{9}{10} \ \blacksquare \ \frac{89}{100}.$$

The LCD is 100.

$$\frac{9}{10} \cdot \frac{10}{10} = \frac{90}{100} \qquad \text{We multiply by } \tfrac{10}{10} \text{ to get the LCD.}$$

Since $90 > 89$, it follows that $\frac{90}{100} > \frac{89}{100}$, so

$$\frac{9}{10} > \frac{89}{100}.$$

Do Exercises 10–12.

 c **SOLVING EQUATIONS**

Now let us solve equations of the form $x + a = b$ or $a + x = b$, where a and b may be fractions. Proceeding as we have before, we subtract a on both sides of the equation.

8. Use < or > for ▮ to write a true sentence:

$$\frac{3}{8} \ \blacksquare \ \frac{5}{8}.$$

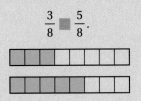

9. Use < or > for ▮ to write a true sentence:

$$\frac{7}{10} \ \blacksquare \ \frac{6}{10}.$$

Use < or > for ▮ to write a true sentence:

10. $\dfrac{2}{3} \ \blacksquare \ \dfrac{5}{8}$

11. $\dfrac{3}{4} \ \blacksquare \ \dfrac{8}{12}$

12. $\dfrac{5}{6} \ \blacksquare \ \dfrac{7}{8}$

Answers on page A-2

Solve.

13. $x + \dfrac{2}{3} = \dfrac{5}{6}$

14. $\dfrac{3}{5} + t = \dfrac{7}{8}$

15. There is $\frac{1}{4}$ cup of olive oil in a measuring cup. How much oil must be added to make a total of $\frac{4}{5}$ cup of oil in the measuring cup?

EXAMPLE 9   Solve: $x + \dfrac{1}{4} = \dfrac{3}{5}$.

$$x + \dfrac{1}{4} - \dfrac{1}{4} = \dfrac{3}{5} - \dfrac{1}{4} \qquad \text{Subtracting } \tfrac{1}{4} \text{ on both sides}$$

$$x + 0 = \dfrac{3}{5} \cdot \dfrac{4}{4} - \dfrac{1}{4} \cdot \dfrac{5}{5} \qquad \begin{array}{l}\text{The LCD is 20. We multiply} \\ \text{by 1 to get the LCD.}\end{array}$$

$$x = \dfrac{12}{20} - \dfrac{5}{20} = \dfrac{7}{20}$$

Do Exercises 13 and 14.

d   SOLVING PROBLEMS

EXAMPLE 10   A jogger has run $\frac{2}{3}$ mi and will stop running when she has run $\frac{7}{8}$ mi. How much farther does the jogger have to go?

1. **Familiarize.**   We first draw a picture or at least visualize the situation. We let $d =$ the distance to go.

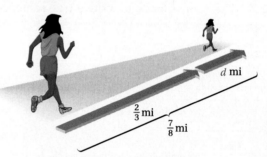

2. **Translate.**   We see that this is a "how much more" situation. Now we translate to an equation.

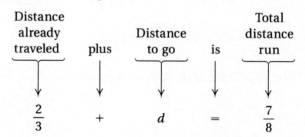

Distance already traveled	plus	Distance to go	is	Total distance run
$\dfrac{2}{3}$	$+$	d	$=$	$\dfrac{7}{8}$

3. **Solve.**   To solve the equation, we subtract $\frac{2}{3}$ on both sides:

$$\dfrac{2}{3} + d - \dfrac{2}{3} = \dfrac{7}{8} - \dfrac{2}{3} \qquad \text{Subtracting } \tfrac{2}{3} \text{ on both sides}$$

$$d + 0 = \dfrac{7}{8} \cdot \dfrac{3}{3} - \dfrac{2}{3} \cdot \dfrac{8}{8} \qquad \begin{array}{l}\text{The LCD is 24. We multiply} \\ \text{by 1 to obtain the LCD.}\end{array}$$

$$d = \dfrac{21}{24} - \dfrac{16}{24} = \dfrac{5}{24}.$$

4. **Check.**   To check, we return to the original problem and add:

$$\dfrac{2}{3} + \dfrac{5}{24} = \dfrac{2}{3} \cdot \dfrac{8}{8} + \dfrac{5}{24} = \dfrac{16}{24} + \dfrac{5}{24} = \dfrac{21}{24} = \dfrac{7}{8} \cdot \dfrac{3}{3} = \dfrac{7}{8}.$$

5. **State.**   The jogger has $\frac{5}{24}$ mi to go.

Do Exercise 15.

Exercise Set 3.3

a Subtract and simplify.

1. $\dfrac{5}{6} - \dfrac{1}{6}$

2. $\dfrac{5}{8} - \dfrac{3}{8}$

3. $\dfrac{11}{12} - \dfrac{2}{12}$

4. $\dfrac{17}{18} - \dfrac{11}{18}$

5. $\dfrac{3}{4} - \dfrac{1}{8}$

6. $\dfrac{2}{3} - \dfrac{1}{9}$

7. $\dfrac{1}{8} - \dfrac{1}{12}$

8. $\dfrac{1}{6} - \dfrac{1}{8}$

9. $\dfrac{4}{3} - \dfrac{5}{6}$

10. $\dfrac{7}{8} - \dfrac{1}{16}$

11. $\dfrac{3}{4} - \dfrac{3}{28}$

12. $\dfrac{2}{5} - \dfrac{2}{15}$

13. $\dfrac{3}{4} - \dfrac{3}{20}$

14. $\dfrac{5}{6} - \dfrac{1}{2}$

15. $\dfrac{3}{4} - \dfrac{1}{20}$

16. $\dfrac{3}{4} - \dfrac{4}{16}$

17. $\dfrac{5}{12} - \dfrac{2}{15}$

18. $\dfrac{9}{10} - \dfrac{11}{16}$

19. $\dfrac{6}{10} - \dfrac{7}{100}$

20. $\dfrac{9}{10} - \dfrac{3}{100}$

21. $\dfrac{7}{15} - \dfrac{3}{25}$

22. $\dfrac{18}{25} - \dfrac{4}{35}$

23. $\dfrac{99}{100} - \dfrac{9}{10}$

24. $\dfrac{78}{100} - \dfrac{11}{20}$

25. $\dfrac{2}{3} - \dfrac{1}{8}$

26. $\dfrac{3}{4} - \dfrac{1}{2}$

27. $\dfrac{3}{5} - \dfrac{1}{2}$

28. $\dfrac{5}{6} - \dfrac{2}{3}$

29. $\dfrac{5}{12} - \dfrac{3}{8}$

30. $\dfrac{7}{12} - \dfrac{2}{9}$

31. $\dfrac{7}{8} - \dfrac{1}{16}$

32. $\dfrac{5}{12} - \dfrac{5}{16}$

33. $\dfrac{17}{25} - \dfrac{4}{15}$

34. $\dfrac{11}{18} - \dfrac{7}{24}$

35. $\dfrac{23}{25} - \dfrac{112}{150}$

36. $\dfrac{89}{90} - \dfrac{53}{120}$

b Use < or > for ▮ to write a true sentence.

37. $\dfrac{5}{8}$ ▮ $\dfrac{6}{8}$

38. $\dfrac{7}{9}$ ▮ $\dfrac{5}{9}$

39. $\dfrac{1}{3}$ ▮ $\dfrac{1}{4}$

40. $\dfrac{1}{8}$ ▮ $\dfrac{1}{6}$

ANSWERS

1.
2.
3.
4.
5.
6.
7.
8.
9.
10.
11.
12.
13.
14.
15.
16.
17.
18.
19.
20.
21.
22.
23.
24.
25.
26.
27.
28.
29.
30.
31.
32.
33.
34.
35.
36.
37.
38.
39.
40.

ANSWERS

41. _____

42. _____

43. _____

44. _____

45. _____

46. _____

47. _____

48. _____

49. _____

50. _____

51. _____

52. _____

53. _____

54. _____

55. _____

56. _____

57. _____

58. _____

59. _____

60. _____

61. _____

62. _____

63. _____

64. _____

65. _____

66. _____

41. $\dfrac{2}{3}$ ▢ $\dfrac{5}{7}$　　**42.** $\dfrac{3}{5}$ ▢ $\dfrac{4}{7}$　　**43.** $\dfrac{4}{5}$ ▢ $\dfrac{5}{6}$　　**44.** $\dfrac{3}{2}$ ▢ $\dfrac{7}{5}$

45. $\dfrac{19}{20}$ ▢ $\dfrac{4}{5}$　　**46.** $\dfrac{5}{6}$ ▢ $\dfrac{13}{16}$　　**47.** $\dfrac{19}{20}$ ▢ $\dfrac{9}{10}$　　**48.** $\dfrac{3}{4}$ ▢ $\dfrac{11}{15}$

49. $\dfrac{31}{21}$ ▢ $\dfrac{41}{13}$　　**50.** $\dfrac{12}{7}$ ▢ $\dfrac{132}{49}$

c Solve.

51. $x + \dfrac{1}{30} = \dfrac{1}{10}$　　　**52.** $y + \dfrac{9}{12} = \dfrac{11}{12}$　　　**53.** $\dfrac{2}{3} + t = \dfrac{4}{5}$

54. $\dfrac{2}{3} + p = \dfrac{7}{8}$　　　**55.** $m + \dfrac{5}{6} = \dfrac{9}{10}$　　　**56.** $x + \dfrac{1}{3} = \dfrac{5}{6}$

d Solve.

57. A parent died and left an estate to four children. One received $\frac{1}{4}$ of the estate, the second received $\frac{1}{16}$, and the third received $\frac{3}{8}$. How much did the fourth receive?

58. A hamburger franchise was owned by three people. One owned $\frac{7}{12}$ of the business and the second owned $\frac{1}{6}$. How much did the third person own?

SKILL MAINTENANCE

59. Divide and simplify:

$$\frac{9}{10} \div \frac{3}{5}.$$

60. A batch of maple fudge requires $\frac{5}{6}$ cup of sugar. How much sugar is needed to make 12 batches?

SYNTHESIS

61. 🖩 In a recent year, Bernard Gilkey of the St. Louis Cardinals got 116 hits in 384 times at bat. Bip Roberts of the Cincinnati Reds got 172 hits in 532 times at bat. Who had the higher batting average?

Simplify. Use the rules for order of operations given in Section 1.9.

62. $\dfrac{2}{5} + \dfrac{1}{6} \div 3$　　**63.** $\dfrac{7}{8} - \dfrac{1}{10} \times \dfrac{5}{6}$　　**64.** $5 \times \dfrac{3}{7} - \dfrac{1}{7} \times \dfrac{4}{5}$　　**65.** $\left(\dfrac{2}{3}\right)^2 + \left(\dfrac{3}{4}\right)^2$

66. A videocassette recorder is purchased that records tapes up to a maximum of 6 hr. It can also record a tape of the same length for either 4 hr or 2 hr—that is, it will fill a 6-hr tape in either 4 hr or 2 hr by running at faster speeds. A 6-hr tape is placed in the machine. It records for $\frac{1}{2}$ hr at the 4-hr speed and $\frac{3}{4}$ hr at the 2-hr speed. How much time is left on the tape to record at the 6-hr speed?

3.4 Mixed Numerals

  WHAT IS A MIXED NUMERAL?

A symbol like $2\frac{3}{4}$ is called a **mixed numeral.**

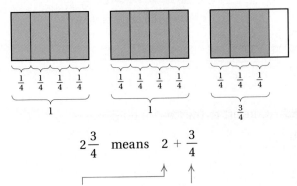

$$2\frac{3}{4} \quad \text{means} \quad 2 + \frac{3}{4}$$

This is a whole number.   This is a fraction less than 1.

EXAMPLES   Convert to a mixed numeral.

1. $7 + \frac{2}{5} = 7\frac{2}{5}$

2. $4 + \frac{3}{10} = 4\frac{3}{10}$

Do Exercises 1–3.

The notation $2\frac{3}{4}$ has a plus sign left out. To aid in understanding, we sometimes write the missing plus sign.

EXAMPLES   Convert to fractional notation.

3. $2\frac{3}{4} = 2 + \frac{3}{4}$ Inserting the missing plus sign

$\quad = \frac{2}{1} + \frac{3}{4}$ $2 = \frac{2}{1}$

$\quad = \frac{2}{1} \cdot \frac{4}{4} + \frac{3}{4}$ Finding a common denominator

$\quad = \frac{8}{4} + \frac{3}{4}$

$\quad = \frac{11}{4}$

4. $4\frac{3}{10} = 4 + \frac{3}{10} = \frac{4}{1} + \frac{3}{10} = \frac{4}{1} \cdot \frac{10}{10} + \frac{3}{10} = \frac{40}{10} + \frac{3}{10} = \frac{43}{10}$

Do Exercises 4 and 5.

1. $1 + \frac{2}{3} = \boxed{}$ Convert to a mixed numeral.

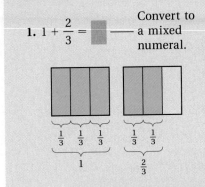

Convert to a mixed numeral.

2. $8 + \frac{3}{4}$ **3.** $12 + \frac{2}{3}$

Convert to fractional notation.

4. $4\frac{2}{5}$ **5.** $6\frac{1}{10}$

Answers on page A-2

Convert to fractional notation. Use the faster method.

6. $4\dfrac{5}{6}$

7. $9\dfrac{1}{4}$

8. $20\dfrac{2}{3}$

Let us now consider a faster method for converting a mixed numeral to fractional notation.

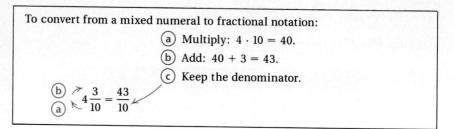

To convert from a mixed numeral to fractional notation:

ⓐ Multiply: $4 \cdot 10 = 40$.

ⓑ Add: $40 + 3 = 43$.

ⓒ Keep the denominator.

$$\begin{array}{c}\text{ⓑ}\\\text{ⓐ}\end{array}\ 4\dfrac{3}{10} = \dfrac{43}{10}$$

EXAMPLES Convert to fractional notation.

5. $6\dfrac{2}{3} = \dfrac{20}{3}$ $6 \cdot 3 = 18,\ 18 + 2 = 20$

6. $8\dfrac{2}{9} = \dfrac{74}{9}$

7. $10\dfrac{7}{8} = \dfrac{87}{8}$

Do Exercises 6–8.

b | WRITING MIXED NUMERALS

We can find a mixed numeral for $\dfrac{5}{3}$ as follows:

$$\dfrac{5}{3} = \dfrac{3}{3} + \dfrac{2}{3} = 1 + \dfrac{2}{3} = 1\dfrac{2}{3}.$$

Fractional symbols like $\dfrac{5}{3}$ also indicate division. Let's divide.

$$\begin{array}{r}1\frac{2}{3}\\3\overline{)5}\\3\\\overline{2}\end{array}\ \leftarrow\ 3\overline{)2}^{\frac{2}{3}}\quad \text{or}\quad 2 \div 3 = \dfrac{2}{3}$$

Thus, $\dfrac{5}{3} = 1\dfrac{2}{3}$.

In terms of objects, we can think of $\dfrac{5}{3}$ as 5 objects, each divided into 3 equal parts, as shown below.

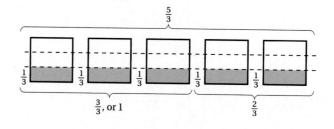

To convert from fractional notation to a mixed numeral, divide.

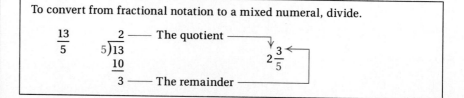

Answers on page A-2

EXAMPLES Convert to a mixed numeral.

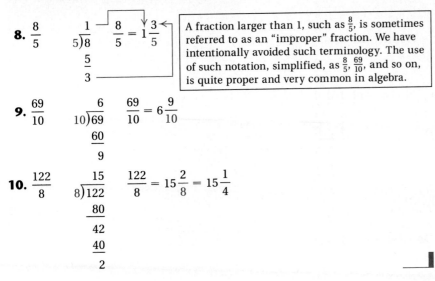

8. $\frac{8}{5}$ $\quad 5\overline{)8}$ $\frac{1}{}$ $\quad \frac{8}{5} = 1\frac{3}{5}$

$\frac{5}{3}$

A fraction larger than 1, such as $\frac{8}{5}$, is sometimes referred to as an "improper" fraction. We have intentionally avoided such terminology. The use of such notation, simplified, as $\frac{8}{5}$, $\frac{69}{10}$, and so on, is quite proper and very common in algebra.

9. $\frac{69}{10}$ $\quad 10\overline{)69}$ $\frac{6}{}$ $\quad \frac{69}{10} = 6\frac{9}{10}$

$\frac{60}{9}$

10. $\frac{122}{8}$ $\quad 8\overline{)122}$ $\frac{15}{}$ $\quad \frac{122}{8} = 15\frac{2}{8} = 15\frac{1}{4}$

$\frac{80}{42}$

$\frac{40}{2}$

Do Exercises 9–11.

c FINDING MIXED NUMERALS FOR QUOTIENTS

It is quite common when dividing whole numbers to write the quotient using a mixed numeral. The remainder is the numerator of the fractional part of the mixed numeral.

EXAMPLE 11 Divide. Write a mixed numeral for the quotient.

$$7\,\overline{)6\ 3\ 4\ 1}$$

We first divide as usual.

$$
\begin{array}{r}
9\ 0\ 5 \\
7\,\overline{)6\ 3\ 4\ 1} \\
6\ 3\ 0\ 0 \\
\hline
4\ 1 \\
3\ 5 \\
\hline
6
\end{array}
$$

The answer is 905 R 6. We write a mixed numeral for the answer as follows:

$$905\frac{6}{7}.$$

The division $6341 \div 7$ can be expressed using fractional notation as follows:

$$\frac{6341}{7} = 905\frac{6}{7}.$$

Convert to a mixed numeral.

9. $\frac{7}{3}$

10. $\frac{11}{10}$

11. $\frac{110}{6}$

Answers on page A-2

Divide. Write a mixed numeral for the answer.

12. 6) 4 8 4 6

EXAMPLE 12 Divide. Write a mixed numeral for the answer.

4 2) 8 9 1 5

We first divide as usual.

$$
\begin{array}{r}
2\ 1\ 2 \\
4\ 2\ \overline{)\ 8\ 9\ 1\ 5} \\
8\ 4\ 0\ 0 \\
\hline
5\ 1\ 5 \\
4\ 2\ 0 \\
\hline
9\ 5 \\
8\ 4 \\
\hline
1\ 1
\end{array}
$$

$$\frac{8915}{42} = 212\frac{11}{42}$$

The answer is $212\frac{11}{42}$.

Do Exercises 12 and 13.

13. 4 5) 6 0 5 3

Exercise Set 3.4

a Convert to fractional notation.

1. $5\frac{2}{3}$

2. $3\frac{4}{5}$

3. $3\frac{1}{4}$

4. $6\frac{1}{2}$

5. $10\frac{1}{8}$

6. $20\frac{1}{5}$

7. $5\frac{1}{10}$

8. $9\frac{1}{10}$

9. $20\frac{3}{5}$

10. $30\frac{4}{5}$

11. $9\frac{5}{6}$

12. $8\frac{7}{8}$

13. $7\frac{3}{10}$

14. $6\frac{9}{10}$

15. $1\frac{5}{8}$

16. $1\frac{3}{5}$

17. $12\frac{3}{4}$

18. $15\frac{2}{3}$

19. $4\frac{3}{10}$

20. $5\frac{7}{10}$

21. $2\frac{3}{100}$

22. $5\frac{7}{100}$

23. $66\frac{2}{3}$

24. $33\frac{1}{3}$

25. $5\frac{29}{50}$

26. $84\frac{3}{8}$

b Convert to a mixed numeral.

27. $\frac{18}{5}$

28. $\frac{17}{4}$

29. $\frac{14}{3}$

30. $\frac{39}{8}$

31. $\frac{27}{6}$

ANSWERS

1.
2.
3.
4.
5.
6.
7.
8.
9.
10.
11.
12.
13.
14.
15.
16.
17.
18.
19.
20.
21.
22.
23.
24.
25.
26.
27.
28.
29.
30.
31.

32. _____

33. _____

34. _____

35. _____

36. _____

37. _____

38. _____

39. _____

40. _____

41. _____

42. _____

43. _____

44. _____

45. _____

46. _____

47. _____

48. _____

49. _____

50. _____

51. _____

52. _____

53. _____

54. _____

55. _____

56. _____

57. _____

58. _____

59. _____

60. _____

61. _____

62. _____

63. _____

64. _____

65. _____

66. _____

32. $\dfrac{30}{9}$ **33.** $\dfrac{57}{10}$ **34.** $\dfrac{89}{10}$ **35.** $\dfrac{53}{7}$ **36.** $\dfrac{59}{8}$

37. $\dfrac{45}{6}$ **38.** $\dfrac{50}{8}$ **39.** $\dfrac{46}{4}$ **40.** $\dfrac{39}{9}$ **41.** $\dfrac{12}{8}$

42. $\dfrac{28}{6}$ **43.** $\dfrac{757}{100}$ **44.** $\dfrac{467}{100}$ **45.** $\dfrac{345}{8}$ **46.** $\dfrac{223}{4}$

c Divide. Write a mixed numeral for the answer.

47. $8\overline{)869}$ **48.** $3\overline{)2126}$ **49.** $5\overline{)3091}$ **50.** $9\overline{)9110}$

51. $21\overline{)852}$ **52.** $85\overline{)7672}$ **53.** $102\overline{)5612}$ **54.** $46\overline{)1081}$

SKILL MAINTENANCE

Multiply and simplify.

55. $\dfrac{6}{5} \cdot 15$ **56.** $\dfrac{5}{12} \cdot 6$ **57.** $\dfrac{7}{10} \cdot \dfrac{5}{14}$ **58.** $\dfrac{1}{10} \cdot \dfrac{20}{5}$

Divide and simplify.

59. $\dfrac{2}{3} \div \dfrac{1}{36}$ **60.** $28 \div \dfrac{4}{7}$ **61.** $200 \div \dfrac{15}{64}$ **62.** $\dfrac{3}{4} \div \dfrac{9}{16}$

SYNTHESIS

Write a mixed numeral.

63. $\dfrac{56}{7} + \dfrac{2}{3}$ **64.** $\dfrac{72}{12} + \dfrac{5}{6}$

Write a mixed numeral for the fraction in the sentence.

65. There are $\frac{366}{7}$ weeks in a leap year.   **66.** There are $\frac{365}{7}$ weeks in a non-leap year.

3.5 *Addition and Subtraction Using Mixed Numerals*

OBJECTIVES

After finishing Section 3.5, you should be able to:

a Add using mixed numerals.

b Subtract using mixed numerals.

c Solve problems involving addition and subtraction with mixed numerals.

FOR EXTRA HELP

TAPE 7   TAPE 6A   MAC: 3A
                            IBM: 3A

a ADDITION

To find the sum $1\frac{5}{8} + 3\frac{1}{8}$, we first add the fractions. Then we add the whole numbers.

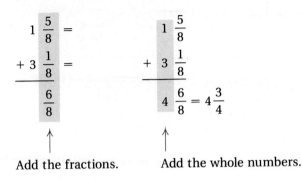

Add the fractions.   Add the whole numbers.

Do Exercise 1.

EXAMPLE 1 Add: $5\frac{2}{3} + 3\frac{5}{6}$. Write a mixed numeral for the answer.

The LCD is 6.

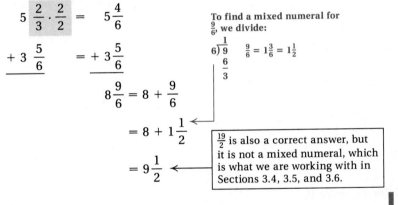

To find a mixed numeral for $\frac{9}{6}$, we divide:

$$6)\overline{9}$$
$$\frac{6}{3}$$

with quotient 1

$$\frac{9}{6} = 1\frac{3}{6} = 1\frac{1}{2}$$

$\frac{19}{2}$ is also a correct answer, but it is not a mixed numeral, which is what we are working with in Sections 3.4, 3.5, and 3.6.

Do Exercise 2.

EXAMPLE 2 Add: $10\frac{5}{6} + 7\frac{3}{8}$.

The LCD is 24.

$$10\,\frac{5}{6} \cdot \frac{4}{4} = \quad 10\frac{20}{24}$$

$$+\; 7\,\frac{3}{8} \cdot \frac{3}{3} = +\; 7\frac{9}{24}$$

$$17\frac{29}{24} = 18\frac{5}{24}$$

Do Exercise 3.

1. Add.

$$2\frac{3}{10}$$
$$+\,5\frac{1}{10}$$

2. Add.

$$8\frac{2}{5}$$
$$+\,3\frac{7}{10}$$

3. Add.

$$9\frac{3}{4}$$
$$+\,3\frac{5}{6}$$

Answers on page A-2

Subtract.

4. $10\dfrac{7}{8}$
$-\ 9\dfrac{3}{8}$
_____

5. $8\dfrac{2}{3}$
$-\ 5\dfrac{1}{2}$
_____

6. Subtract.

$5\dfrac{1}{12}$
$-\ 1\dfrac{3}{4}$
_____

7. Subtract.

5
$-\ 1\dfrac{1}{3}$
_____

Answers on page A-2

---

b SUBTRACTION

EXAMPLE 3   Subtract: $7\frac{3}{4} - 2\frac{1}{4}$.

$$7\ \dfrac{3}{4} \qquad\qquad 7\ \dfrac{3}{4}$$
$$-\ 2\ \dfrac{1}{4} \qquad\qquad -\ 2\ \dfrac{1}{4}$$
$$\overline{\quad\dfrac{2}{4}\quad} \qquad \overline{5\ \dfrac{2}{4} = 5\dfrac{1}{2}}$$

↑ Subtract the fractions.

↑ Subtract the whole numbers.

EXAMPLE 4   Subtract: $9\frac{4}{5} - 3\frac{1}{2}$.

The LCD is 10.

$$9\ \dfrac{4}{5} \cdot \dfrac{2}{2} = \quad 9\ \dfrac{8}{10}$$
$$-\ 3\ \dfrac{1}{2} \cdot \dfrac{5}{5} = -\ 3\ \dfrac{5}{10}$$
$$\overline{\qquad\qquad\qquad\quad 6\ \dfrac{3}{10}}$$

Do Exercises 4 and 5.

EXAMPLE 5   Subtract: $7\frac{1}{6} - 2\frac{1}{4}$.

The LCD is 12.

$$7\ \dfrac{1}{6} \cdot \dfrac{2}{2} = \quad 7\ \dfrac{2}{12}$$
$$-\ 2\ \dfrac{1}{4} \cdot \dfrac{3}{3} = -\ 2\ \dfrac{3}{12}$$

We cannot subtract $\frac{3}{12}$ from $\frac{2}{12}$.
We borrow 1, or $\frac{12}{12}$, from 7:
$7\frac{2}{12} = 6 + 1 + \frac{2}{12} = 6 + \frac{12}{12} + \frac{2}{12} = 6\frac{14}{12}.$

We can write this as

$$7\ \dfrac{2}{12} = \quad 6\ \dfrac{14}{12}$$
$$-\ 2\ \dfrac{3}{12} = -\ 2\ \dfrac{3}{12}$$
$$\overline{\qquad\qquad\qquad 4\ \dfrac{11}{12}}$$

Do Exercise 6.

EXAMPLE 6   Subtract: $12 - 9\frac{3}{8}$.

$$12 \quad = \quad 11\ \dfrac{8}{8} \quad \leftarrow \ 12 = 11 + 1 = 11 + \frac{8}{8} = 11\frac{8}{8}$$
$$-\ 9\ \dfrac{3}{8} = -\ 9\ \dfrac{3}{8}$$
$$\overline{\qquad\qquad\qquad 2\ \dfrac{5}{8}}$$

Do Exercise 7.

c SOLVING PROBLEMS

EXAMPLE 7   On two business days, a salesperson drove $144\frac{9}{10}$ mi and $87\frac{1}{4}$ mi. What was the total distance driven?

1. Familiarize. We let d = the total distance driven.

2. Translate. We translate as follows.

Distance driven first day	+	Distance driven second day	=	Total distance driven
$144\frac{9}{10}$	+	$87\frac{1}{4}$	=	d

3. Solve. The sentence tells us what to do. We add. The LCD is 20.

$$144\frac{9}{10} = \quad 144\,\frac{9}{10}\cdot\frac{2}{2} = \quad 144\frac{18}{20}$$

$$+\ 87\frac{1}{4} = +\ 87\,\frac{1}{4}\cdot\frac{5}{5} = +\ 87\frac{5}{20}$$

$$231\frac{23}{20} = 232\frac{3}{20}$$

Thus, $d = 232\frac{3}{20}$.

4. Check. We check by repeating the calculation. We also note that the answer is larger than either of the distances driven, which means that the answer is reasonable.

5. State. The total distance driven was $232\frac{3}{20}$ mi.

Do Exercise 8.

EXAMPLE 8 Recently, in college football, the distance between goalposts was reduced from $23\frac{1}{3}$ ft to $18\frac{1}{2}$ ft. How much was it reduced?

1. Familiarize. We let d = the amount of reduction and make a drawing to illustrate the situation.

2. Translate. We translate as follows.

Former distance	−	New distance	=	Amount of reduction
$23\frac{1}{3}$	−	$18\frac{1}{2}$	=	d

3. Solve. To solve the equation, we carry out the subtraction. The LCD is 6.

$$23\frac{1}{3} = \quad 23\,\frac{1}{3}\cdot\frac{2}{2} = \quad 23\frac{2}{6} = \quad 22\frac{8}{6}$$

$$-\ 18\frac{1}{2} = -\ 18\,\frac{1}{2}\cdot\frac{3}{3} = -18\frac{3}{6} = -\ 18\frac{3}{6}$$

$$4\frac{5}{6}$$

Thus, $d = 4\frac{5}{6}$ ft.

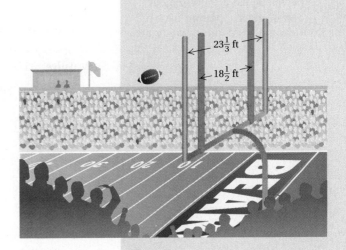

8. A car-seat upholstering company sold two pieces of leather $6\frac{1}{4}$ yd and $10\frac{5}{6}$ yd long. What was the total length of the leather?

Answer on page A-2

9. A $6\frac{1}{2}$-ft pole was set $2\frac{3}{4}$ ft in the ground. How much was left above the ground?

4. Check. To check, we add the reduction to the new distance:

$$18\frac{1}{2} + 4\frac{5}{6} = 18\frac{3}{6} + 4\frac{5}{6} = 22\frac{8}{6} = 23\frac{2}{6} = 23\frac{1}{3}.$$

This checks.

5. State. The reduction in the goalpost distance was $4\frac{5}{6}$ ft.

Do Exercise 9.

MULTISTEP PROBLEMS

EXAMPLE 9 One morning, the stock of Intel Corporation opened at a price of $\$100\frac{3}{8}$ per share. By noon, the price had risen $\$4\frac{7}{8}$. At the end of the day, it had fallen $\$10\frac{3}{4}$ from the price at noon. What was the closing price?

1. Familiarize. We first draw a picture or at least visualize the situation. We let $p =$ the price at noon, after the rise, and $c =$ the price at the close, after the drop.

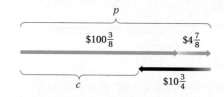

2. Translate. From the figure, we see that the price at the close is the price at noon minus the amount of the drop. Thus,

$$c = p - \$10\frac{3}{4} = \left(\$100\frac{3}{8} + \$4\frac{7}{8}\right) - \$10\frac{3}{4}.$$

3. Solve. This is a two-step problem.

a) We first add $\$4\frac{7}{8}$ to $\$100\frac{3}{8}$ to find the price p of the stock at noon.

$$
\begin{array}{r}
100\dfrac{3}{8} \\[4pt]
+\ \ 4\dfrac{7}{8} \\[2pt]
\hline
104\dfrac{10}{8} = 105\dfrac{1}{4} = p
\end{array}
$$

b) Next we subtract $\$10\frac{3}{4}$ from $\$105\frac{1}{4}$ to find the price c of the stock at closing.

$$
\begin{array}{rcr}
105\dfrac{1}{4} &=& 104\dfrac{5}{4} \\[4pt]
-\ \ 10\dfrac{3}{4} &=& -\ 10\dfrac{3}{4} \\[2pt]
\hline
&& 94\dfrac{2}{4} = 94\dfrac{1}{2} = c
\end{array}
$$

4. Check. We check by repeating the calculation.

5. State. The price of the stock at closing is $\$94\frac{1}{2}$.

10. There are $20\frac{1}{3}$ gal of water in a barrel; $5\frac{3}{4}$ gal are poured out and $8\frac{2}{3}$ gal are poured back in. How many gallons of water are then in the barrel?

Answers on page A-2

Do Exercise 10.

Exercise Set 3.5

a   Add. Write a mixed numeral for the answer.

1. $2\frac{7}{8}$

   $+\ 3\frac{5}{8}$

2. $4\frac{5}{6}$

   $+\ 3\frac{5}{6}$

3. $1\frac{1}{4}$

   $+\ 1\frac{2}{3}$

4. $4\frac{1}{3}$

   $+\ 5\frac{2}{9}$

5. $8\frac{3}{4}$

   $+\ 5\frac{5}{6}$

6. $4\frac{3}{8}$

   $+\ 6\frac{5}{12}$

7. $3\frac{2}{5}$

   $+\ 8\frac{7}{10}$

8. $5\frac{1}{2}$

   $+\ 3\frac{7}{10}$

9. $5\frac{3}{8}$

   $+\ 10\frac{5}{6}$

10. $\frac{5}{8}$

   $+\ 1\frac{5}{6}$

11. $12\frac{4}{5}$

   $+\ \ 8\frac{7}{10}$

12. $15\frac{5}{8}$

   $+\ 11\frac{3}{4}$

13. $14\frac{5}{8}$

   $+\ 13\frac{1}{4}$

14. $16\frac{1}{4}$

   $+\ 15\frac{7}{8}$

15. $7\frac{1}{8}$

   $9\frac{2}{3}$

   $+\ 10\frac{3}{4}$

16. $45\frac{2}{3}$

   $31\frac{3}{5}$

   $+\ 12\frac{1}{4}$

b   Subtract. Write a mixed numeral for the answer.

17. $4\frac{1}{5}$

   $-\ 2\frac{3}{5}$

18. $5\frac{1}{8}$

   $-\ 2\frac{3}{8}$

19. $6\frac{3}{5}$

   $-\ 2\frac{1}{2}$

20. $7\frac{2}{3}$

   $-\ 6\frac{1}{2}$

1. _____
2. _____
3. _____
4. _____
5. _____
6. _____
7. _____
8. _____
9. _____
10. _____
11. _____
12. _____
13. _____
14. _____
15. _____
16. _____
17. _____
18. _____
19. _____
20. _____

21. _____

22. _____

23. _____

24. _____

25. _____

26. _____

27. _____

28. _____

29. _____

30. _____

31. _____

32. _____

33. _____

34. _____

35. _____

36. _____

37. _____

38. _____

21. $34\frac{1}{3}$
$-\ 12\frac{5}{8}$

22. $23\frac{5}{16}$
$-\ 16\frac{3}{4}$

23. 21
$-\ 8\frac{3}{4}$

24. 42
$-\ 3\frac{7}{8}$

25. 34
$-\ 18\frac{5}{8}$

26. 23
$-\ 19\frac{3}{4}$

27. $21\frac{1}{6}$
$-\ 13\frac{3}{4}$

28. $42\frac{1}{10}$
$-\ 23\frac{7}{12}$

29. $14\frac{1}{8}$
$-\ \ \ \frac{3}{4}$

30. $28\frac{1}{6}$
$-\ \ \ \frac{2}{3}$

31. $25\frac{1}{9}$
$-\ 13\frac{5}{6}$

32. $23\frac{5}{16}$
$-\ 14\frac{7}{12}$

c Solve.

33. A butcher sold packages of hamburger weighing $1\frac{2}{3}$ lb and $5\frac{3}{4}$ lb. What was the total weight of the meat?

34. A butcher sold packages of sliced turkey breast weighing $1\frac{1}{3}$ lb and $4\frac{3}{5}$ lb. What was the total weight of the meat?

35. A woman is 66 in. tall and her son is $59\frac{7}{12}$ in. tall. How much taller is the woman?

36. A man is $73\frac{2}{3}$ in. tall and his daughter is $71\frac{5}{16}$ in. tall. How much taller is the man?

37. A plumber uses pipes of lengths $10\frac{5}{16}$ ft and $8\frac{3}{4}$ ft in the installation of a sink. How much pipe was used?

38. The standard pencil is $6\frac{7}{8}$ in. wood and $\frac{1}{2}$ in. eraser. What is the total length of the standard pencil?

$6\frac{7}{8}$ in.

$\frac{1}{2}$ in.

39. One day, a computer technician drove $180\frac{7}{10}$ mi away from Los Angeles for a service call. The next day, the technician drove $85\frac{1}{2}$ mi back toward Los Angeles for another service call. How far was the technician from Los Angeles?

40. A woman is $4\frac{1}{2}$ in. taller than her daughter. The daughter is $66\frac{2}{3}$ in. tall. How tall is the woman?

41. One standard book size is $8\frac{1}{2}$ in. by $9\frac{3}{4}$ in. What is the total distance around (perimeter of) the front cover of such a book?

42. A standard sheet of paper is $8\frac{1}{2}$ in. by 11 in. What is the total distance around (perimeter of) the paper?

43. On a recent day, the stock of Cummins Engine Corporation opened at $\$104\frac{5}{8}$ and dropped $\$1\frac{1}{4}$ during the course of the day. What was the closing price?

44. On a recent day, the stock of Marriott Corporation opened at $\$26\frac{7}{8}$ and closed at $\$27\frac{1}{4}$. How much did it gain that day?

45. When remodeling, a painter used $1\frac{2}{3}$ gal of paint for the family room and $1\frac{1}{2}$ gal for a bedroom. How much paint was used in all?

46. When redecorating, a woman used $1\frac{3}{4}$ gal of paint for the living room and $1\frac{1}{3}$ gal for the family room. How much paint was used in all?

47. A man is $5\frac{1}{4}$ in. taller than his son. The son is $72\frac{5}{6}$ in. tall. How tall is the man?

48. A plane flew 640 mi on a nonstop flight. On the return flight, it landed after having flown $320\frac{3}{10}$ mi. How far was the plane from its original point of departure?

49. An interior designer worked $10\frac{1}{2}$ hr over a three-day period. If the person worked $2\frac{1}{2}$ hr the first day and $4\frac{1}{5}$ hr the second, how many hours were worked the third day?

50. A painter had $3\frac{1}{2}$ gal of paint. It took $2\frac{3}{4}$ gal for a family room. It was estimated that it would take $2\frac{1}{4}$ gal to paint the living room. How much more paint was needed?

39. _____

40. _____

41. _____

42. _____

43. _____

44. _____

45. _____

46. _____

47. _____

48. _____

49. _____

50. _____

Find the distance around (perimeter of) the figure.

51.

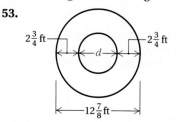

$5\frac{3}{4}$ yd

52.

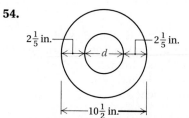

$3\frac{7}{16}$ in.

$6\frac{7}{8}$ in.

$3\frac{7}{16}$ in.

$6\frac{7}{8}$ in.

Find the length d in the figure.

53.

$2\frac{3}{4}$ ft d $2\frac{3}{4}$ ft

$12\frac{7}{8}$ ft

54.

$2\frac{1}{5}$ in. d $2\frac{1}{5}$ in.

$10\frac{1}{2}$ in.

55. Find the smallest length of a bolt that will pass through a piece of tubing with an outside diameter of $\frac{1}{2}$ in., a washer $\frac{1}{16}$ in. thick, a piece of tubing with a $\frac{3}{4}$-in. outside diameter, another washer, and a nut $\frac{3}{16}$ in. thick.

SKILL MAINTENANCE

56. A dairy produced 4578 oz of milk one week. How many 16-oz cartons were filled? How much milk was left over?

57. Divide and simplify: $\frac{12}{25} \div \frac{24}{5}$.

SYNTHESIS

58. A supporting post for a pier is placed through some water into the mud at the bottom of a lake. Half of the post is in the mud, $\frac{1}{3}$ is in the water, and the part above water is $5\frac{1}{2}$ ft long. How long is the post?

59. Solve: $47\frac{2}{3} + n = 56\frac{1}{4}$.

51. _____

52. _____

53. _____

54. _____

55. _____

56. _____

57. _____

58. _____

59. _____

3.6 Multiplication and Division Using Mixed Numerals

OBJECTIVES

After finishing Section 3.6, you should be able to:

a Multiply using mixed numerals.

b Divide using mixed numerals.

c Solve problems involving multiplication and division with mixed numerals.

FOR EXTRA HELP

| TAPE 7 | TAPE 6A | MAC: 3A |
| | | IBM: 3A |

a MULTIPLICATION

Carrying out addition and subtraction with mixed numerals is easier if the numbers are left as mixed numerals. With multiplication and division, however, it is easier to convert the numbers first to fractional notation.

> To multiply using mixed numerals, first convert to fractional notation. Then multiply with fractional notation and convert the answer back to a mixed numeral, if appropriate.

EXAMPLE 1 Multiply: $6 \cdot 2\frac{1}{2}$.

$$6 \cdot 2\frac{1}{2} = \frac{6}{1} \cdot \frac{5}{2} = \frac{6 \cdot 5}{1 \cdot 2} = \frac{2 \cdot 3 \cdot 5}{2 \cdot 1} = \frac{2}{2} \cdot \frac{3 \cdot 5}{1} = 15$$

Here we write fractional notation.

Do Exercise 1.

EXAMPLE 2 Multiply: $3\frac{1}{2} \cdot \frac{3}{4}$.

$$3\frac{1}{2} \cdot \frac{3}{4} = \frac{7}{2} \cdot \frac{3}{4} = \frac{21}{8} = 2\frac{5}{8}$$

> Note that fractional notation is needed to carry out the multiplication.

Do Exercise 2.

EXAMPLE 3 Multiply: $8 \cdot 4\frac{2}{3}$.

$$8 \cdot 4\frac{2}{3} = \frac{8}{1} \cdot \frac{14}{3} = \frac{112}{3} = 37\frac{1}{3}$$

Do Exercise 3.

EXAMPLE 4 Multiply: $2\frac{1}{4} \cdot 3\frac{2}{5}$.

$$2\frac{1}{4} \cdot 3\frac{2}{5} = \frac{9}{4} \cdot \frac{17}{5} = \frac{153}{20} = 7\frac{13}{20}$$

Caution! $2\frac{1}{4} \cdot 3\frac{2}{5} \neq 6\frac{2}{20}$. A common error is to multiply the whole numbers and then the fractions. This does not give the correct answer, $7\frac{13}{20}$, which is found by converting first to fractional notation.

Do Exercise 4.

1. Multiply: $6 \cdot 3\frac{1}{3}$.

2. Multiply: $2\frac{1}{2} \cdot \frac{3}{4}$.

3. Multiply: $2 \cdot 6\frac{2}{5}$.

4. Multiply: $3\frac{1}{3} \cdot 2\frac{1}{2}$.

Answers on page A-2

5. Divide: $84 \div 5\frac{1}{4}$.

6. Divide: $26 \div 3\frac{1}{2}$.

Divide.

7. $2\frac{1}{4} \div 1\frac{1}{5}$

8. $1\frac{3}{4} \div 2\frac{1}{2}$

---

b **DIVISION**

The division $1\frac{1}{2} \div \frac{1}{6}$ is shown here.

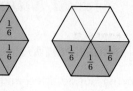

$$1\frac{1}{2} \div \frac{1}{6} = \frac{3}{2} \div \frac{1}{6}$$

$$= \frac{3}{2} \cdot 6 = \frac{3 \cdot 6}{2} = \frac{3 \cdot 3 \cdot 2}{2 \cdot 1} = \frac{3 \cdot 3}{1} \cdot \frac{2}{2} = \frac{3 \cdot 3}{1} \cdot 1 = 9$$

> To divide using mixed numerals, first write fractional notation. Then divide with fractional notation and convert the answer back to a mixed numeral, if appropriate.

EXAMPLE 5 Divide: $32 \div 3\frac{1}{5}$.

$$32 \div 3\frac{1}{5} = \frac{32}{1} \div \frac{16}{5}$$

$$= \frac{32}{1} \cdot \frac{5}{16} = \frac{32 \cdot 5}{1 \cdot 16} = \frac{2 \cdot 16 \cdot 5}{1 \cdot 16} = \frac{16}{16} \cdot \frac{2 \cdot 5}{1} = 10$$

     Remember to multiply by the reciprocal.

Do Exercise 5.

EXAMPLE 6 Divide: $35 \div 4\frac{1}{3}$.

$$35 \div 4\frac{1}{3} = \frac{35}{1} \div \frac{13}{3} = \frac{35}{1} \cdot \frac{3}{13} = \frac{105}{13} = 8\frac{1}{13}$$

Caution! The reciprocal of $4\frac{1}{3}$ is *not* $3\frac{1}{4}$!

Do Exercise 6.

EXAMPLE 7 Divide: $2\frac{1}{3} \div 1\frac{3}{4}$.

$$2\frac{1}{3} \div 1\frac{3}{4} = \frac{7}{3} \div \frac{7}{4} = \frac{7}{3} \cdot \frac{4}{7} = \frac{7 \cdot 4}{7 \cdot 3} = \frac{7}{7} \cdot \frac{4}{3} = 1 \cdot \frac{4}{3} = \frac{4}{3} = 1\frac{1}{3}$$

EXAMPLE 8 Divide: $1\frac{3}{5} \div 3\frac{1}{3}$.

$$1\frac{3}{5} \div 3\frac{1}{3} = \frac{8}{5} \div \frac{10}{3} = \frac{8}{5} \cdot \frac{3}{10} = \frac{2 \cdot 4 \cdot 3}{5 \cdot 2 \cdot 5} = \frac{2}{2} \cdot \frac{4 \cdot 3}{5 \cdot 5} = \frac{4 \cdot 3}{5 \cdot 5} = 1 \cdot \frac{12}{25} = \frac{12}{25}$$

Do Exercises 7 and 8.

C SOLVING PROBLEMS

EXAMPLE 9 A bicycle wheel makes $66\frac{2}{3}$ revolutions per minute. It rotates for 12 min. How many revolutions does it make?

1. Familiarize. We first draw a picture. A rectangular array works well here. We draw a circle for each revolution and a row for each minute. We let n = the number of revolutions.

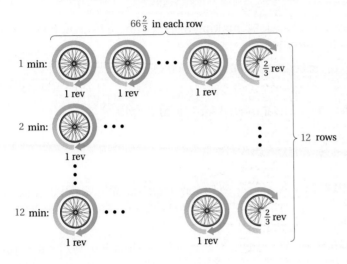

2. Translate. We then translate as follows.

$$
\underbrace{\text{Revolutions per minute}}_{66\frac{2}{3}} \cdot \underbrace{\text{Number of minutes}}_{12} = \underbrace{\text{Total number of revolutions}}_{n}
$$

3. Solve. To solve the equation, we carry out the multiplication:

$$n = 66\frac{2}{3} \cdot 12 = \frac{200}{3} \cdot \frac{12}{1}$$

$$= \frac{200 \cdot 3 \cdot 4}{3 \cdot 1} = \frac{3}{3} \cdot \frac{200 \cdot 4}{1} = 1 \cdot 800 = 800.$$

4. Check. We check by repeating the calculation. We can do a partial check by noting that $66\frac{2}{3} \approx 70$ and $12 \approx 10$. Then the product is about 700. Thus our answer is reasonable.

5. State. The bicycle wheel makes 800 revolutions in 12 min.

Do Exercise 9.

EXAMPLE 10 A bicycle wheel makes 900 revolutions at a rate of $66\frac{2}{3}$ revolutions per minute. How long does the wheel rotate?

1. Familiarize. We first draw a picture as we did in Example 9, using a circle for each revolution and a row for each minute. In this case, we let t = the time the wheel rotates. The last row may be incomplete.

Answer on page A-2

10. A car travels 302 mi on $15\frac{1}{10}$ gal of gas. How many miles per gallon did it get?

11. A room is $22\frac{1}{2}$ ft by $15\frac{1}{2}$ ft. A 9-ft by 12-ft area rug is placed in the center of the room. How much area is not covered by the rug?

2. Translate. The division that corresponds to the situation is

$$t = 900 \div 66\frac{2}{3}.$$

3. Solve. To solve the equation, we carry out the division:

$$t = 900 \div 66\frac{2}{3} = \frac{900}{1} \div \frac{200}{3} = \frac{900}{1} \cdot \frac{3}{200} = \frac{100 \cdot 9 \cdot 3}{2 \cdot 100}$$

$$= \frac{100}{100} \cdot \frac{9 \cdot 3}{2} = 13\frac{1}{2}.$$

4. Check. We check by multiplying the time by the number of revolutions per minute:

$$66\frac{2}{3} \cdot 13\frac{1}{2} = \frac{200}{3} \cdot \frac{27}{2} = \frac{200 \cdot 27}{3 \cdot 2} = \frac{100 \cdot 9 \cdot 3 \cdot 2}{2 \cdot 3} = \frac{100 \cdot 9}{1} \cdot \frac{3 \cdot 2}{3 \cdot 2} = 900.$$

5. State. The wheel rotated for $13\frac{1}{2}$ min.

Do Exercise 10.

EXAMPLE 11 An L-shaped room consists of a rectangle that is $8\frac{1}{2}$ by 11 ft and one that is $6\frac{1}{2}$ by $7\frac{1}{2}$ ft. What is the total area of a carpet that covers the floor?

1. Familiarize. We draw a picture of the situation. We let a = the total area.

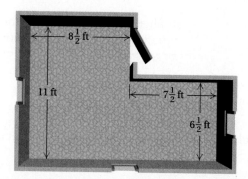

2. Translate. The total area is the sum of the areas of the two rectangles. This gives us the following equation:

$$a = \left(8\frac{1}{2}\right) \cdot (11) + \left(7\frac{1}{2}\right) \cdot \left(6\frac{1}{2}\right).$$

3. Solve. This is a multistep problem. We can carry it out by doing each multiplication and then adding. This follows the rules for order of operations:

$$a = \left(8\frac{1}{2}\right) \cdot (11) + \left(7\frac{1}{2}\right) \cdot \left(6\frac{1}{2}\right) = \frac{17}{2} \cdot 11 + \frac{15}{2} \cdot \frac{13}{2} = \frac{17 \cdot 11}{2} + \frac{15 \cdot 13}{2 \cdot 2}$$

$$= 93\frac{1}{2} + 48\frac{3}{4} = 93\frac{2}{4} + 48\frac{3}{4} = 141\frac{5}{4} = 141 + 1 + \frac{1}{4} + 142\frac{1}{4}.$$

4. Check. We check by repeating the calculation.

5. State. The total area of the carpet is $142\frac{1}{4}$ ft².

Do Exercise 11.

Exercise Set 3.6

a Multiply. Write a mixed numeral for the answer.

1. $8 \cdot 2\frac{5}{6}$

2. $5 \cdot 3\frac{3}{4}$

3. $3\frac{5}{8} \cdot \frac{2}{3}$

4. $6\frac{2}{3} \cdot \frac{1}{4}$

5. $3\frac{1}{2} \cdot 2\frac{1}{3}$

6. $4\frac{1}{5} \cdot 5\frac{1}{4}$

7. $3\frac{2}{5} \cdot 2\frac{7}{8}$

8. $2\frac{3}{10} \cdot 4\frac{2}{5}$

9. $4\frac{7}{10} \cdot 5\frac{3}{10}$

10. $6\frac{3}{10} \cdot 5\frac{7}{10}$

11. $20\frac{1}{2} \cdot 10\frac{1}{5}$

12. $21\frac{1}{3} \cdot 11\frac{1}{3}$

b Divide. Write a mixed numeral for the answer.

13. $20 \div 3\frac{1}{5}$

14. $18 \div 2\frac{1}{4}$

15. $8\frac{2}{5} \div 7$

16. $3\frac{3}{8} \div 3$

17. $4\frac{3}{4} \div 1\frac{1}{3}$

18. $5\frac{4}{5} \div 2\frac{1}{2}$

19. $1\frac{7}{8} \div 1\frac{2}{3}$

20. $4\frac{3}{8} \div 2\frac{5}{6}$

21. $5\frac{1}{10} \div 4\frac{3}{10}$

22. $4\frac{1}{10} \div 2\frac{1}{10}$

23. $20\frac{1}{4} \div 90$

24. $12\frac{1}{2} \div 50$

ANSWERS

1. _____
2. _____
3. _____
4. _____
5. _____
6. _____
7. _____
8. _____
9. _____
10. _____
11. _____
12. _____
13. _____
14. _____
15. _____
16. _____
17. _____
18. _____
19. _____
20. _____
21. _____
22. _____
23. _____
24. _____

c Solve.

25. A water wheel makes $16\frac{3}{4}$ revolutions per minute. If it rotates for 56 min, how many revolutions does it make?

26. A bicycle wheel makes $66\frac{2}{3}$ revolutions per minute. If it rotates for 21 min, how many revolutions does it make?

27. One serving of meat is about $3\frac{1}{2}$ oz. A person needs 2 servings a day for proper nutrition. How many ounces of meat is this?

28. Round steak contains $3\frac{1}{2}$ servings per pound. How many servings are there in 10 lb of round steak?

29. The weight of water is $62\frac{1}{2}$ lb per cubic foot. What is the weight of $5\frac{1}{2}$ cubic feet of water?

30. The weight of water is $62\frac{1}{2}$ lb per cubic foot. What is the weight of $2\frac{1}{4}$ cubic feet of water?

31. Listed below are the ingredients for a low-fat, heart-healthy dish called *Chicken à la King*. What are the ingredients for $\frac{1}{2}$ recipe? for 3 recipes?

CHICKEN À LA KING

- 2 chicken bouillon cubes
- $1^{1}/_{2}$ cups hot water
- 3 tablespoons margarine
- 3 tablespoons flour
- $2^{1}/_{2}$ cups diced cooked chicken
- 1 cup cooked peas
- 1 4-oz can sliced mushrooms, drained
- $^{1}/_{3}$ cup sliced cooked carrots
- $^{1}/_{4}$ cup chopped onions
- 2 tablespoons chopped pimiento
- 1 teaspoon salt

25. _____

26. _____

27. _____

28. _____

29. _____

30. _____

31. _____

32. Listed below are the ingredients for a low-fat, heart-healthy dish called *Italian Stuffed Peppers.* What are the ingredients for $\frac{1}{2}$ recipe? for 3 recipes?

ITALIAN STUFFED PEPPERS
- 1/3 cup Italian dressing
- 4 medium green peppers
- 1 quart water
- 1 1/2 cups cooked brown rice
- 1 16-oz can tomato sauce (no salt or sugar added)
- 1 teaspoon Tamari soy sauce
- 1/2 teaspoon basil
- 1 clove garlic, minced
- 1/3 cup onion, chopped
- 2 15-oz cans dark red kidney beans, rinsed and drained
- 1 tablespoon parsley, chopped
- 4 tablespoons Parmesan cheese

33. Fahrenheit temperature can be obtained from Celsius (centigrade) temperature by multiplying by $1\frac{4}{5}$ and adding 32°. What Fahrenheit temperature corresponds to a Celsius temperature of 20°?

34. Fahrenheit temperature can be obtained from Celsius (centigrade) temperature by multiplying by $1\frac{4}{5}$ and adding 32°. What Fahrenheit temperature corresponds to the Celsius temperature of boiling water, which is 100°?

35. If a water wheel made 469 revolutions at a rate of $16\frac{3}{4}$ revolutions per minute, how long did it rotate?

36. If a bicycle wheel made 480 revolutions at a rate of $66\frac{2}{3}$ revolutions per minute, how long did it rotate?

37. A car traveled 213 mi on $14\frac{2}{10}$ gal of gas. How many miles per gallon did it get?

38. A car traveled 385 mi on $15\frac{4}{10}$ gal of gas. How many miles per gallon did it get?

39. The weight of water is $62\frac{1}{2}$ lb per cubic foot. How many cubic feet would be occupied by 250 lb of water?

40. The weight of water is $62\frac{1}{2}$ lb per cubic foot. How many cubic feet would be occupied by 375 lb of water?

32. _____

33. _____

34. _____

35. _____

36. _____

37. _____

38. _____

39. _____

40. _____

41. _____

42. _____

43. _____

44. _____

45. _____

46. _____

47. _____

48. _____

49. _____

50. _____

51. _____

52. _____

53. _____

54. _____

55. _____

56. _____

57. _____

58. _____

41. Most space shuttles orbit the earth once every $1\frac{1}{2}$ hr. How many orbits are made every 24 hr?

42. Turkey contains $1\frac{1}{3}$ servings per pound. How many pounds are needed for 32 servings?

Find the area of the shaded region.

43.

$\frac{1}{2} \cdot s$

$s = 6\frac{7}{8}$ in.

$\frac{1}{2} \cdot s$

$6\frac{7}{8}$ in.

44.

$10\frac{1}{2}$ ft

$8\frac{1}{2}$ ft

4 ft

$10\frac{1}{2}$ ft

45. A rectangular lot has dimensions of $302\frac{1}{2}$ ft by $205\frac{1}{4}$ ft. A building with dimensions of 100 ft by $25\frac{1}{2}$ ft is built on the lot. How much area is left over?

46. Find the total area of 3 squares of carpeting, each of which is $5\frac{2}{3}$ yd on a side.

SKILL MAINTENANCE

Multiply.

47.
$$\begin{array}{r} 6\ 7\ 0\ 9 \\ \times \quad 2\ 1\ 3 \\ \hline \end{array}$$

48. Round to the nearest hundred: 45,765.

49. Solve: $\frac{5}{7} \cdot t = 420$.

50. Divide and simplify: $\frac{4}{5} \div \frac{6}{5}$.

51. Multiply and simplify: $\frac{3}{8} \cdot \frac{4}{9}$.

52. Round to the nearest ten: 45,765.

SYNTHESIS

Simplify.

53. $8 \div \frac{1}{2} + \frac{3}{4} + \left(5 - \frac{5}{8}\right)^2$

54. $\left(\frac{5}{9} - \frac{1}{4}\right) \times 12 + \left(4 - \frac{3}{4}\right)^2$

55. $\frac{1}{3} \div \left(\frac{1}{2} - \frac{1}{5}\right) \times \frac{1}{4} + \frac{1}{6}$

56. $\frac{7}{8} - 1\frac{1}{8} \times \frac{2}{3} + \frac{9}{10} \div \frac{3}{5}$

57. $4\frac{1}{2} \div 2\frac{1}{2} + 8 - 4 \div \frac{1}{2}$

58. $6 - 2\frac{1}{3} \times \frac{3}{4} + \frac{5}{8} \div \frac{2}{3}$

3.7 Order of Operations

a ORDER OF OPERATIONS: FRACTIONAL NOTATION AND MIXED NUMERALS

The rules for order of operations that we use with whole numbers apply when we are simplifying expressions involving fractional notation and mixed numerals. For review, these rules are listed below.

> **RULES FOR ORDER OF OPERATIONS**
>
> 1. Do all calculations within parentheses before operations outside.
> 2. Evaluate all exponential expressions.
> 3. Do all multiplications and divisions in order from left to right.
> 4. Do all additions and subtractions in order from left to right.

EXAMPLE 1 Simplify: $\frac{2}{3} \div \frac{1}{2} \cdot \frac{5}{8} + \frac{1}{6}$.

$$\frac{2}{3} \div \frac{1}{2} \cdot \frac{5}{8} + \frac{1}{6} = \frac{2}{3} \cdot \frac{2}{1} \cdot \frac{5}{8} + \frac{1}{6}$$
 Doing the division first by multiplying by the reciprocal of $\frac{1}{2}$

$$= \frac{4}{3} \cdot \frac{5}{8} + \frac{1}{6}$$

$$= \frac{4 \cdot 5}{3 \cdot 8} + \frac{1}{6}$$
 Doing the multiplication

$$= \frac{4 \cdot 5}{3 \cdot 4 \cdot 2} + \frac{1}{6}$$
 Factoring in order to simplify

$$= \frac{5}{3 \cdot 2} + \frac{1}{6}$$
 Removing a factor of 1: $\frac{4}{4} = 1$

$$= \frac{5}{6} + \frac{1}{6}$$

$$= \frac{6}{6}, \text{ or } 1$$
 Doing the addition

Do Exercises 1 and 2.

EXAMPLE 2 Simplify: $\frac{2}{3} \cdot 24 - 11\frac{1}{2}$.

$$\frac{2}{3} \cdot 24 - 11\frac{1}{2} = \frac{2 \cdot 24}{3} - 11\frac{1}{2}$$
 Doing the multiplication first

$$= \frac{2 \cdot 3 \cdot 8}{3} - 11\frac{1}{2}$$
 Factoring the numerator

$$= 2 \cdot 8 - 11\frac{1}{2}$$
 Removing a factor of 1: $\frac{3}{3} = 1$

$$= 16 - 11\frac{1}{2}$$
 Completing the multiplication

$$= 4\frac{1}{2}, \text{ or } \frac{9}{2}$$
 Doing the subtraction

Do Exercise 3.

OBJECTIVE

After finishing Section 3.7, you should be able to:

a Simplify expressions using the rules for order of operations.

FOR EXTRA HELP

TAPE 7   TAPE 6B   MAC: 3B
                                 IBM: 3B

Simplify.

1. $\frac{2}{5} \cdot \frac{5}{8} + \frac{1}{4}$

2. $\frac{1}{3} \cdot \frac{3}{4} \div \frac{5}{8} - \frac{1}{10}$

3. Simplify: $\frac{3}{4} \cdot 16 + 8\frac{2}{3}$.

Answers on page A-2

4. Find the average of

$$\frac{1}{2}, \quad \frac{1}{3}, \quad \text{and} \quad \frac{5}{6}.$$

5. Find the average of $\frac{3}{4}$ and $\frac{4}{5}$.

6. Simplify:

$$\left(\frac{2}{3} + \frac{3}{4}\right) \div 2\frac{1}{3} - \left(\frac{1}{2}\right)^3.$$

Answers on page A-2

EXAMPLE 3   To find the **average** of a set of numbers, we add the numbers and then divide by the number of addends. Find the average of $\frac{1}{2}$, $\frac{3}{4}$, and $\frac{7}{8}$.

The average is given by

$$\left(\frac{1}{2} + \frac{3}{4} + \frac{7}{8}\right) \div 3.$$

To find the average, we carry out the computation using the rules for order of operations:

$$\left(\frac{1}{2} + \frac{3}{4} + \frac{7}{8}\right) \div 3 = \left(\frac{4}{8} + \frac{6}{8} + \frac{7}{8}\right) \div 3 \quad \text{Doing the operations inside parentheses first: adding by finding a common denominator}$$

$$= \frac{17}{8} \div 3 \quad \text{Adding}$$

$$= \frac{17}{8} \cdot \frac{1}{3} \quad \text{Dividing by multiplying by the reciprocal}$$

$$= \frac{17}{24} \quad \text{Multiplying}$$

The average is $\frac{17}{24}$.

Do Exercises 4 and 5.

EXAMPLE 4   Simplify: $\left(\frac{7}{8} - \frac{1}{3}\right) \times 48 + \left(13 + \frac{4}{5}\right)^2$.

$$\left(\frac{7}{8} - \frac{1}{3}\right) \times 48 + \left(13 + \frac{4}{5}\right)^2$$

$$= \left(\frac{7}{8} \cdot \frac{3}{3} - \frac{1}{3} \cdot \frac{8}{8}\right) \times 48 + \left(13 \cdot \frac{5}{5} + \frac{4}{5}\right)^2 \quad \text{Carrying out operations inside parentheses first. To do so, we first multiply by 1 to obtain the LCD.}$$

$$= \left(\frac{21}{24} - \frac{8}{24}\right) \times 48 + \left(\frac{65}{5} + \frac{4}{5}\right)^2$$

$$= \frac{13}{24} \times 48 + \left(\frac{69}{5}\right)^2 \quad \text{Completing the operations within parentheses}$$

$$= \frac{13}{24} \times 48 + \frac{4761}{25} \quad \text{Evaluating exponential expressions next}$$

$$= 26 + \frac{4761}{25} \quad \text{Doing the multiplication}$$

$$= 26 + 190\frac{11}{25} \quad \text{Converting to a mixed numeral}$$

$$= 216\frac{11}{25}, \quad \text{or} \quad \frac{5411}{25} \quad \text{Adding}$$

Answers can be given using either fractional notation or mixed numerals as desired. Consult with your instructor.

Do Exercise 6.

Exercise Set 3.7

ANSWERS

1. _____

2. _____

3. _____

4. _____

5. _____

6. _____

7. _____

8. _____

9. _____

10. _____

11. _____

12. _____

13. _____

14. _____

15. _____

16. _____

17. _____

18. _____

19. _____

20. _____

[a] Simplify.

1. $\dfrac{1}{2} \cdot \dfrac{1}{3} \cdot \dfrac{1}{4}$

2. $\dfrac{1}{3} \cdot \dfrac{1}{4} \cdot \dfrac{1}{5}$

3. $6 \div 3 \div 5$

4. $12 \div 4 \div 8$

5. $\dfrac{2}{3} \div \dfrac{4}{3} \div \dfrac{7}{8}$

6. $\dfrac{5}{6} \div \dfrac{3}{4} \div \dfrac{2}{5}$

7. $\dfrac{5}{8} \div \dfrac{1}{4} - \dfrac{2}{3} \cdot \dfrac{4}{5}$

8. $\dfrac{4}{7} \cdot \dfrac{7}{15} + \dfrac{2}{3} \div 8$

9. $\dfrac{1}{2} \cdot \left(\dfrac{3}{4} - \dfrac{2}{3} \right)$

10. $\dfrac{3}{4} \div \left(\dfrac{8}{9} - \dfrac{2}{3} \right)$

11. $28\dfrac{1}{8} - 5\dfrac{1}{4} + 3\dfrac{1}{2}$

12. $10\dfrac{3}{5} - 4\dfrac{1}{10} - 1\dfrac{1}{2}$

13. $\dfrac{7}{8} \div \dfrac{1}{2} \cdot \dfrac{1}{4}$

14. $\dfrac{7}{10} \cdot \dfrac{4}{5} \div \dfrac{2}{3}$

15. $\left(\dfrac{2}{3} \right)^2 - \dfrac{1}{3} \cdot 1\dfrac{1}{4}$

16. $\left(\dfrac{3}{4} \right)^2 + 3\dfrac{1}{2} \div 1\dfrac{1}{4}$

17. $\dfrac{1}{2} - \left(\dfrac{1}{2} \right)^2 + \left(\dfrac{1}{2} \right)^3$

18. $1 + \dfrac{1}{4} + \left(\dfrac{1}{4} \right)^2 - \left(\dfrac{1}{4} \right)^3$

19. Find the average of $\dfrac{2}{3}$ and $\dfrac{7}{8}$.

20. Find the average of $\dfrac{1}{4}$ and $\dfrac{1}{5}$.

21. _____

22. _____

23. _____

24. _____

25. _____

26. _____

27. _____

28. _____

29. _____

30. _____

31. _____

32. _____

33. _____

34. _____

35. a) _____

b) _____

c) _____

21. Find the average of $\frac{1}{6}$, $\frac{1}{8}$, and $\frac{3}{4}$.

22. Find the average of $\frac{4}{5}$, $\frac{1}{2}$, and $\frac{1}{10}$.

23. Find the average of $3\frac{1}{2}$ and $9\frac{3}{8}$.

24. Find the average of $10\frac{2}{3}$ and $24\frac{5}{6}$.

Simplify.

25. $\left(\frac{2}{3} + \frac{3}{4}\right) \div \left(\frac{5}{6} - \frac{1}{3}\right)$

26. $\left(\frac{3}{5} - \frac{1}{2}\right) \div \left(\frac{3}{4} - \frac{3}{10}\right)$

27. $\left(\frac{1}{2} + \frac{1}{3}\right)^2 \cdot 144 - \frac{5}{8} \div 10\frac{1}{2}$

28. $\left(3\frac{1}{2} - 2\frac{1}{3}\right)^2 + 6 \cdot 2\frac{1}{2} \div 32$

SKILL MAINTENANCE

29. Multiply: $27 \cdot 126$.

30. Multiply: $132 \cdot 7865$.

31. Divide: $7865 \div 132$.

32. Multiply: $\frac{2}{3} \cdot 63$.

33. Divide and simplify: $\frac{4}{5} \div \frac{3}{10}$.

34. A 3-oz serving of crabmeat contains 85 mg (milligrams) of cholesterol. A 3-oz serving of shrimp contains 128 mg of cholesterol. How much more cholesterol is in the shrimp?

SYNTHESIS

35. a) Find an expression for the sum of the areas of the two rectangles.
  b) Simplify the expression.
  c) How is the computation in part (b) related to the rules for order of operations?

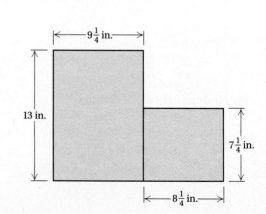

CALCULATOR CONNECTION

Many calculators have the capability of using fractional notation and mixed numerals for computations, giving answers in such notation as well. The following exercises assume that you are using such a calculator.

1. In the sum below, a and b are digits. Find a and b.

$$\frac{a}{17} + \frac{1b}{23} = \frac{35a}{391}$$

2. Consider only the numbers 2, 3, 4, and 5. Assume each is placed in a blank in the following.

$$\frac{\blacksquare}{\blacksquare} + \frac{\blacksquare}{\blacksquare} = ?$$

What placement of the numbers in the blanks yields the largest sum?

3. Consider only the numbers 3, 4, 5, and 6. Assume each can be placed in a blank in the following.

$$\blacksquare + \frac{\blacksquare}{\blacksquare} \cdot \blacksquare = ?$$

What placement of the numbers in the blanks yields the largest number?

4. Use a standard calculator. Arrange the following in order from largest to smallest.

$$\frac{3}{4}, \frac{17}{21}, \frac{13}{15}, \frac{7}{9}, \frac{15}{17}, \frac{13}{12}, \frac{19}{22}$$

EXTENDED SYNTHESIS EXERCISES

1. a) Simplify each of the following, using fractional notation for your answers.

$$\frac{1}{1 \cdot 2}$$

$$\frac{1}{1 \cdot 2} + \frac{1}{2 \cdot 3}$$

$$\frac{1}{1 \cdot 2} + \frac{1}{2 \cdot 3} + \frac{1}{3 \cdot 4}$$

$$\frac{1}{1 \cdot 2} + \frac{1}{2 \cdot 3} + \frac{1}{3 \cdot 4} + \frac{1}{4 \cdot 5}$$

b) Look for a pattern in your answers to part (a). Then find the following without carrying out the computations.

$$\frac{1}{1 \cdot 2} + \frac{1}{2 \cdot 3} + \frac{1}{3 \cdot 4} + \frac{1}{4 \cdot 5} + \frac{1}{5 \cdot 6}$$
$$+ \frac{1}{6 \cdot 7} + \frac{1}{7 \cdot 8} + \frac{1}{8 \cdot 9} + \frac{1}{9 \cdot 10}$$

2. Each of the following represents a portion of the total area of the square shown below.

$$\frac{1}{2}$$

$$\frac{1}{2} + \frac{1}{4}$$

$$\frac{1}{2} + \frac{1}{4} + \frac{1}{8}$$

$$\frac{1}{2} + \frac{1}{4} + \frac{1}{8} + \frac{1}{16}$$

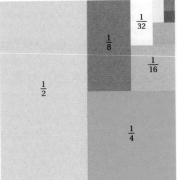

a) Find each of the areas by simplifying the sums. Use fractional notation for your answers.

b) Look for a pattern in your answers to part (a). Make a conjecture about the total area of the square.

3. Yuri and Olga are orangutans who perform in a circus by riding bicycles around a circular track. It takes Yuri 6 min to make one trip around the track and Olga 4 min. Suppose they start at the same point and then complete their act when they again reach the same point. How long is their act?

4. The students in a math class can be organized into study groups of 8 each such that no students are left out. The same class of students can also be organized into groups of 6 such that no students are left out.

a) Find some class sizes for which this will work.

b) Find the smallest such class size.

5. Find r if

$$\frac{1}{r} = \frac{1}{100} + \frac{1}{150} + \frac{1}{200}.$$

(continued)

6. *Estimation with fractions.* A fraction is very close to 0 when the numerator is very small compared to the denominator. For example, 0 is an estimate for $\frac{2}{17}$. A fraction is very close to $\frac{1}{2}$ when the numerator is about half the denominator. For example, $\frac{1}{2}$ is an estimate for $\frac{11}{23}$. A fraction is very close to 1 when the numerator is very close to the denominator. For example, 1 is an estimate for $\frac{37}{38}$ or for $\frac{43}{41}$. Estimate each of the following as either 0, $\frac{1}{2}$, or 1.

a) $\dfrac{2}{47}$ **b)** $\dfrac{4}{5}$ **c)** $\dfrac{1}{13}$ **d)** $\dfrac{7}{8}$

e) $\dfrac{6}{11}$ **f)** $\dfrac{10}{13}$ **g)** $\dfrac{7}{15}$ **h)** $\dfrac{1}{16}$

i) $\dfrac{7}{100}$ **j)** $\dfrac{5}{9}$ **k)** $\dfrac{19}{20}$ **l)** $\dfrac{5}{12}$

7. Find a number for the blank so that the fraction is close to but greater than $\frac{1}{2}$. Answers can vary.

a) $\dfrac{\blacksquare}{11}$ **b)** $\dfrac{\blacksquare}{8}$ **c)** $\dfrac{\blacksquare}{23}$ **d)** $\dfrac{\blacksquare}{35}$

e) $\dfrac{10}{\blacksquare}$ **f)** $\dfrac{7}{\blacksquare}$ **g)** $\dfrac{8}{\blacksquare}$ **h)** $\dfrac{51}{\blacksquare}$

8. Find a number for the blank so that the fraction is close to but less than 1. Answers can vary.

a) $\dfrac{7}{\blacksquare}$ **b)** $\dfrac{11}{\blacksquare}$ **c)** $\dfrac{13}{\blacksquare}$ **d)** $\dfrac{27}{\blacksquare}$

e) $\dfrac{\blacksquare}{15}$ **f)** $\dfrac{\blacksquare}{9}$ **g)** $\dfrac{\blacksquare}{18}$ **h)** $\dfrac{\blacksquare}{100}$

9. Estimate each of the following as a whole number or as a mixed numeral where the fractional part is $\frac{1}{2}$.

a) $2\dfrac{7}{8}$ **b)** $1\dfrac{1}{3}$

c) $12\dfrac{5}{6}$ **d)** $26\dfrac{6}{13}$

10. Estimate the sum.

a) $\dfrac{4}{5} + \dfrac{7}{8}$ **b)** $\dfrac{1}{12} + \dfrac{7}{15}$

c) $\dfrac{2}{3} + \dfrac{7}{13} + \dfrac{5}{9}$ **d)** $\dfrac{8}{9} + \dfrac{4}{5} + \dfrac{11}{12}$

e) $\dfrac{3}{100} + \dfrac{1}{10} + \dfrac{11}{1000}$ **f)** $\dfrac{23}{24} + \dfrac{37}{39} + \dfrac{51}{50}$

EXERCISES FOR THINKING AND WRITING

1. Explain why $2\dfrac{1}{4} \cdot 3\dfrac{2}{5} \neq 6\dfrac{2}{20}$.

2. Explain why $5 \cdot 3\dfrac{2}{5} \neq (5 \cdot 3) \cdot \left(5 \cdot \dfrac{2}{5} \right)$.

3. Discuss the role of least common multiples in adding and subtracting with fractional notation.

4. Find a real-world situation that fits the equation

$$2 \cdot 15\dfrac{3}{4} + 2 \cdot 28\dfrac{5}{8} = 88\dfrac{3}{4}.$$

Summary and Review Exercises: Chapter 3

The review objectives to be tested in addition to the material in this chapter are [1.5b], [1.8a], [2.6a], and [2.7b].

Find the LCM.

1. 12 and 18

2. 18 and 45

3. 3, 6, and 30

Add and simplify.

4. $\dfrac{6}{5} + \dfrac{3}{8}$

5. $\dfrac{5}{16} + \dfrac{1}{12}$

6. $\dfrac{6}{5} + \dfrac{11}{15}$

7. $\dfrac{5}{16} + \dfrac{3}{24}$

Subtract and simplify.

8. $\dfrac{5}{9} - \dfrac{2}{9}$

9. $\dfrac{7}{8} - \dfrac{3}{4}$

10. $\dfrac{11}{27} - \dfrac{2}{9}$

11. $\dfrac{5}{6} - \dfrac{2}{9}$

Use < or > for ▮ to write a true sentence.

12. $\dfrac{4}{7}$ ▮ $\dfrac{5}{9}$

13. $\dfrac{8}{9}$ ▮ $\dfrac{11}{13}$

Solve.

14. $x + \dfrac{2}{5} = \dfrac{7}{8}$

15. $\dfrac{1}{2} + y = \dfrac{9}{10}$

Convert to fractional notation.

16. $7\dfrac{1}{2}$

17. $8\dfrac{3}{8}$

18. $4\dfrac{1}{3}$

19. $10\dfrac{5}{7}$

Convert to a mixed numeral.

20. $\dfrac{7}{3}$

21. $\dfrac{27}{4}$

22. $\dfrac{63}{5}$

23. $\dfrac{7}{2}$

Divide. Write a mixed numeral for the answer.

24. $9\overline{)7896}$

25. $23\overline{)10,493}$

Add. Write a mixed numeral for the answer.

26. $5\dfrac{3}{5}$
$+ 4\dfrac{4}{5}$

27. $8\dfrac{1}{3}$
$+ 3\dfrac{2}{5}$

28. $5\dfrac{5}{6}$
$+ 4\dfrac{5}{6}$

29. $2\dfrac{3}{4}$
$+ 5\dfrac{1}{2}$

Subtract. Write a mixed numeral for the answer.

30. $\begin{array}{r} 12 \\ - \ 4\frac{2}{9} \\ \hline \end{array}$

31. $\begin{array}{r} 9\frac{3}{5} \\ - \ 4\frac{13}{15} \\ \hline \end{array}$

32. $\begin{array}{r} 10\frac{1}{4} \\ - \ 6\frac{1}{10} \\ \hline \end{array}$

33. $\begin{array}{r} 24 \\ - \ 10\frac{5}{8} \\ \hline \end{array}$

Multiply. Write a mixed numeral for the answer.

34. $6 \cdot 2\frac{2}{3}$

35. $5\frac{1}{4} \cdot \frac{2}{3}$

36. $2\frac{1}{5} \cdot 1\frac{1}{10}$

37. $2\frac{2}{5} \cdot 2\frac{1}{2}$

Divide. Write a mixed numeral for the answer.

38. $27 \div 2\frac{1}{4}$

39. $2\frac{2}{5} \div 1\frac{7}{10}$

40. $3\frac{1}{4} \div 26$

41. $4\frac{1}{5} \div 4\frac{2}{3}$

Solve.

42. A curtain requires $2\frac{3}{5}$ yd of material. How many curtains can be made from 39 yd of material?

43. On the first day of trading on the stock market, stock in Alcoa opened at $\$67\frac{3}{4}$ and rose by $\$2\frac{5}{8}$ at the close of trading. What was the stock's closing price?

44. A recipe calls for $1\frac{2}{3}$ cups of sugar. How much is needed for 18 recipes?

45. A wedding cake recipe requires 12 cups of shortening. Being calorie-conscious, the wedding couple decides to reduce the shortening by $3\frac{5}{8}$ cups. How many cups of shortening are used in their new recipe?

46. What is the sum of the areas in the figure?

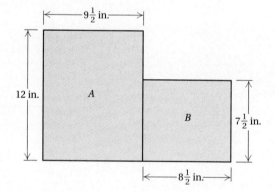

47. In the figure in Exercise 46 how much larger is the area of rectangle A than the area of rectangle B?

48. Simplify: $\frac{1}{8} \div \frac{1}{4} + \frac{1}{2}$.

49. Find the average of $\frac{1}{2}, \frac{1}{4}, \frac{1}{3}$, and $\frac{1}{5}$.

SKILL MAINTENANCE

50. Multiply and simplify: $\frac{9}{10} \cdot \frac{4}{3}$.

51. Divide and simplify: $\frac{5}{4} \div \frac{5}{6}$.

52. Multiply: $\begin{array}{r} 4\ 0\ 2\ 3 \\ \times \ \ \ 1\ 7\ 6 \\ \hline \end{array}$

53. A factory produces 85 digital alarm clocks per day. How long will it take to fill an order for 1445 clocks?

Test: Chapter 3

1. Find the LCM of 12 and 16.

Add and simplify.

2. $\dfrac{1}{2} + \dfrac{5}{2}$

3. $\dfrac{7}{8} + \dfrac{2}{3}$

4. $\dfrac{7}{10} + \dfrac{9}{100}$

Subtract and simplify.

5. $\dfrac{5}{6} - \dfrac{3}{6}$

6. $\dfrac{5}{6} - \dfrac{3}{4}$

7. $\dfrac{17}{24} - \dfrac{5}{8}$

8. Use < or > for ▉ to write a true sentence.

$$\dfrac{6}{7} \ \blacksquare \ \dfrac{21}{25}$$

9. Solve: $x + \dfrac{2}{3} = \dfrac{11}{12}$.

Convert to fractional notation.

10. $3\dfrac{1}{2}$

11. $9\dfrac{7}{8}$

Convert to a mixed numeral.

12. $\dfrac{9}{2}$

13. $\dfrac{74}{9}$

Divide. Write a mixed numeral for the answer.

14. $1\,1\,)\,\overline{1\,7\,8\,9}$

Add. Write a mixed numeral for the answer.

15. $6\dfrac{2}{5}$
$+\ 7\dfrac{4}{5}$

16. $9\dfrac{1}{4}$
$+\ 5\dfrac{1}{6}$

ANSWERS

1.
2.
3.
4.
5.
6.
7.
8.
9.
10.
11.
12.
13.
14.
15.
16.

Subtract. Write a mixed numeral for the answer.

17. $10\frac{1}{6}$

$-\ 5\frac{7}{8}$

18. 14

$-\ 7\frac{5}{6}$

Multiply. Write a mixed numeral for the answer.

19. $9 \cdot 4\frac{1}{3}$

20. $6\frac{3}{4} \cdot \frac{2}{3}$

21. $3\frac{1}{3} \cdot 1\frac{3}{4}$

Divide. Write a mixed numeral for the answer.

22. $33 \div 5\frac{1}{2}$

23. $2\frac{1}{3} \div 1\frac{1}{6}$

24. $2\frac{1}{12} \div 75$

Solve.

25. A low-cholesterol turkey loaf recipe calls for $3\frac{1}{2}$ cups of turkey breast. How much turkey is needed for 5 recipes?

26. An order of books for a math course weighs 220 lb. Each book weighs $2\frac{3}{4}$ lb. How many books are in the order?

27. The weights of two students are $183\frac{2}{3}$ lb and $176\frac{3}{4}$ lb. What is their total weight?

28. A standard piece of paper is $8\frac{1}{2}$ in. by 11 in. By how much does the length exceed the width?

29. Simplify: $\frac{2}{3} + 1\frac{1}{3} \cdot 2\frac{1}{8}$.

30. Find the average of $\frac{2}{5}$, $\frac{3}{4}$, and $\frac{1}{2}$.

SKILL MAINTENANCE

31. Multiply: $\begin{array}{r} 4\ 5\ 6\ 1 \\ \times\quad 7\ 6 \\ \hline \end{array}$

32. Divide and simplify: $\frac{4}{3} \div \frac{5}{6}$.

33. Multiply and simplify: $\frac{4}{3} \cdot \frac{5}{6}$.

34. A container has 8570 oz of beverage with which to fill 16-oz bottles. How many of these bottles can be filled? How much beverage will be left over?

Answers column:

17. _____
18. _____
19. _____
20. _____
21. _____
22. _____
23. _____
24. _____
25. _____
26. _____
27. _____
28. _____
29. _____
30. _____
31. _____
32. _____
33. _____
34. _____

Cumulative Review: Chapters 1–3

1. In the number 2753, what digit names tens?

2. Write expanded notation for 6075.

3. Write a word name for the number in the following sentence: The diameter of Uranus is 29,500 miles.

Add and simplify.

4.
```
  6 2 8
+ 2 7 1
```

5.
```
  3 7 0 4
+ 5 2 7 8
```

6. $\dfrac{3}{8} + \dfrac{1}{24}$

7.
$$2\frac{3}{4}$$
$$+ 5\frac{1}{2}$$

Subtract and simplify.

8.
```
  7 4 6 9
- 2 3 4 5
```

9.
```
  7 6 0 5
- 3 0 8 7
```

10. $\dfrac{3}{4} - \dfrac{1}{3}$

11.
$$2\frac{1}{3}$$
$$- 1\frac{1}{6}$$

Multiply and simplify.

12.
```
  2 7 8
×   1 8
```

13.
```
  8 9 4
× 3 2 8
```

14. $\dfrac{9}{10} \cdot \dfrac{5}{3}$

15. $18 \cdot \dfrac{5}{6}$

16. $2\dfrac{1}{3} \cdot 3\dfrac{1}{7}$

Divide. Write the answer with the remainder in the form 34 R 7.

17. $6 \overline{)\ 4\ 2\ 9\ 0}$

18. $4\ 5 \overline{)\ 2\ 5\ 3\ 1}$

19. In Exercise 18, write a mixed numeral for the answer.

Divide and simplify, where appropriate.

20. $\dfrac{2}{5} \div \dfrac{7}{10}$

21. $2\dfrac{1}{5} \div \dfrac{3}{10}$

22. Round 38,478 to the nearest hundred.

23. Find the LCM of 18 and 24.

24. Determine whether 3718 is divisible by 8.

25. Find all factors of 16.

26. What part is shaded?

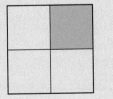

Use $<$, $>$, or $=$ for ▨ to write a true sentence.

27. $\dfrac{4}{5}$ ▨ $\dfrac{4}{6}$

28. $\dfrac{5}{12}$ ▨ $\dfrac{3}{7}$

Simplify.

29. $\dfrac{36}{45}$

30. $\dfrac{320}{10}$

31. Convert to fractional notation: $4\dfrac{5}{8}$.

32. Convert to a mixed numeral: $\dfrac{17}{3}$.

Solve.

33. $x + 24 = 117$

34. $x + \dfrac{7}{9} = \dfrac{4}{3}$

35. $\dfrac{7}{9} \cdot t = \dfrac{4}{3}$

36. $y = 32{,}580 \div 36$

Solve.

37. A jacket costs $87 and a coat costs $148. How much does it cost to buy both?

38. An emergency food pantry fund contains $423. From this fund, $148 and $167 are withdrawn for expenses. How much is left in the fund?

39. A lot measures 27 ft by 11 ft. What is its area?

40. How many people can get equal $16 shares from a total of $496?

41. A recipe calls for $\dfrac{4}{5}$ teaspoon of salt. How much salt should be used in $\dfrac{1}{2}$ recipe?

42. A book weighs $2\dfrac{3}{5}$ lb. How much do 15 books weigh?

43. How many pieces, each $2\dfrac{3}{8}$ ft long, can be cut from a piece of wire 38 ft long?

44. In a walkathon, one person walked $\dfrac{9}{10}$ mi and another walked $\dfrac{75}{100}$ mi. What was the total distance walked?

4

Addition and Subtraction: Decimal Notation

INTRODUCTION

In this chapter, we consider the meaning of decimal notation. Also discussed are order and rounding, addition and subtraction, equation solving, and problem solving involving decimal notation.

• • • • • • • • • • • • • • • • • •

AN APPLICATION

The number of passengers, in millions, who travel through the country's busiest airports annually are: O'Hare International, 59.9; Dallas/Ft. Worth, 48.5; Hartsfield, 48.0; Los Angeles International, 45.8; and San Francisco, 31.1.

THE MATHEMATICS

Let t = the total number of passengers, in millions. The problem then translates to this equation:

$$\underset{\uparrow}{59.9} + 48.5 + 48.0 + 45.8 + 31.1 = t.$$

This is decimal notation.

The review sections to be tested in addition to the material in this chapter are as follows.
[1.2b]   Add whole numbers.
[1.3d]   Subtract whole numbers.
[3.2a, b]   Add using fractional notation.
[3.3a]   Subtract using fractional notation.

Pretest: Chapter 4

1. Write a word name for 2.347.

2. Write a word name, as on a check, for $3264.78.

Write fractional notation.

3. 0.21

4. 5.408

Write decimal notation.

5. $\dfrac{379}{100}$

6. $\dfrac{539}{10,000}$

Which number is larger?

7. 3.2,   0.321

8. 0.099,   0.091

9. 0.54,   0.562

Round 21.0448 to the nearest:

10. Tenth.

11. Hundredth.

12. Thousandth.

Add.

13.
$$\begin{array}{r} 6\ 0\ 1.3 \\ 5.8\ 1 \\ +\quad 0.1\ 0\ 9 \\ \hline \end{array}$$

14.
$$\begin{array}{r} 0.8 \\ 0.0\ 6 \\ 0.0\ 0\ 7 \\ +\ 0.0\ 0\ 1\ 4 \\ \hline \end{array}$$

15. 102.4 + 10.24 + 1.024

16. 4.127 + 0.5 + 2.1167

Subtract.

17.
$$\begin{array}{r} 9\ 4.0\ 6\ 1 \\ -\quad 2.3\ 2\ 9 \\ \hline \end{array}$$

18.
$$\begin{array}{r} 4\ 0.0 \\ -\quad 0.9\ 0\ 9\ 9 \\ \hline \end{array}$$

19. 8 − 0.0049

20. 344.6788 − 91.6851

Solve.

21. $x + 2.33 = 5.6$

22. $54.906 + q = 6400.1177$

Solve.

23. A checking account contained $434.19. After a $148.24 check was drawn, how much was left in the account?

24. On a three-day trip, a traveler drove the following distances: 432.6 mi, 179.2 mi, and 469.8 mi. What is the total number of miles driven?

4.1 *Decimal Notation*

OBJECTIVES

After finishing Section 4.1, you should be able to:

a Given decimal notation, write a word name, and write a word name for an amount of money.

b Convert from decimal notation to fractional notation.

c Convert from fractional notation to decimal notation.

FOR EXTRA HELP

TAPE 7   TAPE 6B   MAC: 4
                                 IBM: 4

The set of **arithmetic numbers,** or **nonnegative rational numbers,** consists of the whole numbers

$$0, 1, 2, 3, 4, 5, 6, 7, 8, 9, 10, \text{ and so on,}$$

and fractions like

$$\frac{1}{2}, \frac{2}{3}, \frac{7}{8}, \frac{17}{10}, \text{ and so on.}$$

We studied the use of fractional notation for arithmetic numbers in Chapters 2 and 3. In Chapters 4 and 5, we will study the use of *decimal notation.* Although we are using different notation, we are still considering the same set of numbers. For example, instead of using fractional notation for $\frac{7}{8}$, we use decimal notation, 0.875.

a **DECIMAL NOTATION AND WORD NAMES**

Decimal notation for the women's shotput record is

$$74.249 \text{ ft.}$$

To understand what 74.249 means, we use a **place-value chart.** The value of each place is $\frac{1}{10}$ as large as the one to its left.

PLACE-VALUE CHART							
Hundreds	Tens	Ones	Ten*ths*	Hundred*ths*	Thousand*ths*	Ten-Thousand*ths*	Hundred-Thousand*ths*
100	10	1	$\frac{1}{10}$	$\frac{1}{100}$	$\frac{1}{1000}$	$\frac{1}{10,000}$	$\frac{1}{100,000}$
7	4 .	2	4	9			

The decimal notation 74.249 means

$$7 \text{ tens} + 4 \text{ ones} + 2 \text{ tenths} + 4 \text{ hundredths} + 9 \text{ thousandths,}$$

or

$$7 \cdot 10 + 4 \cdot 1 + 2 \cdot \frac{1}{10} + 4 \cdot \frac{1}{100} + 9 \cdot \frac{1}{1000},$$

or

$$70 + 4 + \frac{2}{10} + \frac{4}{100} + \frac{9}{1000}.$$

A mixed numeral for 74.249 is $74\frac{249}{1000}$. We read 74.249 as "seventy-four and two hundred forty-nine thousandths." When we come to the decimal point, we read "and." We can also read 74.249 as "seven four *point* two four nine."

Write a word name for the number.

1. Each person in this country consumes an average of 27.3 gallons of coffee per year.

2. The racehorse *Swale* won the Belmont Stakes in a time of 2.4533 minutes.

3. 245.89

4. 31,079.764

Write a word name as on a check.

5. $4217.56

6. $13.98

To write a word name from decimal notation,

a) write a word name for the whole number (the number named to the left of the decimal point),

397.685 $\longrightarrow$ Three hundred ninety-seven

b) write the word "and" for the decimal point, and

397.685 Three hundred ninety-seven and

c) write a word name for the number named to the right of the decimal point, followed by the place value of the last digit.

397.685 Three hundred ninety-seven and six hundred eighty-five *thousandths*

EXAMPLE 1 Write a word name for the number in this sentence: Each person consumes an average of 43.7 gallons of soft drinks per year.

Forty-three and seven tenths

EXAMPLE 2 Write a word name for 413.87.

Four hundred thirteen and eighty-seven hundredths

EXAMPLE 3 Write a word name for the number in this sentence: The world record in the men's marathon is 2.1833 hours.

Two and one thousand eight hundred thirty-three ten-thousandths.

EXAMPLE 4 Write a word name for 1788.405.

One thousand, seven hundred eighty-eight and four hundred five thousandths

Do Exercises 1–4.

Decimal notation is also used with money. It is common on a check to write "and ninety-five cents" as "and $\frac{95}{100}$ dollars."

EXAMPLE 5 Write a word name for the amount on the check, $5876.95.

Five thousand, eight hundred seventy-six and $\frac{95}{100}$ dollars

Do Exercises 5 and 6.

b CONVERTING FROM DECIMAL NOTATION TO FRACTIONAL NOTATION

Write fractional notation.

7. 0.896

We can find fractional notation as follows:

$$9.875 = 9 + \frac{8}{10} + \frac{7}{100} + \frac{5}{1000}$$

$$= 9 \cdot \frac{1000}{1000} + \frac{8}{10} \cdot \frac{100}{100} + \frac{7}{100} \cdot \frac{10}{10} + \frac{5}{1000}$$

$$= \frac{9000}{1000} + \frac{800}{1000} + \frac{70}{1000} + \frac{5}{1000} = \frac{9875}{1000}.$$

Note the following:

$$9.875 \qquad \frac{9875}{1000}$$

3 decimal places   3 zeros

8. 23.78

To convert from decimal to fractional notation,

a) count the number of decimal places,   4.98

                                                  2 places

b) move the decimal point that many   4.98.   Move
   places to the right, and   2 places.

c) write the answer over a denominator $\frac{498}{100}$
   with a 1 followed by that number of zeros.   2 zeros

9. 5.6789

EXAMPLE 6 Write fractional notation for 0.876. Do not simplify.

$$0.876 \qquad 0.876. \qquad 0.876 = \frac{876}{1000}$$

3 places

For a number like 0.876, we generally write a 0 before the decimal to avoid forgetting or omitting the decimal point.

EXAMPLE 7 Write fractional notation for 56.23. Do not simplify.

$$56.23 \qquad 56.23. \qquad 56.23 = \frac{5623}{100}$$

2 places

10. 1.9

EXAMPLE 8 Write fractional notation for 1.5018. Do not simplify.

$$1.5018 \qquad 1.5018. \qquad 1.5018 = \frac{15,018}{10,000}$$

4 places

Do Exercises 7–10.

Answers on page A-2

Write decimal notation.

11. $\dfrac{743}{100}$

12. $\dfrac{406}{1000}$

13. $\dfrac{67,089}{10,000}$

14. $\dfrac{9}{10}$

15. $\dfrac{57}{1000}$

16. $\dfrac{830}{10,000}$

Answers on page A-2

c   CONVERTING FROM FRACTIONAL NOTATION TO DECIMAL NOTATION

If fractional notation has a denominator that is a power of ten, such as 10, 100, 1000, and so on, we reverse the procedure we used before.

To convert from fractional notation to decimal notation when the denominator is 10, 100, 1000, and so on,

a) count the number of zeros, and
$$\dfrac{8679}{1000}$$
3 zeros

b) move the decimal point that number of places to the left. Leave off the denominator.
$$8.679.$$
Move 3 places.
$$\dfrac{8679}{1000} = 8.679$$

EXAMPLE 9 Write decimal notation for $\dfrac{47}{10}$.

$$\dfrac{47}{10}$$
1 zero
$$4.7.$$
$$\dfrac{47}{10} = 4.7$$

EXAMPLE 10 Write decimal notation for $\dfrac{123,067}{10,000}$.

$$\dfrac{123,067}{10,000}$$
4 zeros
$$12.3067.$$
$$\dfrac{123,067}{10,000} = 12.3067$$

EXAMPLE 11 Write decimal notation for $\dfrac{13}{1000}$.

$$\dfrac{13}{1000}$$
3 zeros
$$0.013.$$
$$\dfrac{13}{1000} = 0.013$$

EXAMPLE 12 Write decimal notation for $\dfrac{570}{100,000}$.

$$\dfrac{570}{100,000}$$
5 zeros
$$0.00570.$$
$$\dfrac{570}{100,000} = 0.0057$$

Do Exercises 11–16.

When denominators are numbers other than 10, 100, and so on, we will use another method for conversion. It will be considered in Section 5.3.

Exercise Set 4.1

ANSWERS

1. _____
2. _____
3. _____
4. _____
5. _____
6. _____
7. _____
8. _____
9. _____
10. _____
11. _____
12. _____
13. _____
14. _____
15. _____
16. _____
17. _____
18. _____
19. _____
20. _____
21. _____
22. _____
23. _____
24. _____
25. _____
26. _____
27. _____
28. _____
29. _____
30. _____

a Write a word name for the number in the sentence.

1. The average age of a bride is 23.9 years.

2. The world record in the women's marathon was 2.351 hours.

3. Recently, one dollar was worth 117.65 Japanese yen.

4. Each day, the average person spends $7.89 on health care.

Write a word name.

5. 34.891

6. 27.1245

Write a word name as on a check

7. $326.48

8. $125.99

9. $36.72

10. $0.67

b Write fractional notation.

11. 6.8

12. 8.3

13. 0.17

14. 0.71

15. 8.21

16. 3.56

17. 204.6

18. 203.6

19. 1.509

20. 1.732

21. 46.03

22. 53.81

23. 0.00013

24. 0.0109

25. 20.003

26. 1000.3

27. 1.0008

28. 2.0114

29. 0.5321

30. 4567.2

31.

32.

33.

34.

35.

36.

37.

38.

39.

40.

41.

42.

43.

44.

45.

46.

47.

48.

49.

50.

51.

52.

53.

54.

55.

56.

57.

58.

59.

60.

61.

| c | Write decimal notation.

31. $\dfrac{8}{10}$ **32.** $\dfrac{1}{10}$ **33.** $\dfrac{92}{100}$ **34.** $\dfrac{8}{100}$

35. $\dfrac{93}{10}$ **36.** $\dfrac{51}{10}$ **37.** $\dfrac{889}{100}$ **38.** $\dfrac{776}{100}$

39. $\dfrac{2508}{10}$ **40.** $\dfrac{7803}{10}$ **41.** $\dfrac{3798}{1000}$ **42.** $\dfrac{780}{1000}$

43. $\dfrac{78}{10,000}$ **44.** $\dfrac{9040}{10,000}$ **45.** $\dfrac{56,788}{100,000}$ **46.** $\dfrac{19}{100,000}$

47. $\dfrac{2173}{100}$ **48.** $\dfrac{6743}{100}$ **49.** $\dfrac{66}{100}$ **50.** $\dfrac{178}{100}$

51. $\dfrac{853}{1000}$ **52.** $\dfrac{9563}{1000}$ **53.** $\dfrac{376,193}{1,000,000}$ **54.** $\dfrac{8,953,073}{1,000,000}$

SKILL MAINTENANCE

Round 6172 to the nearest:

55. Ten.   **56.** Hundred.   **57.** Thousand.

Add.

58. 12,342 + 8978   **59.** 138 + 67 + 95 + 79

SYNTHESIS

Write decimal notation.

60. $99\dfrac{44}{100}$ **61.** $4\dfrac{909}{1000}$

4.2 Order and Rounding

a ORDER

To understand how to compare numbers in decimal notation, consider 0.85 and 0.9. First note that $0.9 = 0.90$ because $\frac{9}{10} = \frac{90}{100}$. Then $0.85 = \frac{85}{100}$ and $0.90 = \frac{90}{100}$. Since $\frac{85}{100} < \frac{90}{100}$, it follows that $0.85 < 0.90$. This leads us to a quick way to compare two numbers in decimal notation.

> To compare two numbers in decimal notation, start at the left and compare corresponding digits moving from left to right. If two digits differ, the number with the larger digit is the larger of the two numbers. To ease the comparison, extra zeros can be written to the right of the decimal point, if necessary, so the number of decimal places is the same.

EXAMPLE 1 Which of 2.109 and 2.1 is larger?

2.109 | 2.109 | 2.109 | 2.109
The same | The same | The same | Different; 9 is larger than 0.
2.1 | 2.1 | 2.10 | 2.100

Thus, 2.109 is larger.

EXAMPLE 2 Which of 0.09 and 0.108 is larger?

0.09 | 0.09 | Different; 1 is larger than 0.
The same
0.108 | 0.108

Thus, 0.108 is larger.

Do Exercises 1–6.

b ROUNDING

Rounding is done as for whole numbers. To understand, we first consider an example using a number line.

EXAMPLE 3 Round 0.37 to the nearest tenth.

Here is part of a number line.

0 0.1 0.2 0.3 0.4 0.5 0.6 0.7 0.8 0.9 1

0.30 0.31 0.32 0.33 0.34 0.35 0.36 0.37 0.38 0.39 0.40

We see that 0.37 is closer to 0.40 than to 0.30. Thus, 0.37 rounded to the nearest tenth is 0.4.

OBJECTIVES

After finishing Section 4.2, you should be able to:

a Given a pair of numbers in decimal notation, tell which is larger.

b Round to the nearest thousandth, hundredth, tenth, one, ten, hundred, or thousand.

FOR EXTRA HELP

TAPE 8   TAPE 7A   MAC: 4
                       IBM: 4

Which number is larger?

1. 2.04, 2.039

2. 0.06, 0.008

3. 0.5, 0.58

4. 1, 0.9999

5. 0.8989, 0.09898

6. 21.006, 21.05

Answers on page A-3

Round to the nearest tenth.

7. 2.76   **8.** 13.85

9. 234.448   **10.** 7.009

Round to the nearest hundredth.

11. 0.636   **12.** 7.834

13. 34.675   **14.** 0.025

Round to the nearest thousandth.

15. 0.9434   **16.** 8.0038

17. 43.1119   **18.** 37.4005

Round 7459.3548 to the nearest:

19. Thousandth.

20. Hundredth.

21. Tenth.

22. One.

23. Ten. (*Caution:* "Tens" are not "tenths.")

24. Hundred.

25. Thousand.

Answers on page A-3

To round to a certain place:

a) Locate the digit in that place.

b) Consider the next digit to the right.

c) If the digit to the right is 5 or higher, round up; if the digit to the right is 4 or lower, round down.

EXAMPLE 4   Round 3872.2459 to the nearest tenth.

a) Locate the digit in the tenths place.

3 8 7 2.2 4 5 9
       ↑

b) Consider the next digit to the right.

3 8 7 2.2 4 5 9
         ↑

Caution! 3872.3 is not a correct answer to Example 4. It is incorrect to round from the ten-thousandths digit over to the tenths digit, as follows:

3872.246, 3872.25, 3872.3.

c) Since that digit, 4, is less than 5, round down.

3 8 7 2.2 ← This is the answer.

EXAMPLE 5   Round 3872.2459 to the nearest thousandth, hundredth, tenth, one, ten, hundred, and thousand.

Thousandth:	3872.246	Ten:	3870
Hundredth:	3872.25	Hundred:	3900
Tenth:	3872.2	Thousand:	4000
One:	3872		

EXAMPLE 6   Round 14.8973 to the nearest hundredth.

a) Locate the digit in the hundredths place.   1 4.8 9 7 3
                                                              ↑

b) Consider the next digit to the right.   1 4.8 9 7 3
                                                                ↑

c) Since that digit, 7, is 5 or higher, round up. When we make the hundredths digit a 10, we carry 1 to the tenths place.

The answer is

14.90, or 14.9 .

EXAMPLE 7   Round 0.008 to the nearest tenth.

a) Locate the digit in the tenths place.   0.0 0 8
                                                     ↑

b) Consider the next digit to the right.   0.0 0 8
                                                         ↑

c) Since that digit, 0, is less than 5, round down.

The answer is 0.0, or 0.

Do Exercises 7–25.

Exercise Set 4.2

a Which number is larger?

1. 0.06, 0.58

2. 0.008, 0.8

3. 0.905, 0.91

4. 42.06, 42.1

5. 0.0009, 0.001

6. 7.067, 7.054

7. 234.07, 235.07

8. 0.99999, 1.0

9. 0.4545, 0.05454

10. 0.6, 0.05

11. 0.004, $\dfrac{4}{100}$

12. $\dfrac{43}{10}$, 0.43

13. 0.54, 0.78

14. 0.432, 0.4325

15. 0.8437, 0.84384

16. 0.872, 0.873

17. 0.19, 1.9

18. 0.22, 0.2367

b Round to the nearest tenth.

19. 0.11

20. 0.85

21. 0.16

22. 0.49

23. 0.5794

24. 0.77

25. 2.7449

26. 4.78

27. 13.41

28. 41.23

29. 123.65

30. 36.049

Round to the nearest hundredth.

31. 0.893

32. 0.675

33. 0.666

34. 0.6666

35. 0.4246

36. 6.529

37. 1.435

38. 0.406

ANSWERS

1.
2.
3.
4.
5.
6.
7.
8.
9.
10.
11.
12.
13.
14.
15.
16.
17.
18.
19.
20.
21.
22.
23.
24.
25.
26.
27.
28.
29.
30.
31.
32.
33.
34.
35.
36.
37.
38.

ANSWERS

39. _____

40. _____

41. _____

42. _____

43. _____

44. _____

45. _____

46. _____

47. _____

48. _____

49. _____

50. _____

51. _____

52. _____

53. _____

54. _____

55. _____

56. _____

57. _____

58. _____

59. _____

60. _____

61. _____

62. _____

63. _____

64. _____

65. _____

66. _____

67. _____

68. _____

69. _____

70. _____

71. _____

72. _____

73. _____

74. _____

75. _____

76. _____

77. _____

78. _____

39. 3.581   **40.** 283.1379   **41.** 0.007   **42.** 4.889

43. 37.698   **44.** 207.9976   **45.** 0.995   **46.** 0.094

Round to the nearest thousandth.

47. 0.3246   **48.** 0.4278   **49.** 0.6666   **50.** 7.4294

51. 17.0015   **52.** 2.6776   **53.** 0.0009   **54.** 123.4562

55. 10.1011   **56.** 67.1006   **57.** 0.1161   **58.** 9.9989

Round 809.4732 to the nearest:

59. Hundred.   **60.** Tenth.   **61.** Thousandth.

62. Hundredth.   **63.** One.   **64.** Ten.

Round 34.54389 to the nearest:

65. Ten thousandth.   **66.** Thousandth.   **67.** Hundredth.

68. Tenth.   **69.** One.   **70.** Ten.

SKILL MAINTENANCE

Add.   Subtract.

71. $\begin{array}{r} 6\ 8\ 1 \\ +\ 1\ 4\ 9 \end{array}$ **72.** $\dfrac{681}{1000} + \dfrac{149}{1000}$ **73.** $\begin{array}{r} 2\ 6\ 7 \\ -\ \ \ 8\ 5 \end{array}$ **74.** $\dfrac{267}{100} - \dfrac{85}{100}$

SYNTHESIS

There are other methods of rounding decimal notation. A computer often uses a method called **truncating**. To round using truncating, drop off all decimal places past the rounding place, which is the same as changing all digits to the right to zeros. For example, rounding 6.78093456285102 to the ninth decimal place, using truncating, gives us 6.780934562. Use truncating to round each of the following to the fifth decimal place, that is, the nearest hundred thousandth.

75. 6.78346123   **76.** 6.783461902

77. 99.999999999   **78.** 0.030303030303

4.3 Addition and Subtraction with Decimals

a ADDITION

Adding with decimal notation is similar to adding whole numbers. First we line up the decimal points. Then we add digits from the right. For example, we add the thousandths, then the hundredths, and so on, carrying if necessary. If desired, we can write extra zeros to the right of the decimal point so that the number of places is the same.

OBJECTIVES

After finishing Section 4.3, you should be able to:

a Add using decimal notation.

b Subtract using decimal notation.

c Solve equations of the type $x + a = b$ and $a + x = b$, where a and b may be in decimal notation.

FOR EXTRA HELP

TAPE 8   TAPE 7A   MAC: 4
                            IBM: 4

EXAMPLE 1 Add: 56.314 + 17.78.

$$\begin{array}{r} 5\ 6\ .\ 3\ 1\ 4 \\ +\ 1\ 7\ .\ 7\ 8\ 0 \\ \hline \end{array}$$ Lining up the decimal points in order to add

Writing an extra zero to the right of the decimal point

$$\begin{array}{r} 5\ 6\ .\ 3\ 1\ 4 \\ +\ 1\ 7\ .\ 7\ 8\ 0 \\ \hline 4 \end{array}$$ Adding thousandths

$$\begin{array}{r} 5\ 6\ .\ 3\ 1\ 4 \\ +\ 1\ 7\ .\ 7\ 8\ 0 \\ \hline 9\ 4 \end{array}$$ Adding hundredths

$$\begin{array}{r} \overset{1}{5}\ 6\ .\ 3\ 1\ 4 \\ +\ 1\ 7\ .\ 7\ 8\ 0 \\ \hline .\ 0\ 9\ 4 \end{array}$$ Adding tenths

Write a decimal point in the answer.

We get 10 tenths = 1 one + 0 tenths, so we carry the 1 to the ones column.

$$\begin{array}{r} \overset{1}{5}\ \overset{1}{6}\ .\ 3\ 1\ 4 \\ +\ 1\ 7\ .\ 7\ 8\ 0 \\ \hline 4\ .\ 0\ 9\ 4 \end{array}$$ Adding ones

We get 14 ones = 1 ten + 4 ones, so we carry the 1 to the tens column.

$$\begin{array}{r} \overset{1}{5}\ \overset{1}{6}\ .\ 3\ 1\ 4 \\ +\ 1\ 7\ .\ 7\ 8\ 0 \\ \hline 7\ 4\ .\ 0\ 9\ 4 \end{array}$$ Adding tens

Do Exercises 1 and 2.

Remember, we can write extra zeros to the right of the decimal point to get the same number of decimal places.

EXAMPLE 2 Add: 3.42 + 0.237 + 14.1.

$$\begin{array}{r} 3.4\ 2\ 0 \\ 0.2\ 3\ 7 \\ +\ 1\ 4.1\ 0\ 0 \\ \hline 1\ 7.7\ 5\ 7 \end{array}$$ Lining up the decimal points and writing extra zeros

Adding

Do Exercises 3–5.

Add.

1.
$$\begin{array}{r} 0.8\ 4\ 7 \\ +\ 1\ 0.0\ 7 \\ \hline \end{array}$$

2.
$$\begin{array}{r} 2.1 \\ 0.7\ 3\ 9 \\ +\ 3\ 1.3\ 6\ 8\ 9 \\ \hline \end{array}$$

Add.

3. 0.02 + 4.3 + 0.649

4. 0.12 + 3.006 + 0.4357

5. 0.4591 + 0.2374 + 8.70894

Answers on page A-3

6. 789 + 123.67

Consider the addition 3456 + 19.347. Keep in mind that a whole number, such as 3456, has an "unwritten" decimal point at the right, with 0 fractional parts. When adding, we can always write in that decimal point and extra zeros if desired.

EXAMPLE 3 Add: 3456 + 19.347.

$$
\begin{array}{r}
3\,4\,5\,6.0\,0\,0 \\
+1\,9.3\,4\,7 \\
\hline
3\,4\,7\,5.3\,4\,7
\end{array}
$$

Writing in the decimal point and extra zeros
Lining up the decimal points
Adding

Do Exercises 6 and 7.

7. 45.78 + 2467 + 1.993

b SUBTRACTION

Subtracting with decimal notation is similar to subtracting whole numbers. First we line up the decimal points. Then we subtract digits from the right. For example, we subtract the thousandths, then the hundredths, the tenths, and so on, borrowing if necessary.

EXAMPLE 4 Subtract: 56.314 − 17.78.

$$
\begin{array}{r}
5\,6.3\,1\,4 \\
-1\,7.7\,8\,0 \\
\hline
\end{array}
$$

Lining up the decimal points in order to subtract
Writing an extra 0

Subtract.

8. 37.428 − 26.674

$$
\begin{array}{r}
5\,6.3\,1\,4 \\
-1\,7.7\,8\,0 \\
\hline
4
\end{array}
$$

Subtracting thousandths

$$
\begin{array}{r}
2\;11 \\
5\,6.\cancel{3}\,\cancel{1}\,4 \\
-1\,7.7\,8\,0 \\
\hline
3\,4
\end{array}
$$

Borrowing tenths to subtract hundredths

9.
$$
\begin{array}{r}
0.3\,4\,7 \\
-0.0\,0\,8 \\
\hline
\end{array}
$$

$$
\begin{array}{r}
12 \\
5\;\cancel{2}\;11 \\
5\,\cancel{6}.\cancel{3}\,\cancel{1}\,4 \\
-1\,7.7\,8\,0 \\
\hline
.5\,3\,4
\end{array}
$$

Borrowing ones to subtract tenths

Writing a decimal point

$$
\begin{array}{r}
15\;12 \\
4\;\cancel{5}\;\cancel{2}\;11 \\
\cancel{5}\,\cancel{6}.\cancel{3}\,\cancel{1}\,4 \\
-1\,7.7\,8\,0 \\
\hline
8.5\,3\,4
\end{array}
$$

Borrowing tens to subtract ones

$$
\begin{array}{r}
15\;12 \\
4\;\cancel{5}\;\cancel{2}\;11 \\
\cancel{5}\,\cancel{6}.\cancel{3}\,\cancel{1}\,4 \\
-1\,7.7\,8\,0 \\
\hline
3\,8.5\,3\,4
\end{array}
$$

Subtracting tens

Do Exercises 8 and 9.

Answers on page A-3

EXAMPLE 5 Subtract: 13.07 − 9.205.

$$
\begin{array}{r}
\overset{12}{}\,\overset{2\ 10\ 6\ 10}{\cancel{1}\,\cancel{3}.\cancel{0}\,\cancel{7}\,\cancel{0}} \\
-\ \ 9.2\ 0\ 5 \\
\hline
3.8\ 6\ 5
\end{array}
$$ Writing an extra zero

Subtracting

EXAMPLE 6 Subtract: 23.08 − 5.0053.

$$
\begin{array}{r}
\overset{1\ 13\ \ \ 7\ 9\ 10}{\cancel{2}\,\cancel{3}.0\,\cancel{8}\,\cancel{0}\,\cancel{0}} \\
-\ \ 5.0\ 0\ 5\ 3 \\
\hline
1\ 8.0\ 7\ 4\ 7
\end{array}
$$ Writing two extra zeros

Subtracting

Do Exercises 10–12.

When subtraction involves a whole number, again keep in mind that there is an "unwritten" decimal point that can be written in if desired. Extra zeros can also be written in to the right of the decimal point.

EXAMPLE 7 Subtract: 456 − 2.467.

$$
\begin{array}{r}
\overset{5\ 9\ 9\ 10}{4\ 5\ \cancel{6}.\cancel{0}\,\cancel{0}\,\cancel{0}} \\
-\ \ \ \ 2.4\ 6\ 7 \\
\hline
4\ 5\ 3.5\ 3\ 3
\end{array}
$$ Writing in the decimal point and extra zeros

Subtracting

Do Exercises 13 and 14.

c   SOLVING EQUATIONS

Now let us solve equations $x + a = b$ and $a + x = b$, where a and b may be in decimal notation. Proceeding as we have before, we subtract a on both sides.

EXAMPLE 8 Solve: $x + 28.89 = 74.567$.

We have

$$x + 28.89 - 28.89 = 74.567 - 28.89 \qquad \text{Subtracting 28.89 on both sides}$$
$$x = 45.677.$$

$$
\begin{array}{r}
\overset{6\ 13\ 14\ 16}{\cancel{7}\,\cancel{4}.\cancel{5}\,\cancel{6}\,7} \\
-\ 2\ 8.8\ 9\ 0 \\
\hline
4\ 5.6\ 7\ 7
\end{array}
$$

The solution is 45.677.

Subtract.
10. $1.2345 - 0.7$

11. $0.9564 - 0.4392$

12. $7.37 - 0.00008$

Subtract.
13. $1277 - 82.78$

14. $5 - 0.0089$

Answers on page A-3

Solve.

15. $x + 17.78 = 56.314$

16. $8.906 + t = 23.07$

17. Solve: $241 + y = 2374.5$.

Answers on page A-3

EXAMPLE 9 Solve: $0.8879 + y = 9.0026$.

We have

$$0.8879 + y - 0.8879 = 9.0026 - 0.8879 \qquad \text{Subtracting 0.8879 on both sides}$$
$$y = 8.1147.$$

$$
\begin{array}{r}
{\scriptstyle 8\ \ 9\ \ 9\ \ 11\ \ 16} \\
9.0\,0\,2\,6 \\
-\ 0.8\,8\,7\,9 \\
\hline
8.1\,1\,4\,7
\end{array}
$$

The solution is 8.1147.

Do Exercises 15 and 16.

EXAMPLE 10 Solve: $120 + x = 4380.6$.

We have

$$120 + x - 120 = 4380.6 - 120 \qquad \text{Subtracting 120 on both sides}$$
$$x = 4260.6$$

$$
\begin{array}{r}
4\ 3\ 8\ 0.6 \\
1\ 2\ 0.0 \\
\hline
4\ 2\ 6\ 0.6
\end{array}
$$

The solution is 4260.6.

Do Exercise 17.

S I D E L I G H T S

A NUMBER PUZZLE

Place the numbers 3, 4, 5, 6, 7, 8, 9 in the boxes so that the sum along any line will be 18.

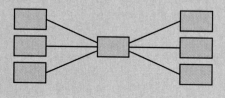

Exercise Set 4.3

a Add.

1.
```
  3 1 6.2 5
+   1 8.1 2
```

2.
```
  6 4 1.8 0 3
+   1 4.9 3 5
```

3.
```
  6 5 9.4 0 3
+ 9 1 6.8 1 2
```

4.
```
  4 2 0 3.2 8
+       3.3 9
```

5.
```
        9.1 0 4
+ 1 2 3.4 5 6
```

6.
```
  6.1 5 2 8
+ 5.2 7 7 7
```

7.
```
  8 1.0 0 8
+   3.4 0 9
```

8. 0.8096 + 0.7856

9. 20.0124 + 30.0124

10. 0.687 + 0.9

11. 39 + 1.007

12. 0.845 + 10.02

13. 0.34 + 3.5 + 0.127 + 768

14. 2.3 + 0.729 + 23

15. 17 + 3.24 + 0.256 + 0.3689

16.
```
      4 7.8
  2 1 9.8 5 2
      4 3.5 9
+ 6 6 6.7 1 3
```

17.
```
        2.7 0 3
      7 8.3 3
    2 8.0 0 0 9
+ 1 1 8.4 3 4 1
```

18.
```
      1 3.7 2
        9.1 1 2
  6 5 4 2.7 9 0 8
+     2 3.9 0 1
```

19. 99.6001 + 7285.18 + 500.042 + 870

20. 65.987 + 9.4703 + 6744.02 + 1.0003 + 200.895

ANSWERS

1. _____

2. _____

3. _____

4. _____

5. _____

6. _____

7. _____

8. _____

9. _____

10. _____

11. _____

12. _____

13. _____

14. _____

15. _____

16. _____

17. _____

18. _____

19. _____

20. _____

$\boxed{\text{c}}$ Solve.

61. $x + 17.5 = 29.15$

62. $t + 50.7 = 54.07$

63. $3.205 + m = 22.456$

64. $4.26 + q = 58.32$

65. $17.95 + p = 402.63$

66. $w + 1.3004 = 47.8$

67. $13{,}083.3 = x + 12{,}500.33$

68. $100.23 = 67.8 + z$

69. $x + 2349 = 17{,}684.3$

70. $1830.4 + t = 23{,}067$

SKILL MAINTENANCE

71. Round 34,567 to the nearest thousand.

72. Round 34,496 to the nearest thousand.

Subtract.

73. $\dfrac{13}{24} - \dfrac{3}{8}$

74. $\dfrac{8}{9} - \dfrac{2}{15}$

75. $8805 - 2639$

76. $8005 - 2639$

SYNTHESIS

77. A student presses the wrong button when using a calculator and adds 235.7 instead of subtracting it. The incorrect answer is 817.2. What is the correct answer?

4.4 Solving Problems

OBJECTIVE

After finishing Section 4.4, you should be able to:

a Solve problems involving addition and subtraction with decimals.

FOR EXTRA HELP

TAPE 8   TAPE 7B   MAC: 4
                                 IBM: 4

a Solving problems with decimals is like solving problems with whole numbers. We translate first to an equation that corresponds to the situation. Then we solve the equation.

EXAMPLE 1 The following bar graph compares the costs of a one-day adult ticket at various Disney recreational parks throughout the world. How much more does it cost for a ticket to Euro Disneyland than for one to Tokyo Disneyland?

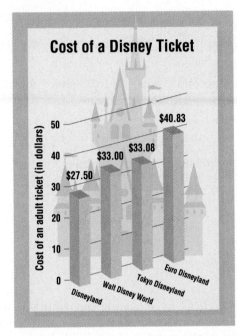

Cost of a Disney Ticket

1. *Familiarize.* We use the graph to help us visualize the situation. We let c = the extra cost at Euro Disneyland over Tokyo Disneyland.

2. *Translate.* This is a "how-much-more" situation. We translate as follows, using the data from the bar graph.

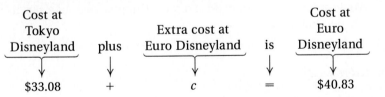

Cost at Tokyo Disneyland	plus	Extra cost at Euro Disneyland	is	Cost at Euro Disneyland
$33.08	+	c	=	$40.83

3. *Solve.* We solve the equation, first subtracting 33.08 from both sides:

$$33.08 + c - 33.08 = 40.83 - 33.08$$
$$c = 7.75 \, .$$

$$
\begin{array}{r}
{\scriptstyle 3 \;\; 10 \;\; 7 \;\; 13} \\
\cancel{4}\,\cancel{0}.\cancel{8}\,\cancel{3} \\
-\; 3\,3.0\,8 \\
\hline
7.7\,5
\end{array}
$$

4. *Check.* We can check by adding 7.75 to 33.08, to get 40.83. This checks.

5. *State.* It costs $7.75 more for a one-day ticket to Euro Disneyland than for one to Tokyo Disneyland.

Do Exercise 1.

1. Normal body temperature is 98.6°F. When fevered, most people will die if their bodies reach 107°F. This is a rise of how many degrees?

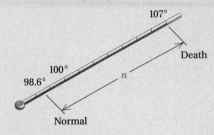

Answer on page A-3

2. Each year, each of us drinks 43.7 gal of soft drinks, 37.3 gal of water, 27.3 gal of coffee, 21.1 gal of milk, and 8.1 gal of fruit juice. What is the total amount that each of us drinks?

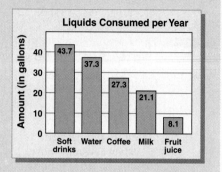

EXAMPLE 2   A patient was given injections of 3.68 mg, 2.7 mg, 3.65 mg, and 5.0 mg over a 24-hr period. What was the total amount of the injections?

1. Familiarize.   We draw a picture or at least visualize the situation. We let t = the amount of the injections.

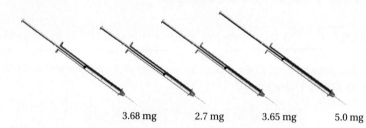

      3.68 mg   2.7 mg   3.65 mg   5.0 mg

2. Translate.   Amounts are being combined. We translate to an equation:

First plus second plus third plus fourth is total.

$$3.68 + 2.7 + 3.65 + 5.0 = t$$

3. Solve.   To solve, we carry out the addition.

$$
\begin{array}{r}
\overset{2}{}\overset{1}{} \\
3.6\,8 \\
2.7\,0 \\
3.6\,5 \\
+\ \ 5.0\,0 \\
\hline
1\,5.0\,3
\end{array}
$$

Thus, $t = 15.03$.

4. Check.   We can check by repeating our addition. We can also see whether our answer is reasonable by first noting that it is indeed larger than any of the numbers being added. We can also check by rounding:

$$3.68 + 2.7 + 3.65 + 5.0 \approx 4.0 + 3.0 + 4.0 + 5.0 = 16 \approx 15.03.$$

If we had gotten an answer like 150.3 or 0.1503, then our estimate, 16, would have told us that we did something wrong, like not lining up the decimal points.

5. State.   The total amount of the injections was 15.03 mg.

Do Exercise 2.

Answer on page A-3

Exercise Set 4.4

a Solve.

1. A consumer bought a CD for $16.99 and paid with a $20 bill. How much change was there?

2. Madeleine buys a book for $44.68 and pays with a $50 bill. How much change does she receive?

3. On a six-day trip, a driver bought the following amounts of gasoline: 23.6 gal, 17.7 gal, 20.8 gal. 17.2 gal, 25.4 gal, and 13.8 gal. How many gallons of gasoline were purchased?

4. One week, a corporate jet pilot flew the following distances: 247.6 mi, 80.5 mi, 536.8 mi, 198.2 mi, and 360 mi. What was the total number of miles that the pilot flew?

5. Normal body temperature is 98.6°F. During an illness, a patient's temperature rose 4.2°. What was the new temperature?

6. A medical assistant draws 17.85 mg of blood and uses 9.68 mg in a blood test. How much is left?

7. A family read the odometer before starting a trip. It read 22,456.8 and they know that they will be driving 234.7 mi. What will the odometer read at the end of the trip?

8. A driver bought gasoline when the odometer read 14,296.3. At the next gasoline purchase, the odometer read 14,515.8. How many miles had been driven?

9. Each year, Americans eat 24.8 billion hamburgers and 15.9 billion hot dogs. How many more hamburgers than hot dogs do they eat?

10. In 1911, Ray Harroun won the first Indianapolis 500-mile race with an average speed of 74.59 mph. In 1990, Arie Luyendyk set a record for the highest average speed with a speed of 185.984 mph. How much faster was Luyendyk than Harroun?

11. _____

12. _____

13. _____

14. _____

15. _____

16. _____

17. a) _____

b) _____

18. a) _____

b) _____

19. _____

20. _____

11. Average annual expenses for golfers who play primarily on public courses are $176.20 for greens fees, $141.87 for cart rentals, and $38.82 for golf balls. What is the total amount spent?

12. A student bought a sweater for $47.95, a pair of jeans for $33.99, and a coat for $88.50. How much was spent?

13. The length of the Panama Canal is 81.6 km (kilometers). The length of the Suez Canal is 175.5 km. How much longer is the Suez Canal?

14. In 1960, the average age of a bride was 20.8. In 1990, the average age was 23.9. How much older was the average bride in 1990 than in 1960?

15. On a 4-mi relay, the times of each member were 4.25 min, 4.86 min, 3.98 min, and 5.0 min. What was their combined time?

16. On a 400-m relay, the times were 10.8 sec, 10.6 sec, 11.1 sec, and 10.2 sec. What was their combined time?

17. Dallas, Texas, receives an average of 34.55 in. (87.757 cm) of rain and 2.3 in. (5.842 cm) of snow each year.
    a) What is the total amount of precipitation in inches?
    b) What is the total amount of precipitation in centimeters?

18. The distance, by air, from New York to St. Louis is 876 mi (1401.6 km) and from St. Louis to Los Angeles is 1562 mi (2499.2 km).
    a) How far, in miles, is it from New York to Los Angeles?
    b) How far, in kilometers, is it from New York to Los Angeles?

19. A driver bought gasoline when the odometer read 28,576.8. At the next gasoline purchase, the odometer read 28,802.6. How many miles had been driven?

20. A family read the odometer at 18,788.9 before starting a trip. They drove 356.4 mi one day and 36.5 mi the next. What did the odometer read at the end of the trip?

21. A businesswoman has $1123.56 in her checking account. She writes checks of $23.82, $507.88, and $98.32 to pay some bills. She then deposits a bonus check of $678.20. How much is in her account after these changes?

22. A student had $185.00 to spend for fall clothes: $44.95 was spent for shoes, $71.95 for a sport coat, and $55.35 for slacks. How much was left?

The following graph shows the number of passengers per year who travel through the country's busiest airports. (Use the graph for Exercises 23–26.)

Busiest Airports in the United States

Airport	Passengers (in millions)
O'Hare International (Chicago)	59.9
Dallas/Ft. Worth	48.5
Hartsfield (Atlanta)	48.0
Los Angeles International	45.8
San Francisco	31.1

Passengers (in millions)

23. How many more passengers does O'Hare handle than San Francisco?

24. How many more passengers does Hartsfield handle than Los Angeles?

25. How many passengers do Dallas/Ft. Worth and San Francisco handle together?

26. How many passengers do all these airports handle together?

27. A pair of cotton jeans costs $39.95. A student bought two pairs. The tax on the purchase was $4.80. How much did the student pay for the jeans, with tax?

28. A pair of running shoes costs $79.95. A jogger buys two pairs. The tax on the purchase was $9.40. How much did the jogger pay for the shoes, with tax?

21. _____

22. _____

23. _____

24. _____

25. _____

26. _____

27. _____

28. _____

29. _____

30. _____

31. _____

32. _____

33. _____

34. _____

35. _____

36. _____

37. _____

38. _____

39. _____

40. _____

41. _____

42. _____

43. _____

44. _____

Find the distance around (perimeter of) the figure.

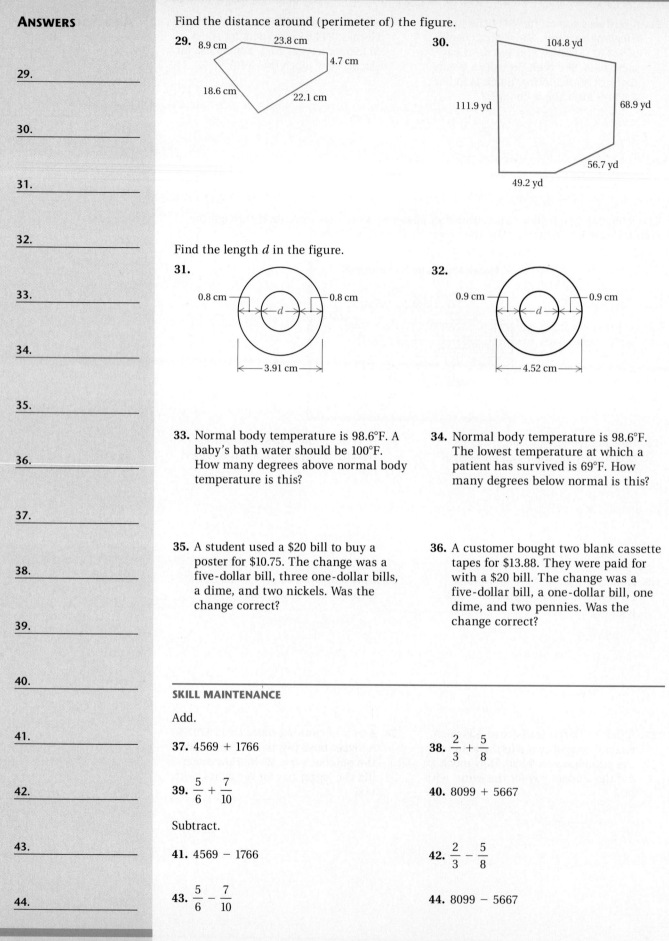

29. 8.9 cm   23.8 cm   4.7 cm   18.6 cm   22.1 cm

30. 104.8 yd   111.9 yd   68.9 yd   56.7 yd   49.2 yd

Find the length *d* in the figure.

31. 0.8 cm — *d* — 0.8 cm   3.91 cm

32. 0.9 cm — *d* — 0.9 cm   4.52 cm

33. Normal body temperature is 98.6°F. A baby's bath water should be 100°F. How many degrees above normal body temperature is this?

34. Normal body temperature is 98.6°F. The lowest temperature at which a patient has survived is 69°F. How many degrees below normal is this?

35. A student used a $20 bill to buy a poster for $10.75. The change was a five-dollar bill, three one-dollar bills, a dime, and two nickels. Was the change correct?

36. A customer bought two blank cassette tapes for $13.88. They were paid for with a $20 bill. The change was a five-dollar bill, a one-dollar bill, one dime, and two pennies. Was the change correct?

SKILL MAINTENANCE

Add.

37. $4569 + 1766$

38. $\dfrac{2}{3} + \dfrac{5}{8}$

39. $\dfrac{5}{6} + \dfrac{7}{10}$

40. $8099 + 5667$

Subtract.

41. $4569 - 1766$

42. $\dfrac{2}{3} - \dfrac{5}{8}$

43. $\dfrac{5}{6} - \dfrac{7}{10}$

44. $8099 - 5667$

CALCULATOR CONNECTION

In the following exercises, use a calculator to make experimental calculations as well as to complete the exercises.

1. In the sum below, *a*, *b*, *c*, and *d* are digits. Find *a*, *b*, *c*, and *d*.

$$\begin{array}{r} 9a.bcd \\ + \ 1a.bcd \\ \hline 105.d7a \end{array}$$

2. Look for a pattern in the following list, and find the missing numbers.

$22.22, $33.34, $44.46, $55.58, ____, ____, ____, ____, ____, ____, ____.

3. Look for a pattern in the following list, and find the missing numbers.

$2344.78, $2266, $2187.22, $2108.44, ____, ____, ____, ____.

4. In the subtraction below, *a* and *b* are digits. Find *a* and *b*.

$$\begin{array}{r} b876.a4321 \\ - \ 1234.a678b \\ \hline 8641.b7a32 \end{array}$$

Each of the following is called a *magic square*. The sum along each row, column, or diagonal is the same. Find the missing numbers.

5.

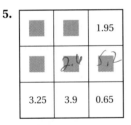

		1.95
	2.4	1.2
3.25	3.9	0.65

Magic sum = ■

6.

2.16		1.08
1.62		0.54

Magic sum = 4.05

7.

6.16		34.72	16.24
38.64	12.32		
	34.16	44.24	
40.32			37.52

Magic sum = 100.24

EXTENDED SYNTHESIS EXERCISES

1. Boneless chicken parts are priced at $2.59 per pound at a local grocery. Chicken parts on the bone cost $1.59 per pound at the same store. What information is needed to determine which is the better buy? How might you go about finding that information?

2. Which of the following is true? Explain your answers.

a) The sum of two numbers greater than 0 and less than 1 is never less than 1.

b) The sum of two numbers greater than 0 and less than 1 is always less than 1.

c) The sum of two numbers greater than 0 and less than 1 is sometimes less than 1.

(continued)

15. $102.4 + 6.1 + 78$

16. $0.93 + 9.3 + 93 + 930$

15. _____

16. _____

Subtract.

17.
$$\begin{array}{r} 5\,2.6\,7\,8 \\ -\quad 4.3\,2\,1 \\ \hline \end{array}$$

18.
$$\begin{array}{r} 2\,0.0 \\ -\quad 0.9\,0\,9\,9 \\ \hline \end{array}$$

17. _____

18. _____

19. $2 - 0.0054$

20. $234.6788 - 81.7854$

19. _____

20. _____

Solve.

21. $x + 0.018 = 9$

22. $34.908 + q = 3400.5677$

21. _____

22. _____

Solve.

23. In 1988, the average speed of the Olympic men's downhill ski winner was 58.848 mph. In the 1948 Olympics, the speed was 29.578 mph. How much faster was the average speed in 1988?

24. A businesswoman wrote checks for $123.89, $56.78, and $3446.98. How much was written in checks altogether?

23. _____

24. _____

25. _____

SKILL MAINTENANCE

Add.

25. $\dfrac{5}{6} + \dfrac{2}{9}$

26.
$$\begin{array}{r} 4\,5\,0\,9 \\ 6\,7\,8\,9 \\ +2\,3\,5\,4 \\ \hline \end{array}$$

26. _____

Subtract.

27. $\dfrac{5}{6} - \dfrac{2}{9}$

28.
$$\begin{array}{r} 4\,5\,0\,9 \\ -2\,3\,5\,4 \\ \hline \end{array}$$

27. _____

28. _____

5

Multiplication and Division: Decimal Notation

INTRODUCTION

In this chapter, we first consider multiplication and division with decimal notation. This will allow us to solve problems like the one below. Also studied are equations and problem solving, as well as estimating sums, differences, products, and quotients. We will also consider conversion from fractional notation to decimal notation in which the decimal notation may be repeating.

5.1 *Multiplication with Decimal Notation*

5.2 *Division with Decimal Notation*

5.3 *Converting from Fractional Notation to Decimal Notation*

5.4 *Estimating*

5.5 *Solving Problems*

AN APPLICATION

The dimensions of a World Cup soccer field are 114.9 yd by 74.4 yd. The dimensions of a standard football field are 120 yd by 53.3 yd. How much greater is the area of a soccer field?

THE MATHEMATICS

This is a multistep problem. First, we find the area of the soccer field. Then we find the area of the football field, and subtract. The area of the soccer field is found as follows:

$$S = \underline{114.9 \times 74.4}.$$

This is multiplication using decimal notation.

The review objectives to be tested in addition to the material in this chapter are as follows.

[2.1d]   Find the prime factorization of a composite number.
[2.5b]   Simplify fractional notation.
[3.5a, b] Add and subtract using mixed numerals.
[3.6a, b] Multiply and divide using mixed numerals.

Pretest: Chapter 5

Multiply.

1. $\begin{array}{r} 4\ 7 \\ \times\ 0.8\ 2 \\ \hline \end{array}$

2. $\begin{array}{r} 0.8\ 3\ 5 \\ \times\ \ \ 0.7\ 4 \\ \hline \end{array}$

3. $\begin{array}{r} 0.4\ 6\ 3 \\ \times\ \ \ 1\ 0\ 0 \\ \hline \end{array}$

4. $\begin{array}{r} 5\ 7.2\ 9\ 9 \\ \times\ \ \ \ \ \ \ 7.6 \\ \hline \end{array}$

5. $\begin{array}{r} 6.8 \\ \times\ 0.5\ 4 \\ \hline \end{array}$

6. $\begin{array}{r} 4.0\ 7 \\ \times\ 0.1\ 0\ 5 \\ \hline \end{array}$

7. $\begin{array}{r} 3\ 2\ 4.5\ 6 \\ \times\ \ \ 0.0\ 0\ 1 \\ \hline \end{array}$

8. $\begin{array}{r} 7\ 3.9\ 6\ 2 \\ \times\ \ \ \ \ \ 1\ 0 \\ \hline \end{array}$

Divide.

9. $8\,\overline{)\,2\ 9}$

10. $2\ 5\,\overline{)\,3\ 3}$

11. $4\ 2\,\overline{)\,2\ 0.1\ 6}$

12. $6.6\,\overline{)\,2\ 0\ 0.6\ 4}$

13. $8\ 2\,\overline{)\,3\ 1.1\ 6}$

14. $\dfrac{576.98}{1000}$

15. $\dfrac{756.89}{0.01}$

16. $\dfrac{0.004653}{100}$

17. Solve: $9.6 \cdot y = 808.896$.

18. Estimate the product 6.92×32.458 by rounding to the nearest one.

19. Estimate the quotient $74.882209 \div 15.03$ by rounding to the nearest ten.

Find decimal notation. Use multiplying by 1.

20. $\dfrac{7}{5}$

21. $\dfrac{23}{16}$

22. $\dfrac{53}{4}$

Find decimal notation Use division.

23. $\dfrac{11}{4}$

24. $\dfrac{7}{9}$

25. $\dfrac{29}{7}$

Round the answer to Exercise 25 to the nearest:

26. Tenth.

27. Hundredth.

28. Thousandth.

Solve.

29. What is the cost of 6 compact discs at $14.95 each?

30. A person walked 10.85 km in 5 hr. How far did the person walk in 1 hr?

31. A developer paid $47,567.89 for 14 acres of land. How much did it cost for 1 acre? Round to the nearest cent.

32. Convert from dollars to cents: $74.96.

33. Convert from cents to dollars: 13,549 cents.

34. Convert to standard notation: 48.6 trillion.

Calculate.

35. $256 \div 3.2 \div 2 - 3.685 + 78.325 \times 0.03$

36. $(1 - 0.06)^2 + \{8[5(12.1 - 7.8) + 20(17.3 - 8.7)]\}$

37. $\dfrac{5}{8} \times 78.95$

38. $\dfrac{2}{3} \times 89.95 - \dfrac{5}{9} \times 3.234$

5.1 *Multiplication with Decimal Notation*

a | MULTIPLICATION

Let us find the product

$$2.3 \times 1.12.$$

To understand how we find such products, we convert each factor to fractional notation. Next, we multiply the whole numbers 23 and 112, and then divide by 1000.

$$2.3 \times 1.12 = \frac{23}{10} \times \frac{112}{100} = \frac{23 \times 112}{10 \times 100} = \frac{2576}{1000} = 2.576$$

Note the number of decimal places.

$$
\begin{array}{r}
1.1\,2 \\
\times \quad 2.3 \\
\hline
2.5\,7\,6
\end{array}
\qquad
\begin{array}{l}
\text{(2 decimal places)} \\
\text{(1 decimal place)} \\
\text{(3 decimal places)}
\end{array}
$$

Now consider

$$0.011 \times 15.0002 = \frac{11}{1000} \times \frac{150{,}002}{10{,}000} = \frac{1{,}650{,}022}{10{,}000{,}000} = 0.1650022.$$

Note the number of decimal places.

$$
\begin{array}{r}
1\,5.0\,0\,0\,2 \\
\times \quad\quad 0.0\,1\,1 \\
\hline
0.1\,6\,5\,0\,0\,2\,2
\end{array}
\qquad
\begin{array}{l}
\text{(4 decimal places)} \\
\text{(3 decimal places)} \\
\text{(7 decimal places)}
\end{array}
$$

To multiply using decimals:

a) Ignore the decimal points and multiply as though both factors were whole numbers.

b) Then place the decimal point in the result. The number of decimal places in the product is the sum of the numbers of places in the factors (count places from the right).

$$0.8 \times 0.43$$

$$
\begin{array}{r}
\overset{2}{} \\
0.4\,3 \\
\times \quad 0.8 \\
\hline
3\,4\,4
\end{array}
\quad
\begin{array}{l}
\text{Ignore the decimal} \\
\text{points for now.}
\end{array}
$$

$$
\begin{array}{r}
0.4\,3 \\
\times \quad 0.8 \\
\hline
0.3\,4\,4
\end{array}
\qquad
\begin{array}{l}
\text{(2 decimal places)} \\
\text{(1 decimal place)} \\
\text{(3 decimal places)}
\end{array}
$$

1. Multiply.

$$\begin{array}{r} 8\,5.4 \\ \times \quad 6.2 \\ \hline \end{array}$$

Multiply.

2.
$$\begin{array}{r} 1\,2\,3\,4 \\ \times\,0.0\,0\,4\,1 \\ \hline \end{array}$$

3.
$$\begin{array}{r} 4\,2.6\,5 \\ \times\,0.8\,0\,4 \\ \hline \end{array}$$

EXAMPLE 1 Multiply: 8.3×74.6.

a) Ignore the decimal points and multiply as though factors were whole numbers:

$$\begin{array}{r} \scriptstyle 3\;4 \\ \scriptstyle 1\;1 \\ 7\,4.6 \\ \times\quad 8.3 \\ \hline 2\,2\,3\,8 \\ 5\,9\,6\,8\,0 \\ \hline 6\,1\,9\,1\,8 \end{array}$$

b) Place the decimal point in the result. The number of decimal places in the product is the sum, $1 + 1$, of the number of places in the factors.

$$\begin{array}{rl} 7\,4.6 & \text{(1 decimal place)} \\ \times\quad 8.3 & \text{(1 decimal place)} \\ \hline 2\,2\,3\,8 & \\ 5\,9\,6\,8\,0 & \\ \hline 6\,1\,9.1\,8 & \text{(2 decimal places)} \end{array}$$

Do Exercise 1.

EXAMPLE 2 Multiply: 0.0032×2148.

As we catch on to the skill, we can combine the two steps.

$$\begin{array}{rl} 2\,1\,4\,8 & \text{(0 decimal places)} \\ \times\,0.0\,0\,3\,2 & \text{(4 decimal places)} \\ \hline 4\,2\,9\,6 & \\ 6\,4\,4\,4\,0 & \\ \hline 6.8\,7\,3\,6 & \text{(4 decimal places)} \end{array}$$

EXAMPLE 3 Multiply: 0.14×0.867.

$$\begin{array}{rl} 0.8\,6\,7 & \text{(3 decimal places)} \\ \times\quad 0.1\,4 & \text{(2 decimal places)} \\ \hline 3\,4\,6\,8 & \\ 8\,6\,7\,0 & \\ \hline 0.1\,2\,1\,3\,8 & \text{(5 decimal places)} \end{array}$$

Do Exercises 2 and 3.

Answers on page A-3

Now let us consider some special kinds of products. The first involves multiplying by a tenth, hundredth, thousandth, or ten-thousandth. Let us look at those products.

$$0.1 \times 38 = \frac{1}{10} \times 38 = \frac{38}{10} = 3.8$$

$$0.01 \times 38 = \frac{1}{100} \times 38 = \frac{38}{100} = 0.38$$

$$0.001 \times 38 = \frac{1}{1000} \times 38 = \frac{38}{1000} = 0.038$$

$$0.0001 \times 38 = \frac{1}{10,000} \times 38 = \frac{38}{10,000} = 0.0038$$

Note in each case that the product is *smaller* than 38.

To multiply any number by a tenth, hundredth, or thousandth,

a) count the number of decimal places in the tenth, hundredth, or thousandth, and

$$0.001 \times 34.45678$$
→ 3 places

b) move the decimal point that many places to the left.

$$0.001 \times 34.45678 = 0.034.45678$$

Move 3 places to the left.

$$0.001 \times 34.45678 = 0.03445678$$

EXAMPLES Multiply.

4. $0.1 \times 14.605 = 1.4605$ 1.4.605

5. $0.01 \times 14.605 = 0.14605$

6. $0.001 \times 14.605 = 0.014605$
     └── We write an extra zero.

7. $0.0001 \times 14.605 = 0.0014605$
     └── We write two extra zeros.

Do Exercises 4–7.

Now let us consider multiplying by a power of ten, such as 10, 100, 1000, and so on. Let us look at those products.

$$10 \times 97.34 = 973.4$$
$$100 \times 97.34 = 9734.$$
$$1000 \times 97.34 = 97,340$$
$$10,000 \times 97.34 = 973,400$$

Note in each case that the product is *larger* than 97.34.

Multiply.

4. 0.1×3.48

5. 0.01×3.48

6. 0.001×3.48

7. 0.0001×3.48

Answers on page A-3

Multiply.

8. 10×3.48

To multiply any number by a power of ten, such as 10, 100, 1000, and so on,

a) count the number of zeros, and

1000×34.45678

$\rightarrow$ 3 zeros

b) move the decimal point that many places to the right.

$1000 \times 34.45678 = 34.456.78$

Move 3 places to the right.

$1000 \times 34.45678 = 34{,}456.78$

9. 100×3.48

EXAMPLES

8. $10 \times 14.605 = 146.05$ 14.6.05

9. $100 \times 14.605 = 1460.5$

10. $1000 \times 14.605 = 14{,}605$

11. $10{,}000 \times 14.605 = 146{,}050$ 14.6050.

Do Exercises 8–11.

10. 1000×3.48

b   APPLICATIONS USING MULTIPLICATION WITH DECIMAL NOTATION

NAMING LARGE NUMBERS

We often see notation like the following in newspapers and magazines and on television.

> O'Hare International Airport handles 59.9 million passengers per year.
>
> Americans drink 17 million gallons of coffee each day.
>
> The population of the world is 5.6 billion.

To understand such notation, it helps to consider the following table.

11. $10{,}000 \times 3.48$

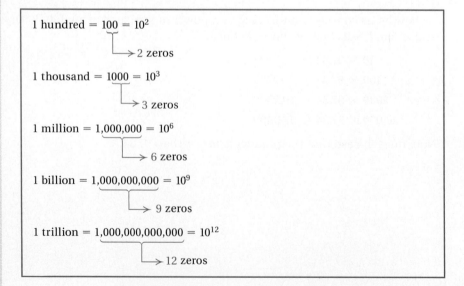

1 hundred = 100 = 10^2

$\rightarrow$ 2 zeros

1 thousand = 1000 = 10^3

$\rightarrow$ 3 zeros

1 million = 1,000,000 = 10^6

$\rightarrow$ 6 zeros

1 billion = 1,000,000,000 = 10^9

$\rightarrow$ 9 zeros

1 trillion = 1,000,000,000,000 = 10^{12}

$\rightarrow$ 12 zeros

Answers on page A-3

To convert to standard notation, we proceed as follows.

EXAMPLE 12 Convert the number in this sentence to standard notation: O'Hare handles 59.9 million passengers per year.

$$59.9 \text{ million} = 59.9 \times 1 \text{ million}$$
$$= 59.9 \times 1,000,000 = 59,900,000$$

Do Exercises 12 and 13.

MONEY CONVERSION

Converting from dollars to cents is like multiplying by 100. To see why, consider $19.43.

$19.43 = 19.43 \times \$1$	We think of $19.43 as 19.43 × 1 dollar, or 19.43 × $1.
$= 19.43 \times 100¢$	Substituting 100¢ for $1: $1 = 100¢
$= 1943¢$	Multiplying

> To convert from dollars to cents, move the decimal point two places to the right and change from the $ sign in front to the ¢ sign at the end.

EXAMPLES Convert from dollars to cents.

13. $189.64 = 18,964¢$
14. $0.75 = 75¢$

Do Exercises 14 and 15.

Converting from cents to dollars is like multiplying by 0.01. To see why, consider 65¢.

$65¢ = 65 \times 1¢$	We think of 65¢ as 65 × 1 cent, or 65 × 1¢.
$= 65 \times \$0.01$	Substituting $0.01 for 1¢: 1¢ = $0.01
$= \$0.65$	Multiplying

> To convert from cents to dollars, move the decimal point two places to the left and change from the ¢ sign at the end to the $ sign in front.

EXAMPLES Convert from cents to dollars.

15. $395¢ = \$3.95$
16. $8503¢ = \$85.03$

Do Exercises 16 and 17.

Convert the number in the sentence to standard notation.

12. Americans drink 17 million gallons of coffee each day.

13. The population of the world is 5.6 billion.

Convert from dollars to cents.
14. $15.69

15. $0.17

Convert from cents to dollars.
16. 35¢

17. 577¢

Answers on page A-3

NUMBER PATTERNS

Look for a pattern in the following. We can use a calculator for the calculations.

$$3 \times 5 = \quad 15$$
$$33 \times 5 = \quad 165$$
$$333 \times 5 = \quad 1665$$
$$3333 \times 5 = 16,665$$

Do you see a pattern? If so, find $33,333 \times 5$ without the use of your calculator.

EXERCISES

In each of the following, do the first four calculations using a calculator. Look for a pattern. Then do the last calculation without your calculator.

1. $\quad 4 \times 6$
    $\quad 44 \times 6$
    $\quad 444 \times 6$
    $\quad 4444 \times 6$
    $\quad 44,444 \times 6$

2. $\quad 9 \times 9$
    $\quad 99 \times 89$
    $\quad 999 \times 889$
    $\quad 9999 \times 8889$
    $\quad 99,999 \times 88,889$

3. $\quad 77 \times 78$
    $\quad 777 \times 78$
    $\quad 7777 \times 78$
    $\quad 77,777 \times 78$
    $\quad 777,777 \times 78$

4. $1 \cdot 13 \cdot 76,923$
    $2 \cdot 13 \cdot 76,923$
    $3 \cdot 13 \cdot 76,923$
    $4 \cdot 13 \cdot 76,923$
    $5 \cdot 13 \cdot 76,923$

Exercise Set 5.1

a Multiply.

1. 8.6
×   7

2. 5.7
× 0.8

3. 0.8 4
×   8

4. 9.4
× 0.6

5.   6.3
× 0.0 4

6.   9.8
× 0.0 8

7.   8 7
× 0.0 0 6

8. 1 8.4
× 0.0 7

9. 10×23.76

10. 100×3.8798

11. 1000×583.686852

12. 0.34×1000

13. 7.8×100

14. 0.00238×10

15. 0.1×89.23

16. 0.01×789.235

17. 0.001×97.68

18. 8976.23×0.001

19. 78.2×0.01

20. 0.0235×0.1

21.   3 2.6
×   1 6

22.   9.2 8
×   8.6

23.   0.9 8 4
×   3.3

24.   8.4 8 9
×   7.4

25.   3 7 4
×   2.4

26.   8 6 5
× 1.0 8

27.   7 4 9
× 0.4 3

28.   9 7 8
× 2 0.5

29.   0.8 7
×   6 4

30.   7.2 5
×   6 0

31.   4 6.5 0
×   7 5

32.   8.2 4
× 7 0 3

33.   8 1.7
× 0.6 1 2

34.   3 1.8 2
×   7.1 5

35. 1 0.1 0 5
× 1 1.3 2 4

36. 1 5 1.2
× 4.5 5 5

ANSWERS

1.
2.
3.
4.
5.
6.
7.
8.
9.
10.
11.
12.
13.
14.
15.
16.
17.
18.
19.
20.
21.
22.
23.
24.
25.
26.
27.
28.
29.
30.
31.
32.
33.
34.
35.
36.

37.   1 2.3
  × 1.0 8

38.   7.8 2
  × 0.0 2 4

39.   3 2.4
  ×   2.8

40.   8.0 9
  × 0.0 0 7 5

41.   0.0 0 3 4 2
  ×   0.8 4

42.   2.0 0 5 6
  ×   3.8

43.   0.3 4 7
  ×   2.0 9

44.   2.5 3 2
  × 1.0 6 7

45.   3.0 0 5
  × 0.6 2 3

46.   1 6.3 4
  × 0.0 0 0 5 1 2

47. 1000×45.678

48. 0.001×45.678

b Convert from dollars to cents.

49. $28.88

50. $67.43

51. $0.66

52. $1.78

Convert from cents to dollars.

53. 34¢

54. 95¢

55. 3445¢

56. 933¢

Convert the number in the sentence to standard notation.

57. The national debt was $4.03 trillion.

58. Americans eat 5.8 million pounds of chocolate each day.

59. It is estimated that in 1995, 4.7 million fax machines will be sold.

60. In a recent year, 4.3 billion CDs were sold.

SKILL MAINTENANCE

61. Multiply: $2\frac{1}{3} \cdot 4\frac{4}{5}$.

62. Divide: $2\frac{1}{3} \div 4\frac{4}{5}$.

Divide.

63. $24 \overline{)8\,2\,0\,8}$

64. $4 \overline{)3\,4\,8}$

65. $7 \overline{)3\,1{,}9\,6\,2}$

66. $18 \overline{)2\,2{,}6\,2\,6}$

SYNTHESIS

Express as a power of 10.

67. (1 trillion) · (1 billion)

68. (1 million) · (1 billion)

5.2 Division with Decimal Notation

a DIVISION

WHOLE-NUMBER DIVISORS

Compare these divisions.

$$\frac{588}{7} = 84$$

$$\frac{58.8}{7} = 8.4$$

$$\frac{5.88}{7} = 0.84$$

$$\frac{0.588}{7} = 0.084$$

The number of decimal places in the *quotient* is the same as the number of decimal places in the *dividend*.

These lead us to the following method for dividing by a whole number.

To divide by a whole number,

a) place the decimal point directly above the decimal point in the dividend, and

b) divide as though dividing whole numbers.

$$
\begin{array}{r}
0.8\,4 \\
7\,\overline{)\,5.8\,8} \\
5\,6\,0 \\
\hline
2\,8 \\
2\,8 \\
\hline
0
\end{array}
$$

EXAMPLE 1 Divide: $379.2 \div 8$.

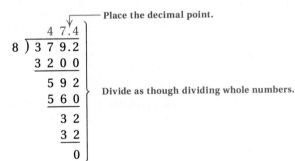

Place the decimal point.

$$
\begin{array}{r}
4\,7.4 \\
8\,\overline{)\,3\,7\,9.2} \\
3\,2\,0\,0 \\
\hline
5\,9\,2 \\
5\,6\,0 \\
\hline
3\,2 \\
3\,2 \\
\hline
0
\end{array}
$$

Divide as though dividing whole numbers.

EXAMPLE 2 Divide: $82.08 \div 24$.

Place the decimal point.

$$
\begin{array}{r}
3.4\,2 \\
2\,4\,\overline{)\,8\,2.0\,8} \\
7\,2\,0\,0 \\
\hline
1\,0\,0\,8 \\
9\,6\,0 \\
\hline
4\,8 \\
4\,8 \\
\hline
0
\end{array}
$$

Divide as though dividing whole numbers.

Do Exercises 1–3.

OBJECTIVES

After finishing Section 5.2, you should be able to:

a Divide using decimal notation.

b Solve equations of the type $a \cdot x = b$, where a and b may be in decimal notation.

c Simplify expressions using the rules for order of operations.

FOR EXTRA HELP

TAPE 9   TAPE 8A   MAC: 5
                              IBM: 5

Divide.

1. $9\,\overline{)\,5.4}$

2. $1\,5\,\overline{)\,2\,2.5}$

3. $8\,2\,\overline{)\,3\,8.5\,4}$

Answers on page A-3

Divide.

4. $2\,5\,\overline{)\,8}$

5. $4\,\overline{)\,1\,5}$

6. $8\,6\,\overline{)\,2\,1.5}$

Sometimes it helps to write some extra zeros to the right of the decimal point. They don't change the number.

EXAMPLE 3   Divide: $30 \div 8$.

$$
\begin{array}{r}
3. \\
8\,\overline{)\,3\,0.} \\
2\,4 \\
\hline
6
\end{array}
$$

Place the decimal point and divide to find how many ones.

$$
\begin{array}{r}
3. \\
8\,\overline{)\,3\,0.0} \\
2\,4\downarrow \\
\hline
6\,0
\end{array}
$$

Write an extra zero.

$$
\begin{array}{r}
3.7 \\
8\,\overline{)\,3\,0.0} \\
2\,4 \\
\hline
6\,0 \\
5\,6 \\
\hline
4
\end{array}
$$

Divide to find how many tenths.

$$
\begin{array}{r}
3.7 \\
8\,\overline{)\,3\,0.0\,0} \\
2\,4 \\
\hline
6\,0 \\
5\,6\downarrow \\
\hline
4\,0
\end{array}
$$

Write an extra zero.

$$
\begin{array}{r}
3.7\,5 \\
8\,\overline{)\,3\,0.0\,0} \\
2\,4 \\
\hline
6\,0 \\
5\,6 \\
\hline
4\,0 \\
4\,0 \\
\hline
0
\end{array}
$$

Divide to find how many hundredths.

EXAMPLE 4   Divide: $4 \div 25$.

$$
\begin{array}{r}
0.1\,6 \\
2\,5\,\overline{)\,4.0\,0} \\
2\,5 \\
\hline
1\,5\,0 \\
1\,5\,0 \\
\hline
0
\end{array}
$$

Do Exercises 4–6.

DIVISORS THAT ARE NOT WHOLE NUMBERS

Consider the division

$$0.2\,4 \overline{)\,8.2\,0\,8}$$

We write the division as $\dfrac{8.208}{0.24}$. Then we multiply by 1 to change to a whole-number divisor:

$$\frac{8.208}{0.24} = \frac{8.208}{0.24} \times \frac{100}{100} = \frac{820.8}{24}.$$

The divisor is now a whole number. The division

$$0.2\,4 \overline{)\,8.2\,0\,8}$$

is the same as

$$2\,4 \overline{)\,8\,2\,0.8}$$

To divide when the divisor is not a whole number,

a) move the decimal point (multiply by 10, 100, and so on) to make the divisor a whole number;

$$0.2\,4 \overline{)\,8.2\,0\,8}$$
Move 2 places to the right.

b) move the decimal point (multiply the same way) in the dividend the same number of places; and

$$0.2\,4 \overline{)\,8.2\,0\,8}$$
Move 2 places to the right.

c) place the decimal point directly above the decimal point in the dividend and divide as though dividing whole numbers.

$$
\begin{array}{r}
3\,4.2 \\
0.2\,4 \overline{)\,8.2\,0\,{}_{\wedge}8} \\
7\,2\,0\,0 \\
\hline
1\,0\,0\,8 \\
9\,6\,0 \\
\hline
4\,8 \\
4\,8 \\
\hline
0
\end{array}
$$

(The new decimal point in the dividend is indicated by a caret.)

EXAMPLE 5 Divide: $5.848 \div 8.6$.

$$8.6 \overline{)\,5.8\,4\,8}$$
Multiply the divisor by 10 (move the decimal point 1 place). Multiply the same way in the dividend (move 1 place).

$$
\begin{array}{r}
0.6\,8 \\
8.6 \overline{)\,5.8\,{}_{\wedge}4\,8} \\
5\,1\,6\,0 \\
\hline
6\,8\,8 \\
6\,8\,8 \\
\hline
0
\end{array}
$$
Then divide.

Note: $\dfrac{5.848}{8.6} = \dfrac{5.848}{8.6} \cdot \dfrac{10}{10} = \dfrac{58.48}{86}.$

Do Exercises 7–9.

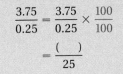

7. a) Complete.

$$
\frac{3.75}{0.25} = \frac{3.75}{0.25} \times \frac{100}{100}
$$
$$
= \frac{(\quad)}{25}
$$

b) Divide.

$$0.2\,5 \overline{)\,3.7\,5}$$

Divide.

8. $0.8\,3 \overline{)\,4.0\,6\,7}$

9. $3.5 \overline{)\,4\,4.8}$

Answers on page A-3

10. Divide.

$$1.6 \overline{)\,2\,5}$$

Answer on page A-3

EXAMPLE 6   Divide: $12 \div 0.64$.

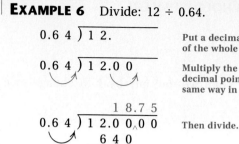

Put a decimal point at the end
of the whole number.

Multiply the divisor by 100 (move the
decimal point 2 places). Multiply the
same way in the dividend (move 2 places).

$$0.6\,4 \overline{)\,1\,2.0\,0\,_{\wedge}0\,0}$$

Then divide.

```
        1 8.7 5
0.6 4 ) 1 2.0 0,0 0
        6 4 0
        5 6 0
        5 1 2
          4 8 0
          4 4 8
            3 2 0
            3 2 0
              0
```

Do Exercise 10.

It is often helpful to be able to divide quickly by a ten, hundred, or thousand, or by a tenth, hundredth, or thousandth. The procedure we use is based on multiplying by 1. Consider the following examples:

$$\frac{23.789}{1000} = \frac{23.789}{1000} \cdot \frac{1000}{1000} = \frac{23{,}789}{1{,}000{,}000} = 0.023789.$$

We are dividing by a number greater than 1: The result is *smaller* than 23.789.

$$\frac{23.789}{0.01} = \frac{23.789}{0.01} \cdot \frac{100}{100} = \frac{2378.9}{1} = 2378.9.$$

We are dividing by a number less than 1: The result is *larger* than 23.789.
We use the following procedure.

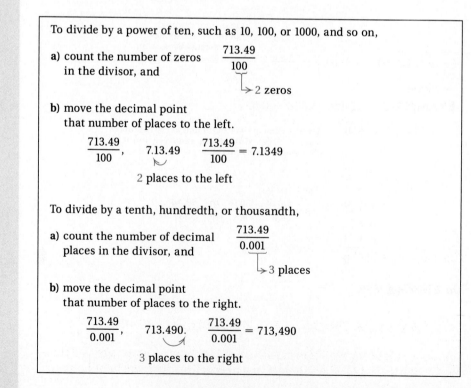

To divide by a power of ten, such as 10, 100, or 1000, and so on,

a) count the number of zeros in the divisor, and

$$\frac{713.49}{100}$$

→ 2 zeros

b) move the decimal point that number of places to the left.

$$\frac{713.49}{100}, \qquad 7.13.49 \qquad \frac{713.49}{100} = 7.1349$$

2 places to the left

To divide by a tenth, hundredth, or thousandth,

a) count the number of decimal places in the divisor, and

$$\frac{713.49}{0.001}$$

→ 3 places

b) move the decimal point that number of places to the right.

$$\frac{713.49}{0.001}, \qquad 713.490. \qquad \frac{713.49}{0.001} = 713{,}490$$

3 places to the right

EXAMPLE 7 Divide: $\dfrac{0.0104}{10}$.

$$\dfrac{0.0104}{\underset{\text{1 zero}}{10}}, \qquad \underset{\substack{\curvearrowleft \\ \text{1 place to the left to change 10 to 1.}}}{0.0.0104,} \qquad \dfrac{0.0104}{10} = 0.00104$$

EXAMPLE 8 Divide: $\dfrac{23.738}{0.001}$.

$$\dfrac{23.738}{\underset{\text{3 places}}{0.001}}, \qquad \underset{\substack{\curvearrowright \\ \text{3 places to the right to change 0.001 to 1.}}}{23.738.} \qquad \dfrac{23.738}{0.001} = 23,738$$

Do Exercises 11–14.

b SOLVING EQUATIONS

Now let us solve equations of the type $a \cdot x = b$, where a and b may be in decimal notation. Proceeding as before, we divide on both sides by a.

EXAMPLE 9 Solve: $8 \cdot x = 27.2$.

We have

$$\dfrac{8 \cdot x}{8} = \dfrac{27.2}{8} \qquad \text{Dividing on both sides by 8}$$

$$x = 3.4. \qquad 8 \overline{)\begin{array}{r} 3.4 \\ 2\ 7.2 \\ \underline{2\ 4\ 0} \\ 3\ 2 \\ \underline{3\ 2} \\ 0 \end{array}}$$

The solution is 3.4. *DIVIDE BY NUMBER NEXT TO VARIABLE*

EXAMPLE 10 Solve: $2.9 \cdot t = 0.14616$.

We have

$$\dfrac{2.9 \cdot t}{2.9} = \dfrac{0.14616}{2.9} \qquad \text{Dividing on both sides by 2.9}$$

$$t = 0.0504. \qquad 2.9 \overline{)\begin{array}{r} 0.0\ 5\ 0\ 4 \\ 0.1_\wedge 4\ 6\ 1\ 6 \\ \underline{1\ 4\ 5\ 0\ 0} \\ 1\ 1\ 6 \\ \underline{1\ 1\ 6} \\ 0 \end{array}}$$

The solution is 0.0504.

Do Exercises 15 and 16.

c ORDER OF OPERATIONS: DECIMAL NOTATION

The same rules for order of operations used with whole numbers and fractional notation apply when simplifying expressions with decimal notation.

Divide.

11. $\dfrac{0.1278}{0.01}$

12. $\dfrac{0.1278}{100}$

13. $\dfrac{98.47}{1000}$

14. $\dfrac{6.7832}{0.1}$

Solve.

15. $100 \cdot x = 78.314$

16. $0.25 \cdot y = 276.4$

Answers on page A-3

Simplify.

17. $0.25 \cdot (1 + 0.08) - 0.0274$

18. $20^2 - 3.4^2 +$
$\{2.5[20(9.2 - 5.6)] + 5(10 - 5)\}$

19. Using the information in the following bar graph, determine the average number of international passengers for the four airlines.

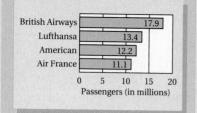

Passengers on International Airlines

British Airways 17.9
Lufthansa 13.4
American 12.2
Air France 11.1

0 5 10 15 20
Passengers (in millions)

Answers on page A-3

> **RULES FOR ORDER OF OPERATIONS**
>
> **1.** Do all calculations within parentheses before operations outside.
> **2.** Evaluate all exponential expressions.
> **3.** Do all multiplications and divisions in order from left to right.
> **4.** Do all additions and subtractions in order from left to right.

EXAMPLE 11 Simplify: $(5 - 0.06) \div 2 + 3.42 \times 0.1$.

$(5 - 0.06) \div 2 + 3.42 \times 0.1 = 4.94 \div 2 + 3.42 \times 0.1$ Carrying out operations inside parentheses

$= 2.47 + 0.342$ Doing all multiplications and divisions in order from left to right

$= 2.812$

EXAMPLE 12 Simplify: $10^2 \times \{[(3 - 0.24) \div 2.4] - (0.21 - 0.092)\}$.

$10^2 \times \{[(3 - 0.24) \div 2.4] - (0.21 - 0.092)\}$

$= 10^2 \times \{[2.76 \div 2.4] - 0.118\}$ Doing the calculations in the innermost parentheses first

$= 10^2 \times \{1.15 - 0.118\}$ Again, doing the calculations in the innermost parentheses

$= 10^2 \times 1.032$ Subtracting inside the parentheses

$= 100 \times 1.032$ Evaluating the exponential expression

$= 103.2$

Do Exercises 17 and 18.

EXAMPLE 13 Aluminum is being used more and more by the American automobile industry because it is lighter than steel, which means better gas mileage, and because it is easy to recycle. The bar graph shows the number of pounds of aluminum used in a particular kind of automobile during four recent years. Find the average amount of aluminum used during those four years.

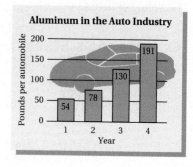

Aluminum in the Auto Industry

Pounds per automobile
200
150
100
50
0

54 78 130 191
1 2 3 4
Year

To find the average of a set of numbers, we add them. Then we divide by the number of addends. We are finding the average of 54, 78, 130, and 191. The average is given by

$$(54 + 78 + 130 + 191) \div 4.$$

Then

$$(54 + 78 + 130 + 191) \div 4 = 453 \div 4 = 113.25.$$

The average amount of aluminum used was 113.25 lb.

Do Exercise 19.

Exercise Set 5.2

a Divide.

1. $2 \overline{)\ 5.9\ 8}$

2. $5 \overline{)\ 1\ 8}$

3. $4 \overline{)\ 9\ 5.1\ 2}$

4. $8 \overline{)\ 2\ 5.9\ 2}$

5. $1\ 2 \overline{)\ 8\ 9.7\ 6}$

6. $2\ 3 \overline{)\ 2\ 5.0\ 7}$

7. $3\ 3 \overline{)\ 2\ 3\ 7.6}$

8. $12.4 \div 4$

9. $9.144 \div 8$

10. $4.5 \div 9$

11. $12.123 \div 3$

12. $7 \overline{)\ 5.6}$

13. $5 \overline{)\ 0.3\ 5}$

14. $0.0\ 4 \overline{)\ 1.6\ 8}$

15. $0.1\ 2 \overline{)\ 8.4}$

16. $0.3\ 6 \overline{)\ 2.8\ 8}$

17. $3.4 \overline{)\ 6\ 8}$

18. $0.2\ 5 \overline{)\ 5}$

19. $1\ 5 \overline{)\ 6}$

20. $1\ 2 \overline{)\ 1.8}$

21. $3\ 6 \overline{)\ 1\ 4.7\ 6}$

ANSWERS

1. _____

2. _____

3. _____

4. _____

5. _____

6. _____

7. _____

8. _____

9. _____

10. _____

11. _____

12. _____

13. _____

14. _____

15. _____

16. _____

17. _____

18. _____

19. _____

20. _____

21. _____

22. $5\,2\,\overline{)\,1\,1\,9.6\,}$

23. $3.2\,\overline{)\,2\,7.2\,}$

24. $8.5\,\overline{)\,2\,7.2\,}$

25. $4.2\,\overline{)\,3\,9.0\,6\,}$

26. $4.8\,\overline{)\,0.1\,1\,0\,4\,}$

27. $8\,\overline{)\,5\,}$

28. $8\,\overline{)\,3\,}$

29. $0.4\,7\,\overline{)\,0.1\,2\,2\,2\,}$

30. $1.0\,8\,\overline{)\,0.5\,4\,}$

31. $4.8\,\overline{)\,7\,5\,}$

32. $0.2\,8\,\overline{)\,6\,3\,}$

33. $0.0\,3\,2\,\overline{)\,0.0\,7\,4\,8\,8\,}$

34. $0.0\,1\,7\,\overline{)\,1.5\,8\,1\,}$

35. $8\,2\,\overline{)\,3\,8.5\,4\,}$

36. $3\,4\,\overline{)\,0.1\,4\,6\,2\,}$

37. $\dfrac{213.4567}{1000}$

38. $\dfrac{213.4567}{100}$

39. $\dfrac{213.4567}{10}$

40. $\dfrac{100.7604}{0.1}$ **41.** $\dfrac{1.0237}{0.001}$ **42.** $\dfrac{1.0237}{0.01}$

40. _____

41. _____

42. _____

43. _____

44. _____

45. _____

46. _____

47. _____

48. _____

49. _____

50. _____

51. _____

52. _____

53. _____

54. _____

55. _____

56. _____

57. _____

58. _____

b Solve.

43. $4.2 \cdot x = 39.06$ **44.** $36 \cdot y = 14.76$

45. $1000 \cdot y = 9.0678$ **46.** $789.23 = 0.25 \cdot q$

47. $1048.8 = 23 \cdot t$ **48.** $28.2 \cdot x = 423$

c Simplify.

49. $14 \times (82.6 + 67.9)$ **50.** $(26.2 - 14.8) \times 12$

51. $0.003 + 3.03 \div 0.01$ **52.** $9.94 + 4.26 \div (6.02 - 4.6) - 0.9$

53. $42 \times (10.6 + 0.024)$ **54.** $(18.6 - 4.9) \times 13$

55. $4.2 \times 5.7 + 0.7 \div 3.5$ **56.** $123.3 - 4.24 \times 1.01$

57. $9.0072 + 0.04 \div 0.1^2$ **58.** $12 \div 0.03 - 12 \times 0.03^2$

59. $(8 - 0.04)^2 \div 4 + 8.7 \times 0.4$

63.3414

15.8404 + 3.48 =

19.3204

60. $(5 - 2.5)^2 \div 100 + 0.1 \times 6.5$

61. $86.13 + 95.7 \div (9.6 - 0.03)$

62. $2.48 \div (1 - 0.504) + 24.3 - 11 \times 2$

63. $(4 \div 0.4) + (0.1 \times 5) - 0.1^2$

64. $6 \times 0.9 + 0.1 \div 4 - 0.2^3$

65. $5.5^2 \times [(6 - 4.2) \div 0.06 + 0.12]$

66. $12^2 \div (12 + 2.4) - [(2 - 1.6) \div 0.8]$

67. $200 \times \{[(4 - 0.25) \div 2.5] - (4.5 - 4.025)\}$

68. $0.03 \times \{1 \times 50.2 - [(8 - 7.5) \div 0.05]\}$

69. Find the average of $1276.59, $1350.49, $1123.78, and $1402.56.

70. Find the average weight of two wrestlers who weigh 308 lb and 296.4 lb.

SKILL MAINTENANCE

71. Add: $10\frac{1}{2} + 4\frac{5}{8}$.

72. Subtract: $10\frac{1}{2} - 4\frac{5}{8}$.

73. Simplify: $\frac{36}{42}$.

74. Find the prime factorization of 162.

75. Find the prime factorization of 684.

76. Simplify: $\frac{56}{64}$.

ANSWERS

59. _____
60. _____
61. _____
62. _____
63. _____
64. _____
65. _____
66. _____
67. _____
68. _____
69. _____
70. _____
71. _____
72. _____
73. _____
74. _____
75. _____
76. _____

5.3 Converting from Fractional Notation to Decimal Notation

a FRACTIONAL NOTATION TO DECIMAL NOTATION

When a denominator has no prime factors other than 2's and 5's, we can find decimal notation by multiplying by 1. We multiply to get a denominator that is a power of ten like 10, 100, or 1000.

EXAMPLE 1 Find decimal notation for $\frac{3}{5}$.

$$\frac{3}{5} = \frac{3}{5} \cdot \frac{2}{2} = \frac{6}{10} = 0.6$$ We use $\frac{2}{2}$ for 1 to get a denominator of 10.

EXAMPLE 2 Find decimal notation for $\frac{7}{20}$.

$$\frac{7}{20} = \frac{7}{20} \cdot \frac{5}{5} = \frac{35}{100} = 0.35$$ We use $\frac{5}{5}$ for 1 to get a denominator of 100.

EXAMPLE 3 Find decimal notation for $\frac{9}{40}$.

$$\frac{9}{40} = \frac{9}{40} \cdot \frac{25}{25} = \frac{225}{1000} = 0.225$$ We use $\frac{25}{25}$ for 1 to get a denominator of 1000.

EXAMPLE 4 Find decimal notation for $\frac{87}{25}$.

$$\frac{87}{25} = \frac{87}{25} \cdot \frac{4}{4} = \frac{348}{100} = 3.48$$ We use $\frac{4}{4}$ for 1 to get a denominator of 100.

Do Exercises 1–4.

We can also divide to find decimal notation.

EXAMPLE 5 Find decimal notation for $\frac{3}{5}$.

$$\frac{3}{5} = 3 \div 5 \qquad 5\overline{)\begin{array}{l} 0.6 \\ 3.0 \\ \underline{3\ 0} \\ 0 \end{array}} \qquad \frac{3}{5} = 0.6$$

EXAMPLE 6 Find decimal notation for $\frac{7}{8}$.

$$\frac{7}{8} = 7 \div 8 \qquad 8\overline{)\begin{array}{l} 0.8\ 7\ 5 \\ 7.0\ 0\ 0 \\ \underline{6\ 4} \\ 6\ 0 \\ \underline{5\ 6} \\ 4\ 0 \\ \underline{4\ 0} \\ 0 \end{array}} \qquad \frac{7}{8} = 0.875$$

Do Exercises 5 and 6.

OBJECTIVES

After finishing Section 5.3, you should be able to:

a Convert from fractional notation to decimal notation.

b Round numbers named by repeating decimals.

c Calculate using fractional and decimal notation together.

FOR EXTRA HELP

TAPE 9   TAPE 8A   MAC: 5
                      IBM: 5

Find decimal notation. Use multiplying by 1.

1. $\frac{4}{5}$

2. $\frac{9}{20}$

3. $\frac{11}{40}$

4. $\frac{33}{25}$

Find decimal notation.

5. $\frac{2}{5}$

6. $\frac{3}{8}$

Answers on page A-3

Find decimal notation.

7. $\dfrac{1}{6}$

8. $\dfrac{2}{3}$

In Examples 5 and 6, the division *terminated,* meaning that eventually we got a remainder of 0. A **terminating decimal** occurs when the denominator has only 2's or 5's, or both, as factors, as in $\frac{17}{25}$, $\frac{5}{8}$, or $\frac{83}{100}$. This assumes that the fractional notation has been simplified.

Consider a different situation:

$$\frac{5}{6} \quad \text{or} \quad \frac{5}{2 \cdot 3}.$$

Since 6 has a 3 as a factor, the division will not terminate. Although we can still use division to get decimal notation, the answer will be a **repeating decimal,** as follows.

EXAMPLE 7   Find decimal notation for $\frac{5}{6}$.

We have

$$\frac{5}{6} = 5 \div 6$$

$$
\begin{array}{r}
0.8\ 3\ 3 \\
6\ )\ \overline{5.0\ 0\ 0} \\
4.8 \\ \hline
2\ 0 \\
1\ 8 \\ \hline
2\ 0 \\
1\ 8 \\ \hline
2
\end{array}
$$

Since 2 keeps reappearing as a remainder, the digits repeat and will continue to do so; therefore,

$$\frac{5}{6} = 0.83333\ldots.$$

The dots indicate an endless sequence of digits in the quotient. When there is a repeating pattern, the dots are often replaced by a bar to indicate the repeating part—in this case, only the 3:

$$\frac{5}{6} = 0.8\overline{3}.$$

Do Exercises 7 and 8.

EXAMPLE 8   Find decimal notation for $\frac{4}{11}$.

$$\frac{4}{11} = 4 \div 11$$

$$
\begin{array}{r}
0.3\ 6\ 3\ 6 \\
1\ 1\ )\ \overline{4.0\ 0\ 0\ 0} \\
3\ 3 \\ \hline
7\ 0 \\
6\ 6 \\ \hline
4\ 0 \\
3\ 3 \\ \hline
7\ 0 \\
6\ 6 \\ \hline
4
\end{array}
$$

Answers on page A-3

Since 7 and 4 keep reappearing as remainders, the sequence of digits "36" repeats in the quotient, and

$$\frac{4}{11} = 0.363636\ldots, \quad \text{or} \quad 0.\overline{36}.$$

Do Exercises 9 and 10.

EXAMPLE 9 Find decimal notation for $\frac{5}{7}$.

We have

```
        0.7 1 4 2 8 5
    7 ) 5 .0 0 0 0 0 0
        4 9
          1 0
            7
          3 0
          2 8
            2 0
            1 4
              6 0
              5 6
                4 0
                3 5
                  5
```

Since 5 appears as a remainder, the sequence of digits "714285" repeats in the quotient, and

$$\frac{5}{7} = 0.714285714285\ldots, \quad \text{or} \quad 0.\overline{714285}.$$

The length of a repeating part can be very long—too long to find on a calculator. An example is $\frac{5}{97}$, which has a repeating part of 96 digits.

Do Exercise 11.

b | ROUNDING IN PROBLEM SOLVING

In applied problems, repeating decimals are rounded to get approximate answers.

EXAMPLES Round each to the nearest tenth, hundredth, and thousandth.

	Nearest tenth	Nearest hundredth	Nearest thousandth
10. $0.8\overline{3} = 0.83333\ldots$	0.8	0.83	0.833
11. $0.\overline{09} = 0.090909\ldots$	0.1	0.09	0.091
12. $0.\overline{714285} = 0.714285714285\ldots$	0.7	0.71	0.714

Do Exercises 12–14.

Find decimal notation.

9. $\dfrac{5}{11}$

10. $\dfrac{12}{11}$

11. Find decimal notation for $\dfrac{3}{7}$.

Round each to the nearest tenth, hundredth, and thousandth.

12. $0.\overline{6}$

13. $0.8\overline{08}$

14. $6.\overline{245}$

Answers on page A-3

15. Calculate: $\dfrac{5}{6} \times 0.864$.

Calculate.

16. $\dfrac{1}{3} \times 0.384 + \dfrac{5}{8} \times 0.6784$

17. $\dfrac{5}{6} \times 0.864 + 14.3 \div \dfrac{8}{5}$

C | CALCULATIONS WITH FRACTIONAL AND DECIMAL NOTATION TOGETHER

In certain kinds of calculations, fractional and decimal notation might occur together. In such cases, there are at least three ways in which we might proceed.

EXAMPLE 13   Calculate: $\frac{2}{3} \times 0.576$.

METHOD 1.   One way to do this calculation is to convert the decimal notation to fractional notation so that both numbers are in fractional notation. The answer can be left in fractional notation and simplified, or we can convert back to decimal notation and round, if appropriate.

$$
\begin{aligned}
\frac{2}{3} \times 0.576 &= \frac{2}{3} \cdot \frac{576}{1000} = \frac{2 \cdot 576}{3 \cdot 1000} \\
&= \frac{2 \cdot 2 \cdot 2 \cdot 2 \cdot 2 \cdot 2 \cdot 2 \cdot 3 \cdot 3}{2 \cdot 2 \cdot 2 \cdot 3 \cdot 5 \cdot 5 \cdot 5} \\
&= \frac{2 \cdot 2 \cdot 2 \cdot 3}{2 \cdot 2 \cdot 2 \cdot 3} \cdot \frac{2 \cdot 2 \cdot 2 \cdot 2 \cdot 3}{5 \cdot 5 \cdot 5} \\
&= 1 \cdot \frac{2 \cdot 2 \cdot 2 \cdot 2 \cdot 3}{5 \cdot 5 \cdot 5} \\
&= \frac{2 \cdot 2 \cdot 2 \cdot 2 \cdot 3}{5 \cdot 5 \cdot 5} = \frac{48}{125}, \text{ or } 0.384
\end{aligned}
$$

METHOD 2.   A second way to do this calculation is to convert the fractional notation to decimal notation so that both numbers are in decimal notation. Since $\frac{2}{3}$ converts to repeating decimal notation, it is first rounded to some chosen decimal place. We choose three decimal places. Then, using decimal notation, we multiply. Note that the answer is not as accurate as that found by method 1, due to the rounding.

$$\frac{2}{3} \times 0.576 = 0.\overline{6} \times 0.576 \approx 0.667 \times 0.576 = 0.384192$$

METHOD 3.   A third way to do this calculation is to treat 0.576 as $\frac{0.576}{1}$. Then we multiply 0.576 by 2, and divide the result by 3.

$$\frac{2}{3} \times 0.576 = \frac{2}{3} \times \frac{0.576}{1} = \frac{2 \times 0.576}{3} = \frac{1.152}{3} = 0.384$$

Do Exercise 15.

EXAMPLE 14   Calculate: $\frac{2}{3} \times 0.576 + 3.287 \div \frac{4}{5}$.

We use the rules for order of operations, doing first the multiplication and then the division. Then we add.

$$
\begin{aligned}
\frac{2}{3} \times 0.576 + 3.287 \div \frac{4}{5} &= 0.384 + 3.287 \cdot \frac{5}{4} \\
&= 0.384 + 4.10875 \\
&= 4.49275
\end{aligned}
$$

Do Exercises 16 and 17.

Exercise Set 5.3

ANSWERS

1. _____

2. _____

3. _____

4. _____

5. _____

6. _____

7. _____

8. _____

9. _____

10. _____

11. _____

12. _____

13. _____

14. _____

15. _____

16. _____

17. _____

18. _____

19. _____

20. _____

21. _____

22. _____

23. _____

24. _____

25. _____

26. _____

27. _____

28. _____

29. _____

30. _____

31. _____

32. _____

33. _____

34. _____

35. _____

36. _____

a Find decimal notation.

1. $\dfrac{3}{5}$

2. $\dfrac{19}{20}$

3. $\dfrac{13}{40}$

4. $\dfrac{3}{16}$

5. $\dfrac{1}{5}$

6. $\dfrac{3}{20}$

7. $\dfrac{17}{20}$

8. $\dfrac{9}{40}$

9. $\dfrac{19}{40}$

10. $\dfrac{81}{40}$

11. $\dfrac{39}{40}$

12. $\dfrac{31}{40}$

13. $\dfrac{13}{25}$

14. $\dfrac{61}{125}$

15. $\dfrac{2502}{125}$

16. $\dfrac{181}{200}$

17. $\dfrac{1}{4}$

18. $\dfrac{1}{2}$

19. $\dfrac{23}{40}$

20. $\dfrac{11}{20}$

21. $\dfrac{18}{25}$

22. $\dfrac{37}{25}$

23. $\dfrac{19}{16}$

24. $\dfrac{5}{8}$

25. $\dfrac{4}{15}$

26. $\dfrac{7}{9}$

27. $\dfrac{1}{3}$

28. $\dfrac{1}{9}$

29. $\dfrac{4}{3}$

30. $\dfrac{8}{9}$

31. $\dfrac{7}{6}$

32. $\dfrac{7}{11}$

33. $\dfrac{4}{7}$

34. $\dfrac{14}{11}$

35. $\dfrac{11}{12}$

36. $\dfrac{5}{12}$

ANSWERS

37. _____
38. _____
39. _____
40. _____
41. _____
42. _____
43. _____
44. _____
45. _____
46. _____
47. _____
48. _____
49. _____
50. _____
51. _____
52. _____
53. _____
54. _____
55. _____
56. _____
57. _____
58. _____
59. _____
60. _____
61. _____
62. _____
63. _____
64. _____
65. _____
66. _____
67. _____
68. _____
69. _____
70. _____
71. _____
72. _____
73. _____
74. _____
75. _____
76. _____
77. _____
78. _____

[b]

37.–47. Round each answer of the odd-numbered Exercises 25–35 to the nearest tenth, hundredth, and thousandth.

38.–48. Round each answer of the even-numbered Exercises 26–36 to the nearest tenth, hundredth, and thousandth.

Round each to the nearest tenth, hundredth, and thousandth.

49. $0.\overline{18}$
50. $0.\overline{83}$
51. $0.2\overline{7}$
52. $3.5\overline{4}$

[c] Calculate.

53. $\dfrac{7}{8} \times 12.64$

54. $\dfrac{4}{5} \times 384.8$

55. $2\dfrac{3}{4} + 5.65$

56. $4\dfrac{4}{5} + 3.25$

57. $\dfrac{47}{9} \times 79.95$

58. $\dfrac{7}{11} \times 2.7873$

59. $\dfrac{1}{2} - 0.5$

60. $3\dfrac{1}{8} - 2.75$

61. $4.875 - 2\dfrac{1}{16}$

62. $55\dfrac{3}{5} - 12.22$

63. $\dfrac{5}{6} \times 0.0765 + \dfrac{5}{4} \times 0.1124$

64. $\dfrac{3}{5} \times 6384.1 - \dfrac{3}{8} \times 156.56$

65. $\dfrac{4}{5} \times 384.8 + 24.8 \div \dfrac{8}{3}$

66. $102.4 \div \dfrac{2}{5} - 12 \times \dfrac{5}{6}$

67. $\dfrac{7}{8} \times 0.86 - 0.76 \times \dfrac{3}{4}$

68. $17.95 \div \dfrac{5}{8} + \dfrac{3}{4} \times 16.2$

69. $3.375 \times 5\dfrac{1}{3}$

70. $2.5 \times 3\dfrac{5}{8}$

71. $6.84 \div 2\dfrac{1}{2}$

72. $8\dfrac{1}{2} \div 2.125$

SKILL MAINTENANCE

73. Multiply: $9 \cdot 2\dfrac{1}{3}$.

74. Divide: $84 \div 8\dfrac{2}{5}$.

75. Subtract: $20 - 16\dfrac{3}{5}$.

76. Add: $14\dfrac{3}{5} + 16\dfrac{1}{10}$.

77. Find the LCM of 25 and 65.

78. Find the prime factorization of 128.

5.4 Estimating

a ESTIMATING SUMS, DIFFERENCES, PRODUCTS, AND QUOTIENTS

Estimating has many uses. It can be done before a problem is even attempted to get an idea of the answer. It can be done afterward as a check, even when we are using a calculator. In many situations, an estimate is all we need. We usually estimate by rounding the numbers so that there are one or two nonzero digits. Consider the following advertisements for Examples 1–4.

Now thru Thursday!

SALE
SALE
SALE $466.95

Monoham®

Monoham Compact Home Fax Machine With Built-In Phone
20-level gray scale, automatic voice switch, and answering machine connection.

SALE $349.95
Reg. 399.95
19" color TV with matrix picture tube, remote control, stereo sound, and on-screen programing.

SALE!
Handy® Vacuum Cleaner with 8 amps of power comes with 9 attachments, and the motorized beater bar has a headlight to aid cleaning.

$219.95
reg. $299.95

1 year warranty parts and service

EXAMPLE 1   Estimate to the nearest ten the total cost of one fax machine and one TV.

We are estimating the sum

$466.95 + $349.95 = Total cost.

The estimate to the nearest ten is

$470 + $350 = $820.   (Estimated total cost)

We rounded $466.95 to the nearest ten and $349.95 to the nearest ten. The estimated sum is $820.

Do Exercise 1.

EXAMPLE 2   About how much more does the fax machine cost than the TV? Estimate to the nearest ten.

We are estimating the difference

$466.95 − $349.95 = Price difference.

The estimate to the nearest ten is

$470 − $350 = $120.   (Estimated price difference)

Do Exercise 2.

1. Estimate to the nearest ten the total cost of one TV and one vacuum cleaner. Which of the following is an appropriate estimate?
a) $5700   b) $570
c) $790   d) $57

2. About how much more does the TV cost than the vacuum cleaner? Estimate to the nearest ten. Which of the following is an appropriate estimate?
a) $130   b) $1300
c) $580   d) $13

Answers on page A-3

3. Estimate the total cost of 6 fax machines. Which of the following is an appropriate estimate?

a) $4400　　b) $300
c) $30,000　d) $3000

4. About how many vacuum cleaners can be bought for $1100? Which of the following is an appropriate estimate?

a) 8　　b) 5
c) 11　d) 124

Estimate the product. Do not find the actual product. Which of the following is an appropriate estimate?

5. 2.4 × 8

a) 16　　b) 34
c) 125　d) 5

6. 24 × 0.6

a) 200　　b) 5
c) 110　d) 20

7. 0.86 × 0.432

a) 0.04　　b) 0.4
c) 1.1　d) 4

8. 0.82 × 0.1

a) 800　　b) 8
c) 0.08　d) 80

9. 0.12 × 18.248

a) 180　　b) 1.8
c) 0.018　d) 18

10. 24.234 × 5.2

a) 200　　b) 125
c) 12.5　d) 234

Answers on page A-3

EXAMPLE 3　Estimate the total cost of 4 vacuum cleaners.

We are estimating the product

$$4 \times \$219.95 = \text{Total cost.}$$

The estimate is found by rounding $219.95 to the nearest ten:

$$4 \times \$220 = \$880.$$

Do Exercise 3.

EXAMPLE 4　About how many fax machines can be bought for $1580?

We estimate the quotient

$$\$1580 \div \$466.95.$$

Since we want a whole-number estimate, we choose our rounding appropriately. Rounding $466.95 to the nearest hundred, we get $500. Since $1580 is close to $1500, which is a multiple of 500, we estimate

$$\$1500 \div \$500,$$

so the answer is 3.

Do Exercise 4.

EXAMPLE 5　Estimate: 4.8 × 52. Do not find the actual product. Which of the following is an appropriate estimate?

a) 25　　　b) 250　　　c) 2500　　　d) 360

We have

$$5 \times 50 = 250. \quad \text{(Estimated product)}$$

We rounded 4.8 to the nearest one and 52 to the nearest ten. Thus an appropriate estimate is (b).

Compare these estimates for the product 4.94 × 38:

$$5 \times 40 = 200, \quad 5 \times 38 = 190, \quad 4.9 \times 40 = 196.$$

The first estimate was the easiest. You could probably do it mentally. The others had more nonzero digits.

Do Exercises 5–10.

EXAMPLE 6 Estimate: 82.08 ÷ 24. Which of the following is an appropriate estimate?

a) 400   b) 16   c) 40   d) 4

This is about 80 ÷ 20, so the answer is about 4. Thus an appropriate estimate is (d).

EXAMPLE 7 Estimate: 94.18 ÷ 3.2. Which of the following is an appropriate estimate?

a) 30   b) 300   c) 3   d) 60

This is about 90 ÷ 3, so the answer is about 30. Thus an appropriate estimate is (a).

EXAMPLE 8 Estimate: 0.0156 ÷ 1.3. Which of the following is an appropriate estimate?

a) 0.2   b) 0.002   c) 0.02   d) 20

This is about 0.02 ÷ 1, so the answer is about 0.02. Thus an appropriate estimate is (c).

Do Exercises 11–13.

In some cases, it is easier to estimate a quotient directly rather than by rounding the divisor and the dividend.

EXAMPLE 9 Estimate: 0.0074 ÷ 0.23. Which of the following is an appropriate estimate?

a) 0.3   b) 0.03   c) 300   d) 3

We estimate 3 for a quotient. We check by multiplying.

$$0.23 \times 3 = 0.69$$

We make the estimate smaller. We estimate 0.3 and check by multiplying.

$$0.23 \times 0.3 = 0.069$$

We make the estimate smaller. We estimate 0.03 and check by multiplying.

$$0.23 \times 0.03 = 0.0069$$

This is about 0.0074, so the quotient is about 0.03. Thus an appropriate estimate is (b).

Do Exercise 14.

Estimate the quotient. Which of the following is an appropriate estimate?

11. 59.78 ÷ 29.1

a) 200   b) 20
c) 2   d) 0.2

12. 82.08 ÷ 2.4

a) 40   b) 4.0
c) 400   d) 0.4

13. 0.1768 ÷ 0.08

a) 8   b) 10
c) 2   d) 20

14. Estimate: 0.0069 ÷ 0.15. Which of the following is an appropriate estimate?

a) 0.5   b) 50
c) 0.05   d) 23.4

Answers on page A-3

13. _____

14. _____

15. _____

16. _____

17. _____

18. _____

19. _____

20. _____

21. _____

22. _____

23. _____

24. _____

13. A family of five can save $6.72 per week by eating cereal to be cooked, such as oatmeal, rather than ready-to-eat cereal. How much will they save in 1 year? Use 52 weeks for 1 year.

14. A medical assistant prepares 200 injections, each with 3.5 mg of penicillin. How much penicillin is used in all?

15. A car loan of $11,178.72 is to be paid off in 24 monthly payments. How much is each payment?

16. A loan of $11,692.50 is to be paid off in 30 monthly payments. How much is each payment?

17. A plane flies 147.9 km/h for 6 hr. How far does it go?

18. Each day, Americans consume 5.8 million lb of chocolate. How much chocolate do we consume in one year? Assume 365 days in 1 year.

19. It costs $24.95 a day plus 27 cents per mile to rent a compact car at Shuttles Rent-a-Car. How much, in dollars, would it cost to drive the car 120 mi in 1 day?

20. A student worked 53 hr during a week one summer. The student earned $6.50 per hour for the first 40 hr and $6.85 per hour for overtime. How much was earned during the week?

21. A driver filled the gasoline tank and noted that the odometer read 26,342.8. After the next filling, the odometer read 26,736.7. It took 19.5 gal to fill the tank. How many miles per gallon did the driver get?

22. A driver filled the gasoline tank and noted that the odometer read 18,943.2. After the next filling, the odometer read 19,305.8. It took 19.6 gal to fill the tank. How many miles per gallon did the driver get?

23. The water in a tank weighs 748.45 lb. One cubic foot of water weighs 62.5 lb. How many cubic feet of water does the tank hold?

24. An investor bought $25 treasury bonds for $18.75 each. She spent $168.75. How many bonds did she buy?

25. The average video game costs 25 cents and runs for 1.5 min. Assuming a player does not win any free games and plays continuously, how much money, in dollars, does it cost to play a video game for 1 hr?

26. A family owns a house with an assessed value of $124,500. For every $1000 of assessed value, they pay $7.68 in taxes. How much in taxes do they pay?

27. A person weighing 170 lb burns 8.6 calories per minute while mowing a lawn. How many calories would be burned in 2 hr of mowing?

28. Lot A measures 250.1 ft by 302.7 ft. Lot B measures 389.4 ft by 566.2 ft. What is the total area of the two lots?

29. At the start of a recent baseball season, Barry Bonds, of the San Francisco Giants, got 58 hits in his first 138 "at bats." What part of the at bats were hits? Give decimal notation. Round to the nearest thousandth. (This is a player's "batting average.")

$\frac{58}{138}$

30. In a recent year, Jose Canseco got 187 hits in 610 at bats. What was his batting average? Round to the nearest thousandth.

31. A person earned $310.37 during a 40-hr week. What was the hourly wage? Round to the nearest cent.

32. A driver wants to estimate gas mileage per gallon. At 36,057.1 mi, the tank is filled with 10.7 gal. At 36,217.6 mi, the tank is filled with 11.1 gal. Find the mileage per gallon. Round to the nearest tenth. 14.5 mpg

33. A student earned $78.27 working part-time for 9 days. How much was earned each day? (Assume that the student worked the same number of hours each day.) Round to the nearest cent.

34. A golfer paid $28.75 for a dozen golf balls. What was the cost of each golf ball? Round to the nearest cent. (*Hint:* 1 dozen = 12.)

35. An 8-lb canned ham costs $31.92. What is the cost per pound?

36. A restaurant owner bought 20 dozen eggs for $13.80. Find the cost of each egg to the nearest tenth of a cent (thousandth of a dollar).

25. _____

26. _____

27. _____

28. _____

29. _____

30. _____

31. _____

32. _____

33. _____

34. _____

35. _____

36. _____

37. _____

38. _____

39. _____

40. _____

41. _____

42. _____

43. _____

44. _____

45. _____

46. _____

47. _____

48. _____

37. A county paid $906,816 for 47.5 acres of land for a new community college. What was the cost per acre? Round to the nearest cent.

38. A person weighing 170 lb burns 8.6 calories per minute while mowing a lawn. One must burn about 3500 calories in order to lose 1 lb. How many pounds would be lost by mowing for 2 hr? Round to the nearest tenth.

39. The dimensions of a World Cup soccer field are 114.9 yd by 74.4 yd. The dimensions of a standard football field are 120 yd by 53.3 yd. How much greater is the area of a soccer field?

40. A construction worker is paid $13.50 per hour for the first 40 hr of work, and time and a half, or $20.25 per hour, for any overtime exceeding 40 hr per week. One week she works 46 hr. How much was her pay?

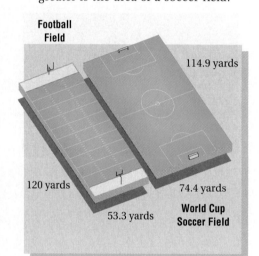

Football Field

114.9 yards

120 yards

74.4 yards

53.3 yards

World Cup Soccer Field

41. In order to make money on loans, financial institutions are paid back more money than they loan. You borrow $120,000 to buy a house and agree to make monthly payments of $880.52 for 30 years. How much do you pay back? How much more do you pay back than the amount of the loan?

42. A car rental agency charges $59.95 per day plus $0.39 per mile for a certain type of car. How much is the rental charge for a 4-day trip of 876 mi?

SKILL MAINTENANCE

43. Subtract: $24\frac{1}{4} - 10\frac{2}{3}$.

44. Divide: $8\frac{1}{2} \div 4\frac{1}{4}$.

45. Add: $24\frac{1}{4} + 10\frac{2}{3}$.

46. Multiply: $8\frac{1}{2} \cdot 4\frac{1}{4}$.

47. Multiply: $24\frac{1}{4} \cdot 10\frac{2}{3}$.

48. Subtract: $8\frac{1}{2} - 4\frac{1}{4}$.

CRITICAL THINKING

CALCULATOR CONNECTION

In Exercises 1–5, use a calculator to find decimal notation.

1. $\dfrac{1}{7}$

2. $\dfrac{2}{7}$

3. $\dfrac{3}{7}$

4. $\dfrac{4}{7}$

5. $\dfrac{5}{7}$

6. From the pattern of Exercises 1–5, guess the decimal notation for $\dfrac{6}{7}$. Then check your guess on a calculator.

In Exercises 7–9, use a calculator to find decimal notation.

7. $\dfrac{1}{9}$

8. $\dfrac{1}{99}$

9. $\dfrac{1}{999}$

10. From the pattern of Exercises 7–9, guess the decimal notation for $\dfrac{1}{9999}$. Then check your guess on a calculator.

In Exercises 11–14, use a calculator to find decimal notation.

11. $\dfrac{1}{81}$ **12.** $\dfrac{10}{81}$ **13.** $\dfrac{100}{81}$ **14.** $\dfrac{1000}{81}$

15. From the pattern of Exercises 11–14, guess the decimal notation for $\dfrac{10,000}{81}$. Then check your guess on a calculator.

16. Find decimal notation for each fraction. Use that notation to arrange the fractions in order from smallest to largest.

$$\dfrac{2}{3}, \ \dfrac{15}{19}, \ \dfrac{11}{13}, \ \dfrac{5}{7}, \ \dfrac{13}{15}, \ \dfrac{17}{20}$$

In Exercises 17 and 18, use one of $+$, $-$, $\times$, and $\div$ in each blank to make a true sentence.

17. $2.56 \ \blacksquare \ 6.4 \ \blacksquare \ 51.2 \ \blacksquare \ 17.4 \ \blacksquare \ 89.7 = 72.62$

18. $(0.37 \ \blacksquare \ 18.78) \ \blacksquare \ 2^{13} = 156{,}876.8$

EXTENDED SYNTHESIS EXERCISES

1. Look over the results of the preceding calculator exercises. Suppose the fractional notation for a/b has been simplified. At most how many digits are in the repeating part of the decimal notation? Explain.

2. The product of two numbers is 322. The sum is 37. Find the numbers.

3. The product of two numbers is 0.15. The sum is 0.8. Find the numbers.

4. You buy a half-dozen packs of basketball cards with a dozen cards in each pack. The cost is a dozen cents for each half-dozen cards. How much do you pay for the cards?

5. Find repeating decimal notation for 1 and explain. Use the following hints:

$$\dfrac{1}{3} = 0.33333333\ldots$$

$$\dfrac{2}{3} = 0.66666666\ldots$$

6. Find repeating decimal notation for 2.

(continued)

7. Use one of the words *sometimes, never,* or *always* to complete each of the following.

    a) The product of two numbers greater than 0 and less than 1 is _____ less than 1.

    b) The product of two numbers greater than 0 and greater than 1 is _____ less than 1.

    c) The product of two numbers greater than 0 if one number is less than 1 and the other is greater than 1 is _____ equal to 1.

    d) The product of two numbers is 1, both are greater than 0, and one of the numbers is less than 1. Then the other number is _____ greater than 1.

EXERCISES FOR THINKING AND WRITING

1. Consider finding decimal notation for $\frac{44}{61}$. Discuss as many ways as you can for finding such notation and give the answer.

2. Discuss how you might use estimating to place the decimal point when calculating with decimal notation.

3. Explain how fractional notation can be used to justify multiplication with decimal notation.

4. Find a real-world situation that fits the equation $8 \cdot 79.95 + 3 \cdot 149.59 = 1088.37$.

Summary and Review Exercises: Chapter 5

The review objectives to be tested in addition to the material in this chapter are [2.1d], [2.5b], [3.5a, b], and [3.6a, b].

Multiply.

1.
$$\begin{array}{r} 4\ 8 \\ \times\ 0.2\ 7 \\ \hline \end{array}$$

2.
$$\begin{array}{r} 0.1\ 7\ 4 \\ \times\ \ 0.8\ 3 \\ \hline \end{array}$$

3.
$$\begin{array}{r} 0.0\ 4\ 3 \\ \times\ \ 1\ 0\ 0 \\ \hline \end{array}$$

4.
$$\begin{array}{r} 0.0\ 8\ 7 \\ \times\ \ \ \ 3.2 \\ \hline \end{array}$$

5.
$$\begin{array}{r} 3.7 \\ \times\ 0.2\ 9 \\ \hline \end{array}$$

6.
$$\begin{array}{r} 2.0\ 8 \\ \times\ 0.1\ 0\ 5 \\ \hline \end{array}$$

7.
$$\begin{array}{r} 2\ 4.6\ 8 \\ \times\ 0.0\ 0\ 1 \\ \hline \end{array}$$

8.
$$\begin{array}{r} 2\ 4.6\ 8 \\ \times 1\ 0\ 0\ 0 \\ \hline \end{array}$$

Divide.

9. $8\ \overline{)\ 6\ 0}$

10. $2\ 5\ \overline{)\ 8\ 0}$

11. $5\ 2\ \overline{)\ 2\ 3.4}$

12. $2.6\ \overline{)\ 1\ 1\ 7.5\ 2}$

13. $7.2\ \overline{)\ 1\ 1.5\ 2}$

14. $2.1\ 4\ \overline{)\ 2.1\ 8\ 7\ 0\ 8}$

15. $\dfrac{276.3}{100}$

16. $\dfrac{276.3}{1000}$

17. $\dfrac{0.1274}{0.1}$

18. $\dfrac{13.892}{0.01}$

Solve.

19. $3 \cdot x = 20.85$

20. $10 \cdot y = 425.4$

21. Estimate the product 7.82×34.487 by rounding to the nearest one.

22. Estimate the quotient $82.304 \div 17.287$ by rounding to the nearest ten.

23. Estimate the difference $219.875 - 4.478$ by rounding to the nearest one.

24. Estimate the sum $\$45.78 + \$78.99$ by rounding to the nearest one.

Find decimal notation. Use multiplying by 1.

25. $\dfrac{13}{5}$

26. $\dfrac{32}{25}$

27. $\dfrac{11}{4}$

Find decimal notation. Use division.

28. $\dfrac{13}{4}$

29. $\dfrac{7}{6}$

30. $\dfrac{17}{11}$

Round the answer to Exercise 30 to the nearest:

31. Tenth.

32. Hundredth.

33. Thousandth.

Solve.

34. A train traveled 496.02 mi in 6 hr. How far did it travel in 1 hr?

35. Four software discs containing video games are purchased, each costing $59.95. How much is spent in total?

36. A person drinks an average of 3.48 cups of tea per day. How many cups of tea are drunk in a week? in a month (30 days)?

37. A wholesale nursery sold 93 potted plants for a total of $423.65. What was the cost of each plant? Round to the nearest cent.

38. A student buys 6 CDs for $95.88. How much was each CD?

Convert from cents to dollars.

39. 8273 cents

40. 487 cents

Convert from dollars to cents.

41. $24.93

42. $9.86

Convert the number in the sentence to standard notation.

43. We breathe 3.4 billion cubic feet of oxygen each day.

44. Your blood travels 1.2 million miles in a week.

Calculate.

45. $(8 - 1.23) \div 4 + 5.6 \times 0.02$

46. $(1 + 0.07)^2 + 10^3 \div 10^2 +$ $[4(10.1 - 5.6) + 8(11.3 - 7.8)]$

47. $\dfrac{3}{4} \times 20.85$

48. $\dfrac{1}{3} \times 123.7 + \dfrac{4}{9} \times 0.684$

SKILL MAINTENANCE

49. Multiply: $8\dfrac{1}{3} \cdot 5\dfrac{1}{4}$.

50. Divide: $20 \div 5\dfrac{1}{3}$.

51. Add: $12\dfrac{1}{2} + 7\dfrac{3}{10}$.

52. Subtract: $24 - 17\dfrac{2}{5}$.

53. Simplify: $\dfrac{28}{56}$.

54. Find the prime factorization of 192.

Test: Chapter 5

Multiply.

1.
$$\begin{array}{r} 3\,2 \\ \times\,0.2\,5 \\ \hline \end{array}$$

2.
$$\begin{array}{r} 0.1\,2\,5 \\ \times\;\;\;0.2\,4 \\ \hline \end{array}$$

3.
$$\begin{array}{r} 0.0\,3\,7 \\ \times\;\;\;1\,0\,0 \\ \hline \end{array}$$

4.
$$\begin{array}{r} 0.0\,9\,9 \\ \times\;\;\;\;\;\;2.1 \\ \hline \end{array}$$

5.
$$\begin{array}{r} 3.4 \\ \times\,0.3\,2 \\ \hline \end{array}$$

6.
$$\begin{array}{r} 3.0\,6 \\ \times\,0.1\,0\,4 \\ \hline \end{array}$$

7.
$$\begin{array}{r} 2\,1\,3.4\,5 \\ \times\;\;\;0.0\,0\,1 \\ \hline \end{array}$$

8.
$$\begin{array}{r} 7\,3.9\,6\,2 \\ \times\;\;\;\;\;\;1\,0 \\ \hline \end{array}$$

Divide.

9. $4\,\overline{)\,1\,9}$

10. $2\,5\,\overline{)\,1\,1}$

11. $4\,2\,\overline{)\,1\,0.0\,8}$

12. $3.3\,\overline{)\,1\,0\,0.3\,2}$

13. $8\,2\,\overline{)\,1\,5.5\,8}$

14. $\dfrac{346.89}{1000}$

15. $\dfrac{346.89}{0.01}$

16. $\dfrac{0.00123}{100}$

17. Solve: $4.8 \cdot y = 404.448$.

18. Estimate the product 8.91×22.457 by rounding to the nearest one.

19. Estimate the quotient $78.2209 \div 16.09$ by rounding to the nearest ten.

Find decimal notation. Use multiplying by 1.

20. $\dfrac{8}{5}$

21. $\dfrac{22}{25}$

22. $\dfrac{21}{4}$

Find decimal notation. Use division.

23. $\dfrac{3}{4}$

24. $\dfrac{11}{9}$

25. $\dfrac{15}{7}$

ANSWERS

1. _____

2. _____

3. _____

4. _____

5. _____

6. _____

7. _____

8. _____

9. _____

10. _____

11. _____

12. _____

13. _____

14. _____

15. _____

16. _____

17. _____

18. _____

19. _____

20. _____

21. _____

22. _____

23. _____

24. _____

25. _____

Round the answer to Exercise 25 to the nearest:

26. Tenth.   **27.** Hundredth.   **28.** Thousandth.

Solve.

29. A government agency bought 6 new flags at $79.95 each. How much was spent?

30. A marathon runner ran 24.85 km in 5 hr. How far did the person run in 1 hr?

31. A rancher paid $23,457 for 14 acres of land adjoining his ranch. How much did he pay for 1 acre? Round to the nearest cent.

32. Convert from dollars to cents: $87.95.

33. Convert from cents to dollars: 949 cents.

34. Convert to standard notation: 38.7 trillion.

Calculate.

35. $256 \div 3.2 \div 2 - 1.56 + 78.325 \times 0.02$

36. $(1 - 0.08)^2 + \{6[5(12.1 - 8.7) + 10(14.3 - 9.6)]\}$

37. $\dfrac{7}{8} \times 345.6$

38. $\dfrac{2}{3} \times 79.95 - \dfrac{7}{9} \times 1.235$

SKILL MAINTENANCE

39. Subtract: $28\dfrac{2}{3} - 2\dfrac{1}{6}$.

40. Add: $2\dfrac{3}{16} + \dfrac{1}{2}$.

41. Divide: $3\dfrac{3}{8} \div 3$.

42. Multiply: $2\dfrac{1}{10} \cdot 6\dfrac{2}{3}$.

43. Simplify: $\dfrac{33}{54}$.

44. Find the prime factorization of 360.

Cumulative Review: Chapters 1–5

Convert to fractional notation.

1. $2\frac{2}{9}$

2. 3.052

Find decimal notation.

3. $\frac{7}{5}$

4. $\frac{6}{11}$

5. Determine whether 43 is prime, composite, or neither.

6. Determine whether 2,053,752 is divisible by 4.

Calculate.

7. $48 + 12 \div 4 - 10 \times 2 + 6892 \div 4$

8. $4.7 - \{0.1[1.2(3.95 - 1.65) + 1.5 \div 2.5]\}$

Round to the nearest hundredth.

9. 584.903

10. $218.\overline{5}$

11. Estimate the product 16.392×9.715 by rounding to the nearest one.

12. Estimate by rounding to the nearest tenth:

$2.714 + 4.562 - 3.31 - 0.0023$.

13. Estimate the product 6418×1984 by rounding to the nearest hundred.

14. Estimate the quotient $717.832 \div 124.998$ by rounding to the nearest ten.

Add and simplify.

15.
$$\begin{array}{r} 2\frac{1}{4} \\ + 3\frac{4}{5} \\ \hline \end{array}$$

16.
$$\begin{array}{r} 3\,4{,}9\,2\,1 \\ 9\,3{,}0\,9\,2 \\ + 1\,1{,}1\,0\,3 \\ \hline \end{array}$$

17. $\frac{1}{6} + \frac{2}{3} + \frac{8}{9}$

18. $143.9 + 2.053$

Subtract and simplify.

19. $723{,}041 - 12{,}904$

20. $19 - 5.903$

21. $5\frac{1}{7} - 4\frac{3}{7}$

22. $\frac{10}{11} - \frac{9}{10}$

Multiply and simplify.

23. $\dfrac{3}{8} \cdot \dfrac{4}{9}$

24.
$$\begin{array}{r} 2\,5\,3\,2 \\ \times\,2\,1\,0\,0 \\ \hline \end{array}$$

25.
$$\begin{array}{r} 2\,3.9 \\ \times\,0.2 \\ \hline \end{array}$$

26.
$$\begin{array}{r} 2\,7.9\,4\,3\,1 \\ \times\,0.0\,0\,1 \\ \hline \end{array}$$

Divide and simplify.

27. $1\,6.5\,\overline{)\,3\,5.0\,1\,3}$

28. $2\,6\,\overline{)\,4\,7{,}9\,1\,8}$

29. $13.8621 \div 0.001$

30. $\dfrac{4}{9} \div \dfrac{8}{15}$

Solve.

31. $8.32 + x = 9.1$

32. $75 \cdot x = 2100$

33. $y \cdot 9.47 = 81.6314$

34. $1062 + y = 368{,}313$

35. $t + \dfrac{5}{6} = \dfrac{8}{9}$

36. $\dfrac{7}{8} \cdot t = \dfrac{7}{16}$

Solve.

37. In a recent year, there were 1368 heart transplants, 8800 kidney transplants, 924 liver transplants, and 130 pancreas transplants. How many transplants of these four organs were performed that year?

38. After making a $150 down payment on a sofa, $\dfrac{3}{10}$ of the total cost was paid. How much did the sofa cost?

39. There are 60 seconds in a minute and 60 minutes in an hour. How many seconds are in a day?

40. A student's tuition was $3600. A loan was obtained for $\dfrac{2}{3}$ of the tuition. How much was the loan?

41. The balance in a checking account is $314.79. After a check is written for $56.02, what is the balance in the account?

42. A clerk in a delicatessen sold $1\frac{1}{2}$ lb of ham, $2\frac{3}{4}$ lb of turkey, and $2\frac{1}{4}$ lb of roast beef. How many pounds of meat were sold?

43. A baker used $\frac{1}{2}$ lb of sugar for cookies, $\frac{2}{3}$ lb of sugar for pie, and $\frac{5}{6}$ lb of sugar for cake. How much sugar was used in all?

44. A rectangular family room measures 19.8 ft by 23.6 ft. Find its area.

SYNTHESIS

45. A box of gelatin mix packages weighs $15\frac{3}{4}$ lb. Each package weighs $1\frac{3}{4}$ oz. How many packages are in the box?

46. A customer in a grocery store used a manufacturer's coupon to buy juice. With the coupon, if 5 cartons of juice were purchased, the sixth carton was free. The price of each carton was $1.09. What was the cost per carton with the coupon? Round to the nearest cent.

6

Ratio and Proportion

INTRODUCTION

The mathematics of the application below shows what is called a *proportion*. The expressions on either side of the equals sign are called *ratios*. In this chapter, we use ratios and proportions to solve problems such as this one. We will also study a topic important to consumers, *unit pricing*.

● ●

AN APPLICATION

To determine the number of deer in a forest, a conservationist catches 612 deer, tags them, and releases them. Later, 244 deer are caught, and it is found that 72 of them are tagged. Estimate how many deer are in the forest.

THE MATHEMATICS

We let D = the number of deer in the forest. Then we translate to a proportion:

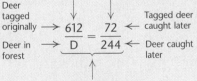

These are ratios.

Deer tagged originally ⟶ $\dfrac{612}{D} = \dfrac{72}{244}$ ← Tagged deer caught later

Deer in forest ⟶ ← Deer caught later

This is a proportion.

OBJECTIVES FOR REVIEW

The review objectives to be tested in addition to the material in this chapter are as follows.

[2.5c]   Test fractions for equality.
[4.4a]   Solve problems involving addition and subtraction with decimals.
[5.1a]   Multiply using decimal notation.
[5.2a]   Divide using decimal notation.

Pretest: Chapter 6

Write fractional notation for the ratio.

1. 35 to 43

2. 0.079 to 1.043

Solve.

3. $\dfrac{5}{6} = \dfrac{x}{27}$

4. $\dfrac{y}{0.25} = \dfrac{0.3}{0.1}$

5. $\dfrac{3\frac{1}{2}}{4\frac{1}{3}} = \dfrac{6\frac{3}{4}}{x}$

6. What is the rate in miles per gallon?

408 miles, 16 gallons

7. A student picked 10 qt (quarts) of strawberries in 45 min. What is the rate in quarts per minute?

8. A 24-oz loaf of bread costs $1.39. Find the unit price in cents per ounce. Round to the nearest hundredth of a cent.

9. Which has the lower unit price?

ORANGE JUICE
Brand A: 93¢ for 12 oz
Brand B: $1.07 for 16 oz

Solve.

10. A person traveled 216 km in 6 hr. At this rate, how far will the person travel in 54 hr?

11. If 4 packs of gum cost $5.16, how many packs of gum can you buy for $28.38?

12. Juan's digital car clock loses 5 min in 10 hr. At this rate, how much will it lose in 24 hr?

13. On a map, 4 in. represents 225 mi. If two cities are 7 in. apart on the map, how far are they apart in reality?

14. These triangles are similar. Find the missing lengths.

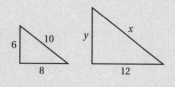

6.1 Introduction to Ratios

a Ratios

OBJECTIVES

After finishing Section 6.1, you should be able to:

a Find fractional notation for ratios.

b Simplify ratios.

FOR EXTRA HELP

TAPE 10   TAPE 9A   MAC: 6
                       IBM: 6

> A *ratio* is the quotient of two quantities.

For example, each day in this country about 5200 people die. Of these, 1070 die of cancer. The *ratio* of those who die of cancer to those who die is shown by the fractional notation

$$\frac{1070}{5200} \quad \text{or by the notation} \quad 1070:5200.$$

We read such notation as "the ratio of 1070 to 5200," listing the numerator first and the denominator second.

> The ratio of a to b is given by $\frac{a}{b}$, where a is the numerator and b is the denominator, or by $a:b$.

In most of our work, we will use fractional notation for ratios.

EXAMPLE 1 Find the ratio of 7 to 8.

The ratio is $\frac{7}{8}$.

EXAMPLE 2 Find the ratio of 31.4 to 100.

The ratio is $\frac{31.4}{100}$.

EXAMPLE 3 Find the ratio of $4\frac{2}{3}$ to $5\frac{7}{8}$.

The ratio is $\dfrac{4\frac{2}{3}}{5\frac{7}{8}}$.

Do Exercises 1–3.

EXAMPLE 4 Hank Aaron hit 755 home runs in 12,364 "at bats." Find the ratio of at bats to home runs.

The ratio is $\frac{12{,}364}{755}$.

EXAMPLE 5 A family earning $21,400 per year allots about $3210 for car expenses. Find the ratio of car expenses to yearly income.

The ratio is $\frac{3210}{21{,}400}$.

Do Exercises 4–6.

1. Find the ratio of 5 to 11.

2. Find the ratio of 57.3 to 86.1.

3. Find the ratio of $6\frac{3}{4}$ to $7\frac{2}{5}$.

4. On average, each of us drinks 182.5 gal of liquid each year. Of this, 21.1 gal is milk. Find the ratio of milk drunk to total amount drunk.

5. Of the 365 days in each year, it takes the average person 107 days of work to pay his or her taxes. Find the ratio of days worked for taxes to total number of days.

6. A pitcher gives up 4 earned runs in $7\frac{2}{3}$ innings of pitching. Find the ratio of earned runs to number of innings pitched.

Answers on page A-3

7. In the triangle below, what is the ratio of the length of the shortest side to the length of the longest side?

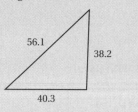

56.1

38.2

40.3

8. Find the ratio of 18 to 27. Then simplify and find two other numbers in the same ratio.

9. Find the ratio of 3.6 to 12. Then simplify and find two other numbers in the same ratio.

10. Find the ratio of 1.2 to 1.5. Then simplify and find two other numbers in the same ratio.

11. In Example 9, what is the ratio of the length of the shortest side of the television screen to the length of the longest side?

Answers on page A-3

EXAMPLE 6   In the triangle at the right:

a) What is the ratio of the length of the longest side to the length of the shortest side?

$$\frac{5}{3}$$

b) What is the ratio of the length of the shortest side to the length of the longest side?

$$\frac{3}{5}$$

3   5   4

Do Exercise 7.

b   SIMPLIFYING NOTATION FOR RATIOS

Sometimes a ratio can be simplified. This provides a means of finding other numbers with the same ratio.

EXAMPLE 7   Find the ratio of 6 to 8. Then simplify and find two other numbers in the same ratio.

We write the ratio in fractional notation and then simplify:

$$\frac{6}{8} = \frac{2 \cdot 3}{2 \cdot 4} = \frac{2}{2} \cdot \frac{3}{4} = 1 \cdot \frac{3}{4} = \frac{3}{4}.$$

Thus, 3 and 4 have the same ratio as 6 and 8. We can express this by saying "6 is to 8" as "3 is to 4."

Do Exercise 8.

EXAMPLE 8   Find the ratio of 2.4 to 10. Then simplify and find two other numbers in the same ratio.

We first write the ratio. Then we multiply by 1 to clear the decimal from the numerator. Then we simplify:

$$\frac{2.4}{10} = \frac{2.4}{10} \cdot \frac{10}{10} = \frac{24}{100} = \frac{4 \cdot 6}{4 \cdot 25} = \frac{4}{4} \cdot \frac{6}{25} = \frac{6}{25}.$$

Thus, 2.4 is to 10 as 6 is to 25.

Do Exercises 9 and 10.

EXAMPLE 9   A standard television screen with a length of 16 in. has a width or height of 12 in. What is the ratio of length to width?

The ratio is $\frac{16}{12} = \frac{4 \cdot 4}{4 \cdot 3} = \frac{4}{4} \cdot \frac{4}{3} = \frac{4}{3}.$

Thus we can say that the ratio of length to width is 4 to 3.

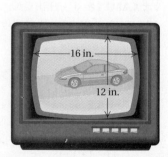

16 in.

12 in.

Do Exercise 11.

Exercise Set 6.1

a Find fractional notation for the ratio. You need not simplify.

1. 4 to 5

2. 3 to 2

3. 178 to 572

4. 329 to 967

5. 0.4 to 12

6. 2.3 to 22

7. 3.8 to 7.4

8. 0.6 to 0.7

9. 56.78 to 98.35

10. 456.2 to 333.1

11. $8\frac{3}{4}$ to $9\frac{5}{6}$

12. $10\frac{1}{2}$ to $43\frac{1}{4}$

13. In Washington, D.C., there are 36.1 lawyers for every 1000 people. What is the ratio of lawyers to people? What is the ratio of people to lawyers?

14. In a bread recipe, there are 2 cups of milk to 12 cups of flour. What is the ratio of cups of milk to cups of flour?

15. Foreign tourists spend $13.1 billion in this country annually. The most money, $2.7 billion, is spent in Florida. What is the ratio of amount spent in Florida to total amount spent? What is the ratio of total amount spent to amount spent in Florida?

16. One person in four plays a musical instrument. In a given group of people, what is the ratio of those who play an instrument to total number of people? What is the ratio of those who do not play an instrument to total number of people?

17. In this rectangle, find the ratios of length to width and of width to length.

478

213

18. In this right triangle, find the ratios of shortest length to longest length and of longest length to shortest length.

47.5
107.3
96.2

b Simplify the ratio. *No Decimals*

19. 4 to 6

20. 6 to 10

21. 18 to 24

22. 28 to 36

23. 4.8 to 10

24. 5.6 to 10

25. 2.8 to 3.6

26. 4.8 to 6.4

ANSWERS

1.
2.
3.
4.
5.
6.
7.
8.
9.
10.
11.
12.
13.
14.
15.
16.
17.
18.
19.
20.
21.
22.
23.
24.
25.
26.

ANSWERS

27. _____

28. _____

29. _____

30. _____

31. _____

32. _____

33. _____

34. _____

35. _____

36. _____

37. _____

38. _____

39. _____

40. _____

41. _____

42. _____

43. _____

44. _____

45. _____

46. _____

27. 20 to 30   **28.** 40 to 60   **29.** 56 to 100   **30.** 42 to 100

31. 128 to 256   **32.** 232 to 116   **33.** 0.48 to 0.64   **34.** 0.32 to 0.96

35. The ratio of Americans aged 18–24 living with their parents is 54 to 100. Find the ratio and simplify.

36. Of every 100 hr, the average mother spends 8 hr cooking. Find the ratio of hours spent cooking to total number of hours and simplify.

37. In this right triangle, find the ratio of shortest length to longest length and simplify.

20.2   6.4

19.2

38. In this rectangle, find the ratio of width to length and simplify.

5.4

8.8

SKILL MAINTENANCE

Use = or ≠ for ▓ to write a true sentence.

39. $\dfrac{12}{8}$ ▓ $\dfrac{6}{4}$

40. $\dfrac{4}{7}$ ▓ $\dfrac{5}{9}$

Divide. Write decimal notation for the answer.

41. 200 ÷ 4   **42.** 95 ÷ 10   **43.** 232 ÷ 16   **44.** 342 ÷ 2.25

SYNTHESIS

45. 🔲 In Australia, there are 13,339,000 people and 145,304,000 sheep. What is the ratio of people to sheep? What is the ratio of sheep to people?

46. Find the ratio of $3\dfrac{3}{4}$ to $5\dfrac{7}{8}$ and simplify.

6.2 Rates

OBJECTIVES

After finishing Section 6.2, you should be able to:

a Give the ratio of two different kinds of measure as a rate.

b Find unit prices and use them to determine which of two possible purchases has the lower unit price.

FOR EXTRA HELP

TAPE 10   TAPE 9A   MAC: 6
                        IBM: 6

a When a ratio is used to compare two different kinds of measure, we call it a **rate**. Suppose that a car is driven 200 km in 4 hr. The ratio

$$\frac{200 \text{ km}}{4 \text{ hr}}, \quad \text{or } 50 \frac{\text{km}}{\text{hr}}, \quad \text{or } 50 \text{ kilometers per hour}, \quad \text{or } 50 \text{ km/h}$$

Recall that "per" means "division," or "for each."

is the rate traveled in kilometers per hour, which is the division of the number of kilometers by the number of hours. A ratio of distance traveled to time is also called **speed**.

EXAMPLE 1 A European driver travels 145 km on 2.5 L of gas. What is the rate in kilometers per liter?

$$\frac{145 \text{ km}}{2.5 \text{ L}}, \quad \text{or} \quad 58 \frac{\text{km}}{\text{L}}.$$

EXAMPLE 2 It takes 60 oz of grass seed to seed 3000 sq ft of lawn. What is the rate in ounces per square foot?

$$\frac{60 \text{ oz}}{3000 \text{ sq ft}} = \frac{1}{50} \frac{\text{oz}}{\text{sq ft}}, \quad \text{or} \quad 0.02 \frac{\text{oz}}{\text{sq ft}}.$$

EXAMPLE 3 A cook buys 10 lb of potatoes for $2.69. What is the rate in cents per pound?

$$\frac{\$2.69}{10 \text{ lb}} = \frac{269 \text{ cents}}{10 \text{ lb}}, \quad \text{or} \quad 26.9 \frac{\text{cents}}{\text{lb}}.$$

EXAMPLE 4 A student nurse working in a health center earned $3690 for working 3 months one summer. What was the rate of pay per month?

The rate of pay is the ratio of money earned per length of time worked, or

$$\frac{\$3690}{3 \text{ mo}} = 1230 \frac{\text{dollars}}{\text{month}}, \quad \text{or} \quad \$1230 \text{ per month}.$$

EXAMPLE 5 There were 9.4 marriages for every 1000 persons in a recent year. What was the marriage rate?

$$\frac{9.4 \text{ marriages}}{1000 \text{ persons}}, \quad \text{or} \quad 0.0094 \frac{\text{marriages}}{\text{person}}$$

Do Exercises 1–8.

What is the rate, or speed, in miles per hour?

1. 45 mi, 9 hr

2. 120 mi, 10 hr

3. 3 mi, 10 hr

What is the rate, or speed, in feet per second?

4. 2200 ft, 2 sec

5. 52 ft, 13 sec

6. 232 ft, 16 sec

7. A well-hit golf ball can travel 500 ft in 2 sec. What is the rate, or speed, of the golf ball in feet per second?

8. A leaky faucet can lose 14 gal of water in a week. What is the rate in gallons per day?

Answers on page A-3

9. A customer bought a 14-oz package of oat bran for $2.89. What is the unit price in cents per ounce? Round to the nearest hundredth of a cent.

.21

2.89/14

10. Which has the lower unit price? [*Note:* 1 qt = 32 fl oz (fluid ounces).]

$1.19

$1.79

Fruit Punch
1qt, 14oz

Fruit Punch
64oz

A

B

.00259   *.00190*

b   UNIT PRICING

A **unit price** is the ratio of price to the number of units.

A unit price is applied so that the price of one unit can be determined.

EXAMPLE 6   A customer bought a 20-lb box of powdered detergent for $19.47. What is the unit price in dollars per pound?

The unit price is the price in dollars for each pound.

$$\text{Unit price} = \frac{\text{Price}}{\text{Number of units}}$$

$$= \frac{\$19.47}{20 \text{ lb}} = \frac{19.47}{20} \cdot \frac{\$}{\text{lb}}$$

$$= 0.9735 \text{ dollars per pound}$$

Do Exercise 9.

For comparison shopping, it helps to find unit prices.

EXAMPLE 7   Which has the lower unit price?

A   B

To find out, we compare the unit prices—in this case, the price per ounce.

For can A: $\dfrac{48 \text{ cents}}{14 \text{ oz}} \approx 3.429 \dfrac{\text{cents}}{\text{oz}}.$

For can B: We need to find the total number of ounces:

1 lb, 15 oz = 16 oz + 15 oz = 31 oz.

Then

$$\frac{99 \text{ cents}}{31 \text{ oz}} \approx 3.194 \frac{\text{cents}}{\text{oz}}.$$

Thus can B has the lower unit price.

In many stores, unit prices are now listed on the items or the shelves.

Do Exercise 10.

Exercise Set 6.2

a In Exercises 1–6, find the rate as a ratio of distance to time.

1. 120 km, 3 hr

2. 18 mi, 9 hr

3. 440 m, 40 sec

4. 200 mi, 25 sec

5. 342 yd, 2.25 days

6. 492 m, 60 sec

7. A car is driven 500 mi in 20 hr. What is the rate in miles per hour? in hours per mile?

8. A student eats 3 hamburgers in 15 min. What is the rate in hamburgers per minute? in minutes per hamburger?

9. To water a lawn adequately requires 623 gal of water for every 1000 sq ft. What is the rate in gallons per square foot?

10. An 8-lb shankless ham contains 36 servings of meat. What is the ratio in servings per pound?

11. A long-distance telephone call between two cities costs $5.75 for 10 min. What is the rate in cents per minute?

12. A car is driven 200 km on 40 L of gasoline. What is the rate in kilometers per liter?

13. Light travels 186,000 mi in 1 sec. What is its rate, or speed, in miles per second?

14. Sound travels 1100 ft in 1 sec. What is its rate, or speed, in feet per second?

1. _____

2. _____

3. _____

4. _____

5. _____

6. _____

7. _____

8. _____

9. _____

10. _____

11. _____

12. _____

13. _____

14. _____

15. Impulses in nerve fibers travel 310 km in 2.5 hr. What is the rate, or speed, in kilometers per hour?

16. A black racer snake can travel 4.6 km in 2 hr. What is its rate, or speed, in kilometers per hour?

15. _____

16. _____

17. A jet flew 2660 mi in 4.75 hr. What was its speed?

18. A turtle traveled 0.42 mi in 2.5 hr. What was its speed?

17. _____

b

19. The fabric for a wedding gown costs $80.75 for 8.5 yd. Find the unit price.

20. An 8-oz bottle of shampoo costs $2.59. Find the unit price.

18. _____

19. _____

20. _____

21. A 2-lb can of decaffeinated coffee costs $8.59. What is the unit price in cents per ounce? Round to the nearest hundredth of a cent.

22. A 24-can package of 12-oz cans of soda costs $5.79. What is the unit price in cents per ounce? Round to the nearest hundredth of a cent.

21. _____

22. _____

23. A $\frac{2}{3}$-lb package of Monterey Jack cheese costs $2.89. Find the unit price in dollars per pound. Round to the nearest hundredth of a dollar.

24. A $1\frac{1}{4}$-lb container of cottage cheese costs $1.35. Find the unit price in dollars per pound.

23. _____

Which has the lower unit price?

24. _____

25.

CHILI SAUCE
Brand A: 18 oz for $1.79
Brand B: 16 oz for $1.65

$\frac{\$1.79}{18} = .099$

$\frac{1.45}{16} . 103$

26.

NAPKINS
Brand A: 140 napkins for 61 cents
Brand B: 125 napkins for 44 cents

$\frac{140}{61} = 2.295$

$\frac{125}{44} 2.840$

25. _____

27.

GRAPEFRUIT JUICE
Brand A: $2.69 for 2 qt
Brand B: $1.97 for 48 oz

28.

EVAPORATED MILK
Brand A: 56 cents for 13 oz
Brand B: $1.99 for 1 qt, 8 oz

26. _____

27. _____

29.

SOAP
Brand A: $3.96 for 3 bars
Brand B: $2.67 for 2 bars

30.

BROCCOLI SOUP
Brand A: 10.75 oz for 79 cents
Brand B: 8.25 oz for 69 cents

28. _____

29. _____

30. _____

ANSWERS

31. _____

32. _____

33. _____

34. _____

35. _____

36. _____

37. _____

38. _____

39. _____

40. _____

41. _____

42. _____

43. a) _____

b) _____

31.

FANCY TUNA
Brand A: $1.29 for 7 oz
Brand B: $1.19 for $6\frac{1}{8}$ oz

32.

FLOUR
Brand A: $1.25 for 3 lb, 2 oz
Brand B: $0.99 for 28 oz

33.

SPARKLING WATER
The same kind of water is sold in two types of bottle. Which type has the lower unit price?
Six 10-oz bottles for $3.09, or Four 12-oz bottles for $2.39

34.

GRAPE JELLY
The same kind of jelly is sold in two sizes. Which size has the lower unit price?
18 oz for $1.33, or 32 oz for $1.59

35.

INSTANT PUDDING (SAME BRAND)
Package A: 3.9 oz for 63¢
Package B: 5.9 oz for 99¢

36.

TACO SHELLS (SAME BRAND)
Package A: 10 in a box for $1.43
Package B: 16 in a box for $2.19

SKILL MAINTENANCE

37. There are 20.6 million people in this country who play the piano and 18.9 million who play the guitar. How many more play the piano than the guitar?

Multiply.

38.
$$\begin{array}{r} 4\,5.6\,7 \\ \times \quad 2.4 \\ \hline \end{array}$$

39.
$$\begin{array}{r} 6\,7\,8.1\,9 \\ \times \quad 1\,0\,0 \\ \hline \end{array}$$

40.
$$\begin{array}{r} 6\,7\,8.1\,9 \\ \times \quad 0.0\,0\,1 \\ \hline \end{array}$$

41. 84.3×69.2

42. 1002.56×465

SYNTHESIS

43. Recently, certain manufacturers have been changing the size of their containers in such a way that the consumer thinks the price of a product has been lowered when in reality, a higher unit price is being charged.

a) Some aluminum juice cans are now concave (curved in) on the bottom. Suppose the volume of the can in the figure has been reduced from a fluid capacity of 6 oz to 5.5 oz, and the price of each can has been reduced from 65¢ to 60¢. Find the unit price of each container in cents per ounce.

b) Suppose at one time the cost of a certain kind of paper towel was $0.89 for a roll containing 78 ft² of absorbent surface. Later the surface area was changed to 65 ft² and the price was decreased to $0.79. Find the unit price of each product in cents per square foot.

6.3 Proportions

a PROPORTION

When two pairs of numbers (such as 3, 2 and 6, 4) have the same ratio, we say that they are **proportional**. The equation

$$\frac{3}{2} = \frac{6}{4}$$

states that the pairs 3, 2 and 6, 4 are proportional. Such an equation is called a **proportion**. We sometimes read $\frac{3}{2} = \frac{6}{4}$ as "3 is to 2 as 6 is to 4."

Since ratios are represented by fractional notation, we can test whether two ratios are the same by using the test for equality discussed in Section 2.5. (It is also a skill to review for the chapter test.)

> **EXAMPLE 1** Determine whether 1, 2 and 3, 6 are proportional.
>
> We can use cross-products:
>
> $1 \cdot 6 = 6$ $\quad \dfrac{1}{2} \dfrac{3}{6} \quad$ $2 \cdot 3 = 6.$
>
> Since the cross-products are the same, $6 = 6$, we know that $\frac{1}{2} = \frac{3}{6}$, so the numbers are proportional.

> **EXAMPLE 2** Determine whether 2, 5 and 4, 7 are proportional.
>
> We can use cross-products:
>
> $2 \cdot 7 = 14$ $\quad \dfrac{2}{5} \dfrac{4}{7} \quad$ $5 \cdot 4 = 20.$
>
> Since the cross-products are not the same, $14 \neq 20$, we know that $\frac{2}{5} \neq \frac{4}{7}$, so the numbers are not proportional.

Do Exercises 1–3.

> **EXAMPLE 3** Determine whether 3.2, 4.8 and 0.16, 0.24 are proportional.
>
> We can use cross-products:
>
> $3.2 \times 0.24 = 0.768$ $\quad \dfrac{3.2}{4.8} \dfrac{0.16}{0.24} \quad$ $4.8 \times 0.16 = 0.768.$
>
> Since the cross-products are the same, $0.768 = 0.768$, we know that $\frac{3.2}{4.8} = \frac{0.16}{0.24}$, so the numbers are proportional.

Do Exercises 4 and 5.

Determine whether the two pairs of numbers are proportional.

1. 3, 4 and 6, 8

2. 1, 4 and 10, 39

3. 1, 2 and 20, 39

Determine whether the two pairs of numbers are proportional.

4. 6.4, 12.8 and 5.3, 10.6

5. 6.8, 7.4 and 3.4, 4.2

Answers on page A-3

6. Determine whether $4\frac{2}{3}$, $5\frac{1}{2}$ and 14, $16\frac{1}{2}$ are proportional.

EXAMPLE 4 Determine whether $4\frac{2}{3}$, $5\frac{1}{2}$ and $8\frac{7}{8}$, $16\frac{1}{3}$ are proportional.

We can use cross-products:

$$4\frac{2}{3} \cdot 16\frac{1}{3} = 76\frac{2}{9} \qquad \boxed{\begin{matrix} 4\frac{2}{3} & 8\frac{7}{8} \\ 5\frac{1}{2} & 16\frac{1}{3} \end{matrix}} \qquad 5\frac{1}{2} \cdot 8\frac{7}{8} = 48\frac{13}{16}.$$

Since the cross-products are not the same, $76\frac{2}{9} \neq 48\frac{13}{16}$, we know that the numbers are not proportional.

Do Exercise 6.

b SOLVING PROPORTIONS

Let us now see how to solve proportions. Consider the proportion

$$\frac{x}{8} = \frac{3}{5}.$$

One way to solve a proportion is to use cross-products. Then we can divide on both sides to get the variable alone:

$$5 \cdot x = 8 \cdot 3 \qquad \text{Finding cross-products}$$

$$x = \frac{8 \cdot 3}{5} \qquad \text{Dividing by 5 on both sides}$$

$$x = \frac{24}{5} \qquad \text{Multiplying}$$

$$x = 4.8. \qquad \text{Dividing}$$

7. Solve: $\frac{x}{63} = \frac{2}{9}$.

We can check that 4.8 is the solution by replacing x by 4.8 and using cross-products:

$$4.8 \cdot 5 = 24 \qquad \boxed{\begin{matrix} 4.8 & 3 \\ 8 & 5 \end{matrix}} \qquad 8 \cdot 3 = 24.$$

Since the cross-products are the same, it follows that $\frac{4.8}{8} = \frac{3}{5}$, so the numbers 4.8, 8 and 3, 5 are proportional, and 4.8 is the solution of the equation.

> To solve $\frac{x}{a} = \frac{c}{d}$, find cross-products and divide on both sides to get x alone.

Do Exercise 7.

EXAMPLE 5 *Estimating wildlife populations.* To determine the number of deer in a forest, a conservationist catches 612 deer, tags them, and releases them. Later, 244 deer are caught, and it is found that 72 of them are tagged. Estimate how many deer are in the forest.

5. To determine the number of fish in a lake, a conservationist catches 225 fish, tags them, and throws them back into the lake. Later, 108 fish are caught, and it is found that 15 of them are tagged. Estimate how many fish are in the lake.

We let D = the number of deer in the forest. Then we translate to a proportion.

$$\text{Deer tagged originally} \rightarrow \frac{612}{D} = \frac{72}{244} \leftarrow \text{Tagged deer caught later}$$
$$\text{Deer in forest} \longrightarrow \phantom{\frac{612}{D}} \phantom{\frac{72}{244}} \leftarrow \text{Deer caught later}$$

Solve:

$$244 \cdot 612 = D \cdot 72 \qquad \text{Finding cross-products}$$

$$\frac{244 \cdot 612}{72} = D \qquad \text{Dividing by 72 on both sides}$$

$$\frac{2 \cdot 2 \cdot 2 \cdot 2 \cdot 3 \cdot 3 \cdot 17 \cdot 61}{2 \cdot 2 \cdot 2 \cdot 3 \cdot 3} = D \qquad \text{Factoring}$$

$$2 \cdot 17 \cdot 61 = D \qquad \text{Simplifying}$$

$$2074 = D$$

Thus we estimate that there are 2074 deer in the forest.

Do Exercise 5.

6. In Example 6, how much seed would be needed for 7000 sq ft of lawn?

EXAMPLE 6 It takes 60 oz of grass seed to seed 3000 ft² of lawn. At this rate, how much would be needed for 5000 ft² of lawn?

We let g = the number of ounces of grass seed. Then we translate to a proportion.

$$\text{Grass seed needed} \rightarrow \frac{g}{5000} = \frac{60}{3000} \leftarrow \text{Grass seed needed}$$
$$\text{Amount of lawn} \longrightarrow \phantom{\frac{g}{5000}} \phantom{\frac{60}{3000}} \leftarrow \text{Amount of lawn}$$

Solve:

$$3000 \cdot g = 5000 \cdot 60 \qquad \text{Finding cross-products}$$

$$g = \frac{5000 \cdot 60}{3000} \qquad \text{Dividing by 3000 on both sides}$$

$$g = \frac{1000 \cdot 5 \cdot 3 \cdot 20}{1000 \cdot 3} \qquad \text{Factoring}$$

$$g = 5 \cdot 20 \qquad \text{Simplifying}$$

$$g = 100$$

Thus, 100 oz are needed for 5000 ft² of lawn.

Do Exercise 6.

Answers on page A-3

APPLICATIONS: QUARTERBACK COMPARISONS

Troy Aikman of the Dallas Cowboys and Jim Kelly of the Buffalo Bills faced each other recently in a Super Bowl. At right are many of the statistics used to compare them before the game. Many such comparisons can be done using ratio and proportions.

Use ratio and proportions to analyze which, if either, of the quarterbacks seems to be the better player before that Super Bowl.

Find decimal notation for each ratio.

The Quarterback Showdown*

Comparing the starting quarterbacks for a recent Super Bowl—Troy Aikman of the Dallas Cowboys and Jim Kelly of the Buffalo Bills.

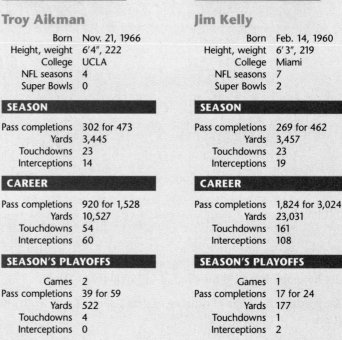

Troy Aikman		**Jim Kelly**	
Born	Nov. 21, 1966	Born	Feb. 14, 1960
Height, weight	6'4", 222	Height, weight	6'3", 219
College	UCLA	College	Miami
NFL seasons	4	NFL seasons	7
Super Bowls	0	Super Bowls	2

SEASON		**SEASON**	
Pass completions	302 for 473	Pass completions	269 for 462
Yards	3,445	Yards	3,457
Touchdowns	23	Touchdowns	23
Interceptions	14	Interceptions	19

CAREER		**CAREER**	
Pass completions	920 for 1,528	Pass completions	1,824 for 3,024
Yards	10,527	Yards	23,031
Touchdowns	54	Touchdowns	161
Interceptions	60	Interceptions	108

SEASON'S PLAYOFFS		**SEASON'S PLAYOFFS**	
Games	2	Games	1
Pass completions	39 for 59	Pass completions	17 for 24
Yards	522	Yards	177
Touchdowns	4	Touchdowns	1
Interceptions	0	Interceptions	2

TROY AIKMAN

1. Pass completion that season
2. Pass completion for his career
3. Pass completion for the playoffs
4. Career interceptions for total passes thrown
5. Career touchdowns for total passes thrown

JIM KELLY

6. Pass completion that season
7. Pass completion for his career
8. Pass completion for the playoffs
9. Career interceptions for total passes thrown
10. Career touchdowns for total passes thrown

11. On the basis of your ratio comparisons, who, if either, do you think would be the winning quarterback in the Super Bowl?

12. On the basis of his career records, how many completions do you think Troy Aikman is likely to throw in his first 100 passes next season? How many touchdowns?

13. On the basis of his career records, how many completions do you think Jim Kelly is likely to throw in his first 100 passes next season? How many touchdowns?

*From *USA Today*, January 29, 1993. Copyright 1993 USA TODAY. Reprinted with permission.

Exercise Set 6.4

a Solve.

1. A car travels 234 km in 14 days. At this rate, how far would it travel in 42 days?

2. A car traveled 84 mi on 6.5 gal of gasoline. At this rate, how many gallons would be needed to travel 126 mi?
$$\frac{84}{6.5} = \frac{126}{N}$$

3. If 2 sweatshirts cost $18.80, how much would 9 sweatshirts cost?

4. If 2 bars of soap cost $0.49, how many bars of soap can you buy for $6.37?

5. Jim Abbott, a pitcher for the New York Yankees, gave up 73 earned runs in 211 innings. What was his earned run average? Round to the nearest hundredth.

6. Scott Erikson, a pitcher for the Minnesota Twins, gave up 86 earned runs in 212 innings. What was his earned run average? Round to the nearest hundredth.

7. To determine the number of deer in a game preserve, a forest ranger catches 318 deer, tags them, and releases them. Later, 168 deer are caught, and it is found that 56 of them are tagged. Estimate how many deer there are in the game preserve.

8. To determine the number of trout in a lake, a conservationist catches 112 trout, tags them, and throws them back into the lake. Later, 82 trout are caught, and it is found that 32 of them are tagged. Estimate how many trout there are in the lake.

9. A bookstore manager knows that 24 books weigh 37 lb. How much do 40 books weigh?

10. An 8-lb shankless ham contains 36 servings of meat. How many pounds of ham would be needed for 54 servings?

11. The coffee beans from 14 trees are required to produce the 17 lb of coffee that each person in the United States drinks, on average, each year. How many trees are required to produce 391 lb of coffee?

12. A student bought a car. In the first 8 months, it was driven 10,000 km. At this rate, how many kilometers would the car be driven in 1 year? (Use 12 months for 1 year.)

13. In a metal alloy, the ratio of zinc to copper is 3 to 13. If there are 520 lb of copper, how many pounds of zinc are there?

14. A contractor finds that 6 gal of paint will cover 1650 ft^2 of wall area. How much paint would be needed for a job of 5775 ft^2?

15.

16.

17.

18.

19.

20.

21.

22. a)

b)

23.

15. A quality-control inspector examined 200 lightbulbs and found 18 of them to be defective. At this rate, how many defective bulbs will there be in a lot of 22,000?

16. A professor must grade 32 essays in a literature class. She can grade 5 essays in 40 min. At this rate, how long will it take her to grade all 32 essays?

17. On a National Geographic map, 1 in. represents 16.6 mi. If two cities are 3.5 in. apart on the map, how far are they apart in reality?

18. On a map, $\frac{1}{4}$ in. represents 50 mi. If two cities are $3\frac{1}{4}$ in. apart on the map, how far are they apart in reality?

19. Under typical conditions, $1\frac{1}{2}$ ft of snow will melt to 2 in. of water. To how many inches of water will $5\frac{1}{2}$ ft of snow melt?

20. Tires are often priced according to the number of miles they are expected to be driven. Suppose a tire priced at $39.76 is expected to be driven 35,000 mi. How much would you expect to pay for a tire that is expected to be driven 40,000 mi?

21. A student attends a university whose academic year consists of two 16-week semesters. She budgets $800 for incidental expenses for the academic year. After 3 weeks, she has spent $80 for incidental expenses. Assuming the student continues to spend at the same rate, will the budget for incidental expenses be adequate? If not, when will the money be exhausted and how much more will be needed to complete the year?

22. A basic sound system consists of a CD player, a receiver–amplifier, and two speakers. A rule of thumb used to estimate the relative investment in these components is 1 : 3 : 2. That is, the receiver–amplifier should cost three times the amount spent on the CD player and the speakers should cost twice as much as the amount spent on the CD player.

a) You have $1800 to spend. How should you allocate the funds if you use this rule of thumb?

b) How should you allocate a budget of $3000?

SYNTHESIS

23. Cy Young, one of the greatest baseball pitchers of all time, had an earned run average of 2.63. He pitched more innings, 7356, than anyone in the history of baseball. How many earned runs did he give up?

6.5 *Similar Triangles*

a | PROPORTIONS AND SIMILAR TRIANGLES

Look at the pair of triangles below. Note that they appear to have the same shape, but their sizes are different. These are examples of **similar triangles**. By using a magnifying glass, you could imagine enlarging the smaller triangle to get the larger. This process works because the corresponding sides of each triangle have the same ratio. That is, the following proportion is true.

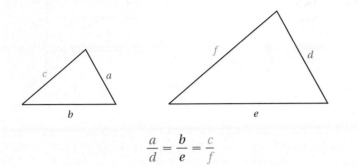

$$\frac{a}{d} = \frac{b}{e} = \frac{c}{f}$$

> *Similar triangles* have the same shape. Their corresponding sides have the same ratio—that is, they are proportional.

EXAMPLE 1   The triangles below are similar triangles. Find the missing length x.

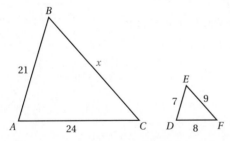

The ratio of x to 9 is the same as the ratio of 24 to 8. We get the proportion

$$\frac{x}{9} = \frac{24}{8}.$$

We solve the proportion:

$8 \cdot x = 9 \cdot 24$ **Finding cross-products**

$x = \dfrac{9 \cdot 24}{8}$ **Dividing by 8 on both sides**

$x = 27.$ **Simplifying**

The missing length x is 27. We could have also used $\frac{x}{9} = \frac{21}{7}$ to find x. ____

Do Exercise 1.

OBJECTIVE

After finishing Section 6.5, you should be able to:

a   Find lengths of sides of similar triangles using proportions.

FOR EXTRA HELP

TAPE 11   TAPE 10A   MAC: 6
                                   IBM: 6

1. This pair of triangles is similar. Find the missing length x.

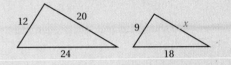

Answer on page A-3

2. How high is a flagpole that casts a 45-ft shadow at the same time that a 5.5-ft woman casts a 10-ft shadow?

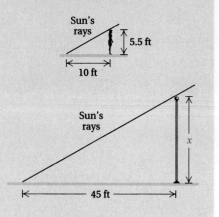

Similar triangles and proportions can often be used to find lengths that would ordinarily be difficult to measure. For example, we could find the height of a flagpole without climbing it or the distance across a river without crossing it.

EXAMPLE 2   How high is a flagpole that casts a 56-ft shadow at the same time that a 6-ft man casts a 5-ft shadow?

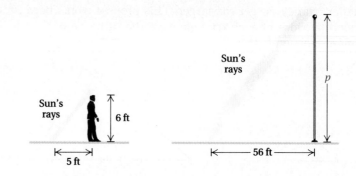

If we use the sun's rays to represent the third side of the triangle in our drawing of the situation, we see that we have similar triangles. Let p = the height of the flagpole. The ratio of 6 to p is the same as the ratio of 5 to 56. Thus we have the proportion

$$\frac{6}{p} = \frac{5}{56}.$$

Solve: $6 \cdot 56 = 5 \cdot p$ Finding cross-products

$\quad\quad\dfrac{6 \cdot 56}{5} = p$ Dividing by 5 on both sides

$\quad\quad\quad 67.2 = p$ Simplifying

The height of the flagpole is 67.2 ft.

Do Exercise 2.

Answer on page A-3

Exercise Set 6.5

a The triangles in each exercise are similar. Find the missing lengths.

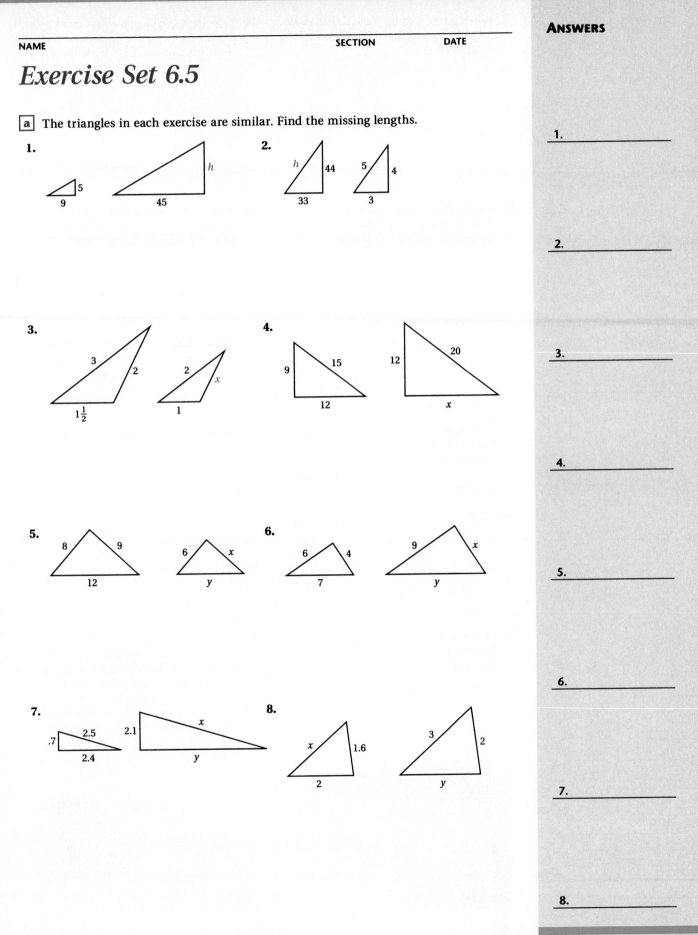

1.

5

9

h

45

2.

h

44

33

5

4

3

3.

3

2

$1\frac{1}{2}$

2

x

1

4.

9

15

12

12

20

x

5.

8   9

12

6   x

y

6.

6   4

7

9   x

y

7.

.7

2.5

2.4

2.1

x

y

8.

x

1.6

2

3

2

y

9. When a tree 8 m high casts a shadow 5 m long, how long a shadow is cast by a person 2 m tall?

10. How high is a flagpole that casts a 42-ft shadow at the same time that a $5\frac{1}{2}$-ft woman casts a 7-ft shadow?

11. How high is a tree that casts a 27-ft shadow at the same time that a 4-ft fence post casts a 3-ft shadow?

12. How high is a tree that casts a 32-ft shadow at the same time that an 8-ft light pole casts a 9-ft shadow?

$$\frac{x}{8} = \frac{32}{9}$$

$$9x = 8 \cdot 32 \qquad \frac{8 \times 32}{9} =$$

SKILL MAINTENANCE

13. A student has $34.97 to spend for a book at $49.95, a CD at $14.88, and a sweatshirt at $29.95. How much more money does the student need to make these purchases?

14. Divide: $80.892 \div 8.4$.

Multiply.

15. 8.4×80.892

16. 0.01×274.568

17. 100×274.568

18. 0.002×274.568

SYNTHESIS

19. Find the distance across the river. Assume that the ratio of d to 25 ft is the same as the ratio of 40 ft to 10 ft.

20. To measure the height of a hill, a string is drawn tight from level ground to the top of the hill. A 3-ft yardstick is placed under the string, touching it at point P, a distance of 5 ft from point G, where the string touches the ground. The string is then detached and found to be 120 ft long. How high is the hill?

ANSWERS

9. _____

10. _____

11. _____

12. _____

13. _____

14. _____

15. _____

16. _____

17. _____

18. _____

19. _____

20. _____

CALCULATOR CONNECTION

Solve. Round the answer to the nearest thousandth.

1. $\dfrac{8664.3}{10{,}344.8} = \dfrac{x}{9776.2}$ **2.** $\dfrac{12.0078}{56.0115} = \dfrac{789.23}{y}$

3. $\dfrac{840}{2100} = \dfrac{420}{m}$ **4.** $\dfrac{24.62}{t} = \dfrac{122\frac{5}{6}}{645\frac{7}{8}}$

The triangles in each exercise are similar triangles. Find the lengths not given.

5.

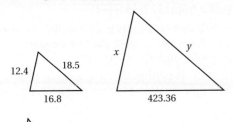

6.

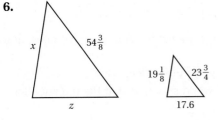

7.

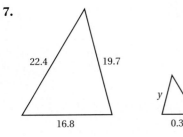

8.

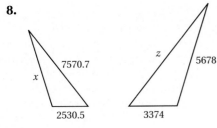

EXTENDED SYNTHESIS EXERCISES

1. The score of an NBA basketball game at the end of the first quarter is the New York Knickerbockers 28, the Portland Trail Blazers 23. Assuming the teams continue to score at this rate, what will be the final score?

2. Nancy Morano-Smith wants to win a season football ticket from the local bookstore. Her goal is to guess the number of marbles in an 8-gal jar. She knows that there are 128 oz in a gallon. She goes home and fills an 8-oz jar with 46 marbles. How many marbles should she guess are in the jar?

3. It takes Yancy Martinez 10 min to type two-thirds of a page of his term paper. At this rate, how long will it take him to type a 7-page term paper?

(continued)

State Lottery Profits The chart below shows the profits of state lotteries in a recent year. Use the information to do the exercises.

STATE	PROFIT (IN MILLIONS)
Arizona	$ 22.0
Colorado	32.0
Connecticut	148.8
Delaware	15.0
Illinois	517.8
Maine	4.4
Maryland	263.7
Massachusetts	284.0
Michigan	320.0
New Hampshire	4.3
New Jersey	388.0
New York	615.0
Ohio	338.0
Pennsylvania	572.6
Rhode Island	18.6
Vermont	1.2
Washington	58.8

4. Which state profited the most from lotteries?

5. Which state profited the least from lotteries?

6. How much more did the state with the most lottery profit make than the state with the least profit?

7. How much, in billions, did these states profit together from lotteries?

8. What is the ratio of the lottery profits of New York to the entire profits taken in by lotteries? Use a calculator.

9. How much did New England profit from lotteries?

10. How much more did Ohio make than Maryland?

11. The population of Washington is 4,300,000. At what rate, in dollars per person, did the people of Washington profit from their lottery?

12. The population of New York is 17,667,000. At what rate, in dollars per person, did the people of New York profit from their lottery?

13. The population of Illinois is 11,486,000. At what rate, in dollars per person, did the people of Illinois profit from their lottery?

14. Which state, Washington, New York, or Illinois, has the highest ratio of lottery profits per person?

EXERCISES FOR THINKING AND WRITING

1. Discuss the relationships among ratios, rates, and proportions.

2. In what way is a unit price a rate?

3. Find a real-world situation that fits the equation

$$\frac{18}{128.95} = \frac{y}{789.89}.$$

Summary and Review Exercises: Chapter 6

The review objectives to be tested in addition to the material in this chapter are [2.5c], [4.4a], [5.1a], and [5.2a].

Write fractional notation for the ratio. Do not simplify.

1. 47 to 84

$$\frac{47}{84}$$

2. 46 to 1.27

$$\frac{46}{1.27}$$

3. 83 to 100

4. 0.72 to 197

$$\frac{.72}{197}$$

5. Each day in the United States, 5200 people die. Of these, 1070 die of cancer. Write fractional notation for the ratio of the number of people who die to the number of people who die of cancer.

Simplify the ratio.

6. 9 to 12 $\frac{3}{4}$

7. 3.6 to 6.4 $\frac{3.6}{4.4} \text{ or } \frac{10}{10} = \frac{36}{64} = \frac{9}{16}$

8. What is the rate in miles per hour?

117.7 miles,   5 hours

9. A lawn requires 319 gal of water for every 500 ft². What is the rate in gallons per square foot?

10. What is the rate in dollars per kilogram?

$355.04,   14 kilograms

11. A 25-lb turkey serves 18 people. What is the rate in servings per pound?

$$\frac{18}{25} = \frac{x}{1} = .72$$

12. A 1 lb, 7 oz package of flour costs $1.30. Find the unit price in cents per ounce. Round to the nearest tenth of a cent.

13. It costs 79 cents for a $14\frac{1}{2}$-oz can of tomatoes. Find the unit price in cents per ounce. Round to the nearest hundredth of a cent.

$$\frac{79}{14^{1}/_{2}} = 5.45\cent \text{ per oz}$$

What has the lower unit price?

14.

WHITE BREAD
Brand A: 16 oz for 89 cents
Brand B: 12 oz for 65 cents

15.

CANNED PINEAPPLE JUICE
Brand A: 12 oz for 99 cents
Brand B: 18 oz for $1.26

Determine whether the two pairs of numbers are proportional

16. 9, 15 and 36, 59   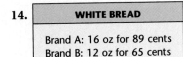 531 540

No

17. 24, 37 and 40, 46.25

$$\frac{24}{37} = \frac{40}{46.25}$$ 1480   No
                              1110

Solve.

18. $\dfrac{8}{9} = \dfrac{x}{36}$

19. $\dfrac{120}{\frac{3}{7}} = \dfrac{7}{x}$

$120 \cdot x$

$7 \times \frac{3}{7}$

$\frac{3}{120} = \frac{1}{40}$

20. $\dfrac{6}{x} = \dfrac{48}{56}$

21. $\dfrac{4.5}{120} = \dfrac{0.9}{x}$

$$\frac{120 \times 0.9}{4.5} = 24$$

Solve.

22. If 3 dozen eggs cost $2.67, how much will 5 dozen eggs cost?

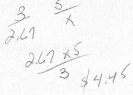

$$\frac{2.67 \times 5}{3} \quad \$4.45$$

23. A factory manufacturing computer circuits found 39 defective circuits in a lot of 65 circuits. At this rate, how many defective circuits can be expected in a lot of 585 circuits?

24. A train travels 448 mi in 7 hr. At this rate, how far will it travel in 13 hr?

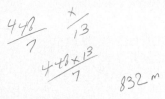

$$\frac{448 \times 13}{7} \quad 832 \, m$$

25. Fifteen acres are required to produce 54 bushels of tomatoes. At this rate, how many acres are required to produce 97.2 bushels of tomatoes?

26. It is known that 5 people produce 13 kg of garbage in one day. San Diego, California, has 920,000 people. How many kilograms of garbage are produced in San Diego in one day?

27. Under typical conditions, $1\frac{1}{2}$ ft of snow will melt to 2 in. of water. To how many inches of water will $4\frac{1}{2}$ ft of snow melt?

28. In Michigan, there are 2.3 lawyers for each 1000 people. The population of Detroit is 1,140,000. How many lawyers are there in Detroit?

29. These triangles are similar. Find x and y.

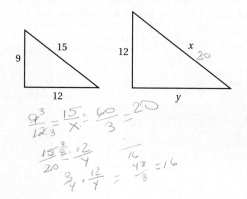

SKILL MAINTENANCE

30. A family has $2347.89 in its checking account. Someone writes checks for $678.95 and $38.54. How much is left in the checking account?

Use = or ≠ for ▨ to write a true sentence.

31. $\frac{5}{2}$ ▨ $\frac{10}{4}$

32. $\frac{4}{6}$ ▨ $\frac{8}{10}$

33. Multiply.

$$\begin{array}{r} 4\ 5\ 6.1 \\ \times\quad 2\ 3.4 \\ \hline \end{array}$$

34. Divide. Write decimal notation for the answer.

$$5.6\,\overline{)\,2\ 5\ 4.8}$$

Test: Chapter 6

Write fractional notation for the ratio. Do not simplify.

1. 85 to 97

2. 0.34 to 124

Simplify the ratio.

3. 18 to 20

4. 0.75 to 0.96

5. What is the rate in feet per second?

  10 feet,   16 seconds

6. A 12-lb shankless ham contains 16 servings. What is the rate in servings per pound?

7. A 1-lb, 2-oz package of Mahi Mahi fish costs $3.49. Find the unit price in cents per ounce. Round to the nearest hundredth of a cent.

8. Which has the lower unit price?

ORANGE JUICE
Brand A: $1.19 for 12 oz
Brand B: $1.33 for 16 oz

Determine whether the two pairs of numbers are proportional.

9. 7, 8   and   63, 72

10. 1.3, 3.4   and   5.6, 15.2

Solve.

11. $\dfrac{9}{4} = \dfrac{27}{x}$

12. $\dfrac{150}{2.5} = \dfrac{x}{6}$

Solve.

13. A person traveled 432 km in 12 hr. At this rate, how far would the person travel in 42 hr?

14. If 2 cans of apricots cost $3.39, how many cans of apricots can you buy for $74.58?

15. A watch loses 2 min in 10 hr. At this rate, how much will it lose in 24 hr?

16. On a map, 3 in. represents 225 mi. If two cities are 7 in. apart on the map, how far are they apart in reality?

17. A tower 3 m high casts a shadow 5 m long. At the same time, the shadow of a second tower is 110 m long. How high is the second tower?

h

3 m

5 m

110 m

18. In a recent year, Kellogg sold 146.2 million lb of Corn Flakes and 120.4 million lb of Frosted Flakes. How many more pounds of Corn Flakes did they sell than Frosted Flakes?

19. Use = or ≠ for ▇ to write a true sentence:

$$\frac{6}{5} \; ▇ \; \frac{11}{9}.$$

20. Multiply:
$$\begin{array}{r} 2\ 3\ 4.1\ 1 \\ \times \qquad 7\ 4 \\ \hline \end{array}$$

21. Divide: $\dfrac{99.44}{100}$.

13. _____

14. _____

15. _____

16. _____

17. _____

18. _____

19. _____

20. _____

21. _____

Cumulative Review: Chapters 1–6

Add and simplify.

1.
```
   2 7.6 8
      3.0 1 9
+ 4 8 3.2 9 7
```

2. $2\frac{1}{3}$
$+\ 4\frac{5}{12}$

3. $\dfrac{6}{35} + \dfrac{5}{28}$

Subtract and simplify.

4.
```
  4 0.2
-   9.7 0 9
```

5. $73.82 - 0.908$

6. $\dfrac{4}{15} - \dfrac{3}{20}$

Multiply and simplify.

7.
```
  3 7.6 4
×     5.9
```

8. 5.678×100

9. $2\dfrac{1}{3} \cdot 1\dfrac{2}{7}$

Divide and simplify.

10. $2.3\,\overline{)\,9\,8.9}$

11. $5\,4\,\overline{)\,4\,8{,}5\,4\,6}$

12. $\dfrac{7}{11} \div \dfrac{14}{33}$

13. Write expanded notation: 30,074.

14. Write a word name for 120.07.

Which number is larger?

15. 0.7, 0.698

16. 0.799, 0.8

17. Find the prime factorization of 144.

18. Find the LCM of 28 and 35.

19. What part is shaded?

20. Simplify: $\dfrac{90}{144}$.

Calculate.

21. $\dfrac{3}{5} \times 9.53$

22. $\dfrac{1}{3} \times 0.645 - \dfrac{3}{4} \times 0.048$

23. Write fractional notation for the ratio 0.3 to 15.

24. Determine whether the pairs 3, 9 and 25, 75 are proportional.

25. What is the rate in meters per second?

660 meters,   12 seconds

26. A 14-oz jar of applesauce costs $0.39. A 30-oz jar of applesauce costs $0.99. Which has the lower unit price?

Solve.

27. $\dfrac{14}{25} = \dfrac{x}{54}$

28. $423 = 16 \cdot t$

29. $\dfrac{2}{3} \cdot y = \dfrac{16}{27}$

30. $\dfrac{7}{16} = \dfrac{56}{x}$

31. $34.56 + n = 67.9$

32. $t + \dfrac{7}{25} = \dfrac{5}{7}$

Solve.

33. A particular kind of fettucini alfredo has 520 calories in 1 cup. How many calories are there in $\frac{3}{4}$ cup?

34. The four largest hotel chains in the United States are Holiday Inn with 325,848 rooms, Hospitality Inns with 300,000 rooms, Best Western with 266,123 rooms, and Choice hotels with 262,811 rooms. How many rooms are there in all?

35. A Greyhound tour bus traveled 347.6 mi, 249.8 mi, and 379.5 mi on three separate trips. What was the total mileage of the bus?

36. A machine can stamp out 925 washers in 5 min. The company owning the machine needs 1295 washers by the end of the morning. How long will it take to stamp them out?

37. A 46-oz juice can contains $5\frac{3}{4}$ cups of juice. A recipe calls for $3\frac{1}{2}$ cups of juice. How many cups are left over?

38. It takes a carpenter $\frac{2}{3}$ hr to hang a door. How many doors can the carpenter hang in 8 hr?

39. A car travels 337.62 mi in 8 hr. How far does it travel in 1 hr?

40. A recent space shuttle made 16 orbits a day during an 8.25-day mission. How many orbits were made during the entire mission?

SYNTHESIS

41. A car travels 88 ft in 1 sec. What is the rate in miles per hour?

42. A 12-oz bag of shredded mozzarella cheese costs $2.07. Blocks of mozzarella cheese are sold for $2.79 per pound. Which is the better buy?

7
Percent Notation

INTRODUCTION

This chapter introduces a new kind of notation for numbers: percent notation. We will see that $\frac{1}{2}$, 0.5, and 50% are all names for the same number. Then we will use percent notation and equations to solve problems.

●●●●●●●●●●●●●●●●●●●●●

AN APPLICATION

It has been determined by sociologists that 17% of the population is left-handed. Each tournament conducted by the Professional Bowlers Association typically has 160 bowlers. How many would you expect to be left-handed?

THE MATHEMATICS

We let x = the number you would expect to be left-handed. We then translate the problem to an equation and solve as follows.

This is percent notation.
↓

Restate: What number is 17% of 160?

Translate: $x = 17\% \times 160$
↑

This is the type of equation we will consider.

OBJECTIVES FOR REVIEW

The review objectives to be tested in addition to the material in this chapter are as follows.

[3.4b]   Convert from fractional notation to mixed numerals.
[5.2b]   Solve equations of the type $a \cdot x = b$, where a and b may be in decimal notation.
[5.3a]   Convert from fractional notation to decimal notation.
[6.3b]   Solve proportions.

Pretest: Chapter 7

1. Find decimal notation for 87%.

2. Find percent notation for 0.537.

3. Find percent notation for $\frac{3}{4}$.

4. Find fractional notation for 37%.

5. Translate to an equation. Then solve.

   What is 60% of 75?

6. Translate to a proportion. Then solve.

   What percent of 50 is 35?

Solve.

7. The weight of muscles in a human body is 40% of total body weight. A person weighs 225 lb. What do the muscles weigh?

8. The population of a town increased from 3000 to 3600. Find the percent of increase in population.

9. The sales tax rate in Massachusetts is 5%. How much tax is charged on a purchase of $286? What is the total price?

10. A salesperson's commission rate is 28%. What is the commission from the sale of $18,400 worth of merchandise?

11. The marked price of a stereo is $450. The stereo is on sale at Lowland Appliances for 25% off. What are the discount and the sale price?

12. What is the simple interest on $1200 principal at the interest rate of 8.3% for one year?

13. What is the simple interest on $500 at 8% for $\frac{1}{2}$ year?

14. Interest is compounded annually. Find the amount in an account if $6000 is invested at 9% for 2 years.

7.1 *Percent Notation*

a UNDERSTANDING PERCENT NOTATION

Of all drivers, 59% use their seat belt. What does this mean? It means that, on the average, of every 100 people, 59 fasten their seat belt when driving. Thus, 59% is a ratio of 59 to 100, or $\frac{59}{100}$.

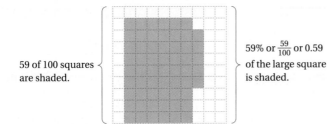

59 of 100 squares are shaded.

59% or $\frac{59}{100}$ or 0.59 of the large square is shaded.

Percent notation is used extensively in our lives. Here are some examples:

24% of us go to the movies once a month.

95% of hair spray is alcohol.

67% of all contact lens users are female.

62.4% of all aluminum cans were recycled in a recent year.

51.6% of all new marriages will end in divorce.

45.8% of us sleep between 7 and 8 hours per night.

74% of the times a major-league baseball player strikes out swinging, the pitch was out of the strike zone.

Percent notation is often represented by pie charts to show how the parts of a quantity are related. For example, the chart below relates the length of time couples were engaged before marrying.

Engagement Times of Married Couples

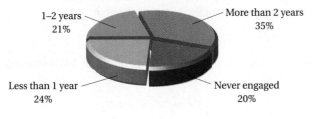

1–2 years
21%

More than 2 years
35%

Less than 1 year
24%

Never engaged
20%

> The notation $n\%$ means "n per hundred."

This definition leads us to the following equivalent ways of defining percent notation.

> Percent notation, $n\%$, is defined using:
>
> ratio ➤ $n\%$ = the ratio of n to 100 = $\dfrac{n}{100}$;
>
> fractional notation ➤ $n\% = n \times \dfrac{1}{100}$;
>
> decimal notation ➤ $n\% = n \times 0.01$.

OBJECTIVES

After finishing Section 7.1, you should be able to:

a Write three kinds of notation for a percent.

b Convert from percent notation to decimal notation.

c Convert from decimal notation to percent notation.

FOR EXTRA HELP

TAPE 11   TAPE 10A   MAC: 7
                                   IBM: 7

40% of those who run marathons say they do so for a sense of accomplishment.

Write three kinds of notation as in Examples 1 and 2.

1. 70%

2. 23.4%

3. 100%

EXAMPLE 1 Write three kinds of notation for 35%.

Using ratio: $35\% = \dfrac{35}{100}$ A ratio of 35 to 100

Using fractional notation: $35\% = 35 \times \dfrac{1}{100}$ Replacing % with $\times \dfrac{1}{100}$

Using decimal notation: $35\% = 35 \times 0.01$ Replacing % with $\times 0.01$

EXAMPLE 2 Write three kinds of notation for 67.8%.

Using ratio: $67.8\% = \dfrac{67.8}{100}$ A ratio of 67.8 to 100

Using fractional notation: $67.8\% = 67.8 \times \dfrac{1}{100}$ Replacing % with $\times \dfrac{1}{100}$

Using decimal notation: $67.8\% = 67.8 \times 0.01$ Replacing % with $\times 0.01$

Do Exercises 1–3.

b **CONVERTING FROM PERCENT NOTATION TO DECIMAL NOTATION**

Consider 78%. To convert to decimal notation, we can think of percent notation as a ratio and write

$78\% = \dfrac{78}{100}$ Using the definition of percent as a ratio

$ = 0.78.$ Converting to decimal notation

Similarly,

$4.9\% = \dfrac{4.9}{100}$ Using the definition of percent as a ratio

$ = 0.049.$ Converting to decimal notation

We could also convert 78% to decimal notation by replacing "%" with "$\times 0.01$" and write

$78\% = 78 \times 0.01$ Replacing % with $\times 0.01$

$ = 0.78.$ Multiplying

Similarly,

$4.9\% = 4.9 \times 0.01$ Replacing % with $\times 0.01$

$ = 0.049.$ Multiplying

It is thought that the Roman Emperor Augustus began percent notation by taxing goods sold at a rate of $\frac{1}{100}$. In time, the symbol "%" evolved by interchanging the parts of the symbol "100" to "0/0" and then to "%".

Answers on page A-3

Dividing by 100 amounts to moving the decimal point two places to the left. This is the same as multiplying by 0.01. Thus a quick way to convert from percent notation to decimal notation is to drop the percent symbol and move the decimal point two places to the left.

To convert from percent notation to decimal notation,	36.5%
a) replace the percent symbol % with × 0.01, and	36.5 × 0.01
b) multiply by 0.01, which means move the decimal point two places to the left.	0.36.5⤸ Move 2 places to the left. 36.5% = 0.365

EXAMPLE 3 Find decimal notation for 99.44%.

a) Replace the percent symbol with × 0.01.   99.44 × 0.01

b) Move the decimal point two places to the left.   0.99.44⤸

Thus, 99.44% = 0.9944.

EXAMPLE 4 The population growth rate of Europe is 1.1%. Find decimal notation for 1.1%.

a) Replace the percent symbol with × 0.01.   1.1 × 0.01

b) Move the decimal point two places to the left.   0.01.1⤸

Thus, 1.1% = 0.011.

Do Exercises 4–7.

CONVERTING FROM DECIMAL NOTATION TO PERCENT NOTATION

To convert 0.38 to percent notation, we can first write fractional notation, as follows:

$$0.38 = \frac{38}{100} \qquad \text{Converting to fractional notation}$$

$$= 38\%. \qquad \text{Using the definition of percent as a ratio}$$

Note that 100% = 100 × 0.01 = 1. Thus to convert 0.38 to percent notation, we can multiply by 1, using 100% as a symbol for 1. Then

$$0.38 = 0.38 \times 1$$
$$= 0.38 \times 100\%$$
$$= 0.38 \times 100 \times 0.01$$
$$= 38 \times 0.01$$
$$= 38\%.$$

Even more quickly, since 0.38 = 0.38 × 100%, we can simply multiply 0.38 by 100 and write the % symbol.

Find decimal notation.

4. 34%

5. 78.9%

6. One year the rate of inflation was 12.08%. Find decimal notation for 12.08%.

7. The present world population growth rate is 2.1% per year. Find decimal notation for 2.1%.

Answers on page A-3

Find percent notation.

8. 0.24

9. 3.47

10. 1

11. Contact lenses are worn by 0.1 of the population. Find percent notation for 0.1.

12. Of all the sports trading cards sold, 0.52 are of baseball players. Find percent notation for 0.52.

To convert from decimal notation to percent notation, multiply by 100%—that is, move the decimal point two places to the right and write a percent symbol.

To convert from decimal notation to percent notation, multiply by 100%. That is,	$0.675 = 0.675 \times 100\%$
a) move the decimal point two places to the right, and	0.67.5 Move 2 places to the right.
b) write a % symbol.	67.5% $0.675 = 67.5\%$

EXAMPLE 5 Find percent notation for 1.27.

a) Move the decimal point two places to the right.   1.27.

b) Write a % symbol.   127%

Thus, $1.27 = 127\%$.

EXAMPLE 6 Television sets are on 0.25 of the time. Find percent notation for 0.25.

a) Move the decimal point two places to the right.   0.25.

b) Write a % symbol.   25%

Thus, $0.25 = 25\%$.

Do Exercises 8–12.

Answers on page A-3

Exercise Set 7.1

a Write three kinds of notation as in Examples 1 and 2 on p. 310.

1. 90%

2. 58.7%

1) $\frac{58.7}{100}$

2) $58.7 \times \frac{1}{100}$

3) 58.7×0.01

3. 12.5%

4. 130%

b Find decimal notation.

5. 67%

.67

6. 17%

.17

7. 45.6%

.454

8. 76.3%

.743

9. 59.01%

10. 30.02%

11. 10%

12. 40%

13. 1%

14. 100%

15. 200%

16. 300%

17. 0.1%

18. 0.4%

19. 0.09%

20. 0.12%

21. 0.18%

22. 5.5%

.055

23. 23.19%

24. 87.99%

25. Blood is 90% water. Find decimal notation for 90%.

.90

26. Of all college football players, 2.6% go on to play professional football. Find decimal notation for 2.6%.

.026

27. Of those accidents requiring medical attention, 10.8% of them occur on roads. Find decimal notation for 10.8%.

.108

28. Of all CDs purchased, 58.1% are pop/rock. Find decimal notation for 58.1%.

.581

ANSWERS

1. _____
2. _____
3. _____
4. _____
5. _____
6. _____
7. _____
8. _____
9. _____
10. _____
11. _____
12. _____
13. _____
14. _____
15. _____
16. _____
17. _____
18. _____
19. _____
20. _____
21. _____
22. _____
23. _____
24. _____
25. _____
26. _____
27. _____
28. _____

ANSWERS

29. _____

30. _____

31. _____

32. _____

33. _____

34. _____

35. _____

36. _____

37. _____

38. _____

39. _____

40. _____

41. _____

42. _____

43. _____

44. _____

45. _____

46. _____

47. _____

48. _____

49. _____

50. _____

51. _____

52. _____

53. _____

54. _____

55. _____

56. _____

57. _____

58. _____

59. _____

60. _____

61. _____

62. _____

63. _____

64. _____

29. It is known that 45.8% of us sleep between 7 and 8 hours. Find decimal notation for 45.8%.

30. It is known that 24% of us go to the movies once a month. Find decimal notation for 24%.

$\boxed{\text{c}}$ Find percent notation.

31. 0.47

32. 0.87

33. 0.03

34. 0.01

35. 1.00

36. 4.00

37. 0.334

38. 0.889

39. 0.75

40. 0.99

41. 0.4

42. 0.5

43. 0.006

44. 0.008

45. 0.017

46. 0.024

47. 0.2718

48. 0.8911

49. 0.0239

50. 0.00073

51. A person's brain is 0.025 of the body weight. Find percent notation for 0.025.

52. It is known that 0.16 of all dessert orders in restaurants are for pie. Find percent notation for 0.16.

53. Of all 18-year-olds, 0.684 have a driver's license. Find percent notation for 0.684.

54. It is known that 0.622 of us think Monday is the worst day of the week. Find percent notation for 0.622.

SKILL MAINTENANCE

Convert to a mixed numeral.

55. $\dfrac{100}{3}$

56. $\dfrac{75}{2}$

57. $\dfrac{75}{8}$

58. $\dfrac{297}{16}$

Convert to decimal notation.

59. $\dfrac{2}{3}$

60. $\dfrac{1}{3}$

61. $\dfrac{5}{6}$

62. $\dfrac{17}{12}$

SYNTHESIS

63. $\boxed{\blacksquare}$ What would you do to an entry on a calculator in order to get percent notation?

64. $\boxed{\blacksquare}$ What would you do to percent notation on a calculator in order to get decimal notation?

7.2 *Percent Notation and Fractional Notation*

a CONVERTING FROM FRACTIONAL NOTATION TO PERCENT NOTATION

Consider the fractional notation $\frac{7}{8}$. To convert to percent notation, we use two skills we already have. We first find decimal notation by dividing:

$$\frac{7}{8} = 0.875$$

$$
\begin{array}{r}
0.8\,7\,5 \\
8\,)\,\overline{7.0\,0\,0} \\
\underline{6\,4} \\
6\,0 \\
\underline{5\,6} \\
4\,0 \\
\underline{4\,0} \\
0
\end{array}
$$

Then we convert the decimal notation to percent notation. We move the decimal point two places to the right

$$0.8\,7.5 \curvearrowright$$

and write a % symbol:

$$\frac{7}{8} = 87.5\%, \text{ or } 87\frac{1}{2}\%.$$

To convert from fractional notation to percent notation,	$\frac{3}{5}$ Fractional notation
a) find decimal notation by division, and	$\begin{array}{r} 0.6 \\ 5\,)\,\overline{3.0} \\ \underline{3\,0} \\ 0 \end{array}$
b) convert the decimal notation to percent notation.	$0.6 = 0.60 = 60\%$ Percent notation $\frac{3}{5} = 60\%$

EXAMPLE 1   Find percent notation for $\frac{3}{8}$.

a) Find decimal notation by division.

$$
\begin{array}{r}
0.3\,7\,5 \\
8\,)\,\overline{3.0\,0\,0} \\
\underline{2\,4} \\
6\,0 \\
\underline{5\,6} \\
4\,0 \\
\underline{4\,0} \\
0
\end{array}
\qquad \frac{3}{8} = 0.375
$$

OBJECTIVES

After finishing Section 7.2, you should be able to:

a Convert from fractional notation to percent notation.

b Convert from percent notation to fractional notation.

FOR EXTRA HELP

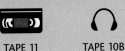

TAPE 11   TAPE 10B   MAC: 7
                           IBM: 7

Find percent notation.

1. $\frac{1}{4}$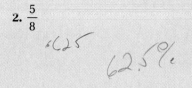

2. $\frac{5}{8}$ .625   62.5%

3. The human body is $\frac{2}{3}$ water. Find percent notation for $\frac{2}{3}$.

.66%

4. Find percent notation: $\frac{5}{6}$.

$83.\overline{3}$ %

Find percent notation.

5. $\frac{57}{100}$ 51%

6. $\frac{19}{25}$ 76%

b) Convert the decimal notation to percent notation. Move the decimal point two places to the right, and write a % symbol.

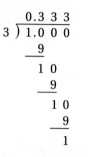

$$\frac{3}{8} = 37.5\%, \text{ or } 37\frac{1}{2}\%$$

Don't forget the % symbol.

Do Exercises 1 and 2.

EXAMPLE 2   Of all meals, $\frac{1}{3}$ are eaten outside the home. Find percent notation for $\frac{1}{3}$.

a) Find decimal notation by division.

$$\begin{array}{r} 0.3\ 3\ 3 \\ 3\overline{)1.0\ 0\ 0} \\ \underline{9} \\ 1\ 0 \\ \underline{9} \\ 1\ 0 \\ \underline{9} \\ 1 \end{array}$$

We get a repeating decimal: $0.33\overline{3}$.

b) Convert the answer to percent notation.

$$\frac{1}{3} = 33.\overline{3}\%, \text{ or } 33\frac{1}{3}\%$$

Do Exercises 3 and 4.

In some cases, division is not the easiest way to convert. The following are some optional ways conversion might be done.

EXAMPLE 3   Find percent notation for $\frac{69}{100}$.

We use the definition of percent as a ratio.

$$\frac{69}{100} = 69\%$$

EXAMPLE 4   Find percent notation for $\frac{17}{20}$.

We multiply by 1 to get 100 in the denominator. We think of what we have to multiply 20 by in order to get 100. That number is 5, so we multiply by 1 using $\frac{5}{5}$.

$$\frac{17}{20} \cdot \frac{5}{5} = \frac{85}{100} = 85\%$$

Do Exercises 5 and 6.

b CONVERTING FROM PERCENT NOTATION TO FRACTIONAL NOTATION

To convert from percent notation to fractional notation,	30% Percent notation
a) use the definition of percent as a ratio, and	$\dfrac{30}{100}$
b) simplify, if possible.	$\dfrac{3}{10}$ Fractional notation

EXAMPLE 5   Find fractional notation for 75%.

$$75\% = \frac{75}{100} \qquad \text{Using the definition of percent}$$

$$= \frac{3 \cdot 25}{4 \cdot 25} = \frac{3}{4} \cdot \frac{25}{25}$$

$$= \frac{3}{4}$$

       Simplifying

EXAMPLE 6   Find fractional notation for 62.5%.

$$62.5\% = \frac{62.5}{100} \qquad \text{Using the definition of percent}$$

$$= \frac{62.5}{100} \times \frac{10}{10} \qquad \text{Multiplying by 1 to eliminate the decimal point in the numerator}$$

$$= \frac{625}{1000}$$

$$= \frac{5 \cdot 125}{8 \cdot 125} = \frac{5}{8} \cdot \frac{125}{125}$$

$$= \frac{5}{8}$$

       Simplifying

EXAMPLE 7   Find fractional notation for $16\frac{2}{3}\%$.

$$16\frac{2}{3}\% = \frac{50}{3}\% \qquad \text{Converting from the mixed numeral to fractional notation}$$

$$= \frac{50}{3} \times \frac{1}{100} \qquad \text{Using the definition of percent}$$

$$= \frac{50 \cdot 1}{3 \cdot 50 \cdot 2} = \frac{1}{6} \cdot \frac{50}{50}$$

$$= \frac{1}{6}$$

       Simplifying

Do Exercises 7–10.

     The table on the inside back cover lists decimal, fractional, and percent equivalents used so often that it would speed up your work if you learned them. For example, $\frac{1}{3} = 0.\overline{3}$, so we say that the **decimal equivalent** of $\frac{1}{3}$ is $0.\overline{3}$, or that $0.\overline{3}$ has the **fractional equivalent** $\frac{1}{3}$.

Find fractional notation.

7. 60%

$\dfrac{60}{100}$

8. 3.25%

$\dfrac{3.25}{100} \times \dfrac{10}{10} = \dfrac{325}{1000}$

$13/40$

9. $66\frac{2}{3}\%$

$\dfrac{2}{3}$

$\dfrac{833}{1000}$

10. Complete this table.

FRACTIONAL NOTATION	$\dfrac{1}{5}$	$\dfrac{5}{6}$	$3/8$
DECIMAL NOTATION	.2	$0.83\overline{3}$	.375
PERCENT NOTATION	20%	$83.\overline{3}\%$	$37\frac{1}{2}\%$

Answers on page A-3

APPLICATIONS OF RATIO AND PERCENT:
THE PRICE–EARNINGS RATIO AND STOCK YIELDS

THE PRICE–EARNINGS RATIO

If the total earnings of a company one year were $5,000,000 and 100,000 shares of stock were issued, the earnings per share was $50. At one time, the price per share of Apple Computer was 64\frac{1}{4}$ and the earnings per share was $4.33. The **price–earnings ratio, P/E**, is the price of the stock divided by the earnings per share. For the Apple Computer stock, the price–earnings ratio, *P/E*, is given by

$$\frac{P}{E} = \frac{64\frac{1}{4}}{4.33}$$

$$= \frac{64.25}{4.33} \quad \text{Converting to decimal notation}$$

$$\approx 14.8. \quad \text{Dividing, using a calculator, and rounding to the nearest tenth}$$

STOCK YIELDS

At one time, the price per share of Apple Computer stock was 64\frac{1}{4}$ and the company was paying a yearly dividend of $0.48 per share. It is helpful to those interested in stocks to know what percent the dividend is of the price of the stock. The percent is called the **yield**. For the Apple Computer stock, the yield is given by

$$\text{Yield} = \frac{\text{Dividend}}{\text{Price per share}}$$

$$= \frac{0.48}{64\frac{1}{4}}$$

$$= \frac{0.48}{64.25} \quad \text{Converting to decimal notation}$$

$$\approx 0.007 \quad \text{Dividing and rounding to the nearest thousandth}$$

$$\approx 0.7\%. \quad \text{Converting to percent notation}$$

EXERCISES

Compute the price–earnings ratio and the yield for each stock.

	Stock	Price per Share	Earnings	Dividend
1.	Monsanto	52\frac{1}{4}$	$3.24	$2.24
2.	K-Mart	23	1.99	0.92
3.	Rubbermaid	30$\frac{1}{2}$	1.01	0.39
4.	AT&T	51$\frac{5}{8}$	2.60	1.32

Exercise Set 7.2

a Find percent notation.

1. $\frac{41}{100}$ **2.** $\frac{36}{100}$ **3.** $\frac{5}{100}$ **4.** $\frac{1}{100}$ **5.** $\frac{2}{10}$ **6.** $\frac{7}{10}$

7. $\frac{3}{10}$ **8.** $\frac{9}{10}$ **9.** $\frac{1}{2}$ **10.** $\frac{3}{4}$ **11.** $\frac{5}{8}$ **12.** $\frac{1}{8}$

13. $\frac{4}{5}$ **14.** $\frac{2}{5}$ **15.** $\frac{2}{3}$ **16.** $\frac{1}{3}$ **17.** $\frac{1}{6}$ **18.** $\frac{5}{6}$

19. $\frac{4}{25}$ **20.** $\frac{17}{25}$ **21.** $\frac{1}{20}$ **22.** $\frac{31}{50}$ **23.** $\frac{17}{50}$ **24.** $\frac{3}{20}$

25. Bread is $\frac{9}{25}$ water. Find percent notation for $\frac{9}{25}$.

26. Milk is $\frac{7}{8}$ water. Find percent notation for $\frac{7}{8}$.

1. _____

2. _____

3. _____

4. _____

5. _____

6. _____

7. _____

8. _____

9. _____

10. _____

11. _____

12. _____

13. _____

14. _____

15. _____

16. _____

17. _____

18. _____

19. _____

20. _____

21. _____

22. _____

23. _____

24. _____

25. _____

26. _____

ANSWERS

27. _____

28. _____

29. _____

30. _____

31. _____

32. _____

33. _____

34. _____

35. _____

36. _____

37. _____

38. _____

39. _____

40. _____

41. _____

42. _____

43. _____

44. _____

45. _____

46. _____

47. _____

48. _____

49. _____

50. _____

51. _____

52. _____

53. _____

54. _____

Write percent notation for the fractions in this pie chart.

Types of Hotels

Moderate
$\frac{16}{25}$

First class
$\frac{11}{50}$

Deluxe
$\frac{3}{100}$

Economy
$\frac{11}{100}$

27. $\frac{3}{100}$

28. $\frac{11}{50}$

29. $\frac{16}{25}$

30. $\frac{11}{100}$

b Find fractional notation.

31. 85% **32.** 55% **33.** 62.5% **34.** 12.5%

35. $33\frac{1}{3}\%$ **36.** $83\frac{1}{3}\%$ **37.** $16.\overline{6}\%$ **38.** $66.\overline{6}\%$

39. 7.25% **40.** 4.85% **41.** 0.8% **42.** 0.2%

43. $25\frac{3}{8}\%$ **44.** $48\frac{7}{8}\%$ **45.** $78\frac{2}{9}\%$ **46.** $16\frac{5}{9}\%$

47. $64\frac{7}{11}\%$ **48.** $73\frac{3}{11}\%$ **49.** 150% **50.** 110%

51. 0.0325% **52.** 0.419% **53.** $33.\overline{3}\%$ **54.** $83.\overline{3}\%$

Find fractional notation for the percents in this bar graph.

Popularity of Ethnic Foods

Percent who "greatly enjoyed"

60 — 55%
50
40 — 39% 38%
30
20
10 — 12% 11% 8%

Italian Mexican Chinese German French Other

Type of food

Note that there can be multiple responses.

55. 55%

56. 39%

57. 38%

58. 12%

59. 11%

60. 8%

55. _____

56. _____

57. _____

58. _____

59. _____ 24/... 5/...

60. _____ 1000/...

61. The United States uses 24% of the world's energy. Find fractional notation for 24%. 6/25

62. The United States has 4.8% of the world's population. Find fractional notation for 4.8%. 6/125

63. Of all 18-year-olds, 27.5% are registered to vote. Find fractional notation for 27.5%. 11/40

64. Of all those who buy CDs, 57% are in the 20–39 age group. Find fractional notation for 57%. 57/100

61. _____

62. _____

63. _____

64. _____

Complete the table.

65.

FRACTIONAL NOTATION	DECIMAL NOTATION	PERCENT NOTATION
$\frac{1}{8}$	.125	$12\frac{1}{2}\%$, or 12.5%
$\frac{1}{6}$	.16$\overline{6}$	14.6%
$\frac{1}{5}$	.20	20%
$\frac{1}{4}$	0.25	25%
$33\frac{3}{10}$	.333	$33\frac{1}{3}\%$, or 33.3%
$\frac{3}{8}$	.375	$37\frac{1}{2}\%$, or 37.5%
$\frac{2}{5}$	.40	40%
$\frac{1}{2}$	0.5	50%

66.

FRACTIONAL NOTATION	DECIMAL NOTATION	PERCENT NOTATION
$\frac{3}{5}$	0.60	60%
$\frac{5}{8}$	$\frac{625}{1000}$ 0.625	62.5%
$\frac{2}{3}$	.6$\overline{6}$	66.6%
	0.75	75%
$\frac{4}{5}$	.8	80%
$\frac{5}{6}$	.83	$83\frac{1}{3}\%$, or 83.3%
$\frac{7}{8}$	0.875	$87\frac{1}{2}\%$, or 87.5%
		100%

65. See table.

66. See table.

67. See table.

68. See table.

69.

70.

71.

72.

73.

74.

75.

76.

77.

78.

67.

FRACTIONAL NOTATION	DECIMAL NOTATION	PERCENT NOTATION
	0.5	
$\frac{1}{3}$	.33	$33\frac{1}{3}\%$
$\frac{1}{4}$	.15	25%
$\frac{1}{6}$	.164	$16\frac{2}{3}\%$, or $16.\overline{6}\%$
	0.125	12.5%
$\frac{3}{4}$	.75	75%
	$0.8\overline{3}$	83%
$\frac{3}{8}$	.375	37.5

68.

FRACTIONAL NOTATION	DECIMAL NOTATION	PERCENT NOTATION
		40%
		$62\frac{1}{2}\%$, or 62.5%
	0.875	
$\frac{1}{1}$		
	0.6	
	$0.\overline{6}$	
$\frac{1}{5}$		

SKILL MAINTENANCE

Solve.

69. $10 \cdot x = 725$

70. $15 \cdot y = 75$

71. $0.05 \times b = 20$

72. $3 = 0.16 \times b$

73. $\dfrac{24}{37} = \dfrac{15}{x}$

74. $\dfrac{17}{18} = \dfrac{x}{27}$

SYNTHESIS

Find percent notation.

75. $\dfrac{41}{369}$

76. $\dfrac{54}{999}$

Find decimal notation.

77. $\dfrac{14}{9}\%$

78. $\dfrac{19}{12}\%$

7.3 Solving Percent Problems Using Equations

a TRANSLATING TO EQUATIONS

To solve a problem involving percents, it is helpful to translate first to an equation.

EXAMPLE 1 Translate:

$$
\begin{array}{ccccc}
23\% & \text{of} & 5 & \text{is} & \text{what?} \\
\downarrow & \downarrow & \downarrow & \downarrow & \downarrow \\
23\% & \cdot & 5 & = & a
\end{array}
$$

"Of" translates to "·", or "×".	"Is" translates to "=".
"What" translates to any letter.	% translates to "× $\frac{1}{100}$" or "× 0.01".

EXAMPLE 2 Translate:

$$
\begin{array}{ccccc}
\text{What} & \text{is} & 11\% & \text{of} & 49? \\
\downarrow & \downarrow & \downarrow & \downarrow & \downarrow \\
a & = & 11\% & \cdot & 49
\end{array}
$$

Any letter can be used.

Do Exercises 1 and 2.

EXAMPLE 3 Translate:

$$
\begin{array}{ccccc}
3 & \text{is} & 10\% & \text{of} & \text{what?} \\
\downarrow & \downarrow & \downarrow & \downarrow & \downarrow \\
3 & = & 10\% & \cdot & b
\end{array}
$$

EXAMPLE 4 Translate:

$$
\begin{array}{ccccc}
45\% & \text{of} & \text{what} & \text{is} & 23? \\
\downarrow & \downarrow & \downarrow & \downarrow & \downarrow \\
45\% & \times & b & = & 23
\end{array}
$$

Do Exercises 3 and 4.

EXAMPLE 5 Translate:

$$
\begin{array}{ccccc}
10 & \text{is} & \text{what percent} & \text{of} & 20? \\
\downarrow & \downarrow & \downarrow & \downarrow & \downarrow \\
10 & = & n & \times & 20
\end{array}
$$

EXAMPLE 6 Translate:

$$
\begin{array}{ccccc}
\text{What percent} & \text{of} & 50 & \text{is} & 7? \\
& \downarrow & \downarrow & \downarrow & \downarrow \\
n & \cdot & 50 & = & 7
\end{array}
$$

Do Exercises 5 and 6.

OBJECTIVES

After finishing Section 7.3, you should be able to:

a Translate percent problems to equations.

b Solve basic percent problems.

FOR EXTRA HELP

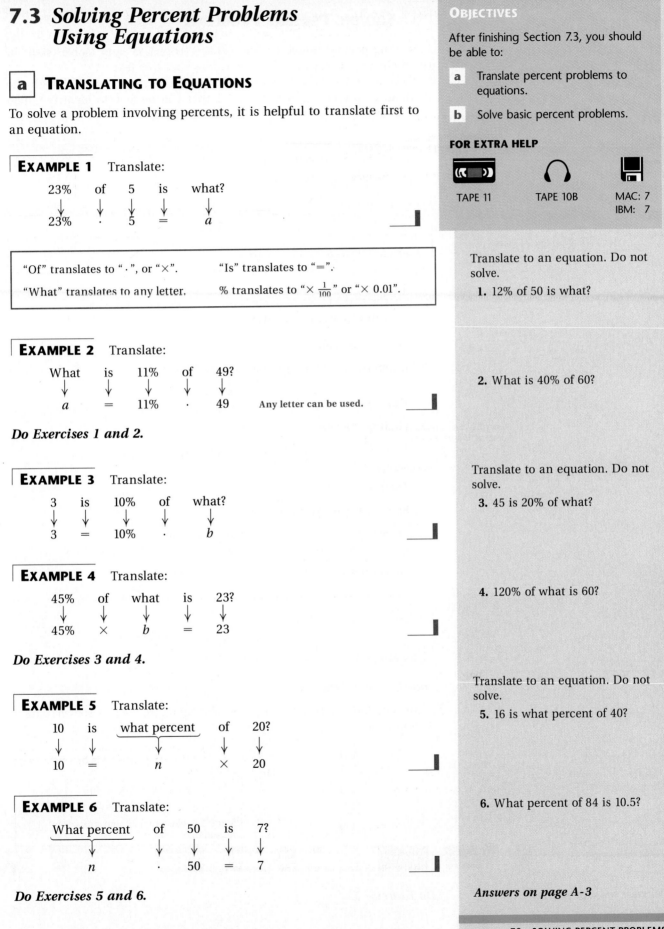

TAPE 11   TAPE 10B   MAC: 7
IBM: 7

Translate to an equation. Do not solve.

1. 12% of 50 is what?

2. What is 40% of 60?

Translate to an equation. Do not solve.

3. 45 is 20% of what?

4. 120% of what is 60?

Translate to an equation. Do not solve.

5. 16 is what percent of 40?

6. What percent of 84 is 10.5?

Answers on page A-3

7. Solve:

What is 12% of 50?

b | SOLVING PERCENT PROBLEMS

In solving percent problems, we use the "translate" and "solve" steps in the problem-solving strategy used throughout this text.

Percent problems are actually of three different types. Although the method we present does *not* require that you be able to identify which type we are studying, it is helpful to know them.

We know that

15 is 25% of 60, or

$15 = 25\% \times 60.$

We can think of this as:

Amount = Percent number × Base.

Each of the three types of percent problems depends on which of the three pieces of information is missing.

1. Finding the amount

Example:　What　is　25%　of　60?

Translation:　　y　　=　25%　·　60

2. Finding the base

Example:　15　is　25%　of　what number?

Translation:　15　=　25%　·　　　y

3. Finding the percent number

Example:　15　is　what percent　of　60?

Translation:　15　=　　　y　·　60

FINDING THE AMOUNT

⌐ **EXAMPLE 7**　What is 11% of 49?

Translate: $a = 11\% \times 49.$

Solve: The letter is by itself. To solve the equation, we just convert 11% to decimal notation and multiply.

$$\begin{array}{r} 4\ 9 \\ \times\ 0.1\ 1 \\ \hline 4\ 9 \\ 4\ 9\ 0 \\ \hline a = 5.3\ 9 \end{array}$$

$11\% = 0.11$

A way of checking answers is by estimating as follows:

$$11\% \times 49 \approx 10\% \times 50$$
$$= 0.10 \times 50 = 5.$$

Since 5 is close to 5.39, our answer is reasonable.

Thus, 5.39 is 11% of 49. The answer is 5.39.

Do Exercise 7.

EXAMPLE 8   120% of $42 is what?

Translate: 120% × 42 = *a*.

Solve: The letter is by itself. To solve the equation, we carry out the calculation.

$$
\begin{array}{r}
4\ 2 \\
\times\ 1.2 \\
\hline
8\ 4 \\
4\ 2\ 0 \\
\hline
a = 5\ 0.4
\end{array}
$$

120% = 1.20 = 1.2

Thus, 120% of $42 is $50.40. The answer is $50.40.

Do Exercise 8.

FINDING THE BASE

EXAMPLE 9   5% of what is 20?

Translate: 5% × *b* = 20.

Solve: This time the letter is *not* by itself. To solve the equation, we divide on both sides by 5%:

$b = 20 ÷ 5\%$ Dividing on both sides by 5%

$b = 20 ÷ 0.05$ 5% = 0.05

$b = 400.$

$$
\begin{array}{r}
4\ 0\ 0. \\
0.0\ 5\ \overline{)\ 2\ 0 0.0\ 0_\wedge} \\
2\ 0\ 0\ 0 \\
\hline
0
\end{array}
$$

Thus, 5% of 400 is 20. The answer is 400.

EXAMPLE 10 $3 is 16% of what?

Translate:
$$
\begin{array}{ccccc}
\$3 & \text{is} & 16\% & \text{of} & \text{what?} \\
\downarrow & \downarrow & \downarrow & \downarrow & \downarrow \\
3 & = & 16\% & \times & b.
\end{array}
$$

Solve: Again, the letter is not by itself. To solve the equation, we divide on both sides by 16%:

$3 ÷ 16\% = b$ Dividing on both sides by 16%

$3 ÷ 0.16 = b$ 16% = 0.16

$18.75 = b.$

$$
\begin{array}{r}
1\ 8.7\ 5 \\
0.1\ 6\ \overline{)\ 3.0\ 0_\wedge 0\ 0} \\
1\ 6 \\
\hline
1\ 4\ 0 \\
1\ 2\ 8 \\
\hline
1\ 2\ 0 \\
1\ 1\ 2 \\
\hline
8\ 0 \\
8\ 0 \\
\hline
0
\end{array}
$$

Thus, $3 is 16% of $18.75. The answer is $18.75.

Do Exercises 9 and 10.

Answers on page A-3

11. Solve:

16 is what percent of 40?

12. Solve:

What percent of $84 is $10.50?

FINDING THE PERCENT NUMBER

In solving these problems, you *must* remember to convert to percent nota-tion after you have solved the equation.

EXAMPLE 11   10 is what percent of 20?

Translate: 10   is   what percent   of   20?

$$10 = n \times 20.$$

Solve: To solve the equation, we divide on both sides by 20 and convert the result to percent notation:

$$n \cdot 20 = 10$$

$$\frac{n \cdot 20}{20} = \frac{10}{20} \qquad \text{Dividing on both sides by 20}$$

$$n = 0.50 = 50\%. \qquad \text{Converting to percent notation}$$

Thus, 10 is 50% of 20. The answer is 50%.

Do Exercise 11.

EXAMPLE 12   What percent of $50 is $16?

Translate: What percent   of $50   is $16?

$$n \times 50 = 16.$$

Solve: To solve the equation, we divide on both sides by 50 and convert the answer to percent notation:

$$n = 16 \div 50 \qquad \text{Dividing on both sides by 50}$$

$$n = \frac{16}{50}$$

$$n = \frac{16}{50} \cdot \frac{2}{2}$$

$$n = \frac{32}{100}$$

$$n = 32\%. \qquad \text{Converting to percent notation}$$

Thus, 32% of $50 is $16. The answer is 32%.

Do Exercise 12.

Caution!   Remember to convert the solution to percent notation when completing problems such as these.

Exercise Set 7.3

a Translate to an equation. Do not solve.

1. What is 32% of 78?

2. 98% of 57 is what?

3. 89 is what percent of 99?

4. What percent of 25 is 8?

5. 13 is 25% of what?

6. 21.4% of what is 20?

b Solve.

7. What is 85% of 276?

8. What is 74% of 53?

9. 150% of 30 is what?

10. 100% of 13 is what?

11. What is 6% of $300?

12. What is 4% of $45?

13. 3.8% of 50 is what?

14. $33\frac{1}{3}$% of 480 is what?
(*Hint:* $33\frac{1}{3}\% = \frac{1}{3}$.)

15. $39 is what percent of $50?

16. $16 is what percent of $90?

17. 20 is what percent of 10?

18. 60 is what percent of 20?

1. _____

2. _____

3. _____

4. _____

5. _____

6. _____

7. _____

8. _____

9. _____

10. _____

11. _____

12. _____

13. _____

14. _____

15. _____

16. _____

17. _____

18. _____

19. _____

20. _____

21. _____

22. _____

23. _____

24. _____

25. _____

26. _____

27. _____

28. _____

29. _____

30. _____

31. _____

32. _____

33. _____

34. _____

35. _____

36. _____

37. _____

38. _____

39. _____

40. _____

41. _____

42. _____

43. _____

44. _____

19. What percent of $300 is $150?

20. What percent of $50 is $40?

21. What percent of 80 is 100?

22. What percent of 60 is 15?

23. 20 is 50% of what?

24. 57 is 20% of what?

25. 40% of what is $16?

26. 100% of what is $74?

27. 56.32 is 64% of what?

28. 71.04 is 96% of what?

29. 70% of what is 14?

30. 70% of what is 35?

31. What is $62\frac{1}{2}$% of 10?

32. What is $35\frac{1}{4}$% of 1200?

33. What is 8.3% of $10,200?

34. What is 9.2% of $5600?

SKILL MAINTENANCE

Write fractional notation.

35. 0.09

36. 1.79

37. 0.875

38. 0.9375

Write decimal notation.

39. $\dfrac{89}{100}$

40. $\dfrac{7}{100}$

41. $\dfrac{3}{10}$

42. $\dfrac{17}{1000}$

SYNTHESIS

Solve.

43. 🖩 What is 7.75% of $10,880?

Estimate _____

Calculate _____

44. 🖩 50,951.775 is what percent of 78,995?

Estimate _____

Calculate _____

7.4 Solving Percent Problems Using Proportions*

a TRANSLATING TO PROPORTIONS

A percent is a ratio of some number to 100. For example, 75% is the ratio $\frac{75}{100}$. We also know that 3 and 4 have the same ratio as 75 and 100. Thus,

$$75\% = \frac{75}{100} = \frac{3}{4}.$$

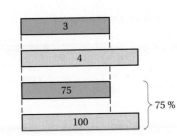

To solve a percent problem using a proportion, we translate as follows:

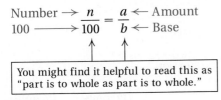

$$\text{Number} \rightarrow \frac{n}{100} \xrightarrow{} = \frac{a}{b} \leftarrow \text{Amount} \atop \leftarrow \text{Base}$$

You might find it helpful to read this as "part is to whole as part is to whole."

For example,

36 is 75% of 48

translates to

$$\frac{75}{100} = \frac{36}{48}.$$

A clue in translating is that the base, b, corresponds to 100 and usually follows the wording "percent of." Also, $n\%$ always translates to $n/100$. Another aid in translating is to make a comparison drawing. Beginning with the percent side, we list 0% at the top and 100% at the bottom. Then we estimate where the 75% would be located. The numbers, or quantities, that correspond are then filled in. The base—in this case, 48—always corresponds to 100% and the amount—in this case, 36—corresponds to 75%.

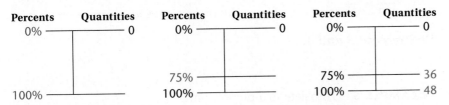

The proportion can then be read easily from the drawing.

TAPE 12	TAPE 11A	MAC: 7
		IBM: 7

*Note: This section presents an alternative method for solving basic percent problems. You can use either equations or proportions to solve percent problems, but you might prefer one method over the other, or your instructor may direct you to use one method over the other.

OBJECTIVES

After finishing Section 7.4, you should be able to:

a Translate percent problems to proportions.

b Solve basic percent problems.

FOR EXTRA HELP

Translate to a proportion. Do not solve.

1. 12% of 50 is what?

2. What is 40% of 60?

Translate to a proportion. Do not solve.

3. 45 is 20% of what?

4. 120% of what is 60?

Answers on page A-3

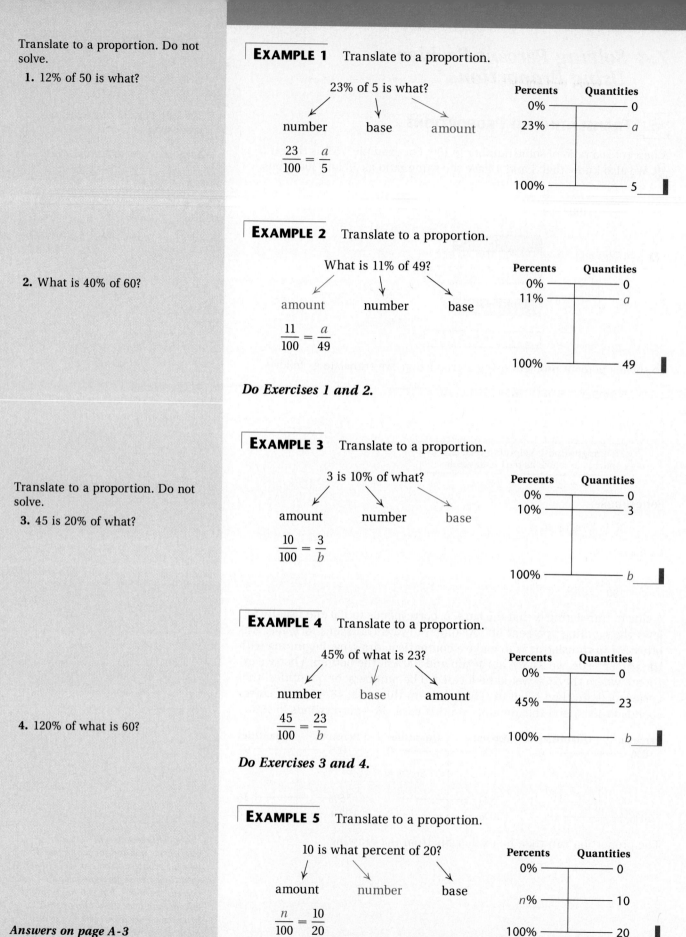

EXAMPLE 1   Translate to a proportion.

23% of 5 is what?

number   base   amount

$$\frac{23}{100} = \frac{a}{5}$$

Percents	Quantities
0%	0
23%	a
100%	5

EXAMPLE 2   Translate to a proportion.

What is 11% of 49?

amount   number   base

$$\frac{11}{100} = \frac{a}{49}$$

Percents	Quantities
0%	0
11%	a
100%	49

Do Exercises 1 and 2.

EXAMPLE 3   Translate to a proportion.

3 is 10% of what?

amount   number   base

$$\frac{10}{100} = \frac{3}{b}$$

Percents	Quantities
0%	0
10%	3
100%	b

EXAMPLE 4   Translate to a proportion.

45% of what is 23?

number   base   amount

$$\frac{45}{100} = \frac{23}{b}$$

Percents	Quantities
0%	0
45%	23
100%	b

Do Exercises 3 and 4.

EXAMPLE 5   Translate to a proportion.

10 is what percent of 20?

amount   number   base

$$\frac{n}{100} = \frac{10}{20}$$

Percents	Quantities
0%	0
n%	10
100%	20

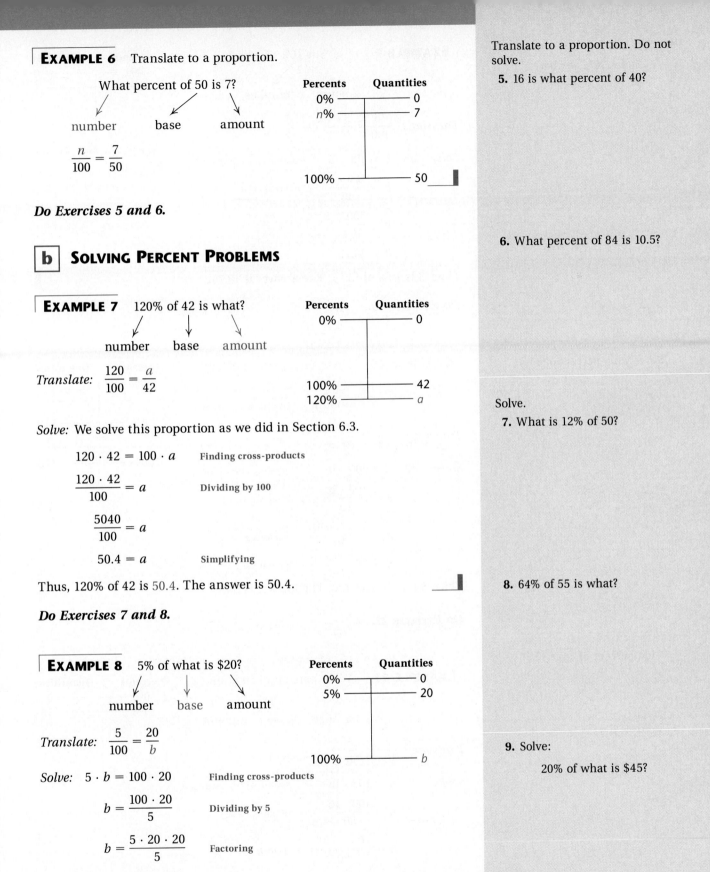

EXAMPLE 6 Translate to a proportion.

What percent of 50 is 7?

number · base · amount

$$\frac{n}{100} = \frac{7}{50}$$

Percents   Quantities

0% — 0
n% — 7

100% — 50

Do Exercises 5 and 6.

b | SOLVING PERCENT PROBLEMS

EXAMPLE 7 120% of 42 is what?

number · base · amount

Translate: $\frac{120}{100} = \frac{a}{42}$

Percents   Quantities

0% — 0

100% — 42
120% — a

Solve: We solve this proportion as we did in Section 6.3.

$120 \cdot 42 = 100 \cdot a$ **Finding cross-products**

$\frac{120 \cdot 42}{100} = a$ **Dividing by 100**

$\frac{5040}{100} = a$

$50.4 = a$ **Simplifying**

Thus, 120% of 42 is 50.4. The answer is 50.4.

Do Exercises 7 and 8.

EXAMPLE 8 5% of what is $20?

number · base · amount

Translate: $\frac{5}{100} = \frac{20}{b}$

Percents   Quantities

0% — 0
5% — 20

100% — b

Solve: $5 \cdot b = 100 \cdot 20$ **Finding cross-products**

$b = \frac{100 \cdot 20}{5}$ **Dividing by 5**

$b = \frac{5 \cdot 20 \cdot 20}{5}$ **Factoring**

$b = 400$ **Simplifying**

Thus, 5% of $400 is $20. The answer is $400.

Do Exercise 9.

Translate to a proportion. Do not solve.

5. 16 is what percent of 40?

6. What percent of 84 is 10.5?

Solve.

7. What is 12% of 50?

8. 64% of 55 is what?

9. Solve:

20% of what is $45?

Answers on page A-3

10. Solve:

60 is 120% of what?

11. Solve:

$16 is what percent of $40?

12. Solve:

What percent of 84 is 10.5?

Answers on page A-3

EXAMPLE 9   3 is 16% of what?

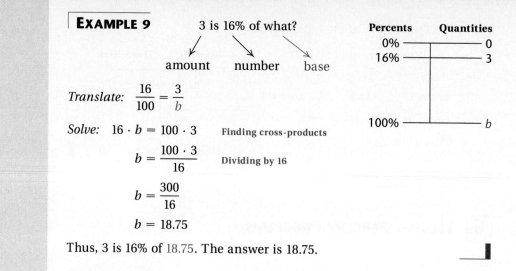

amount   number   base

Translate: $\dfrac{16}{100} = \dfrac{3}{b}$

Solve: $16 \cdot b = 100 \cdot 3$ **Finding cross-products**

$b = \dfrac{100 \cdot 3}{16}$ **Dividing by 16**

$b = \dfrac{300}{16}$

$b = 18.75$

Thus, 3 is 16% of 18.75. The answer is 18.75.

Do Exercise 10.

EXAMPLE 10 $10 is what percent of $20?

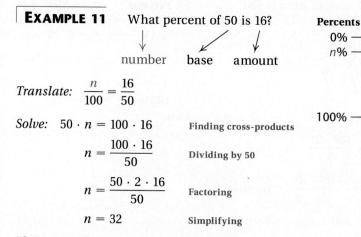

amount   number   base

Translate: $\dfrac{n}{100} = \dfrac{10}{20}$

Solve: $20 \cdot n = 100 \cdot 10$ **Finding cross-products**

$n = \dfrac{100 \cdot 10}{20}$ **Dividing by 20**

$n = \dfrac{20 \cdot 5 \cdot 10}{20}$ **Factoring**

$n = 50$ **Simplifying**

Thus, $10 is 50% of $20. The answer is 50%.

Do Exercise 11.

EXAMPLE 11   What percent of 50 is 16?

number   base   amount

Translate: $\dfrac{n}{100} = \dfrac{16}{50}$

Solve: $50 \cdot n = 100 \cdot 16$ **Finding cross-products**

$n = \dfrac{100 \cdot 16}{50}$ **Dividing by 50**

$n = \dfrac{50 \cdot 2 \cdot 16}{50}$ **Factoring**

$n = 32$ **Simplifying**

Thus, 32% of 50 is 16. The answer is 32%.

Do Exercise 12.

Exercise Set 7.4

a Translate to a proportion. Do not solve.

1. What is 37% of 74?

2. 66% of 74 is what?

3. 4.3 is what percent of 5.9?

4. What percent of 6.8 is 5.3?

5. 14 is 25% of what?

6. 22.3% of what is 40?

b Solve.

7. What is 76% of 90?

8. What is 32% of 70?

9. 70% of 660 is what?

10. 80% of 920 is what?

11. What is 4% of 1000?

12. What is 6% of 2000?

13. 4.8% of 60 is what?

14. 63.1% of 80 is what?

15. $24 is what percent of $96?

16. $14 is what percent of $70?

17. 102 is what percent of 100?

18. 103 is what percent of 100?

19. What percent of $480 is $120?

20. What percent of $80 is $60?

1.
2.
3.
4.
5.
6.
7.
8.
9.
10.
11.
12.
13.
14.
15.
16.
17.
18.
19.
20.

21. What percent of 160 is 150?

22. What percent of 33 is 11?

23. $18 is 25% of what?

24. $75 is 20% of what?

25. 60% of what is 54?

26. 80% of what is 96?

27. 65.12 is 74% of what?

28. 63.7 is 65% of what?

29. 80% of what is 16?

30. 80% of what is 10?

31. What is $62\frac{1}{2}$% of 40?

32. What is $43\frac{1}{4}$% of 2600?

33. What is 9.4% of $8300?

34. What is 8.7% of $76,000?

SKILL MAINTENANCE

Solve.

35. $\dfrac{x}{188} = \dfrac{2}{47}$

36. $\dfrac{15}{x} = \dfrac{3}{800}$

37. $\dfrac{4}{7} = \dfrac{x}{14}$

38. $\dfrac{612}{t} = \dfrac{72}{244}$

39. $\dfrac{5000}{t} = \dfrac{3000}{60}$

40. $\dfrac{75}{100} = \dfrac{n}{20}$

SYNTHESIS

Solve.

41. ▦ What is 8.85% of $12,640?
Estimate _____
Calculate _____

42. ▦ 78.8% of what is 9809.024?
Estimate _____
Calculate _____

7.5 Applications of Percent

a PERCENT PROBLEMS

Problems involving percent are not always stated in a manner easily translated to an equation. In such cases, it is helpful to restate the problem before translating. Sometimes it also helps to draw a picture.

EXAMPLE 1 If a restaurant sells 220 desserts in an evening, it is typical that 44 of them will be chocolate cake. What percent of the desserts sold will be chocolate cake?

METHOD 1 Solve using an equation.

Restate: 44   is   what percent   of   220?

Translate: 44   = n ×   220

Solve: We divide on both sides by 220:

$$44 \div 220 = n$$
$$0.2 = n$$
$$20\% = n.$$

Of the desserts, 20% will be chocolate cake.

METHOD 2* Solve using a proportion.

Restate: 44 is what percent of 220?

amount   number   base

Translate: $\dfrac{n}{100} = \dfrac{44}{220}$

Percents	Quantities
0%	0
n%	44
100%	220

Solve: $220 \cdot n = 100 \cdot 44$ **Finding cross-products**

$n = \dfrac{100 \cdot 44}{220}$ **Dividing by 220**

$n = \dfrac{4400}{220}$

$n = 20$

Of the desserts, 20% will be chocolate cake.

Do Exercise 1.

Note: If you skipped Section 7.4, then you should ignore Method 2.

OBJECTIVES

After finishing Section 7.5, you should be able to:

a Solve applied percent problems.

b Solve percent problems involving percent increase or decrease.

FOR EXTRA HELP

TAPE 12   TAPE 11A   MAC: 7
                          IBM: 7

1. A college basketball team won 11 of its 25 games. What percent of its games did it win?

Answer on page A-4

2. The weight of a human brain is 2.5% of total body weight. A person weighs 200 lb. What does the brain weigh?

Brain Weight vs. Body Weight

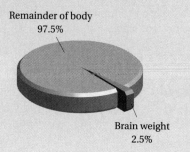

Remainder of body
97.5%

Brain weight
2.5%

EXAMPLE 2 Have you ever wondered why you receive so much junk mail? One reason offered by the U.S. Postal Service is that we open and read 78% of the advertising we receive in the mail. Suppose that a business sends out 9500 advertising brochures. How many of them can it expect to be opened and read?

Mail Advertising Opened

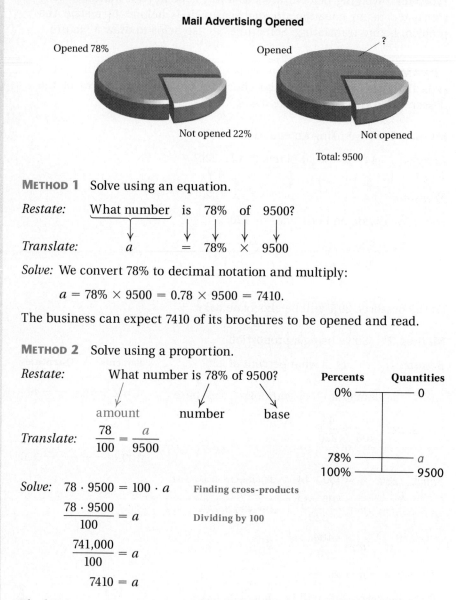

Opened 78%

Not opened 22%

Opened

?

Not opened

Total: 9500

METHOD 1 Solve using an equation.

Restate: What number is   78% of   9500?

Translate: a $=$ 78% $\times$ 9500

Solve: We convert 78% to decimal notation and multiply:

$$a = 78\% \times 9500 = 0.78 \times 9500 = 7410.$$

The business can expect 7410 of its brochures to be opened and read.

METHOD 2 Solve using a proportion.

Restate: What number is 78% of 9500?

amount   number   base

Translate: $\dfrac{78}{100} = \dfrac{a}{9500}$

Percents	Quantities
0%	0
78%	a
100%	9500

Solve: $78 \cdot 9500 = 100 \cdot a$ **Finding cross-products**

$\dfrac{78 \cdot 9500}{100} = a$ **Dividing by 100**

$\dfrac{741{,}000}{100} = a$

$7410 = a$

The business can expect 7410 of its brochures to be opened and read.

Do Exercise 2.

Answer on page A-4

b | PERCENT INCREASE OR DECREASE

Percent is often used to state increases or decreases. For example, sales of frozen yogurt have recently *increased* 44%. This means that the increase was 44% of the former sales. Sales of frozen yogurt were 82.5 million gallons in 1989. By 1991, sales had increased 44%. Thus the increase was 44% of 82.5, or 36.3 million gallons. The new sales were 82.5 + 36.3, or 118.8 million gallons, as illustrated below.

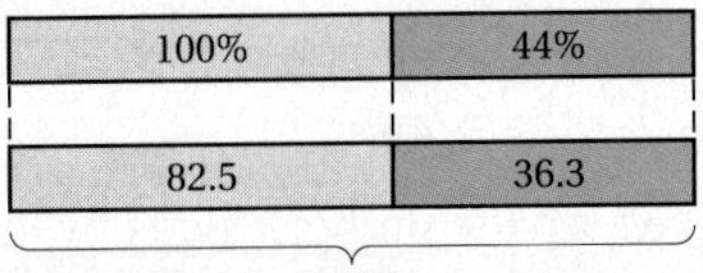

What do we mean when we say that the price of Swiss cheese has decreased 8%? If the price was $1.00 a pound and it went down to $0.92 a pound, then the decrease is $0.08, which is 8% of the original price. We can see this in the following figure.

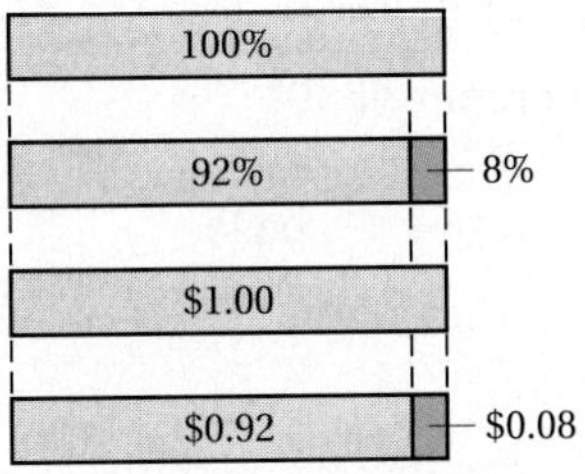

> To find a percent of increase or decrease,
> a) find the amount of increase or decrease, and
> b) then determine what percent this is of the original amount.

EXAMPLE 3 The price of fruit punch increased from 40 cents per quart to 45 cents per quart. What was the percent of increase?

We make a drawing.

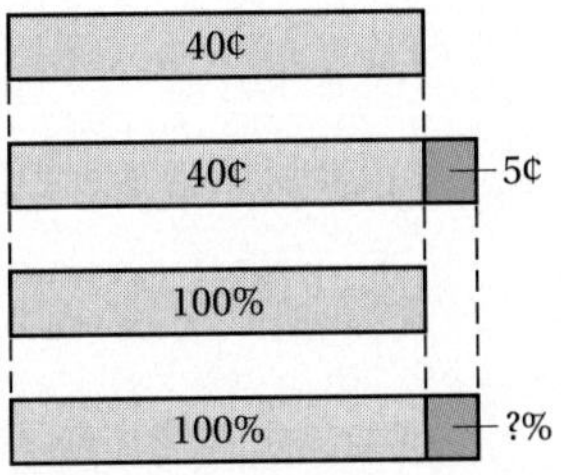

a) First, we find the increase by subtracting.

$$
\begin{array}{rl}
4\ 5 & \text{New price} \\
-\ 4\ 0 & \text{Original price} \\
\hline
5 & \text{Increase}
\end{array}
$$

The increase is 5 cents.

3. The price of an automobile increased from \$15,800 to \$17,222. What was the percent of increase?

b) Now we ask:

5 is what percent of 40 (the original price)?

> *Caution!* A common error is to use 45 instead of 40, the *original* amount.

To find out, we use either of our two methods.

METHOD 1 Solve using an equation.

5 is what percent of 40?

$$5 = n \times 40$$

To solve the equation, we divide on both sides by 40:

$$5 \div 40 = n \qquad \text{Dividing by 40}$$
$$0.125 = n$$
$$12.5\% = n.$$

The percent of increase was 12.5%.

METHOD 2 Solve using a proportion.

5 is what percent of 40?

amount   number   base

$$\frac{n}{100} = \frac{5}{40}$$
$$40 \cdot n = 100 \cdot 5 \qquad \text{Finding cross-products}$$
$$n = \frac{100 \cdot 5}{40} \qquad \text{Dividing by 40}$$
$$n = \frac{500}{40}$$
$$n = 12.5$$

The percent of increase was 12.5%.

Do Exercise 3.

Answer on page A-4

EXAMPLE 4 By proper furnace maintenance, a family that pays a monthly fuel bill of $78.00 can reduce their bill to $70.20. What is the percent of decrease?

We make a drawing.

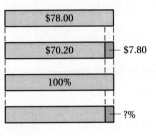

a) First, we find the decrease.

$$
\begin{array}{rl}
7\;8.0\;0 & \text{Original bill} \\
-\;7\;0.2\;0 & \text{New bill} \\
\hline
7.8\;0 & \text{Decrease}
\end{array}
$$

The decrease is $7.80.

b) Now we ask:

7.80 is what percent of 78.00 (the original bill)?

> *Caution!* A common error is to use $70.20 instead of $78.00, the *original* amount.

METHOD 1 Solve using an equation.

7.80	is	what percent	of	78.00?
↓	↓	↓	↓	↓
7.80	=	n	×	78.00

To solve the equation, we divide on both sides by 78:

$$7.8 \div 78 = n \qquad \text{Dividing by 78}$$
$$0.1 = n$$
$$10\% = n.$$

The percent of decrease is 10%.

METHOD 2 Solve using a proportion.

7.80 is what percent of 78.00?

↓ amount   ↓ number   ↓ base

$$\frac{n}{100} = \frac{7.80}{78.00}$$
$$78.00 \times n = 100 \times 7.80 \qquad \text{Finding cross-products}$$
$$n = \frac{100 \times 7.80}{78.00} \qquad \text{Dividing by 78.00}$$
$$n = \frac{780}{78}$$
$$n = 10$$

The percent of decrease is 10%.

Do Exercise 4.

4. By using only cold water in the washing machine, a family that pays a monthly fuel bill of $78.00 can reduce their bill to $74.88. What is the percent of decrease?

Answer on page A-4

5. A part-time salesperson earns $9800 one year and gets a 9% raise the next. What is the new salary?

EXAMPLE 5 A part-time teacher's aide earns $9700 one year and receives a 6% raise the next. What is the new salary?

We make a drawing.

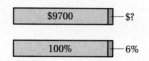

a) First, we find the increase. We ask:

What is 6% of 9700?

METHOD 1 Solve using an equation.

$$\begin{array}{ccccc} \text{What} & \text{is} & 6\% & \text{of} & 9700? \\ \downarrow & \downarrow & \downarrow & \downarrow & \downarrow \\ a & = & 6\% & \times & 9700 \end{array}$$

This tells us what to do. We convert 6% to decimal notation and multiply:

$$a = 0.06 \times 9700 = 582.$$

The increase is $582.00.

METHOD 2 Solve using a proportion.

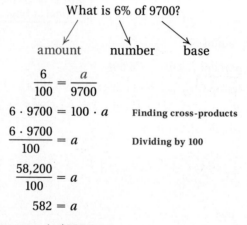

What is 6% of 9700?

amount   number   base

$$\frac{6}{100} = \frac{a}{9700}$$

$$6 \cdot 9700 = 100 \cdot a \qquad \text{Finding cross-products}$$

$$\frac{6 \cdot 9700}{100} = a \qquad \text{Dividing by 100}$$

$$\frac{58,200}{100} = a$$

$$582 = a$$

The increase is $582.00.

b) The new salary is

$$\$9700 + \$582 = \$10,282.$$

Do Exercise 5.

Exercise Set 7.5

a Solve.

1. It has been determined by sociologists that 17% of the population is left-handed. Each tournament conducted by the Professional Bowlers Association has 160 entrants. How many would you expect to be left-handed? not left-handed? Round to the nearest one.

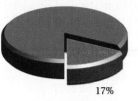

17%

Total: 160

2. A common guideline for businesses is to use 5% of their operating budget for advertising. Ariel Electronics has an operating budget of $8000 per week. How much should it spend each week for advertising? for other expenses?

5%

Total: $8000

3. Of all moviegoers, 67% are in the 12–29 age group. A theater contained 800 people for a showing of *Home Alone-18*. How many were in the 12–29 age group? not in this age group?

4. Deming, New Mexico, claims to have the purest drinking water in the world. It is 99.9% pure. If you had 240 L of water from Deming, how much of it, in liters, would be pure? impure?

5. A baseball player gets 13 hits in 40 at bats. What percent are hits? not hits?

6. On a test of 80 items, Erika had 76 correct. What percent were correct? incorrect?

7. A lab technician has 680 mL of a solution of water and acid; 3% is acid. How many milliliters are acid? water?

8. A lab technician has 540 mL of a solution of alcohol and water; 8% is alcohol. How many milliliters are alcohol? water?

9. Of the 8760 hr in a year, most television sets are on for 2190 hr. What percent is this?

10. In a medical study, it was determined that if 800 people kiss someone else who has a cold, only 56 will actually catch a cold. What percent is this?

11. A nut dealer has 1800 lb of peanuts, 1500 lb of cashews, and 700 lb of almonds. What percent of the total is peanuts? cashews? almonds?

12. It costs an oil company $40,000 a day to operate two refineries. Refinery A takes 37.5% of the cost, and refinery B takes the rest of the cost.

   a) What percent of the cost does it take to run refinery B?
   b) What is the cost of operating refinery A? refinery B?

b Solve.

13. The amount in a savings account increased from $200 to $216. What was the percent of increase?

14. The population of a small mountain town increased from 840 to 882. What was the percent of increase?

15. During a sale, a dress decreased in price from $70 to $56. What was the percent of decrease?

16. A person on a diet goes from a weight of 125 lb to a weight of 110 lb. What is the percent of decrease?

17. A person earns $18,600 one year and receives a 5% raise in salary. What is the new salary?

18. A person earns $20,400 one year and receives an 8% raise in salary. What is the new salary?

19. The value of a car typically decreases by 30% in the first year. A car is bought for $18,000. What is its value one year later?

20. One year the pilots of an airline shocked the business world by taking an 11% pay cut. The former salary was $55,000. What was the reduced salary?

21. World population is increasing by 1.6% each year. In 1995, it was 5.6 billion. How much will it be in 1996? 1997? 1998?

22. By increasing the thermostat from 72° to 78°, a family can reduce its cooling bill by 50%. If the cooling bill was $106.00, what would the new bill be? By what percent has the temperature been increased?

23. A car normally depreciates 30% of its original value in the first year. A car is worth $11,480 after the first year. What was its original cost?

$$11\,480$$
$$3\,444$$
$$14\,924$$

24. A cross-section of a standard or nominal "two by four" board actually measures $1\frac{1}{2}$ in. by $3\frac{1}{2}$ in. The rough board is 2 in. by 4 in. but is planed and dried to the finished size. What percent of the wood is removed in planing and drying?

25. *Treadmill test.* Treadmill tests are often administered to diagnose heart ailments. A guideline in such a test is to try to get you to reach your *maximum heart rate,* in beats per minute. The maximum heart rate is found by subtracting a person's age from 220 and then multiplying by 85%. What is the maximum heart rate of a person of age 25? 36? 48? 60? 76? Round to the nearest one.

$$16\,400$$

26. *Car depreciation.* Given normal use, an American-made car will depreciate 30% of its original cost the first year and 14% of its remaining value in the second year. What is the value of a car at the end of the second year if its original cost was $16,500? $18,400? $20,800?

$$X = 11\,480(.30)x$$

ANSWERS

19. _____

20. _____

21. _____

22. _____

23. _____

24. _____

25. _____

26. _____

27. In baseball, the *strike zone* is normally a 15-in. by 40-in. rectangle. Some batters give the pitcher the advantage by swinging at pitches thrown out of the strike zone. By what percent is the area of the strike zone increased if a 2-in. border is added to the outside?

28. Carlos is planting grass on a 24-ft by 36-ft area in his back yard. He installs a 6-ft by 8-ft garden. By what percent has he reduced the area he has to mow?

27. _____

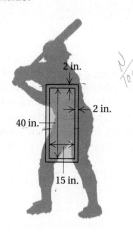

2 in.

2 in.

40 in.

15 in.

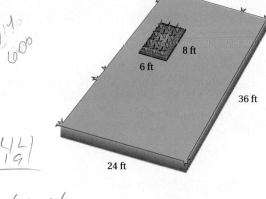

8 ft

6 ft

36 ft

24 ft

28. _____

(handwritten) 2:40 600

N/100

44/19

.600 IN

836 IN

836/600

N/100   234/600

29. _____

30. _____

(handwritten) 23600 = 600w

39,3

31. _____

SYNTHESIS

29. Which is higher, if either?

a) $1000 increased by 15%, then that amount decreased by 15%, or,

b) $1000 decreased by 15%, then that amount increased by 15%.

30. If *p* is 120% of *q*, *q* is what percent of *p*?

31. It has been determined that at the age of 10, a girl has reached 84.4% of her final adult growth. A girl is 4 ft, 8 in. at the age of 10. What will be her final adult height?

32. It has been determined that at the age of 15, a boy has reached 96.1% of his final adult height. A boy is 6 ft, 4 in. at the age of 15. What will be his final adult height?

32. _____

7.6 Consumer Applications: Sales Tax

OBJECTIVE

After finishing Section 7.6, you should be able to:

a   Solve problems involving percent and sales tax.

FOR EXTRA HELP

TAPE 12   TAPE 11B   MAC: 7
                            IBM: 7

a   Percent is used in sales tax computations. The sales tax rate in Arkansas is 3%. This means that the tax is 3% of the purchase price. Suppose the purchase price on a coat is $124.95. The sales tax is then

$$3\% \text{ of } \$124.95, \quad \text{or} \quad 0.03 \times 124.95,$$

or

$$3.7485, \quad \text{or about } \$3.75.$$

The total that you pay is the price plus the sales tax:

$$\$124.95 + \$3.75, \quad \text{or } \$128.70.$$

$124.95
+ 3% sales tax

Bill:		
Purchase price	=	$124.95
Sales tax (3% of $124.95)	=	+ 3.75
Final price		$128.70

Sales tax = Sales tax rate × Purchase price

Total price = Purchase price + Sales tax

---

1. The sales tax rate in Connecticut is 8%. How much tax is charged on the purchase of a refrigerator that sells for $668.95? What is the total price?

EXAMPLE 1   The sales tax rate in Florida is 6%. How much tax is charged on the purchase of a Florida Marlins baseball jacket that sells for $124.95? What is the total price?

a) We find the sales tax by first writing an equation.

$$\underbrace{\text{Sales tax}} = \underbrace{\text{Sales tax rate}} \times \underbrace{\text{Purchase price}}$$
$$t \quad = \quad 6\% \quad \times \quad 124.95$$

This tells us what to do. We multiply.

$$
\begin{array}{r}
1\ 2\ 4.9\ 5 \\
\times \quad\ \ 0.0\ 6 \\
\hline
7.4\ 9\ 7\ 0
\end{array}
$$

The sales tax is 7.497, or about $7.50.

b) The total price is given by the purchase price plus the sales tax, or

$$
\begin{array}{r}
1\ 2\ 4.9\ 5 \\
+ \quad\ \ \ 7.5\ 0 \\
\hline
1\ 3\ 2.4\ 5
\end{array}
$$

The total price is $132.45.

Do Exercise 1.

Answer on page A-4

2. The sales tax is $33 on the purchase of a washing machine that sells for $550. What is the sales tax rate?

EXAMPLE 2   The sales tax is $32 on the purchase of a sofa that sells for $800. What is the sales tax rate?

Restate: Sales tax   is   what percent   of   purchase price?

Translate: 32   =   r   ×   800

Solve: We divide on both sides by 800:

$$32 \div 800 = r$$
$$0.04 = r$$
$$4\% = r.$$

The sales tax rate is 4%.

Do Exercise 2.

3. The sales tax on a stereo is $25.20 and the sales tax rate is 6%. Find the purchase price (the price before taxes are added).

EXAMPLE 3   The sales tax on a laser printer is $31.74 and the sales tax rate is 5%. Find the purchase price (the price before taxes are added).

Restate: Sales tax   is   5%   of   what?

Translate: 31.74   =   5%   ×   p,   or 31.74 = 0.05 × p.

Solve: We divide on both sides by 0.05:

$$31.74 \div 0.05 = p$$
$$634.8 = p.$$

$$
\begin{array}{r}
6\ 3\ 4.8 \\
0.0\ 5\ )\overline{3\ 1.7\ 4\ 0} \\
3\ 0\ 0\ 0 \\
\hline
1\ 7\ 4 \\
1\ 5\ 0 \\
\hline
2\ 4 \\
2\ 0 \\
\hline
4\ 0 \\
4\ 0 \\
\hline
0
\end{array}
$$

The purchase price is $634.80.

Do Exercise 3.

Exercise Set 7.6

a Solve.

1. The sales tax rate in New York City is 8.25%. How much tax is charged on a purchase of $248? What is the total price?

   $248 + 8.25% of $248

2. The sales tax rate in Indiana is 5%. How much tax is charged on a purchase of $586? What is the total price?

   $586 + 5% of $586

3. The sales tax rate in Kentucky is 6%. How much tax is charged on a purchase of $189.95? What is the total price?

4. The sales tax rate in Illinois is 6.25%. How much tax is charged on a purchase of $265? What is the total price?

5. The sales tax is $48 on the purchase of a dining room set that sells for $960. What is the sales tax rate?

6. The sales tax is $15 on the purchase of a diamond ring that sells for $500. What is the sales tax rate?

7. The sales tax is $35.80 on the purchase of a refrigerator–freezer that sells for $895. What is the sales tax rate?

8. The sales tax is $9.12 on a purchase of $456. What is the sales tax rate?

9. The sales tax on a used car is $100 and the sales tax rate is 5%. Find the purchase price (the price before taxes are added).

10. The sales tax on a purchase is $112 and the sales tax rate is 2%. Find the purchase price.

1. _____

2. _____

3. _____

4. _____

5. _____

6. _____

7. _____

8. _____

9. _____

10. _____

11. _____

12. _____

13. _____

14. _____

15. _____

16. _____

17. _____

18. _____

19. _____

20. _____

21. _____

22. _____

23. _____

24. _____

11. The sales tax on a purchase is $28 and the sales tax rate is 3.5%. Find the purchase price.

12. The sales tax on a purchase is $66 and the sales tax rate is 5.5%. Find the purchase price.

13. The sales tax rate in Dallas is 1% for the city and 6% for the state. How much tax is charged on a purchase of $665?

14. The sales tax rate in Omaha is 1.5% for the city and 5% for the state. How much tax is charged on a purchase of $780?

15. The sales tax is $1030.40 on an automobile purchase of $18,400. What is the sales tax rate?

16. The sales tax is $979.60 on an automobile purchase of $15,800. What is the sales tax rate?

SKILL MAINTENANCE

Solve.

17. $2.3 \times y = 85.1$

18. $\dfrac{5}{x} = \dfrac{9}{11}$

19. $4.3 \times t = 34.4$

20. $256.8 = 10.7 \times x$

21. Convert to decimal notation: $\dfrac{13}{11}$.

22. Convert to a mixed numeral: $\dfrac{29}{11}$.

SYNTHESIS

23. ▦ The sales tax rate on a purchase is 5.4%. How much tax is charged on a purchase of $96,568.95?

24. ▦ The sales tax is $3811.88 on a purchase of $58,644.24. What is the sales tax rate?

7.7 Consumer Applications: Commission and Discount

a COMMISSION

When you work for a **salary**, you receive the same amount of money each week or month. When you work for a **commission**, you are paid a percentage of the amount that you sell. To find commission, take a certain percentage of sales.

Commission = Commission rate × Sales

EXAMPLE 1 A salesperson's commission rate is 20%. What is the commission from the sale of $25,560 worth of boom boxes?

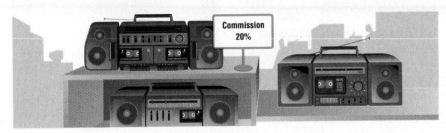

$$Commission = Commission\ rate \times Sales$$
$$C = 20\% \times 25{,}560$$

This tells us what to do. We multiply.

$$
\begin{array}{r}
2\,5{,}5\,6\,0 \\
\times \qquad 0.2 \\
\hline
5\,1\,1\,2.0
\end{array}
$$

20% = 0.20 = 0.2

The commission is $5112.

Do Exercise 1.

EXAMPLE 2 A salesperson earns a commission of $3000 selling $60,000 worth of farm machinery. What is the commission rate?

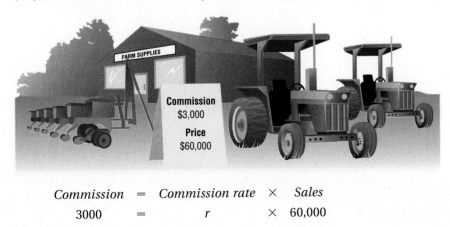

$$Commission = Commission\ rate \times Sales$$
$$3000 = r \times 60{,}000$$

1. A salesperson's commission rate is 30%. What is the commission from the sale of $18,760 worth of air conditioners?

Answer on page A-4

2. A salesperson earns a commission of $6000 selling $24,000 worth of refrigerators. What is the commission rate?

To solve this equation, we divide on both sides by 60,000:

$$3000 \div 60{,}000 = r.$$

We can divide, but this time we simplify by removing a factor of 1:

$$r = \frac{3000}{60{,}000} = \frac{1}{20} \cdot \frac{3000}{3000} = \frac{1}{20} = 0.05 = 5\%.$$

The commission rate is 5%.

Do Exercise 2.

EXAMPLE 3 A motorcycle salesperson's commission rate is 25%. He receives a commission of $425 on the sale of a motorbike. How much did the motorbike cost?

$$\begin{array}{ccccc} Commission & = & Commission\ rate & \times & Sales \\ 425 & = & 25\% & \times & S \end{array}$$

To solve this equation, we divide on both sides by 25%:

$$425 \div 25\% = S$$
$$425 \div 0.25 = S$$
$$1700 = S.$$

$$0.25\overline{\smash{\big)}\,425.00_\wedge} \begin{array}{r} 1\ 7\ 0\ 0. \\ \end{array}$$

$$\begin{array}{r} 2\ 5\ 0 \\ \hline 1\ 7\ 5 \\ 1\ 7\ 5 \\ \hline 0 \end{array}$$

The motorbike cost $1700.

Do Exercise 3.

3. A clothing salesperson's commission is 16%. She receives a commission of $268. How many dollars worth of clothing were sold?

Answers on page A-4

b | DISCOUNT

The regular price of a rug is $60. It is on sale at 25% off. Since 25% of $60 is $15, the sale price is $60 − $15, or $45. We call $60 the **marked price,** 25% the **rate of discount,** $15 the **discount,** and $45 the **sale price.** These are related as follows.

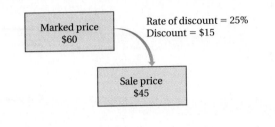

Rate of discount = 25%
Discount = $15

Discount = Rate of discount × Marked price

Sale price = Marked price − Discount

EXAMPLE 4 A rug is marked $240 and is on sale at 25% off. What is the discount? the sale price?

a) *Discount* = *Rate of discount* × *Marked price*

$$D = 25\% \times 240$$

This tells us what to do. We convert 25% to decimal notation and multiply.

```
    2 4 0
×   0.2 5       25% = 0.25
  1 2 0 0
  4 8 0 0
  6 0.0 0
```

The discount is $60.

4. A suit is marked $140 and is on sale at 24% off. What is the discount? the sale price?

b) *Sale price = Marked price − Discount*
$$S = 240 - 60$$

This tells us what to do. We subtract.

$$\begin{array}{r} 2\ 4\ 0 \\ -\ \ \ 6\ 0 \\ \hline 1\ 8\ 0 \end{array}$$

The sale price is $180.

Do Exercise 4.

Answer on page A-4

SIDELIGHTS

AN APPLICATION: WATER LOSS

The human body is $\frac{2}{3}$ water. If you lose

    1% of your body water, you will be thirsty;

    8% of your body water, you will almost collapse;

    10% of your body water, you will be unconscious; and

    20% of your body water, you will die.

EXERCISES
You weigh 180 lb.

1. How much of your body weight, in pounds, is water?
2. A loss of how many pounds of water would make you thirsty?
3. A loss of how many pounds of water would make you almost collapse?
4. A loss of how many pounds of water would make you lose consciousness?
5. A loss of how many pounds of water would cause you to die?

Exercise Set 7.7

a Solve.

1. A salesperson's commission rate is 6%. What is the commission from the sale of $45,000 worth of furnaces?

2. A salesperson's commission rate is 32%. What is the commission from the sale of $12,500 worth of atlases?

3. A salesperson earns $120 selling $2400 worth of television sets. What is the commission rate?

4. A salesperson earns $408 selling $3400 worth of stereos. What is the commission rate?

5. A dishwasher salesperson's commission rate is 40%. He receives a commission of $392. How many dollars worth of dishwashers were sold?

6. A real estate agent's commission rate is 7%. She receives a commission of $2800 on the sale of a home. How much did the home sell for?

7. A real estate commission is 7%. What is the commission on the sale of a $98,000 home?

8. A real estate commission is 8%. What is the commission on the sale of a piece of land for $68,000?

9. An encyclopedia salesperson earns a salary of $500 a month, plus a 2% commission on sales. One month $990 worth of encyclopedias were sold. What were the wages that month?

10. Some salespersons have their commission increased according to how much they sell. A salesperson gets a commission of 5% for the first $2000 and 8% on the amount over $2000. What is the total commission on sales of $6000?

b Find what is missing.

11.

Marked price	Rate of discount	Discount	Sale price
$300	10%	$30	270

12.

$2000	40%	8''	120'

13.

Marked price	Rate of discount	Discount	Sale price
$350	15%		

14.

$20.00	25%		

15.

~~$125~~	10%	$12.50	

$D = M \times R$

16.

	15%	$65.70	

17.

$600		$240	

18.

$12,800		$1920	

19. Find the discount and the rate of discount for the ring in this ad.

20. What is the mathematical error in this ad?

1/2 CARAT T.W.
DIAMOND, 14K GOLD
LADY'S BRIDAL SET
WAS $1275.00
$888

Water-Resistant Watch
With 24-Hour Alarm

Cut
30%

6⁹⁵
Reg.
9.95

Resists water to 100 feet!
Calendar, chime. #63-5058

SKILL MAINTENANCE

Find decimal notation.

21. $\frac{5}{9}$ .55   **22.** $\frac{23}{11}$ 2.09   **23.** $\frac{11}{12}$ **24.** $\frac{13}{7}$ 1.857142   **25.** $\frac{15}{7}$ **26.** $\frac{19}{12}$

SYNTHESIS

27. 🖩 A real estate commission rate is 7.5%. A house sells for $78,990. What is the commission? How much does the seller get for the house after paying the commission?

28. In a recent subscription drive, *People* offered a subscription of 104 weekly issues for a price of $1.29 per issue. They advertised that this was a savings of 27.9% off the newsstand price. What was the newsstand price?

7.8 Consumer Applications: Interest

a | SIMPLE INTEREST

You put $100 in a savings account for 1 year. The $100 is called the **principal**. The **interest rate** is 8%. This means you get back 8% of the principal, which is

$$8\% \text{ of } \$100, \quad \text{or} \quad 0.08 \times 100, \quad \text{or} \quad \$8.00,$$

in addition to the principal. The $8.00 is called the **interest**.

EXAMPLE 1   What is the interest on $2500 principal at the interest rate of 6% for 1 year?

We take 6% of $2500:

$$6\% \times 2500 = 0.06 \times 2500$$
$$= 150.$$

$$\begin{array}{r} 2\ 5\ 0\ 0 \\ \times \quad 0.0\ 6 \\ \hline 1\ 5\ 0.0\ 0 \end{array}$$

The interest for 1 year is $150.

Do Exercise 1.

To find interest for a fraction t of a year, we compute the interest for 1 year and multiply by t.

EXAMPLE 2   What is the interest on $2500 principal at the interest rate of 6% for $\frac{1}{4}$ year?

a) We find the interest for 1 year. We take 6% of $2500:

$$6\% \times 2500 = 0.06 \times 2500 = 150.$$

b) We multiply by $\frac{1}{4}$:

$$\frac{1}{4} \times 150 = \frac{150}{4} = 37.50.$$

$$\begin{array}{r} 3\ 7.5 \\ 4\overline{)1\ 5\ 0.0} \\ 1\ 2\ 0 \\ \hline 3\ 0 \\ 2\ 8 \\ \hline 2\ 0 \\ 2\ 0 \\ \hline 0 \end{array}$$

The interest for $\frac{1}{4}$ year is $37.50.

Do Exercise 2.

1. What is the interest on $4300 principal at the interest rate of 8% for 1 year?

2. What is the interest on $4300 principal at the interest rate of 8% for $\frac{3}{4}$ year?

Answers on page A-4

3. What is the interest on $4800 at 7% for 60 days?

Money is often borrowed for 30, 60, or 90 days even though the interest rate is given **per year.** To simplify calculations, businesspeople consider there to be 360 days in a year. If a loan is for 30 days, it is for 30/360 of a year. The actual interest is found by finding interest for 1 year and taking 30/360 of it.

EXAMPLE 3 What is the interest on $400 at 8% for 30 days?

We convert 30 days to a fractional part of one year.

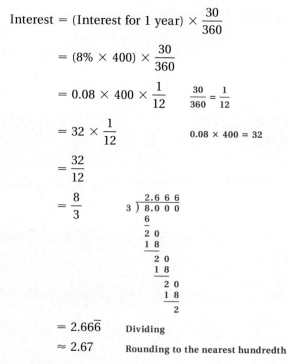

$$\text{Interest} = (\text{Interest for 1 year}) \times \frac{30}{360}$$

$$= (8\% \times 400) \times \frac{30}{360}$$

$$= 0.08 \times 400 \times \frac{1}{12} \qquad \frac{30}{360} = \frac{1}{12}$$

$$= 32 \times \frac{1}{12} \qquad 0.08 \times 400 = 32$$

$$= \frac{32}{12}$$

$$= \frac{8}{3}$$

$$
\begin{array}{r}
2.6\;6\;6 \\
3\,\overline{)\,8.0\;0\;0} \\
6 \\
\overline{2\;0} \\
1\;8 \\
\overline{2\;0} \\
1\;8 \\
\overline{2\;0} \\
1\;8 \\
\overline{2}
\end{array}
$$

$$= 2.66\overline{6} \qquad \textbf{Dividing}$$

$$\approx 2.67 \qquad \textbf{Rounding to the nearest hundredth}$$

The interest for 30 days is $2.67.

A general formula for interest is as follows.

Interest = Rate · Principal · Time (expressed in some part of a year), or

$I = (r \cdot P) \cdot t,$ or, more commonly, $I = P \cdot r \cdot t.$

Interest computed in this way is called **simple interest.**

Caution! Remember that time is to be expressed in years or a fractional part of a year.

Do Exercise 3.

Answer on page A-4

b | COMPOUND INTEREST

When interest is paid *on interest*, we call it **compound interest.** This type of interest is usually paid on savings accounts. Suppose you have $100 in a savings account at 6%. In 1 year, you earn

6% of $100,   or $6 interest.

Then you have $106. If you leave the interest in your account, the next year you earn interest on $106, which is

6% of $106,   or $0.06 × 106,   or $6.36.

You then have $106 + $6.36, or $112.36 in your account. When this happens, we say that interest is **compounded annually.** The interest of $6 the first year earned $0.36 the second year.

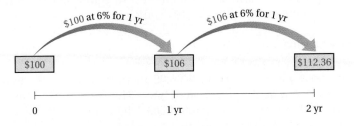

| $100 | $106 | $112.36 |

| 0 | 1 yr | 2 yr |

EXAMPLE 4   Interest is compounded annually. Find the amount in an account if $2000 is invested at 8% for 2 years.

a) We find the interest at the end of 1 year:

$$I = (r \times P) \times t$$
$$= (8\% \times \$2000) \times 1$$
$$= 0.08 \times \$2000$$
$$= \$160.$$

b) We then find the new principal after 1 year:

$$\$2000 + \$160 = \$2160.$$

c) Going into the second year, the principal is $2160. We now find the interest for 1 year after that:

$$I = (r \times P) \times t$$
$$= (8\% \times \$2160) \times 1$$
$$= 0.08 \times \$2160$$
$$= \$172.80.$$

d) Next, we find the new principal after 2 years:

$$\$2160 + \$172.80 = \$2332.80.$$

The amount in the account after 2 years is $2332.80.

Do Exercise 4.

Answer on page A-4

5. Interest is compounded
semiannually. Find the
amount in an account if $2000
is invested at 5% for 1 year.

Interest added to an account every half year is **compounded semi-annually.** Suppose you have $100 in a savings account at 6%. In $\frac{1}{2}$ year, you earn

$$6\% \times \$100 \times \frac{1}{2}, \quad \text{or} \quad \$3 \text{ interest.}$$

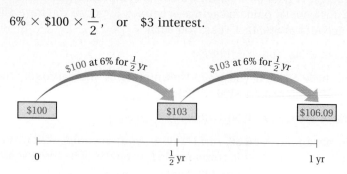

Then you have $103 if you leave the interest in your account. The last half of the year, you earn interest on $103, which is

$$6\% \times \$103 \times \frac{1}{2}, \quad \text{or} \quad \$3.09.$$

You then have $103 + $3.09, or $106.09 in your account. If interest were compounded annually, you would have $106, but with interest compounded semiannually, you have $106.09. The more often interest is compounded, the more interest your money earns.

EXAMPLE 5 Interest is compounded semiannually. Find the amount in an account if $2000 is invested at 8% for 1 year.

a) We find the interest at the end of $\frac{1}{2}$ year:

$$I = (r \times P) \times t$$

$$= (8\% \times \$2000) \times \frac{1}{2}$$

$$= (0.08 \times \$2000) \times \frac{1}{2}$$

$$= \$160 \times \frac{1}{2} = \$80.$$

b) We then find the new principal after $\frac{1}{2}$ year:

$$\$2000 + \$80 = \$2080.$$

c) Going into the second half of the year, the principal is $2080. Now we find the interest for $\frac{1}{2}$ year after that:

$$I = (r \times P) \times t$$

$$= (8\% \times \$2080) \times \frac{1}{2}$$

$$= (0.08 \times \$2080) \times \frac{1}{2}$$

$$= \$166.40 \times \frac{1}{2} = \$83.20.$$

d) Next, we find the new principal after 1 year:

$$\$2080 + \$83.20 = \$2163.20.$$

The amount in the account after 1 year is $2163.20.

Answer on page A-4

Do Exercise 5.

Exercise Set 7.8

a Find the *simple* interest.

	Principal	Rate of interest	Time
1.	$200	13%	1 year
2.	$450	18%	1 year
3.	$2000	12.4%	1 year
4.	$200	7.7%	$\frac{1}{2}$ year
5.	$4300	14%	$\frac{1}{4}$ year
6.	$2000	15%	30 days
7.	$5000	14.5%	60 days
8.	$3300	12%	90 days
9.	$4400	9.4%	1 year
10.	$8800	8.8%	1 year

1. _____

2. _____

3. _____

4. _____

5. 15050 _____

6. _____

7. 10012 _____

8. $99 _____

9. _____

10. _____

21. In a certain state, a sales tax of $378 is collected on the purchase of a car for $7560. What is the sales tax rate?

$$\frac{378}{7560} = \frac{.05}{5\%}$$

22. A salesperson earns $753.50 selling $6850 worth of televisions. What is the commission rate?

23. An item has a marked price of $350. It is placed on sale at 12% off. What are the discount and the sale price?

24. An item priced at $305 is discounted at the rate of 14%. What are the discount and the sale price?

25. An insurance salesperson receives a 7% commission. If $420 worth of insurance is sold, what is the commission?

26. What is the simple interest on $180 at 6% for $\frac{1}{3}$ year?

$$I = (180)(0.06)\left(\frac{1}{3}\right)$$
$$\$3.60$$

27. What is the simple interest on $180 principal at the interest rate of 7% for 1 year?

28. What is the simple interest on $220 principal at the interest rate of 5.5% for 1 year?

29. What is the simple interest on $250 at 4.2% for $\frac{1}{2}$ year?

30. Interest is compounded semiannually. Find the amount in an account if $200 is invested at 8% for 1 year.

$$200 \times (.08)\left(\frac{1}{2}\right) = 8$$
$$208 \times (0.08)\left(\frac{1}{2}\right) = 8.32$$
$$\$216.32$$

31. Interest is compounded annually. Find the amount in an account if $150 is invested at 9% for 2 years.

SKILL MAINTENANCE

Solve.

32. $\dfrac{3}{8} = \dfrac{7}{x}$

33. $10.4 \times y = 665.6$

34. $100 \times x = 761.23$

35. $\dfrac{1}{6} = \dfrac{7}{x}$

Convert to decimal notation.

36. $\dfrac{11}{3}$

37. $\dfrac{11}{7}$

Convert to a mixed numeral.

38. $\dfrac{11}{3}$

39. $\dfrac{121}{7}$

Test: Chapter 7

1. Find decimal notation for 89%.

.89

2. Find percent notation for 0.674.

67.4%

3. Find percent notation for $\frac{7}{8}$.

87.5%

4. Find fractional notation for 65%.

$\frac{13}{20}$

5. Translate to an equation. Then solve.

What is 40% of 55?

$X = 40 \times 55$

$X = 40\% \times 55$ $X = 22$

$X = 22$

6. Translate to a proportion. Then solve.

What percent of 80 is 65?

$\frac{N}{100} = \frac{A}{B}$

$\frac{N}{100} = \frac{4}{80}$

$\frac{65 \times 100}{80}$

81.25%

Solve.

7. The weight of muscles in a human body is 40% of total body weight. A person weighs 125 lb. What do the muscles weigh?

$\frac{40}{1} \cdot \frac{50}{125}$

$.40 \times 125 = 50$

8. The population of a town increased from 2000 to 2400. Find the percent of increase in population.

$\frac{400}{2400}$

$\frac{400}{2000}$ 20%

$\frac{N}{100} = \frac{2000}{}$

10%

ANSWERS

1. _____

2. _____

3. _____

4. _____

$\frac{80}{100} = \frac{65}{B}$

81.25

5. _____

6. _____

7. _____

8. _____

9.

10.

11.

12.

13.

14.

15.

16.

17.

18.

9. The sales tax rate in Arizona is 5%. How much tax is charged on a purchase of $324? What is the total price?

$.05 \times 324 = 14.20$

340.20

10. A salesperson's commission rate is 15%. What is the commission from the sale of $4200 worth of merchandise?

$.15 \times 4200 =$

630

11. The marked price of an item is $200 and the item is on sale at 20% off. What are the discount and the sale price?

$.20 \times 200 = 40$

40

12. What is the simple interest on $120 principal at the interest rate of 7.1% for 1 year?

$P \times R \times S$

8.52

13. What is the simple interest on $100 at 8.6% for $\frac{1}{2}$ year?

$100 \times .086 \times .5$

4.3

14. Interest is compounded annually. Find the amount in an account if $100 is invested at 5% for 2 years.

105

110.25

SKILL MAINTENANCE

Solve.

15. $8.4 \times y = 1864.8$

16. $\dfrac{5}{8} = \dfrac{10}{x}$

17. Convert to decimal notation: $\dfrac{17}{12}$.

1.416

18. Convert to a mixed numeral: $\dfrac{153}{44}$.

$3\frac{21}{44}$

Cumulative Review: Chapters 1–7

1. Find fractional notation for 0.091.

2. Find decimal notation for $\frac{13}{6}$.

3. Find decimal notation for 3%.

4. Find percent notation for $\frac{9}{8}$.

5. Write fractional notation for the ratio 5 to 0.5.

6. Find the rate in kilometers per hour.

350 km,   15 hr

Use $<$, $>$, or $=$ for ▨ to write a true sentence.

7. $\frac{5}{7}$ ▨ $\frac{6}{8}$

8. $\frac{6}{14}$ ▨ $\frac{15}{25}$

Estimate the sum or difference by rounding to the nearest hundred.

9. $263{,}961 + 32{,}090 + 127.89$

10. $73{,}510 - 23{,}450$

Calculate.

11. $46 - [4(6 + 4 \div 2) + 2 \times 3 - 5]$

12. $[0.8(1.5 - 9.8 \div 49) + (1 + 0.1)^2] \div 1.5$

Add and simplify.

13. $\frac{6}{5} + 1\frac{5}{6}$

14. $46.9 + 2.84$

15.
$$
\begin{array}{r}
4\,8\,7{,}0\,9\,4 \\
6{,}9\,3\,6 \\
+\ \ \ 2\,1{,}1\,2\,0 \\
\hline
\end{array}
$$

Subtract and simplify.

16. $35 - 34.98$

17. $3\frac{1}{3} - 2\frac{2}{3}$

18. $\frac{8}{9} - \frac{6}{7}$

Multiply and simplify.

19. $\frac{7}{9} \cdot \frac{3}{14}$

20.
$$
\begin{array}{r}
2\,3\,6{,}9\,8\,4 \\
\times\ \ \ \ \ 3{,}6\,0\,0 \\
\hline
\end{array}
$$

21.
$$
\begin{array}{r}
4\,6.0\,1\,2 \\
\times\ \ \ \ \ 0.0\,3 \\
\hline
\end{array}
$$

Divide and simplify.

22. $6\frac{3}{5} \div 4\frac{2}{5}$

23. $431.2 \div 35.2$

24. $15 \overline{)\,1\,8\,5\,0}$

Solve.

25. $36 \cdot x = 3420$

26. $y + 142.87 = 151$

27. $\frac{2}{15} \cdot t = \frac{6}{5}$

28. $\frac{3}{4} + x = \frac{5}{6}$

29. $\frac{y}{25} = \frac{24}{15}$

30. $\frac{16}{n} = \frac{21}{11}$

Solve.

31. On a checking account of $7428.63, a check was drawn for $549.79. What was left in the account?

32. A total of $57.50 was paid for 5 neckties. How much did each cost?

33. A 12-oz box of crackers costs $3.69. Find the unit price in cents per ounce.

34. A bus travels 456 km in 6 hr. At this rate, how far would the bus travel in 8 hr?

35. In a recent year, Americans threw away 50 million lb of paper. It is projected that this will increase to 65 million lb in the year 2000. Find the percent of increase.

36. The state of Utah has an area of 1,722,850 mi^2. Of this area, 60% is owned by the government. How many square miles are owned by the government?

37. In a recent year, there were 9145 McDonald's restaurants and 7258 Pizza Hut restaurants in the United States. How many more McDonald's restaurants were there?

38. How many pieces of ribbon $1\frac{4}{5}$ yd long can be cut from a length of ribbon 9 yd long?

39. A student walked $\frac{7}{10}$ mi to school and then $\frac{8}{10}$ mi to the library. How far did the student walk?

40. On a map, 1 in. represents 80 mi. How much does $\frac{3}{4}$ in. represent?

SYNTHESIS

41. First National Bank offers 10% simple interest. Saturn Bank offers 9.75% interest compounded semiannually. Which bank offers the highest rate of return on the investment?

42. On a trip through the mountains, a car traveled 240 mi on $7\frac{1}{2}$ gal of gasoline. On a trip across the plains, the same car traveled 351 mi on $9\frac{3}{4}$ gal of gasoline. What was the percent of increase or decrease in miles per gallon?

8

Descriptive Statistics

INTRODUCTION

There are many ways in which to analyze or describe data. One is to look at certain numbers, or *statistics*, related to the data. We will consider three kinds of statistics: the *average*, the *median*, and the *mode*. Another way is to create graphs. We will consider several kinds of graphs: pictographs, bar graphs, line graphs, and circle graphs.

AN APPLICATION

The circle graph on page 401 shows music preferences of customers on the basis of music store sales, according to the National Association of Recording Merchandisers. A music store sells 3000 CDs a month. How many of these are country?

THE MATHEMATICS

We find from the circle graph that 9% of all recordings are country. We then take 9% of 3000:

$$9\% \cdot 3000 = 0.09 \cdot 3000 = 270.$$

6. According to recent EPA estimates, a Honda Civic was expected to travel 700 mi (city) on 25 gal of gasoline. What was the average number of miles expected per gallon?

7. A student obtained the following grades one semester.

GRADE	NUMBER OF CREDIT HOURS IN COURSE
B	3
C	4
C	4
A	2

What was the student's grade point average? Assume that the grade point values are 4.00 for an A, 3.00 for a B, and so on.

EXAMPLE 3   According to recent EPA estimates, a Chevrolet Corvette was expected to travel 375 mi (city) on 25 gal of gasoline. What was the average number of miles expected per gallon?

We divide the total number of miles, 375, by the number of gallons, 25:

$$\frac{375}{25} = 15.$$

The average was 15 miles per gallon.

Do Exercise 6.

EXAMPLE 4   *Grade point average, GPA.* In most colleges, students are assigned grade point values for grades obtained. The **grade point average,** or **GPA,** is the average of the grade point values for each hour taken. Suppose that at a certain college, grade point values are assigned as follows:

A: 4.00
B: 3.00
C: 2.00
D: 1.00
F: 0.00

A student obtained the following grades for one semester. What was the student's grade point average?

COURSE	GRADE	NUMBER OF CREDIT HOURS IN COURSE
Accounting	B	4
Calculus	A	5
English	A	5
French	C	3
Physical education	F	1

To find the GPA, we first add all the grade point values for each hour taken. We do this by first multiplying the grade point value (in color below) by the number of hours in the course and then adding as follows:

Accounting	$3.00 \cdot 4 = 12$
Calculus	$4.00 \cdot 5 = 20$
English	$4.00 \cdot 5 = 20$
French	$2.00 \cdot 3 = 6$
Physical education	$0.00 \cdot 1 = \underline{0}$
	58 (Total)

The total number of hours taken is $4 + 5 + 5 + 3 + 1$, or 18. We divide 58 by 18 and round to the nearest hundredth:

$$\frac{58}{18} \approx 3.22.$$

The student's grade point average was 3.22.

Do Exercise 7.

EXAMPLE 5 To get a B in math, a student must score an average of 80 on the tests. On the first four tests, the scores were 79, 88, 64, and 78. What is the lowest score that the student can get on the last test and still get a B?

We can find the total of the five scores needed as follows:

$$80 + 80 + 80 + 80 + 80 = 5 \cdot 80, \quad \text{or} \quad 400.$$

The total of the scores on the first four tests is

$$79 + 88 + 64 + 78 = 309.$$

Thus the student needs to get at least

$$400 - 309, \quad \text{or} \quad 91$$

in order to get a B. We can check this as follows:

$$\frac{79 + 88 + 64 + 78 + 91}{5} = \frac{400}{5}, \quad \text{or} \quad 80.$$

Do Exercise 8.

b MEDIANS

Another kind of center point statistic is a *median*. Suppose a student made the following scores on five tests.

Test 1: 78	Test 4: 76
Test 2: 81	Test 5: 84
Test 3: 82	

Let's first list the scores in order from smallest to largest:

76, 78, 81, 82, 84.

↑

Middle score

The middle score is called the **median.** Thus, 81 is the median of the scores.

EXAMPLE 6 What is the median of this set of numbers?

99, 870, 91, 98, 106, 90, 98

We first rearrange the numbers in order from smallest to largest. Then we locate the middle number, 98.

90, 91, 98, 98, 99, 106, 870

↑

Middle number

The median is 98.

Do Exercises 9–11.

MEDIAN

The *median* of a set of data is the middle number if there is an odd number of numbers. If there is an even number of numbers, then there are two numbers in the middle and the *median* is the number that is halfway between the two middle numbers.

---

8. To get an A in math, a student must score an average of 90 on the tests. On the first three tests, the scores were 80, 100, and 86. What is the lowest score that the student can get on the last test and still get an A?

Find the median.

9. 17, 13, 18, 14, 19

10. 20, 14, 13, 19, 16, 18, 17

11. 78, 81, 83, 91, 103, 102, 122, 119, 88

Answers on page A-4

Find the median.

12. 13, 20, 19, 16, 18, 14

13. 68, 34, 67, 69, 34, 70

Find the mode.

14. 23, 45, 45, 45, 78

15. 34, 34, 67, 67, 68, 70

16. 13, 24, 27, 28, 67, 89

17. George received the following test scores:

74, 86, 96, 67, 82.

a) What is the median score?
b) What is the average?
c) What is the mode?

Answers on page A-4

EXAMPLE 7   What is the median of this set of numbers?

69, 80, 61, 63, 62, 65

We first rearrange the numbers in order from smallest to largest. There is an even number of numbers. We look for the middle two, which are 63 and 65. The median is halfway between 63 and 65. It is 64, and is not in the set.

61, 62, 63, 65, 69, 80

⎵— Median 64

Note that the number halfway between two numbers is their average. In this example, the number halfway between 63 and 65 is found as follows:

$$\text{Median} = \frac{63 + 65}{2} = \frac{128}{2} = 64.$$

EXAMPLE 8   What is the median of this set of numbers?

25, 26, 24, 23

We first rearrange the numbers in order from smallest to largest. There is an even number of numbers. The two middle numbers are 24 and 25. Thus the median is halfway between 24 and 25. We find it as follows:

23, 24, 25, 26

⎵— Median 24.5

$$\text{Median} = \frac{24 + 25}{2} = \frac{49}{2} = 24.5.$$

Do Exercises 12 and 13.

c   MODES

The final type of center point statistic is the **mode**.

> **MODE**
>
> The *mode* of a set of data is the number or numbers that occur most often.

EXAMPLE 9   Find the mode of this set of data.

13, 14, 17, 17, 18, 19

The number that occurs most often is 17. Thus the mode is 17.

A set of data has just one average (mean) and just one median, but it can have more than one mode. If no number repeats, then each number in the set is a mode.

EXAMPLE 10   Find the mode, or modes, of this set of data.

33, 34, 34, 34, 35, 36, 37, 37, 37, 38, 39, 40

There are two numbers that occur most often, 34 and 37. Thus the modes are 34 and 37.

Do Exercises 14–17.

Exercise Set 8.1

a, **b**, **c** For each set of numbers, find the average, the median, and the mode.

1. 16, 18, 29, 14, 29, 19, 15

2. 72, 83, 85, 88, 92

3. 5, 30, 20, 20, 35, 5, 25

4. 13, 32, 25, 27, 13

5. 1.2, 4.3, 5.7, 7.4, 7.4

6. 13.4, 13.4, 12.6, 42.9

7. 234, 228, 234, 229, 234, 278

8. $29.95, $28.79, $30.95, $29.95

The following are the weights of the defensive linemen of the Dallas Cowboys. Use the data for Exercises 9 and 10.

Weight (lb)	Weight (kg)
250	113
255	116
260	118
260	118

9. What are the average, the median, and the mode of the weights in pounds?

10. What are the average, the median, and the mode of the weights in kilograms?

11. The following temperatures were recorded for seven days in Hartford:

43°, 40°, 23°, 38°, 54°, 35°, 47°.

What was the average temperature? the median? the mode?

12. Lauri Merten, a professional golfer, scored 71, 71, 70, and 68 to win the U.S. Women's Open in a recent year. What was the average score? the median? the mode? 68 70 72 72

13. According to recent EPA estimates, a Geo Prizm was expected to get 522 mi (highway) on 18 gal of gasoline. What was the average number of miles per gallon?

14. According to recent EPA estimates, a Saturn was expected to get 840 mi (highway) on 24 gal of gasoline. What was the average number of miles per gallon?

In Exercises 15 and 16 are the grades of a student for one semester. In each case, find the grade point average. Assume that the grade point values are 4.00 for an A, 3.00 for a B, and so on.

15.

GRADES	NUMBER OF CREDIT HOURS IN COURSE
B	4
B	5
B	3
C	4

G.PA 2.75

16.

GRADES	NUMBER OF CREDIT HOURS IN COURSE
A 4	20 5
B 3	12 4
B 3	9 3
C 2	10 5
	51 17

51÷17= 3.00 gpa

1. _____

2. _____

3. _____

4. _____

5. _____

6. _____

7. _____

8. _____

9. _____

10. _____

11. _____

12. _____

13. _____

14. _____

15. _____

16. _____

17. The following prices per pound of filet mignon were found at five supermarkets:

$9.79, $9.59, $9.69, $9.79, $9.89.

What was the average price per pound of filet mignon? the median price? the mode?

18. The following prices per pound of ground beef were found at five supermarkets:

$2.39, $2.29, $2.49, $2.09, $1.99.

What was the average price per pound of ground beef? the median price? the mode?

19. To get a B in math, Alexander Pappas must average 80 on five tests. Scores on the first four tests were 80, 74, 81, and 75. What is the lowest score that he can get on the last test and still get a B?

$80 \times 5 = 400$
400
-310
90

20. To get an A in math, Herb Cohen must average 90 on five tests. Scores on the first four tests were 90, 91, 81, and 92. What is the lowest score that he can get on the last test and still get an A?

354

$90 \times 5 = 450$
-354
96

21. The following are the salaries of the employees at the Raggs, Ltd. Clothing Store. What is the average salary?

Number	Type	Salary
1	Owner	$29,200
5	Salesperson	19,600
3	Secretary	14,800
1	Custodian	13,000

SKILL MAINTENANCE

Multiply.

22. $14 \cdot 14$

23. $\frac{2}{3} \cdot \frac{2}{3}$

24. 1.4×1.4

25. 1.414×1.414

SYNTHESIS

🖩 *Bowling averages.* Computing a bowling average involves a special kind of rounding. In effect, we never round up. For example, suppose a bowler gets a total of 599 for 3 games. To find the average, we divide 599 by 3 and drop the amount to the right of the decimal point:

$$\frac{599}{3} \approx 199.67. \qquad \text{The bowler's average is 199.}$$

In each case, find the bowling average.

26. 547 pins in 3 games

27. 4621 in 27 games

8.2 Tables and Pictographs

a READING AND INTERPRETING TABLES

A **table** is often used to present data in rows and columns of words and numbers.

EXAMPLE 1 The following table lists information about certain statistics of five men considered to be the best first basemen in the National League in a recent year.

PLAYER	TEAM	BATTING AVERAGE	HOME RUNS	ERRORS
Will Clark	Giants	0.300	16	10
Mark Grace	Cubs	0.307	9	4
Eric Karros	Dodgers	0.257	20	9
Fred McGriff	Braves	0.286	35	12
Hal Morris	Reds	0.271	6	1

a) Which player had the highest batting average? Which team did he play for?

b) Which player had the lowest batting average? Which team did he play for?

c) Which player had the most home runs?

d) Which player had the least home runs?

e) What was the average number of home runs hit by these first basemen?

Careful attention to the table will give us the answers.

a) To find the highest batting average, we look down the column headed "Batting average" until we find the highest number. That number is 0.307. Then we look across that row to the left to find the player, who is Mark Grace. We look back to the second column and find that he played for the Cubs.

b) To find the lowest batting average, we look down the column headed "Batting average" until we find the lowest number. That number is 0.257. Then we look across that row to the left to find the player, who is Eric Karros. We look back to the second column and find that he played for the Dodgers.

c) To find which player hit the most home runs, we look down the column headed "Home runs" until we find the highest number. That number is 35. Then we look across that row to the left to find the player, who is Fred McGriff.

d) To find which player hit the least home runs, we look down the column headed "Home runs" until we find the lowest number. That number is 6. Then we look across that row to the left to find the player, who is Hal Morris.

Use the table in Example 1 to answer each of the following.

1. Which of the first basemen had the most errors?

2. Which of the first basemen had the least number of errors?

3. What was the average number of errors made by these first basemen?

4. Of all the first basemen, who had a batting average of 0.280 or better?

Answers on page A-4

Use the table in Example 2 to answer each of the following.

5. In which cities will you have to pay $50 or more, per week, for family day care?

6. If you live in St. Louis, what will be the combined weekly cost range to place your 1-year-old and your 3-year-old in a day-care center?

7. In which cities are you sure to spend under $150 per week for your child to be cared for in your own home?

e) To find the average number of home runs hit by these first basemen, we proceed as in Section 8.1. We add the numbers of home runs, and then divide by the number of addends, 5:

$$\frac{16 + 9 + 35 + 20 + 6}{5} = \frac{86}{5} = 17.2.$$

Thus the average number of home runs hit by these first basemen is 17.2.

Do Exercises 1–4 on the preceding page.

EXAMPLE 2   According to the U.S. Bureau of Labor Statistics, 56% of women with children under the age of 6, or about 9 million women, were employed outside the home in a recent year. This table shows the weekly cost ranges, per child, for day-care programs in seven major U.S. cities.

	FAMILY DAY CARE		DAY-CARE CENTER		CAREGIVER COMES TO CHILD'S HOME
	Age	Cost	Age	Cost	
BOSTON	0–2	$45–160	0–2	$90–150	$260–340
	2–5	$40–160	2–5	$75–110	
NEW YORK	0–2	$35–140	0–2	$60–150	$165–300
	2–5	$40–160	2–5	$75–110	
ATLANTA	0–2	$30–60	0–2	$35–70	$165–230
	2–5	$30–55	2–5	$50–70	
ST. LOUIS	0–2	$45–50	0–2	$65–80	$165 and up
	2–5	$40–160	2–5	$75–110	
DALLAS	0–2	$50–70	0–2	$60–90	$165–200
	2–5	$50–70	2–5	$50–70	
DENVER	0–2	$65–105	0–2	$65–105	$165–200
	2–5	$55–105	2–5	$55–105	
SAN FRANCISCO	0–2	$55–90	0–2	$90–120	$165–200
	2–5	$55–85	2–5	$65–90	

a) In which city is it most expensive to have the caregiver come to your home?

b) How much will it cost you each week to have your 4-year-old child cared for in a day-care center in New York?

c) What is the maximum, per child, that you would expect to pay for a 4-year-old child cared for in family day care in Atlanta?

We look at the table to answer the questions.

a) We go to the last column, since the caregiver will come to your home, and read down the column, looking for the greatest entry. When we find it ($260–$340), we read back across, all the way to the left of that entry, and find that that is the rate in Boston.

b) This time we find New York in the first column. We then read across to the column under the heading "Day-care center." Since your child is 4 years old, we select the 2–5 age range, and the corresponding cost entry is $75–$110.

c) Locating Atlanta on the left, we read across to the column headed "Family day care." Since the range indicated is $30–$55, the maximum that you would expect to pay is $55.

Do Exercises 5–7.

b READING AND INTERPRETING PICTOGRAPHS

Pictographs (or *picture graphs*) are another way to show information. Instead of actually listing the amounts to be considered, a **pictograph** uses symbols to represent the amounts. In addition, a *key* is given telling what each symbol represents.

EXAMPLE 3 This pictograph shows approximately how many productive oil wells there are in eight Middle Eastern countries. Just below the graph, there is a key that tells you that each symbol represents 50 wells.

Productive Oil Wells in the Middle East

= 50 wells

a) How many productive oil wells are there in Saudi Arabia?

b) How many more productive oil wells are there in Oman than in Turkey?

c) What is the total number of productive oil wells in the top four oil-producing countries?

We can compute the answers by first reading the pictograph.

a) Saudi Arabia's wells are represented by 11 symbols. Since each symbol stands for 50 wells, we can multiply 11 × 50 and get 550 oil wells.

b) Oman has 9 symbols representing about 450 oil wells (9 × 50). Turkey has about 400 oil wells, as shown by 8 symbols (8 × 50). Therefore, Oman has approximately 50 more wells than Turkey does.

c) By counting the symbols, you can see that Iran (11 symbols), Kuwait (11 symbols), Saudi Arabia (11 symbols), and Syria (12 symbols) are the top four oil-producing countries. The total number of symbols for these countries is 45, which represents a total of 2250 oil wells (45 × 50).

Do Exercises 8–10.

Use the pictograph in Example 3 to answer each of the following.

8. How many productive oil wells are there in Syria?

9. Turkey has more productive oil wells than which two countries combined?

10. What is the average number of productive oil wells for these 8 countries?

Answers on page A-4

Use the graph in Example 4 to answer each of the following.

11. India has a greater population than the combined populations of which other countries?

You should realize by now that, although they seem to be very easy to read, pictographs are difficult to draw accurately because whole symbols reflect loose approximations due to significant rounding. In pictographs, you also need to use some mathematics to find the actual amounts.

EXAMPLE 4 This pictograph shows recent estimates of the population of five countries.

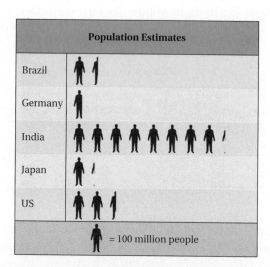

12. Which two countries are closest in population?

Give the approximate population of each country.

Brazil's population is represented by 1 whole symbol (100 million people) and about $\frac{1}{2}$ of another symbol (50 million people) for a total of 150 million people.

Germany's population is shown by about $\frac{3}{4}$ of a symbol, representing a total of 75 million people.

For India, there are 8 whole symbols (8×100 million people) and $\frac{1}{4}$ of another (25 million people), giving a total of 825 million people.

The population of Japan is represented by 1 whole symbol (100 million people) and about $\frac{1}{4}$ of another (25 million people) for a total of 125 million people.

13. Which two countries have the greatest difference in population?

The population of the United States is shown by 2 whole symbols (2×100 million people) and about $\frac{3}{5}$ of another (60 million people), giving a total of 260 million people.

One advantage of pictographs is that the appropriate choice of a symbol will tell you, at a glance, the kind of measurement being made. Another advantage is that the comparison of amounts represented in the graph can be expressed more easily by just counting symbols. For instance, in Example 3, the ratio of Iraq's oil wells to Saudi Arabia's oil wells is 6:11.

One disadvantage of pictographs is that, to make a pictograph easy to read, the amounts must be rounded significantly to the unit that a symbol represents. This makes it difficult to accurately represent an amount. Another problem is that it is difficult to determine very accurately how much a partial symbol represents. A third disadvantage is that you must use some mathematics to finally compute the amount represented, since there is usually no explicit statement of the amount.

Answers on page A-4

Do Exercises 11–13.

C | DRAWING PICTOGRAPHS

EXAMPLE 5 Draw a pictograph showing the U.S. production of apples in five different years. Let the apple symbol represent 50,000,000 bushels of apples and use the following information.

a) 1970: 150,000,000 bushels

b) 1975: 175,000,000 bushels

c) 1980: 200,000,000 bushels

d) 1985: 190,000,000 bushels

e) 1990: 225,000,000 bushels

Some computation is necessary.

a) Since 3 × 50 million is 150 million, three apple symbols will show the bushels for 1970.

b) Since 3 × 50 million is only 150 million, we need not only these three apple symbols, but also $\frac{1}{2}$ of another $\left(\frac{1}{2}\text{ of 50 million is 25 million}\right)$ to show 175 million bushels for 1975.

c) Four apple symbols will be used to show the bushels for 1980, since 4 × 50 million is 200 million.

d) We will need three whole apple symbols to represent 150 million bushels plus $\frac{4}{5}$ of another to show an additional 40 million bushels. This will give a total of 190 million bushels for 1985.

e) Since 4 × 50 million is only 200 million, we need not only these four apple symbols, but also $\frac{1}{2}$ of another $\left(\frac{1}{2}\text{ of 50 million is 25 million}\right)$ to show 225 million bushels for 1990.

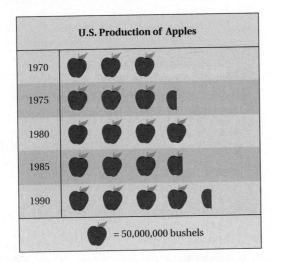

Do Exercise 14.

14. Draw a pictograph to represent the following data regarding the number of television sets in various cities.

a) Honolulu, Hawaii: 200,000

b) Columbus, Ohio: 400,000

c) Houston, Texas: 550,000

d) Pittsburgh, Pennsylvania: 467,000

Answers on page A-4

CAREERS INVOLVING MATHEMATICS

If you have done well in this course in basic mathematics, you might be considering a career involving mathematics. If such is the case, the following information may be valuable to you.

CAREERS INVOLVING MATHEMATICS

The following is the result of a survey conducted by *The Jobs Related Almanac*, published by American References Inc., of Chicago. It used the criteria of salary, stress, work environment, outlook, security, and physical demands to rate the desirability of 250 jobs. The top 10 of the 250 jobs listed were:

1. Actuary
2. Computer programmer   } The
3. Computer systems analyst   } top 5 involve
4. Mathematician   } mathematics.
5. Statistician
6. Hospital administrator
7. Industrial engineer   } These also
8. Physicist   } involve
9. Astrologer   } mathematics.
10. Paralegal

Note that eight of the top ten rated professions involve a heavy use of mathematics. The top, actuary, involves the application of mathematics to insurance. Note also that although choices like doctor, lawyer, and astronaut are *not* in the top ten, perhaps due to physical demands and stress, they also require the use of mathematics.

Perhaps you are interested in a career in teaching mathematics. This profession will be expanding in the next ten years. The field of mathematics will need well-qualified teachers in all areas from elementary to junior high to secondary to two-year college to college instruction. Some questions you might ask yourself in making a decision about a career in math teaching are the following.

1. Do you find yourself carefully observing the strengths and weaknesses of your teachers?
2. Are you deeply interested in mathematics?
3. Are you interested in the ways of learning? If a student is struggling with a topic, would it be challenging for you to discover two or three other ways to present the material so that the student might better understand?

4. Are you able to put yourself in the place of students, in order to help them be more successful in learning mathematics?

If you are interested in a career involving mathematics, the next courses you should take are *introductory algebra, intermediate algebra, precalculus algebra and trigonometry,* and *calculus.* You might want to seek out a counselor in the mathematics department at your college for further assistance.

WHAT KIND OF SALARIES ARE THERE IN VARIOUS FIELDS?

In a recent year, The College Placement Council published the following comparisons of the average salaries of graduating students with a bachelor's degree who were taking the following jobs:

SUBJECT AREA	ANNUAL SALARY
All engineering	$27,800
Computer science	$26,400
Mathematics	$25,900
Sciences other than math and computer science	$22,200
Humanities and social science	$21,800
Accounting	$21,700
All business	$21,300

Many people choose to go on to earn a master's degree. Here are salaries in the same fields for students just graduating with a master's degree:

SUBJECT AREA	ANNUAL SALARY
All engineering	$34,000
Computer science	$33,800
Mathematics	$27,900
Sciences other than math and computer science	$27,400
Humanities and social science	$22,300
Accounting	$26,000
Business administration	$30,700

Exercise Set 8.2

a This table lists information about various types of nails. Use the table for Exercises 1–10.

NAIL SIZES				
		APPROXIMATE NUMBER PER POUND		
PENNY NUMBER	**LENGTH (INCHES)**	**Common Nails**	**Box Nails**	**Finishing Nails**
4	$1\frac{1}{2}$	316	437	548
6	2	181	236	309
8	$2\frac{1}{2}$	106	145	189
10	3	69	94	121
12	$3\frac{1}{4}$	64	87	113
16	$3\frac{1}{2}$	49	71	90
20	4	31	52	62
30	$4\frac{1}{2}$	20		
40	5			
50	$5\frac{1}{2}$			

1. How long is a 40-penny nail?

2. What penny number is given to a nail that is $2\frac{1}{2}$ in. long? 8

3. How many 10-penny box nails are there in a pound?

4. What type of nail comes 309 to the pound? 6

5. How many more 16-penny finishing nails can you get in a pound than 10-penny common nails?

6. How many fewer 8-penny common nails will you get in a pound than 6-penny common nails? 181
−106
75

7. How many 20-penny box nails will you get in a pound?

8. What type of nail comes 52 to the pound? Box Nails, 20 penny

9. How many nails will you get if you buy 5 lb of 4-penny finishing nails?

10. You need approximately 448 12-penny common nails. How many pounds should you buy?
448 ÷ 64 = 7

ANSWERS

1.

2.

3.

4.

5.

6.

7.

8.

9.

10.

This table lists the number of calories burned in 30 min of exercise for various types of activities and several weight categories. Use the table for Exercises 11–22.

ACTIVITY	CALORIES BURNED IN 30 MIN		
	110 lb	132 lb	154 lb
Aerobic dance	201	237	282
Calisthenics	216	261	351
Racquetball	213	252	294
Tennis	165	192	222
Moderate bicycling	138	171	198
Moderate jogging	321	378	453
Moderate walking	111	132	159

11. How many calories are burned by a 154-lb person after 30 min of racquetball?

12. How many calories are burned by a 132-lb person after 30 min of moderate jogging?

13. What activity burns 216 calories in 30 min for a 110-lb person?

14. What activity burns calories at the rate of 132 every 30 min for a 132-lb person?

15. Which burns more calories in 30 min for a 154-lb person: aerobic dance or tennis?

16. Which burns more calories in 30 min: moderate bicycling or moderate walking?

17. How many calories will have been burned by a 110-lb person after 2 hr of tennis?

18. How many minutes of moderate walking will it take for a 110-lb person to burn as many calories as a 154-lb person will burn playing 30 min of tennis?

19. Which activity burns the least number of calories for a 132-lb person?

20. To reach her weight goal of 110 lb, a woman must burn at least 215 calories every 30 min. What activities will provide at least that rate of burn?

21. How many calories would you expect a 120-lb person to burn during 30 min of moderate walking?

22. How many calories would you expect a 143-lb person to burn during 30 min of calisthenics?

b This pictograph shows sales of shampoo for a soap company for six consecutive years. Use the pictograph for Exercises 23–30.

Shampoo Sales

1995	
1994	
1993	
1992	
1991	
1990	

= 1000 bottles sold

23. In which year was the greatest number of bottles sold?

24. Between which two consecutive years was there the greatest growth? 94 - 95

25. Between which two years was the amount of growth the least?

26. How many sales does one bottle represent? 1 m

27. How many bottles were sold in 1992?

28. How many more bottles were sold in 1995 than in 1990?

29. In which year was there actually a decline in the number of bottles sold?

30. The sales for 1995 were how many times the sales for 1990?

9000/2000 = 4 1/2 times

23. _____

24. _____

25. _____

26. _____

27. _____

28. _____

29. _____

30. _____

This pictograph shows a baseball player's at-bats in one month. Use the pictograph for Exercises 31–38.

_____

At-Bat Record	
Home runs	⚾ ⚾
Triples	⚾
Doubles	⚾ ⚾ ⚾
Singles	⚾ ⚾ ⚾ ⚾ ⚾ ⚾
Walks	⚾ ⚾
Outs	⚾ ⚾ ⚾ ⚾
⚾ = 3 times at bat	

31. How many times at bat does one baseball symbol represent?

32. How many of the player's hits were home runs? 6

33. How many hits were singles?

34. How many more doubles than home runs did the player hit? 4

35. How many fewer triples than singles did the player hit?

36. What was the total number of hits for the month? 6+3+9+14=33

37. What happened exactly 12 times to the batter?

38. What did the player do most during the month? Singles

_____

_____

_____

_____

c

39. Draw a pictograph representing the calories per tablespoon in various tablespreads. Be sure to put in all of the appropriate labels. Use a flame to represent 10 calories.

Jam: 54 per tablespoon
Mayonnaise: 51 per tablespoon
Peanut Butter: 94 per tablespoon
Honey: 64 per tablespoon
Syrup: 60 per tablespoon

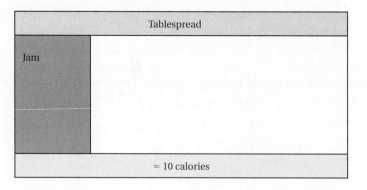

Tablespread	
Jam	
= 10 calories	

_____

_____

_____

_____

_____

36

SKILL MAINTENANCE

Solve.

40. A football team has won 3 of its first 4 games. At this rate, how many games will it win in a 16-game season?

41. The state of Maine is 90% forest. The area of Maine is 30,955 mi^2. How many square miles of Maine are forest?

_____

8.3 Bar Graphs and Line Graphs

A **bar graph** is convenient for showing comparisons because you can tell at a glance which amount represents the largest or smallest quantity. Of course, since bar graphs are a more abstract form of pictographs, this is true of pictographs as well. However, with bar graphs, a *second scale* is usually included so that a more accurate determination of the amount can be made.

a READING AND INTERPRETING BAR GRAPHS

EXAMPLE 1 A recent National Assessment of Educational Progress Survey showed these reasons given by students for dropping out of high school.

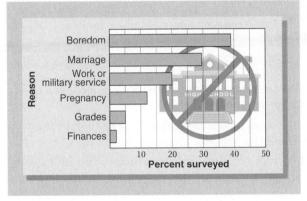

a) Approximately how many in the survey dropped out because of pregnancy?

b) What reason was given least often for dropping out?

c) What reason was given by about 30% for dropping out?

We look at the graph to answer the questions.

a) We go to the right end of the bar representing pregnancy and then go down to the percent scale. We can read, fairly accurately, that approximately 12% dropped out because of pregnancy.

b) We look for the shortest bar and find that it represents finances.

c) We go to the right on the percent scale to find the 30% mark and then up until we reach a bar that ends at approximately 30%. We then go across to the left and read the reason. The reason given by about 30% was marriage.

Do Exercises 1–3.

OBJECTIVES

After finishing Section 8.3, you should be able to:

a Read and interpret data from bar graphs.

b Draw bar graphs.

c Read and interpret data from line graphs.

d Draw simple line graphs.

FOR EXTRA HELP

TAPE 14   TAPE 12B   MAC: 8
                                   IBM: 8

Use the bar graph in Example 1 to answer each of the following.

1. What reason was given most often for dropping out?

2. What reason was given by about 20% for dropping out?

3. How many in the survey dropped out because of grades?

Answers on page A-4

4. What is the average annual income for a person who has completed only high school?

5. What is the greatest average annual income for a person who doesn't finish high school?

6. How much more can you expect to earn annually if you complete high school than if you complete only the 8th grade?

Of course, the bars can be drawn vertically as well.

EXAMPLE 2   A recent survey of 2000 individuals produced the following information on average income based on years of schooling.

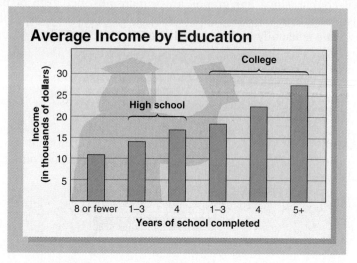

a) What is the average annual income for a person who has completed two years of high school?

b) How many years of schooling does it take to expect an average annual income of at least $20,000?

c) How much more income, on the average, can be expected after completing three years of college than after completing only one year of college?

Interpreting the graph carefully will give us the answers.

a) We go to the right, across the bottom, to the bar representing income for a person with 1–3 years of high school. We then go up to the top of the bar and, from there, back to the left to read approximately $14,000 on the income scale.

b) We go up the left-hand scale of the graph to the $20,000 mark and read to the right, until we come to a bar crossing our path. Moving down on that bar, we find that at least 4 years of college are needed.

c) There is only one bar representing 1–3 years of college. Therefore, this graph shows no difference in income between the two groups, though it may exist.

Do Exercises 4–6.

Answers on page A-4

b DRAWING BAR GRAPHS

EXAMPLE 3 Make a vertical bar graph to show the following information about degree expectations of incoming college freshmen.

Law:	4.4%	Bachelor's:	28.2%
Ph.D. or Ed.D.:	12.5%	Associate (Two-year):	7.3%
Master's:	35.5%	None:	12.1%

First, we indicate on the base or horizontal scale the different degrees. (See the figure on the left below.) Then we label the marks on the vertical scale appropriately by 10's to represent the percent who expect to receive a particular kind of degree. Finally, we draw vertical bars to show the degree expectation of incoming freshmen, as shown in the figure on the right below.

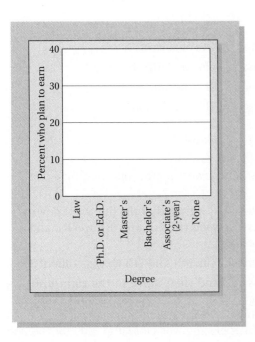

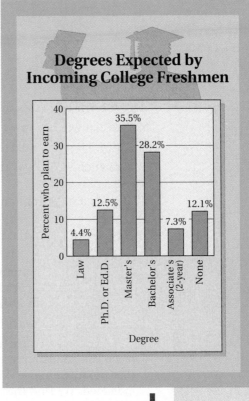

Degrees Expected by Incoming College Freshmen

Do Exercise 7.

7. Make a horizontal bar graph to show the loudness of various sounds, as listed below. (*Hint:* See Example 1.) A decibel is a measure of the loudness of sounds.

SOUND	LOUDNESS (IN DECIBELS)
Whisper	15
Tick of watch	30
Speaking aloud	60
Noisy factory	90
Moving car	80
Car horn	98
Subway	104

Answer on page A-4

Use the line graph in Example 4 to answer each of the following.

8. For which week was the DJIA closing the highest?

9. For which week was the DJIA closing about 3150?

10. About how many points did the DJIA increase between weeks 1 and 7?

Line graphs are often used to show a change over time as well as to indicate patterns or trends.

EXAMPLE 4 This line graph shows the closing Dow Jones Industrial Average (DJIA) for each of six weeks. The jagged line at the base of the vertical scale indicates an unnecessary portion of the scale. Note that the vertical scale differs from the horizontal scale so that numbers fit reasonably.

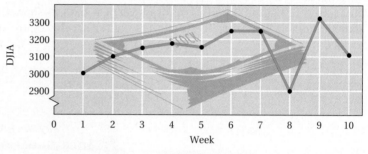

a) For which week was the closing DJIA the lowest?

b) Between which two weeks did the closing DJIA decrease?

c) For which weeks was the closing DJIA the same?

d) For which weeks was the closing DJIA about 3100?

We look at the graph to find the answers.

a) For the 8th week, the line is at its lowest point, representing a close of about 2900.

b) Reading the graph from left to right, we see that the line went down between the 4th and 5th weeks, between the 7th and 8th weeks, and between the 9th and 10th weeks.

c) The closing DJIA was the same for the 6th and 7th weeks—about 3250.

d) We locate 3100 on the DJIA scale and then move to the right until we reach the points representing a closing that is closest to our position. At that point, we move down to the "Week" scale and see which week is indicated. We find that the DJIA closing was closest to 3100 for the 2nd and 10th weeks.

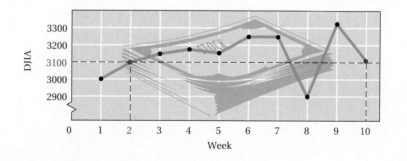

Do Exercises 8–10.

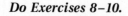

EXAMPLE 5 This line graph gives information regarding the cost of the first year of college based on the present age of a child.

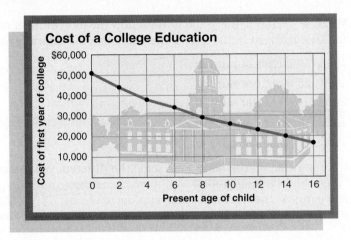

Cost of a College Education

Cost of first year of college

Present age of child

a) Estimate the average cost for a child who is now 16 years old.

b) Give the age of a child for which the cost will be about $29,500.

c) How much more is the cost for a newborn than the cost for a 16-year-old?

We read the graph to find the answers.

a) We find age 16 on the bottom scale and move up from that point to the line. We then go straight across to the left from the line and find that we are about seven-tenths of the way between $10,000 and $20,000. We estimate the cost to be about $17,000.

b) Going up the left scale to a point slightly below $30,000, we move straight across to the right until we cross the line. At that point, we go down to the scale for "Present age" on the bottom and find that this cost occurs at age 8.

c) The graph shows the cost for a 16-year-old to be about $17,000 and for a newborn to be about $50,100. This gives an increase of about $33,100.

Do Exercises 11–13.

Use the line graph in Example 5 to answer each of the following.

11. Estimate the average cost for a child who is now 14 years old.

12. Give the age of a child for which the cost will be about $38,000.

13. About how much does the cost rise for an 8-year-old over that of a 14-year-old?

Answers on page A-4

14. Draw a line graph to show how the average SAT score changed in four years. Use the following information.

    1989: 903

    1990: 898

    1991: 896

    1992: 899

Answer on page A-4

d | DRAWING LINE GRAPHS

EXAMPLE 6   Draw a line graph to show how the total number of inches of rainfall has changed in five years. Use the following information.

    1992: 30 inches of rainfall

    1993: 28 inches of rainfall

    1994: 25 inches of rainfall

    1995: 30 inches of rainfall

    1996: 27 inches of rainfall

First, we indicate on the horizontal scale the different years and title it "Year." (See the following graph.)

Then we mark the vertical scale appropriately by 5's to show the number of inches of rainfall (see the graph below) and title it "Inches of rainfall."

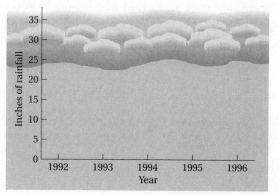

Now, we mark the points above each of the years at the appropriate level to indicate the number of inches of rainfall, and draw line segments connecting them to show the change.

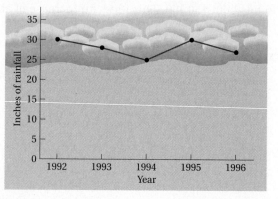

Do Exercise 14.

Exercise Set 8.3

a This horizontal bar graph shows the average length, in weeks, of the growing season for eight U.S. cities.

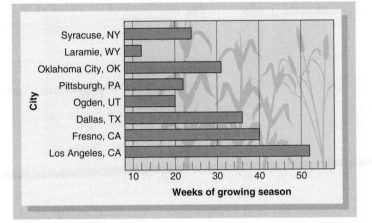

1. _____

2. _____

3. _____

1. Which city has the longest growing season?

2. Which city has the shortest growing season?

4. _____

3. How many weeks long is the growing season in Pittsburgh?

4. How many weeks long is the growing season in Dallas?

$$\frac{3x}{12} = 3x$$

5. _____

5. How many times longer is the growing season in Fresno than in Ogden?

6. How many times longer is the growing season in Dallas than in Laramie?

6. _____

7. Which city most closely approximates one-half of the growing season of Los Angeles?

8. Which city has approximately $2\frac{1}{2}$ times the growing season of Laramie?

$12 \times 2.5 = 30$
Oklahoma

7. _____

8. _____

This vertical bar graph shows the average daily expenses for lodging, food, and rental car for traveling executives.

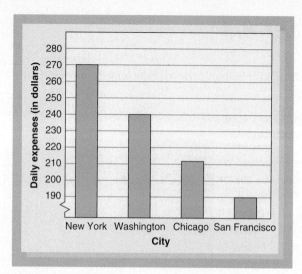

9. _____

10. _____

11. _____

12. _____

13. _____

14. _____

15. See graph. _____

16. _____

17. _____

18. _____

9. What are the average daily expenses in New York?

10. What are the average daily expenses in Chicago?

11. Which city is the least expensive of the four?

12. Which city is the most expensive of the four?

13. How much more are the average daily expenses in Washington than in San Francisco?

14. How much less are the average daily expenses in Chicago than in Washington? $28

b

15. Use the following information to make a bar graph showing the number of calories burned during each activity by a person weighing 152 lb. Use the blank graph below.

Tennis: 420 calories per hour

Jogging: 650 calories per hour

Hiking: 590 calories per hour

Office work: 180 calories per hour

Sleeping: 70 calories per hour

16. What is the difference in the number of calories burned per hour between sleeping and jogging?

17. Suppose you were trying to lose weight by exercising and had to choose one of these exercises. If the doctor told you not to jog, what would be the most beneficial exercise?

18. What is the average of the number of calories burned in these exercises?

[C] This line graph reflects the hourly readings of the outside temperature during one fall day.

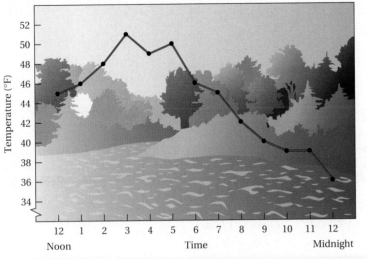

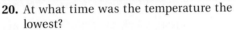

19. At what time was the temperature the highest?

20. At what time was the temperature the lowest?

21. What was the difference in temperature between the highest and lowest readings?

22. Between which two hours did the temperature increase the most?

23. Between which two hours did the temperature decrease the most?

24. How much colder was it at midnight than at noon?

This line graph shows the predicted sales (in millions of dollars) for several years for a company. Again, note the jagged line at the base of the vertical scale.

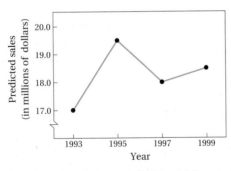

25. In which year are predicted sales the greatest?

26. In which year are predicted sales the least?

27. What are the predicted sales for 1993?

28. What are the predicted sales for 1999?

ANSWERS

19. _____

20. _____

21. _____

22. _____

23. _____

24. _____

25. _____

26. _____

27. _____

28. _____

29. How much greater are predicted sales for 1995 than those for 1997?

30. How much less are predicted sales for 1993 than those for 1999?

[d]

31. A rural intersection is being considered for an automatic traffic signal. A traffic survey, recording the number of cars passing through the intersection, gave the following results. Use these facts to make a line graph showing the number of cars counted for each hour. Be sure to label the two scales appropriately.

12 noon: 50 cars	5 P.M.: 100 cars
1 P.M.: 40 cars	6 P.M.: 112 cars
2 P.M.: 60 cars	7 P.M.: 88 cars
3 P.M.: 65 cars	8 P.M.: 70 cars
4 P.M.: 77 cars	9 P.M.: 35 cars

32. Between which two hours did traffic increase the most?

33. Between which two hours did traffic decrease the most?

34. How much did traffic increase from 1 P.M. to 6 P.M.?

35. What was the average number of cars passing through the intersection in a 1-hr period?

SKILL MAINTENANCE

36. A clock loses 3 min every 12 hr. At this rate, how much time will the clock lose in 72 hr?

37. It is known to operators of pizza restaurants that if 50 pizzas are ordered in an evening, people will request extra cheese on 9 of them. What percent of the pizzas sold are ordered with extra cheese?

38. 110% of 75 is what?

39. 34 is what percent of 51?

8.4 Circle Graphs

We often use **circle graphs,** also called *pie charts,* to show the percent of a quantity used in different categories. Circle graphs can also be used very effectively to show visually the *ratio* of one category to another. In either case, it is quite often necessary to use mathematics to find the actual amounts represented for each specific category.

a READING AND INTERPRETING CIRCLE GRAPHS

EXAMPLE 1   This circle graph shows expenses as a percent of income in a family of four, according to a recent study of the Bureau of Labor Statistics. (*Note:* Due to rounding, the sum of the percents is 101% instead of 100%.)

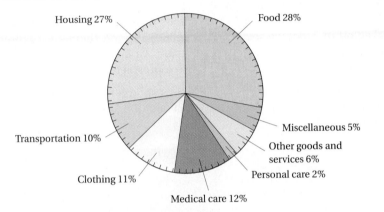

a) Which item accounts for the greatest expense?

b) For a family with a $2000 monthly income, how much is spent for transportation?

c) Some surveys combine medical care with personal care. What percent would be spent on those two items combined?

We look at the sections of the graph to find the answers.

a) It is immediately apparent that there are two sections that are larger than the rest. Of those two sections, the one representing food is the larger, at 28%.

b) The section of the circle representing transportation shows a 10% expense; 10% of $2000 is $200.

c) In a circle graph, we can add percents safely for problems of this type. Therefore, 12% (medical care) + 2% (personal care) = 14%.

Do Exercises 1–3.

OBJECTIVES

After finishing Section 8.4, you should be able to:

a Read and interpret data from circle graphs.

b Draw circle graphs.

FOR EXTRA HELP

TAPE 14   TAPE 13A   MAC: 8
IBM: 8

Consider a family with a $2000 monthly income and use the circle graph in Example 1 to answer each of the following.

1. How much would this family typically spend on housing each month?

2. What percent of the income is spent on housing and clothing combined?

3. Compare the amount spent on medical care with the amount spent on personal care. What is the ratio?

Answers on page A-4

b | DRAWING CIRCLE GRAPHS

EXAMPLE 2   In a quick inventory, it was found that the types of books listed below made up the indicated percent of available books in a library. Use this information to draw a circle graph reflecting the different types of books available.

a) History books: 25%

b) Science books: 10%

c) Fiction: 45%

d) Reference books: 5%

e) Other: 15%

We will first draw each section in a separate working circle to illustrate more clearly how each is made. We will then combine them in a single circle to show the complete graph.

Remember that there are 360 degrees in a circle. Our circles will be marked off in 5-degree sections. Since $360°/5° = 72$, there are 72 intervals in the entire circle and 18 in each quarter. We are providing you with circles marked off, but you could also use a protractor.

a) History books account for 25%, so they should be shown by 25% of the circle. Mathematically, this is 25% of 360° (0.25 × 360), or 90° ($90°/5° = 18$ intervals).

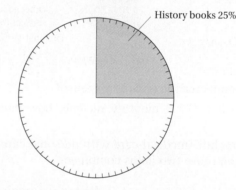

History books 25%

b) Science books account for 10% of the books; 10% of 360° (0.10 × 360) is 36° more ($36°/5° = 7.2$ intervals).

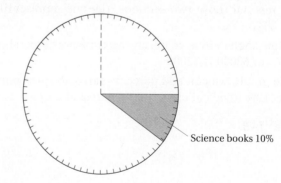

Science books 10%

c) Fiction accounts for 45% of the books; 45% of 360° (0.45 × 360) is another 162°, or 32.4 intervals.

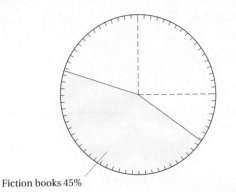

Fiction books 45%

d) Reference books account for 5% of the books; 5% of 360° (0.05 × 360) is another 18°, or 3.6 intervals.

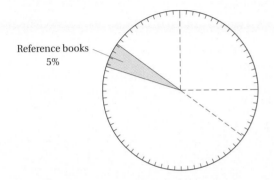

Reference books
5%

e) Other books make up the remaining 15%. We find that 15% of 360° (0.15 × 360) is 54°, or 10.8 intervals. Note that this last section accounts exactly for the remainder of the circle.

Other books 15%

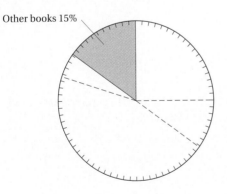

4. For each of the following uses of paper, find the number of degrees needed, to the nearest degree, to draw a circle graph. Then draw the graph.

a) Packaging: 48%

b) Writing paper: 30%

c) Tissues: 8%

d) Other: 14%

e) Draw the graph.

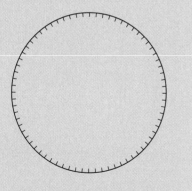

Answers on page A-4

Now we can combine all of these sections in a single circle, which results in the circle graph below.

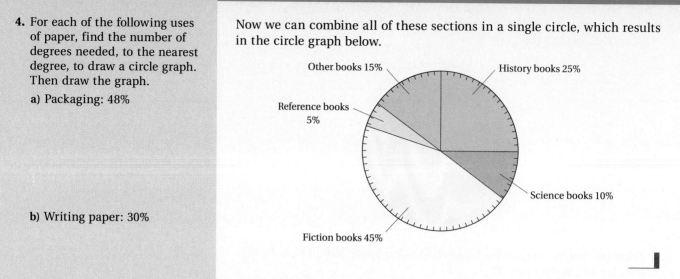

Do Exercise 4.

Exercise Set 8.4

a This circle graph, in the shape of a CD, shows music preferences of customers on the basis of music store sales, according to the National Association of Recording Merchandisers.

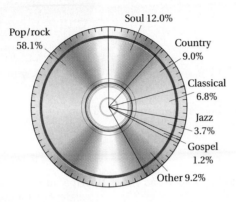

Soul 12.0%

Pop/rock 58.1%

Country 9.0%

Classical 6.8%

Jazz 3.7%

Gospel 1.2%

Other 9.2%

1. What percent of all recordings sold are jazz?

2. Together, what percent of all recordings sold are either soul or pop/rock?

$$\begin{array}{r} 58.1 \\ +12.0 \\ \hline 70.1 \% \end{array}$$

3. A music store sells 3000 recordings a month. How many are country?

4. A music store sells 2500 recordings a month. How many are gospel?

1.2%

.012 × 2500 = 30

5. What percent of all recordings sold are classical?

6. Together, what percent of all recordings sold are either classical or jazz?

$$\begin{array}{r} 6.8 \\ 3.7 \\ \hline 10.5 \end{array}$$

1. _____

2. _____

3. _____

4. _____

5. _____

6. _____

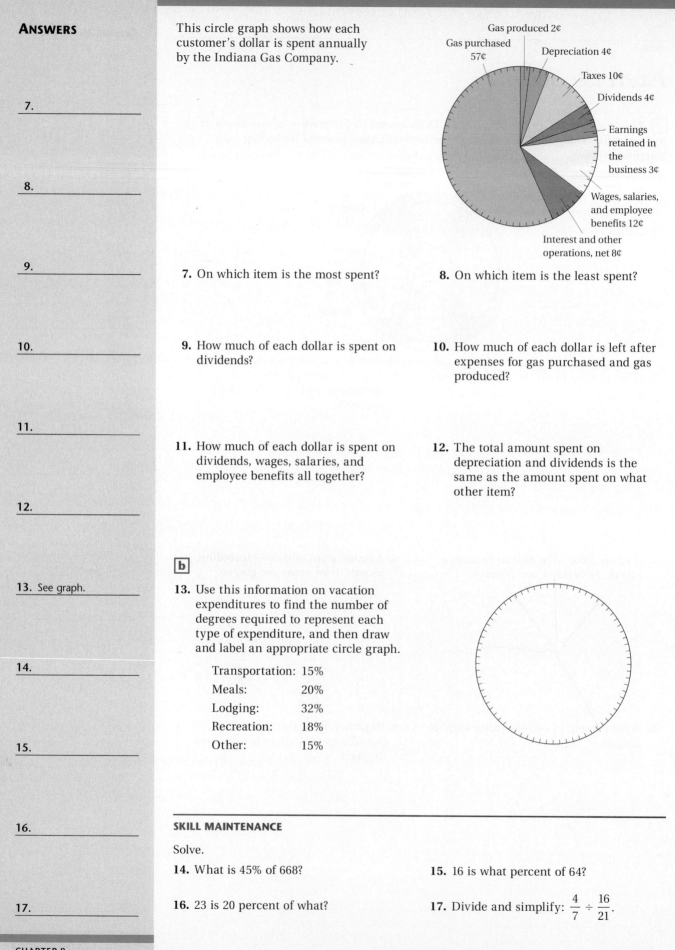

This circle graph shows how each customer's dollar is spent annually by the Indiana Gas Company.

Gas produced 2¢

Gas purchased 57¢

Depreciation 4¢

Taxes 10¢

Dividends 4¢

Earnings retained in the business 3¢

Wages, salaries, and employee benefits 12¢

Interest and other operations, net 8¢

7.
8.
9.
10.
11.
12.
13. See graph.
14.
15.
16.
17.

7. On which item is the most spent?

8. On which item is the least spent?

9. How much of each dollar is spent on dividends?

10. How much of each dollar is left after expenses for gas purchased and gas produced?

11. How much of each dollar is spent on dividends, wages, salaries, and employee benefits all together?

12. The total amount spent on depreciation and dividends is the same as the amount spent on what other item?

b

13. Use this information on vacation expenditures to find the number of degrees required to represent each type of expenditure, and then draw and label an appropriate circle graph.

Transportation: 15%

Meals: 20%

Lodging: 32%

Recreation: 18%

Other: 15%

SKILL MAINTENANCE

Solve.

14. What is 45% of 668?

15. 16 is what percent of 64?

16. 23 is 20 percent of what?

17. Divide and simplify: $\frac{4}{7} \div \frac{16}{21}$.

EXTENDED SYNTHESIS EXERCISES

1. Examine the line graph of the Nike data in the test at the end of this chapter. Use the data to predict the revenue of Nike in 1995. Then find the actual revenue and compare your prediction with the actual amount. Then estimate the revenue in 2000.

This table lists maximum running speeds of various animals, in miles per hour, compared to the speed of the fastest human, 28 mph. Use it for Exercises 2–9.

ANIMAL	SPEED (IN MILES PER HOUR)	ANIMAL	SPEED (IN MILES PER HOUR)
Antelope	61	Elephant	25
Bear	30	Elk	45
Cat	30	Fox	42
Cheetah	70	Jackal	35
Chicken	9	Giraffe	32
Coyote	43	Greyhound	39
Deer	30	Horse	48
Fastest		Lion	50
human	28	Rabbit	35
Pig	11	Zebra	40
Turkey	15		

Source: Natural History, March 1974. Most measurements are for maximum speeds over a quarter-mile distance. For humans, the speed was found over a 15-yd segment of a 100-yd run, where the overall speed was 13.6 sec.

2. Make a vertical bar graph of the data.
3. Find the average of all the speeds.

4. Find the average speed of all the animals except humans. Compare that speed to that of humans.
5. Which animals have speeds below 20 mph? of 20–30 mph? of 31–40 mph? of 41–50 mph? above 50 mph? What, if anything, does each group of animals have in common?
6. Which animals have speeds that differ considerably from the rest? Do these animals have anything in common?
7. Find out the kinds of food the animals eat. Then use the results to separate the animals into groups and find the average speed in each group. What conclusions can you draw?
8. Compare the average speed of all the two-legged animals with that of the four-legged animals. What conclusions can you draw?

This line graph compares the averages for different kinds of business-travel costs for several recent years. Use it for Exercises 9–17.

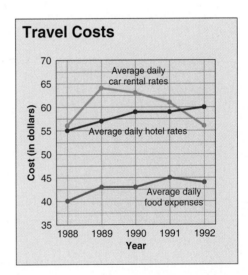

9. Find the average daily hotel rate for all the years.
10. Find the average daily car rental rate for all the years.
11. Find the average daily food expense for all the years.
12. In which year did car rental rates begin to decrease?

(continued)

13. Why do you think car rental rates have been decreasing?

14. Over what period did food expenses stay the same?

15. About what time were hotel rates the same as car rental rates?

16. Use the graphs to estimate each of the daily rates in 1993. Explain your reasons for the estimates.

17. In which year do you think food expenses and car rental rates will be the same?

18. Choose a stock that interests you. Make a line graph of the value of the stock over a period of eight days. Analyze the results and try to make predictions. Would you recommend the stock to a friend?

EXERCISES FOR THINKING AND WRITING

1. Compare averages, medians, and modes. Discuss why you might use one over the others to analyze a set of data.

2. Compare the use of bar graphs and line graphs. Discuss why you might use one over the other to graph a set of data.

3. Compare bar graphs and circle graphs. Discuss why you might use one over the other to graph a set of data.

4. Find a real-world situation that fits the equation

$$\frac{(20{,}500 + 22{,}800 + 23{,}400 + 26{,}000)}{4} = 23{,}175.$$

Summary and Review Exercises: Chapter 8

The review objectives to be tested in addition to the material in this chapter are [2.7b], [6.4a], [7.3b], [7.4b], and [7.5a].

Find the average.

1. 26, 34, 43, 51

2. 7, 11, 14, 17, 18

3. 0.2, 1.7, 1.9, 2.4

4. 700, 900, 1900, 2700, 3000

5. $2, $14, $17, $17, $21, $29

6. 20, 190, 280, 470, 470, 500

Find the median.

7. 26, 34, 43, 51

8. 7, 11, 14, 17, 18

9. 0.2, 1.7, 1.9, 2.4

10. 700, 900, 1900, 2700, 3000

11. $2, $14, $17, $17, $21, $29

12. 20, 190, 280, 470, 470, 500

Find the mode.

13. 26, 34, 43, 26, 51

14. 7, 11, 11, 14, 17, 17, 18

15. 0.2, 0.2, 0.2, 1.7, 1.9, 2.4

16. 700, 700, 800, 2700, 800

17. $2, $14, $17, $17, $21, $29

18. 20, 20, 20, 20, 20, 500

19. One summer, a student earned the following amounts over a four-week period: $102, $112, $130, and $98. What was the average amount earned per week? the median?

20. The following temperatures were recorded in St. Louis every four hours on a certain day in June: 63°, 58°, 66°, 72°, 71°, 67°. What was the average temperature for that day?

21. To get an A in math, a student must average 90 on four tests. Scores on the first three tests were 94, 78, and 92. What is the lowest score that the student can make on the last test and still get an A?

This table illustrates desirable weight recommendations for men and women of various heights. Use it for Exercises 22–27.

DESIRABLE WEIGHT OF MEN			
HEIGHT	SMALL FRAME (IN POUNDS)	MEDIUM FRAME (IN POUNDS)	LARGE FRAME (IN POUNDS)
5 ft, 9 in.	144	151	165
5 ft, 11 in.	150	159	171
6 ft, 1 in.	157	166	180
6 ft, 3 in.	161	174	183

DESIRABLE WEIGHT OF WOMEN			
HEIGHT	SMALL FRAME (IN POUNDS)	MEDIUM FRAME (IN POUNDS)	LARGE FRAME (IN POUNDS)
5 ft, 3 in.	118	128	140
5 ft, 5 in.	124	134	147
5 ft, 7 in.	130	140	154
5 ft, 9 in.	136	146	160

22. What is the recommended weight for a 6 ft, 1 in. man with a medium frame?

23. What is the recommended weight for a 5 ft, 5 in. woman with a small frame?

24. What size woman has a recommended weight of 154 lb?

25. What size man has a recommended weight of 171 lb?

26. How much more is the recommended weight of a 5 ft, 9 in. man than that of a 5 ft, 9 in. woman, each with a large frame?

27. What is the average recommended weight of a 5 ft, 3 in. woman?

This pictograph shows the number of television sets sold by a company. Use it for Exercises 28–31.

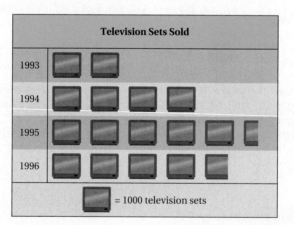

28. About how many TV sets were sold in 1995?

29. In which year did the company sell the least number of TV sets?

30. In which year did the company sell the most TV sets?

31. Estimate the average number of TV sets sold over the four years.

This bar graph shows the Fast Food Price Index for several major cities. By definition for the purpose of this comparison, "fast food" consists of a quarter-pound cheeseburger, a large order of fries, and a medium-sized soft drink. Use it for Exercises 32–37.

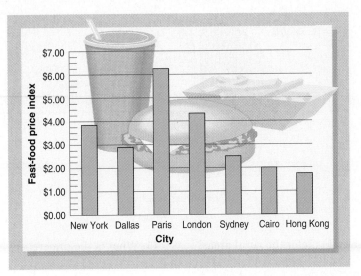

32. What is the most that you will pay for this meal?

33. Where will you pay the least for this meal?

34. In which of the given U.S. cities will you spend the most for this meal?

35. What is the least that you will spend for this meal considering the given U.S. cities?

36. How much more will you pay for this meal in Paris than in Hong Kong?

37. Where will you pay close to the same price for this meal that you would pay in Dallas?

This line graph shows the number of accidents per 100 drivers, by age. Use it for Exercises 38–43.

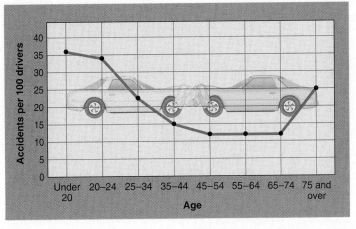

38. Which age group has the most accidents per 100 drivers?

39. What is the fewest number of accidents per 100 in any age group?

40. How many more accidents do people over 75 years of age have than those in the age range of 65–74?

41. Between what ages does the number of accidents stay basically the same?

42. How many fewer accidents do people 25–34 years of age have than those 20–24 years of age?

43. Which age group has accidents more than three times as often as people 55–64 years of age?

This circle graph shows the percent of homebuyers preferring various locations for their homes. Use it for Exercises 44–47.

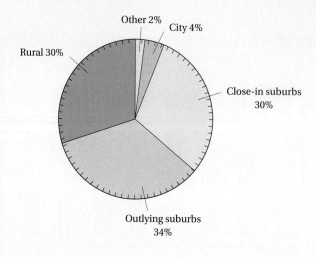

Other 2%
City 4%
Rural 30%
Close-in suburbs 30%
Outlying suburbs 34%

44. What percent of homebuyers prefer the rural areas?

45. What is the preference of 34% of the homebuyers?

46. What percent of homebuyers prefer to live somewhere in the suburbs?

47. Homebuyers prefer a rural area how many times more than the city?

$$\frac{30}{4} = 7\frac{1}{2}x$$

SKILL MAINTENANCE

Solve.

48. A company car was driven 4200 mi in the first 4 months of a year. At this rate, how far will it be driven in 12 months?

49. 92% of the world population does not have a telephone. The population is about 5.4 billion. How many do not have a telephone?

50. 789 is what percent of 355.05?

51. What percent of 98 is 49?

Divide and simplify.

52. $\frac{3}{4} \div \frac{5}{6}$

53. $\frac{5}{8} \div \frac{3}{2}$

Test: Chapter 8

Find the average.

1. 45, 49, 52, 54   **2.** 1, 2, 3, 4, 5   **3.** 3, 17, 17, 18, 18, 20

Find the median and the mode.

4. 45, 49, 52, 54   **5.** 1, 2, 3, 4, 5   **6.** 3, 17, 17, 18, 18, 20

7. A car traveled 754 km in 13 hr. What was the average number of kilometers per hour?

8. To get a C in chemistry, a student must average 70 on four tests. Scores on the first three tests were 68, 71, and 65. What is the lowest score that the student can make on the last test and still get a C?

This table lists the number of calories burned during various walking activities. Use it for Questions 9–12.

WALKING ACTIVITY	CALORIES BURNED IN 30 MIN		
	110 lb	132 lb	154 lb
Walking			
Fitness (5 mph)	183	213	246
Mildly energetic (3.5 mph)	111	132	159
Strolling (2 mph)	69	84	99
Hiking			
3 mph with 20-lb load	210	249	285
3 mph with 10-lb load	195	228	264
3 mph with no load	183	213	246

9. Which activity provides the greatest benefit in burned calories for a person who weighs 132 lb?

10. What is the least strenuous activity you must perform if you weigh 154 lb and you want to burn at least 250 calories every 30 min?

1. _____

2. _____

3. _____

4. _____

5. _____

6. _____

7. _____

8. _____

9. _____

10. _____

11. How is "mildly energetic walking" defined?

12. What type of walking can a person who weighs 110 lb do that will give the same benefit as some type of hiking?

13. Draw a vertical bar graph using an appropriate set of scales, showing the recent starting salary of college graduates in various fields. Be sure to label the scales properly.

Accounting:	$27,400
Advertising:	$22,000
Elementary education:	$23,000
Computer science:	$32,000
Mathematics:	$31,000

$20,000

Accounting

This pictograph shows the number of hits in a season for several professional baseball players. Use it for Questions 14–18.

Number of Hits in a Season for 5 Professional Players

Tony Gwynn	⚾⚾⚾⚾⚾⚾⚾ (
Don Mattingly	⚾⚾⚾⚾⚾⚾⚾⚾
Mark Carreon	⚾⚾⚾⚾ (
Terry Pendleton	⚾⚾⚾⚾⚾⚾⚾⚾
Dave Justice	⚾⚾⚾⚾⚾

⚾ = 25 hits

14. How many hits did Dave Justice have?

15. Who had the most hits?

16. Who had the least number of hits?

17. Who had 200 hits?

18. How many more hits did Don Mattingly have than Tony Gwynn?

This line graph shows the revenues of Nike, Inc. Use it for Questions 19–26.

Revenue of Nike, Inc.

19. What trend, if any, is shown in this graph?

20. In which year was the increase the greatest?

21. How much revenue was earned in 1991?

22. How much more revenue was earned in 1991 than in 1987?

23. What was the average of the revenues for the seven years?

24. What was the median of the revenues for the seven years?

16. _____

17. _____

18. _____

19. _____

20. _____

21. _____

22. _____

23. _____

24. _____

25. In which year did revenue decrease?

26. In which year was the revenue about $2.2 billion?

27. Use the following information to make a circle graph showing the percent of available money that people invest in certain ways. Be sure to label each section appropriately. (*Note:* The given circle is divided into 5-degree sections.)

Savings accounts: 48%

Stocks: 12%

Mutual funds: 16%

Retirement funds: 24%

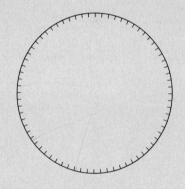

SKILL MAINTENANCE

28. Divide and simplify: $\dfrac{3}{5} \div \dfrac{12}{125}$.

29. 17 is 25% of what number?

30. On a particular Sunday afternoon, 78% of the television sets that were on were tuned to one of the major networks. Suppose 20,000 TV sets in a town are being watched. How many are tuned to a major network?

31. A baseball player gets 7 hits in the first 20 times at bat. At this rate, how many times at bat will it take to get 119 hits?

Cumulative Review: Chapters 1–8

1. In 402,513, what does the digit 5 mean?

2. Evaluate: $3 + 5^3$.

3. Find all the factors of 60.

4. Round 52.045 to the nearest tenth.

5. Convert to fractional notation: $3\frac{3}{10}$.

6. Convert from cents to dollars: 210¢.

7. Convert to standard notation: $3.25 billion.

8. Determine whether 11, 30 and 4, 12 are proportional.

Add and simplify.

9. $2\frac{2}{5} + 4\frac{3}{10}$

10. $41.063 + 3.5721$

Subtract and simplify.

11. $\frac{14}{15} - \frac{3}{5}$

12. $350 - 24.57$

Multiply and simplify.

13. $3\frac{3}{7} \cdot 4\frac{3}{8}$

14. $12,456 \times 220$

Divide and simplify.

15. $\dfrac{13}{15} \div \dfrac{26}{27}$

16. $104{,}676 \div 24$

Solve.

17. $\dfrac{5}{8} = \dfrac{6}{x}$

18. $\dfrac{2}{5} \cdot y = \dfrac{3}{10}$

19. $21.5 \cdot y = 146.2$

20. $x = 398{,}112 \div 26$

Solve.

21. Tortilla chips cost $2.99 for 14.5 oz. Find the unit price in cents per ounce.

22. A college has a student body of 6000 students. Of these, 55.4% own a car. How many students own a car?

23. In any given year, the average American eats 2.7 lb of peanut butter, 1.5 lb of salted peanuts, 1.2 lb of peanut candy, 0.7 lb of in-shell peanuts, and 0.1 lb of peanuts in other forms. How many pounds of peanuts and products containing peanuts does the average American eat in one year?

24. A piece of fabric $1\frac{3}{4}$ yd long is cut into 7 equal strips. What is the width of each strip?

25. In a recent year, American utility companies generated 1464 billion kilowatt-hours of electricity using coal, 455 billion using nuclear power, 273 billion using natural gas, 250 billion using hydroelectric plants, 118 billion using petroleum, and 12 billion using geothermal technology and other methods. How many kilowatt-hours of electricity were produced that year?

26. A recipe calls for $\frac{3}{4}$ cup of sugar. How much sugar should be used for an amount that is $\frac{1}{2}$ of the recipe?

27. A business is owned by four people. One owns $\frac{1}{3}$, the second owns $\frac{1}{4}$, and the third owns $\frac{1}{6}$. How much does the fourth person own?

28. In manufacturing valves for engines, a factory was discovered to have made 4 defective valves in a lot of 18 valves. At this rate, how many defective valves can be expected in a lot of 5049 valves?

29. A landscaper bought 22 evergreen trees for $210. What was the cost of each tree? Round to the nearest cent.

30. A salesperson earns $182 selling $2600 worth of electronic equipment. What is the commission rate?

This circle graph shows the percent of 18-year-olds surveyed who planned to vote in an upcoming presidential election. Use it for Exercises 31–33.

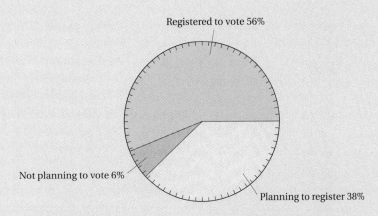

Registered to vote 56%

Not planning to vote 6%

Planning to register 38%

31. What percent of all 18-year-olds were registered to vote?

32. What percent of all 18-year-olds were planning to vote in the election?

33. In a class of 250 18-year-old freshmen, how many were not planning to vote?.

34. Draw a vertical bar graph, using an appropriate set of scales, showing the percentage of people in a National Geographic Society survey who said it is absolutely necessary to know something about the following listed subjects. Be sure to label the scales properly.

Math: 83%

Computer skills: 64%

Science: 38%

Geography: 37%

History: 36%

Foreign languages: 20%

35. Find the mode of this set of numbers:

3, 5, 2, 5, 1, 3, 5, 2 .

36. Find the median of this set of numbers:

61, 67, 60, 63 .

SYNTHESIS

37. A photography club meets four times a month. In September, the attendance figures were 28, 23, 26, and 23. In October, the attendance figures were 26, 20, 14, and 28. What was the percent increase or decrease in average attendance from September to October?

9
Geometry and Measures: Length and Area

INTRODUCTION

In this chapter, we introduce American and metric systems used to measure length, and we present conversion from one unit to another within as well as between each system. These concepts are then applied to finding areas of squares, rectangles, triangles, parallelograms, and circles. We then study right triangles using square roots and the Pythagorean theorem.

AN APPLICATION

A fence is to be built around a 173-m by 240-m field. What is the perimeter of the field?

THE MATHEMATICS

We find the perimeter by finding the distance around the rectangle:

$$\underbrace{240 + 240 + 173 + 173}.$$

↑

This is the perimeter.

The review objectives to be tested in addition to the material in this chapter are as follows.

[7.1b]   Convert from percent notation to decimal notation.
[7.1c]   Convert from decimal notation to percent notation.
[7.2a]   Convert from fractional notation to percent notation.
[7.2b]   Convert from percent notation to fractional notation.

Pretest: Chapter 9

Complete.

1. 8 ft = _____ in.

2. 5 in. = _____ ft

3. 8.46 km = _____ m

4. 9.2 mm = _____ cm

5. Find the perimeter.

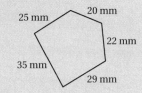

6. Find the area of a square whose sides have length 10 ft.

Find the area.

7.

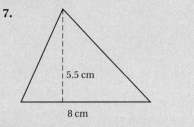

8.

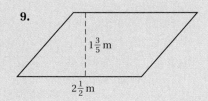

9.

10. Find the length of a diameter of a circle with a radius of 4.8 m.

11. Find the circumference of the circle in Exercise 10. Use 3.14 for π.

12. Find the area of the circle in Exercise 10. Use 3.14 for π.

13. Find the area of the shaded region.

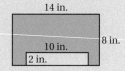

14. Simplify: $\sqrt{81}$.

15. Approximate to three decimal places: $\sqrt{97}$.

In a right triangle, find the length of the side not given. Find an exact answer and an approximation to three decimal places.

16.

c = ?
a = 12
b = 16

17.

c = 7
b = ?
a = 2

9.1 *Linear Measures: American Units*

Length, or distance, is one kind of measure. To find lengths, we start with some **unit segment** and assign to it a measure of 1. Suppose $\overline{AB}$ below is a unit segment.

Let us measure segment $\overline{CD}$ below, using $\overline{AB}$ as our unit segment.

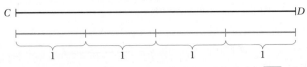

Since we can place 4 unit segments end to end along $\overline{CD}$, the measure of $\overline{CD}$ is 4.

Sometimes we have to use parts of units, called **subunits**. For example, the measure of the segment $\overline{MN}$ below is $1\frac{1}{2}$. We place one unit segment and one half-unit segment end to end.

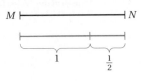

Do Exercises 1–4.

a | AMERICAN MEASURES

American units of length are related as follows.

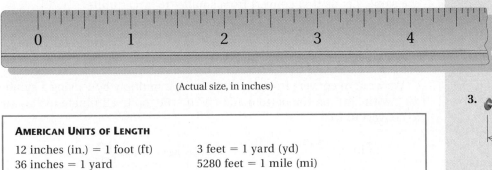

(Actual size, in inches)

AMERICAN UNITS OF LENGTH	
12 inches (in.) = 1 foot (ft)	3 feet = 1 yard (yd)
36 inches = 1 yard	5280 feet = 1 mile (mi)

These American units have also been called "English," or "British–American," because at one time they were used by both countries. Today, both Canada and England have officially converted to the metric system. However, if you travel in England, you will still see units such as "miles" on road signs.

OBJECTIVE

After finishing Section 9.1, you should be able to:

a Convert from one American unit of length to another.

FOR EXTRA HELP

TAPE 14   TAPE 13A   MAC: 9
                         IBM: 9

Use the unit below to measure the length of each segment or object.

1.

2.

3.

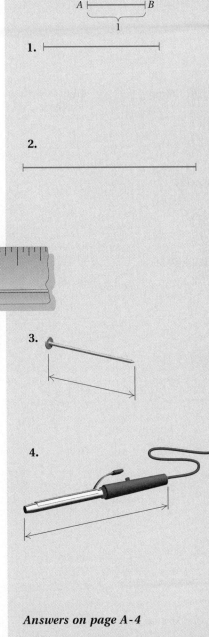

4.

Answers on page A-4

Complete.

5. 8 yd = _____ in.

To change from certain American units to others, we make substitutions. Such a substitution is usually helpful when we are converting from a larger unit to a smaller one.

> **EXAMPLE 1** Complete: 1 yd = _____ in.
>
> $$1 \text{ yd} = 3 \text{ ft}$$
> $$= 3 \times 1 \text{ ft} \qquad \text{We think of 3 ft as } 3 \times \text{ ft, or } 3 \times 1 \text{ ft.}$$
> $$= 3 \times 12 \text{ in.} \qquad \text{Substituting 12 in. for 1 ft}$$
> $$= 36 \text{ in.} \qquad \text{Multiplying}$$

6. 14.5 yd = _____ ft

> **EXAMPLE 2** Complete: 7 yd = _____ in.
>
> $$7 \text{ yd} = 7 \times 1 \text{ yd}$$
> $$= 7 \times 3 \text{ ft} \qquad \text{Substituting 3 ft for 1 yd}$$
> $$= 7 \times 3 \times 1 \text{ ft}$$
> $$= 7 \times 3 \times 12 \text{ in.} \qquad \begin{array}{l}\text{Substituting 12 in. for 1 ft;}\\ 7 \times 3 = 21; 21 \times 12 = 252\end{array}$$
> $$= 252 \text{ in.}$$

7. 3.8 mi = _____ in.

Do Exercises 5–7.

Sometimes it helps to use multiplying by 1 in making conversions. For example, 12 in. = 1 ft, so

$$\frac{12 \text{ in.}}{1 \text{ ft}} = 1 \quad \text{and} \quad \frac{1 \text{ ft}}{12 \text{ in.}} = 1.$$

If we divide 12 in. by 1 ft or 1 ft by 12 in., we get 1 because the lengths are the same. Let us first convert from smaller to larger units.

Complete.

8. 72 in. = _____ ft

> **EXAMPLE 3** Complete: 48 in. = _____ ft.
>
> We want to convert from "in." to "ft." We multiply by 1 using a symbol for 1 with "in." on the bottom and "ft" on the top to eliminate inches and to convert to feet:
>
> $$48 \text{ in.} = \frac{48 \text{ in.}}{1} \times \frac{1 \text{ ft}}{12 \text{ in.}} \qquad \text{Multiplying by 1 using } \frac{1 \text{ ft}}{12 \text{ in.}} \text{ to eliminate in.}$$
> $$= \frac{48 \text{ in.}}{12 \text{ in.}} \times 1 \text{ ft}$$
> $$= \frac{48}{12} \times \frac{\text{in.}}{\text{in.}} \times 1 \text{ ft}$$
> $$= 4 \times 1 \text{ ft} \qquad \text{The } \frac{\text{in.}}{\text{in.}} \text{ acts like 1, so we can omit it.}$$
> $$= 4 \text{ ft.}$$

9. 17 in. = _____ ft

> We can also look at this conversion as "canceling" units:
>
> $$48 \text{ in.} = \frac{48 \text{ in.}}{1} \times \frac{1 \text{ ft}}{12 \text{ in.}} = \frac{48}{12} \times 1 \text{ ft} = 4 \text{ ft.}$$

Answers on page A-4

Do Exercises 8 and 9.

EXAMPLE 4 Complete: 25 ft = _____ yd.

Since we are converting from "ft" to "yd," we choose a symbol for 1 with "yd" on the top and "ft" on the bottom:

$$25 \text{ ft} = 25 \text{ ft} \times \frac{1 \text{ yd}}{3 \text{ ft}}$$

3 ft = 1 yd, so $\frac{3 \text{ ft}}{1 \text{ yd}} = 1$, and $\frac{1 \text{ yd}}{3 \text{ ft}} = 1$. We use $\frac{1 \text{ yd}}{3 \text{ ft}}$ to eliminate ft.

$$= \frac{25}{3} \times \frac{\text{ft}}{\text{ft}} \times 1 \text{ yd}$$

$$= 8\frac{1}{3} \times 1 \text{ yd} \qquad \text{The } \frac{\text{ft}}{\text{ft}} \text{ acts like 1, so we can omit it.}$$

$$= 8\frac{1}{3} \text{ yd, or } 8.\overline{3} \text{ yd.}$$

Again, in this example, we can consider conversion from the point of view of canceling:

$$25 \text{ ft} = 25 \, \cancel{\text{ft}} \times \frac{1 \text{ yd}}{3 \, \cancel{\text{ft}}}$$

$$= \frac{25}{3} \times 1 \text{ yd} = 8\frac{1}{3} \text{ yd, or } 8.\overline{3} \text{ yd.}$$

Do Exercises 10 and 11.

EXAMPLE 5 Complete: 23,760 ft = _____ mi.

We choose a symbol for 1 with "mi" on the top and "ft" on the bottom:

$$23,760 \text{ ft} = 23,760 \text{ ft} \times \frac{1 \text{ mi}}{5280 \text{ ft}}$$

5280 ft = 1 mi, so $\frac{1 \text{ mi}}{5280 \text{ ft}} = 1$.

$$= \frac{23,760}{5280} \times \frac{\text{ft}}{\text{ft}} \times 1 \text{ mi}$$

$$= 4.5 \times 1 \text{ mi} \qquad \text{Dividing}$$

$$= 4.5 \text{ mi.}$$

Let us also consider this example using canceling:

$$23,760 \text{ ft} = 23,760 \, \cancel{\text{ft}} \times \frac{1 \text{ mi}}{5280 \, \cancel{\text{ft}}}$$

$$= \frac{23,760}{5280} \times 1 \text{ mi}$$

$$= 4.5 \times 1 \text{ mi} = 4.5 \text{ mi.}$$

Do Exercises 12 and 13.

We can also use multiplying by 1 to convert from larger to smaller units. Let us redo Example 2.

EXAMPLE 6 Complete: 7 yd = _____ in.

$$7 \text{ yd} = \frac{7 \, \cancel{\text{yd}}}{1} \times \frac{3 \, \cancel{\text{ft}}}{1 \, \cancel{\text{yd}}} \times \frac{12 \text{ in.}}{1 \, \cancel{\text{ft}}}$$

$$= 7 \times 3 \times 12 \times 1 \text{ in.} = 252 \text{ in.}$$

Do Exercise 14.

Complete.

10. 24 ft = _____ yd

11. 35 ft = _____ yd

Complete.

12. 26,400 ft = _____ mi

13. 2640 ft = _____ mi

14. Complete. Use multiplying by 1.

8 yd = _____ in.

Answers on page A-4

9.1 LINEAR MEASURES: AMERICAN UNITS

421

APPLICATIONS TO BASEBALL

There are many applications of mathematics to baseball. We studied one when we considered *earned run average* in Example 3 of Section 6.4. Here we consider several more.

EXERCISES

1. The *batting average* of a player is the number of hits divided by the number of times at bat. The result is usually rounded to three decimal places.

   a) Ty Cobb holds the record for the highest career batting average. He had 4191 hits in 11,429 at bats. What was his batting average?

   b) Ted Williams won the batting title in 1941 with a batting average of 0.406. He was also the last major league player to win a batting title with a 0.400 average. He had 456 at bats. How many hits did he have? Round to the nearest one.

   c) What is the highest batting average that a player can have?

2. The *slugging average* of a player is the total number of bases divided by the total number of at bats. The result is usually rounded to three decimal places.

   a) Rogers Hornsby holds the single season record for slugging average. This happened in 1925 when he got 381 total bases in 504 at bats. What was his slugging average?

   b) Willie Mays got 1960 singles, 523 doubles, 140 triples, and 660 home runs in 10,881 at bats. What was his career slugging average? (*Hint:* Total bases = $1960 \cdot 1 + 523 \cdot 2 + 140 \cdot 3 + 660 \cdot 4$.)

   c) What is the highest slugging average that a player can have?

3. Many excellent baseball players have had their career statistics lessened because of military service. Examples are such Hall of Fame players as Bob Feller, Ted Williams, and Joe DiMaggio. Use ratio and proportion to answer the following questions.

   a) Bob Feller won 266 games in 16 actual seasons of major league pitching. He would have pitched 4 more seasons had he not served in World War II. Estimate how many career wins he would have had if he had played all 20 years. Round to the nearest one.

   b) Feller had 3 no-hitters in his career. Estimate how many he would have had if he had played 4 more seasons. Round to the nearest one.

   c) In Ted Williams' 19-year career in the major league, he had 1839 runs batted in (RBIs). He missed 3 years of playing due to his military service. Estimate how many career RBIs he would have had if he had played the additional 3 years. Round to the nearest one.

4. Hank Aaron holds the major league career record for home runs, with 755. This was 41 more than Babe Ruth had in his career. Ruth hit 8.5 home runs for every 100 times at bat, but Aaron hit 6.5 home runs for every 100 times at bat. The difference is that Aaron took better care of his physical health and had 3965 more at bats than Ruth did.

   a) Ruth had 8399 at bats. How many did Aaron have?

   b) If Ruth had had the same number of at bats as Aaron, how many career home runs would he have hit? Round to the nearest one.

   c) Would Ruth have had the major league home run record if he had had the same number of at bats as Aaron?

5. The *designated hitter rule,* used in the American League but not in the National League, has been a source of great controversy since its inception. A poll was recently conducted among 234,832 fans to see whether they liked the rule. Of these, 41% were for the rule and 59% were against the rule. How many were for the rule? How many against?

Exercise Set 9.1

a Complete.

1. 1 ft = _12_ in.

2. 1 yd = _____ ft

3. 1 in. = _1/12_ ft

4. 1 mi = _____ yd

5. 1 mi = _5280_ ft

6. 1 ft = _____ yd

7. 3 yd = _36/108_ in.

8. 10 yd = _____ ft

9. 84 in. = _7_ ft

10. 48 ft = _____ yd

11. 18 in. = _1 1/2_ ft

12. 29 ft = _____ yd

13. 5 mi = _26400_ ft

14. 5 mi = _____ yd

15. 36 in. = _3_ ft

16. 11,616 ft = _____ mi

17. 10 ft = _3 1/3_ yd

18. 9.6 yd = _____ ft

19. 10 mi = _52800_ ft

20. 31,680 ft = _____ mi

21. $4\frac{1}{2}$ ft = _1 1/2_ yd

ANSWERS

1. _____

2. _____

3. _____

4. _____

5. _____

6. _____

7. _____

8. _____

9. _____

10. _____

11. _____

12. _____

13. _____

14. _____

15. _____

16. _____

17. _____

18. _____

19. _____

20. _____

21. _____

ANSWERS

22. _____

23. _____

24. _____

25. _____

26. _____

27. _____

28. _____

29. _____

30. _____

31. _____

32. _____

33. _____

34. _____

35. _____

36. _____

37. _____

38. _____

39. _____

40. _____

41. _____

42. _____

43. _____

22. 48 in. = _____ ft **23.** 36 in. = 3 yd **24.** 20 yd = _____ in.

25. 330 ft = 110 yd **26.** 5280 yd = _____ mi **27.** 3520 yd = 2 mi

28. 25 mi = _____ ft **29.** 100 yd = 300 ft **30.** 480 in. = _____ ft

31. 360 in. = 30 ft **32.** 720 in. = _____ yd **33.** 1 in. = $\frac{1}{36}$ yd

34. 25 in. = _____ ft **35.** 2 mi = 126,720 in. **36.** 63,360 in. = _____ mi

SKILL MAINTENANCE

Convert to fractional notation.

37. 9.25% $\frac{9.25}{100} = \frac{925}{10,000}$ $\frac{37}{400}$ **38.** $87\frac{1}{2}$%

Convert to percent notation.

39. $\frac{11}{8}$ 137% **40.** $\frac{2}{3}$ **41.** $\frac{1}{4}$ 25 **42.** $\frac{7}{16}$

SYNTHESIS

43. ▦ Recently the national debt was $4.265 trillion. To get an idea of this amount, picture that if that many $1 bills were stacked on top of each other, they would reach 1.169 times the distance to the moon. The distance to the moon is 238,866 mi. How thick, in inches, is a $1 bill?

9.2 *Linear Measures: The Metric System*

The **metric system** is used in most countries of the world, and the United States is now making greater use of it as well. The metric system does not use inches, feet, pounds, and so on, although units for time and electricity are the same as those you use now.

An advantage of the metric system is that it is easier to convert from one unit to another. That is because the metric system is based on the number 10.

The basic unit of length is the **meter**. It is just over a yard. In fact, 1 meter ≈ 1.1 yd.

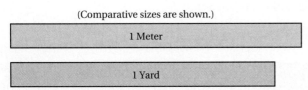

(Comparative sizes are shown.)

1 Meter

1 Yard

The other units of length are multiples of the length of a meter:

10 times a meter,   100 times a meter,   1000 times a meter,   and so on,

or fractions of a meter:

$\frac{1}{10}$ of a meter, $\frac{1}{100}$ of a meter, $\frac{1}{1000}$ of a meter,   and so on.

METRIC UNITS OF LENGTH

1 *kilo*meter (km) = 1000 meters (m)

1 *hecto*meter (hm) = 100 meters (m)

1 *deka*meter (dam) = 10 meters (m)

 1 meter (m)   *dam* and *dm* are not used much.

1 *deci*meter (dm) = $\frac{1}{10}$ meter (m)

1 *centi*meter (cm) = $\frac{1}{100}$ meter (m)

1 *milli*meter (mm) = $\frac{1}{1000}$ meter (m)

You should memorize these names and abbreviations. Think of *kilo-* for 1000, *hecto-* for 100, *deka-* for 10, *deci-* for $\frac{1}{10}$, *centi-* for $\frac{1}{100}$, and *milli-* for $\frac{1}{1000}$. We will also use these prefixes when considering units of area, capacity, and mass (weight).

FOR EXTRA HELP

TAPE 15   TAPE 13B   MAC: 9
                                   IBM: 9

THINKING METRIC

To familiarize yourself with metric units, consider the following.

1 kilometer (1000 meters)   is slightly more than $\frac{1}{2}$ mile (0.6 mi).

1 meter   is just over a yard (1.1 yd).

1 centimeter (0.01 meter)   is a little more than the width of a paperclip (about 0.3937 inch).

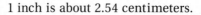

1 cm

1 cm

1 inch is about 2.54 centimeters.

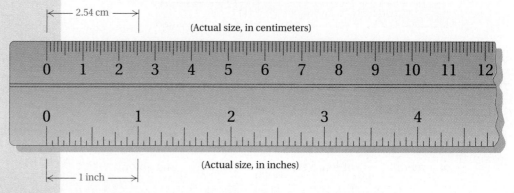

2.54 cm

(Actual size, in centimeters)

0   1   2   3   4   5   6   7   8   9   10   11   12

0   1   2   3   4

(Actual size, in inches)

1 inch

1 millimeter is about the diameter of a paperclip wire.

1 mm

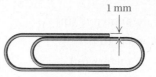

The millimeter (mm) is used to measure small distances, especially in industry.

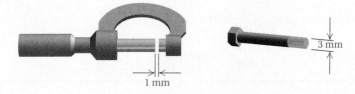

3 mm

1 mm

In many countries, the centimeter (cm) is used for body dimensions and clothing sizes.

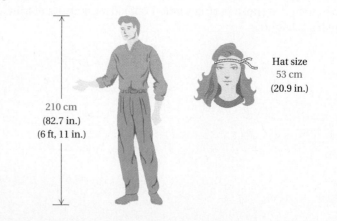

210 cm
(82.7 in.)
(6 ft, 11 in.)

Hat size
53 cm
(20.9 in.)

The meter (m) is used for expressing dimensions of larger objects—say, the length of a building—and for shorter distances, such as the length of a rug.

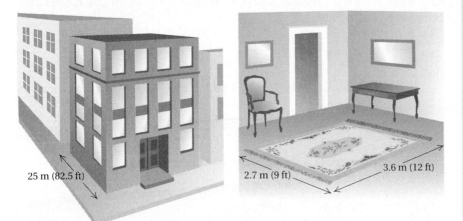

25 m (82.5 ft)

2.7 m (9 ft)

3.6 m (12 ft)

The kilometer (km) is used for longer distances, mostly in places where miles are now being used.

Atlanta
80 MI

Atlanta
128 KM

Do Exercises 1–6.

a | CHANGING METRIC UNITS

EXAMPLE 1   Complete: 4 km = _____ m.

$$4 \text{ km} = 4 \times 1 \text{ km}$$
$$\qquad = 4 \times 1000 \text{ m} \qquad \text{Substituting 1000 m for 1 km}$$
$$\qquad = 4000 \text{ m}$$

Do Exercises 7 and 8.

Since

$$\frac{1}{10} \text{ m} = 1 \text{ dm}, \qquad \frac{1}{100} \text{ m} = 1 \text{ cm}, \quad \text{and} \quad \frac{1}{1000} \text{ m} = 1 \text{ mm},$$

it follows that

1 m = 10 dm,   1 m = 100 cm,   and   1 m = 1000 mm.

Memorizing these will help you to write forms of 1 when making conversions.

Complete with mm, cm, m, or km.

1. A stick of gum is 7 _____ long.

2. Minneapolis is 3213 _____ from San Francisco.

3. A penny is 1 _____ thick.

4. The halfback ran 7 _____ .

5. The book is 3 _____ thick.

6. The desk is 2 _____ long.

Complete.

7. 23 km = _____ m

8. 4 hm = _____ m

Answers on page A-4

Complete.

9. 1.78 m = _____ cm

10. 9.04 m = _____ mm

Complete.

11. 7814 m = _____ km

12. 7814 m = _____ dam

EXAMPLE 2 Complete: 93.4 m = _____ cm.

We want to convert from "m" to "cm." We multiply by 1 using a symbol for 1 with "m" on the bottom and "cm" on the top to eliminate meters and convert to centimeters:

$$93.4 \text{ m} = 93.4 \text{ m} \times \frac{100 \text{ cm}}{1 \text{ m}} \qquad \text{Multiplying by 1 using } \frac{100 \text{ cm}}{1 \text{ m}}$$

$$= 93.4 \times 100 \times \frac{\text{m}}{\text{m}} \times 1 \text{ cm} \qquad \text{The } \frac{\text{m}}{\text{m}} \text{ acts like 1, so we omit it.}$$

$$= 9340 \text{ cm}. \qquad \text{Multiplying by 100 moves the decimal point two places to the right.}$$

We can also work this example by canceling:

$$93.4 \text{ m} = 93.4 \text{ m} \times \frac{100 \text{ cm}}{1 \text{ m}}$$

$$= 93.4 \times 100 \times 1 \text{ cm}$$

$$= 9340 \text{ cm}.$$

EXAMPLE 3 Complete: 0.248 m = _____ mm.

We are converting from "m" to "mm," so we choose a symbol for 1 with "mm" on the top and "m" on the bottom:

$$0.248 \text{ m} = 0.248 \text{ m} \times \frac{1000 \text{ mm}}{1 \text{ m}} \qquad \text{Multiplying by 1 using } \frac{1000 \text{ mm}}{1 \text{ m}}$$

$$= 0.248 \times 1000 \times \frac{\text{m}}{\text{m}} \times 1 \text{ mm} \qquad \text{The } \frac{\text{m}}{\text{m}} \text{ acts like 1, so we omit it.}$$

$$= 248 \text{ mm}. \qquad \text{Multiplying by 1000 moves the decimal point three places to the right.}$$

Using canceling, we can work this example as follows:

$$0.248 = 0.248 \text{ m} \times \frac{1000 \text{ mm}}{1 \text{ m}}$$

$$= 0.248 \times 1000 \times 1 \text{ mm} = 248 \text{ mm}.$$

Do Exercises 9 and 10.

EXAMPLE 4 Complete: 2347 m = _____ km.

$$2347 \text{ m} = 2347 \text{ m} \times \frac{1 \text{ km}}{1000 \text{ m}} \qquad \text{Multiplying by 1 using } \frac{1 \text{ km}}{1000 \text{ m}}$$

$$= \frac{2347}{1000} \times \frac{\text{m}}{\text{m}} \times 1 \text{ km} \qquad \text{The } \frac{\text{m}}{\text{m}} \text{ acts like 1, so we omit it.}$$

$$= 2.347 \text{ km} \qquad \text{Dividing by 1000 moves the decimal point three places to the left.}$$

Using canceling, we can work this example as follows:

$$2347 \text{ m} = 2347 \text{ m} \times \frac{1 \text{ km}}{1000 \text{ m}} = \frac{2347}{1000} \times 1 \text{ km} = 2.347 \text{ km}.$$

Do Exercises 11 and 12.

Sometimes we multiply by 1 more than once.

EXAMPLE 5 Complete: 8.42 mm = _____ cm.

$$8.42 \text{ mm} = 8.42 \text{ mm} \times \frac{1 \text{ m}}{1000 \text{ mm}} \times \frac{100 \text{ cm}}{1 \text{ m}}$$

Multiplying by 1 using $\frac{1 \text{ m}}{1000 \text{ mm}}$ and $\frac{100 \text{ cm}}{1 \text{ m}}$

$$= \frac{8.42 \times 100}{1000} \times \frac{\text{mm}}{\text{mm}} \times \frac{\text{m}}{\text{m}} \times 1 \text{ cm}$$

$$= \frac{842}{1000} \text{ cm}$$

$$= 0.842 \text{ cm}$$

Using canceling, we can work this example as follows:

$$8.42 \text{ mm} = 8.42 \text{ mm} \times \frac{1 \text{ m}}{1000 \text{ mm}} \times \frac{100 \text{ cm}}{1 \text{ m}}$$

$$= \frac{8.42 \times 100}{1000} \times 1 \text{ cm} = 0.842 \text{ cm}.$$

Do Exercises 13 and 14.

MENTAL CONVERSION

Look back over the examples and exercises done so far and you will see that changing from one unit to another in the metric system amounts to only the movement of a decimal point. That is because the metric system is based on 10. Let's find a faster way to convert. Look at the following table.

1000	100	10	1	0.1	0.01	0.001
km	hm	dam	m	dm	cm	mm

Each place in the table has a value $\frac{1}{10}$ that to the left or 10 times that to the right. Thus moving one place in the table corresponds to one decimal place. Let us convert mentally.

EXAMPLE 6 Complete: 8.42 mm = _____ cm.

Think: To go from mm to cm in the table is a move of one place to the left. Thus we move the decimal point one place to the left.

8.42   0.8.42   8.42 mm = 0.842 cm

EXAMPLE 7 Complete: 1.886 km = _____ cm.

Think: To go from km to cm is a move of five places to the right. Thus we move the decimal point five places to the right.

1.886   1.88600.   1.886 km = 188,600 cm

Complete.

13. 9.67 mm = _____ cm

14. 89 km = _____ cm

Answers on page A-4

Complete. Try to do this mentally using the table.

15. 6780 m = _____ km

16. 9.74 cm = _____ mm

17. 1 mm = _____ cm

18. 845.1 mm = _____ dm

Complete.

19. 100 yd = _____ m
(The length of a football field)

20. 500 mi = _____ km
(The Indianapolis 500-mile race)

21. 2383 km = _____ mi
(The distance from St. Louis to Phoenix)

Answers on page A-4

EXAMPLE 8   Complete: 3 m = _____ cm.

Think: To go from m to cm in the table is a move of two places to the right. Thus we move the decimal point two places to the right.

3   3.00.   3 m = 300 cm

You should try to make metric conversions mentally as much as possible.

The fact that conversions can be done so easily is an important advantage of the metric system. The most commonly used metric units of length are km, m, cm, and mm. We have purposely used these more often than the others in the exercises.

Do Exercises 15–18.

b CONVERTING BETWEEN AMERICAN AND METRIC UNITS

We can make conversions between American and metric units by using the following table. Again, we either make a substitution or multiply by 1 appropriately.

METRIC	AMERICAN
1 m	39.37 in.
1 m	3.3 ft
2.54 cm	1 in.
1 km	0.621 mi
1.609 km	1 mi

EXAMPLE 9   Complete: 26.2 mi = _____ km. (This is the length of the Olympic marathon.)

$$26.2 \text{ mi} = 26.2 \times 1 \text{ mi}$$
$$\approx 26.2 \times 1.609 \text{ km}$$
$$\approx 42.1558 \text{ km}$$

EXAMPLE 10   Complete: 100 m = _____ yd. (This is the length of a dash in track.)

$$100 \text{ m} = 100 \times 1 \text{ m} \approx 100 \times 3.3 \text{ ft} \approx 330 \text{ ft}$$
$$\approx 330 \text{ ft} \times \frac{1 \text{ yd}}{3 \text{ ft}} \approx \frac{330}{3} \text{ yd} \approx 110 \text{ yd}$$

Do Exercises 19–21.

Exercise Set 9.2

a Complete. Do as much as possible mentally.

1. a) 1 km = _1000_ m   **2. a)** 1 hm = _____ m   **3. a)** 1 dam = _100_ m

b) 1 m = _$\frac{1}{1000}$ .001_ km   **b)** 1 m = _____ hm   **b)** 1 m = _$\frac{1}{100}$ .01_ dam

4. a) 1 dm = _$\frac{1}{100}$_ m   **5. a)** 1 cm = _100_ m   **6. a)** 1 mm = _____ m

b) 1 m = _10_ dm   **b)** 1 m = _$\frac{1}{100}$_ cm   **b)** 1 m = _____ mm

7. 6.7 km = _670_ m   **8.** 27 km = _____ m   **9.** 98 cm = _.98_ m

10. 0.789 cm = _____ m   **11.** 8921 m = _.8921_ km   **12.** 8664 m = _____ km

13. 56.66 m = _.056_ km   **14.** 4.733 m = _____ km   **15.** 5666 m = _566.00_ cm

16. 869 m = _____ cm   **17.** 477 cm = _.00477_ m   **18.** 6.27 mm = _.00627_ m

19. 6.88 m = _688_ cm   **20.** 6.88 m = _____ dm   **21.** 1 mm = _.01_ cm

22. 1 cm = _____ km   **23.** 1 km = _100,000_ cm   **24.** 2 km = _200,000_ cm

ANSWERS

1. a) _____
b) _____
2. a) _____
b) _____
3. a) _____
b) _____
4. a) _____
b) _____
5. a) _____
b) _____
6. a) _____
b) _____
7. _____
8. _____
9. _____
10. _____
11. _____
12. _____
13. _____
14. _____
15. _____
16. _____
17. _____
18. _____
19. _____
20. _____
21. _____
22. _____
23. _____
24. _____

25. _____

26. _____

27. _____

28. _____

29. _____

30. _____

31. _____

32. _____

33. _____

34. _____

35. _____

36. _____

37. _____

38. _____

39. _____

40. _____

41. _____

42. _____

43. _____

44. _____

45. _____

46. _____

47. _____

48. _____

49. _____

50. _____

51. _____

25. 14.2 cm = _142_ mm

26. 25.3 cm = _____ mm

27. 8.2 mm = _.82_ cm

28. 9.7 mm = _____ cm

29. 4500 mm = _450_ cm

30. 8,000,000 m = _____ km

31. 0.024 mm = _0000024_ m

32. 60,000 mm = _____ dam

33. 6.88 m = _688_ dam

34. 7.44 m = _____ hm

35. 2.3 dam = _230_ dm

36. 9 km = _____ hm

37. 392 dam = _39.2_ km

38. 0.056 mm = _____ dm

[b] Complete.

39. 330 ft = _100_ m
(The length of most baseball foul lines)

40. 12 in. = _____ cm

1171.352 ÷ 1.609
727.409

41. 1171.352 km = _____ mi
(The distance from Cleveland to Atlanta)

42. 2 m = _____ ft
(The length of a desk)

65 × 1.609
104.585

43. 65 mph = _104.585_ km/h
(A common speed limit in the U.S.)

44. 100 km/h = _____ mph
(A common speed limit in Canada)

SKILL MAINTENANCE

Divide. Find decimal notation for the answer.

45. 21 ÷ 12 1.75

46. 23.4 ÷ 10

47. 23.4 ÷ 100 .234

48. 23.4 ÷ 1000

49. Multiply 3.14 × 4.41. Round to the nearest hundredth. 13.85

50. Multiply: $4 \times 20\frac{1}{8}$.

51. Multiply: $48 \times \frac{1}{12}$. 4.0

9.3 *Perimeter*

a | FINDING PERIMETERS

> A *polygon* is a geometric figure with three or more sides. The *perimeter* of a polygon is the distance around it, or the sum of the lengths of its sides.

| **EXAMPLE 1** Find the perimeter of this polygon.

We add the lengths of the sides. Since all the units are the same, we add the numbers, keeping meters (m) as the unit.

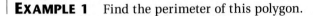

Perimeter = 6 m + 5 m + 4 m + 5 m + 9 m

= (6 + 5 + 4 + 5 + 9) m

= 29 m

Do Exercises 1 and 2.

A **rectangle** is a figure with sides and 90°-angles like the one in Example 2.

| **EXAMPLE 2** Find the perimeter of a rectangle that is 3 cm by 4 cm.

Perimeter = 3 cm + 3 cm + 4 cm + 4 cm

= (3 + 3 + 4 + 4) cm

= 14 cm

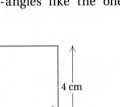

Do Exercise 3.

> The *perimeter of a rectangle* is twice the sum of the length and the width, or 2 times the length plus 2 times the width:
> $$P = 2 \cdot (l + w), \quad \text{or} \quad P = 2 \cdot l + 2 \cdot w.$$
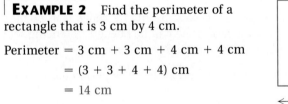

| **EXAMPLE 3** Find the perimeter of a rectangle that is 4.3 ft by 7.8 ft.

$$P = 2 \cdot (l + w) = 2 \cdot (4.3 \text{ ft} + 7.8 \text{ ft}) = 2 \cdot (12.1 \text{ ft}) = 24.2 \text{ ft}$$

Do Exercises 4 and 5.

A **square** is a rectangle with all sides the same length.

| **EXAMPLE 4** Find the perimeter of a square whose sides are 9 mm long.

$$P = 9 \text{ mm} + 9 \text{ mm} + 9 \text{ mm} + 9 \text{ mm}$$

$$= (9 + 9 + 9 + 9) \text{ mm} = 36 \text{ mm}$$

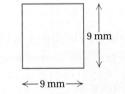

OBJECTIVES

After finishing Section 9.3, you should be able to:

a Find the perimeter of a polygon.

b Solve problems involving perimeter.

FOR EXTRA HELP

TAPE 15   TAPE 13B   MAC: 9
                               IBM: 9

Find the perimeter of the polygon.

1.

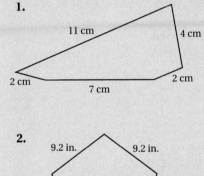

2.

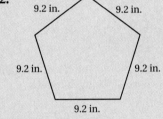

3. Find the perimeter of a rectangle that is 2 cm by 4 cm.

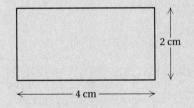

4. Find the perimeter of a rectangle that is 5.25 yd by 3.5 yd.

5. Find the perimeter of a rectangle that is 8 km by 8 km.

Answers on page A-4

6. Find the perimeter of a square whose sides have length 10 km.

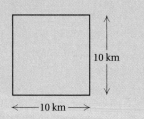

7. Find the perimeter of a square whose sides have length $5\frac{1}{4}$ yd.

8. Find the perimeter of a square whose sides have length 7.8 km.

9. A play area is 25 ft by 10 ft. A fence is to be built around the play area. How many feet of fencing will be needed? If fencing costs $4.95 per foot, what will the fencing cost?

Answers on page A-4

Do Exercise 6.

The *perimeter of a square* is four times the length of a side:

$$P = 4 \cdot s.$$

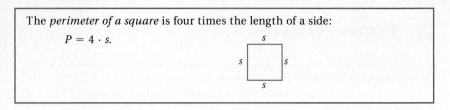

EXAMPLE 5   Find the perimeter of a square whose sides are $20\frac{1}{8}$ in. long.

$$P = 4 \cdot s = 4 \cdot 20\frac{1}{8} \text{ in.}$$

$$= 4 \cdot \frac{161}{8} \text{ in.} = \frac{4 \cdot 161}{4 \cdot 2} \text{ in.}$$

$$= \frac{161}{2} \cdot \frac{4}{4} \text{ in.} = 80\frac{1}{2} \text{ in.}$$

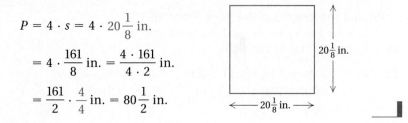

Do Exercises 7 and 8.

b   SOLVING PROBLEMS

EXAMPLE 6   A vegetable garden is 15 ft by 20 ft. A fence is to be built around the garden. How many feet of fence will be needed? If fencing sells for $2.95 per foot, what will the fencing cost?

1. *Familiarize.* We make a drawing and let P = the perimeter.

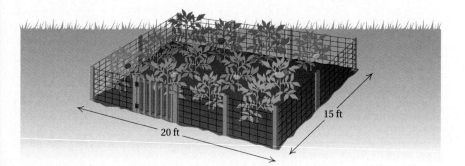

2. *Translate.* The perimeter of the garden is given by

$$P = 2 \cdot (l + w) = 2 \cdot (15 \text{ ft} + 20 \text{ ft}).$$

3. *Solve.* We calculate the perimeter as follows:

$$P = 2 \cdot (15 \text{ ft} + 20 \text{ ft}) = 2 \cdot (35 \text{ ft}) = 70 \text{ ft}.$$

Then we multiply by $2.95 to find the cost of the fencing:

$$\text{Cost} = \$2.95 \times \text{Perimeter} = \$2.95 \times 70 \text{ ft} = \$206.50.$$

4. *Check.* The check is left to the student.

5. *State.* The 70 ft of fencing will cost $206.50.

Do Exercise 9.

Exercise Set 9.3

a Find the perimeter of the polygon.

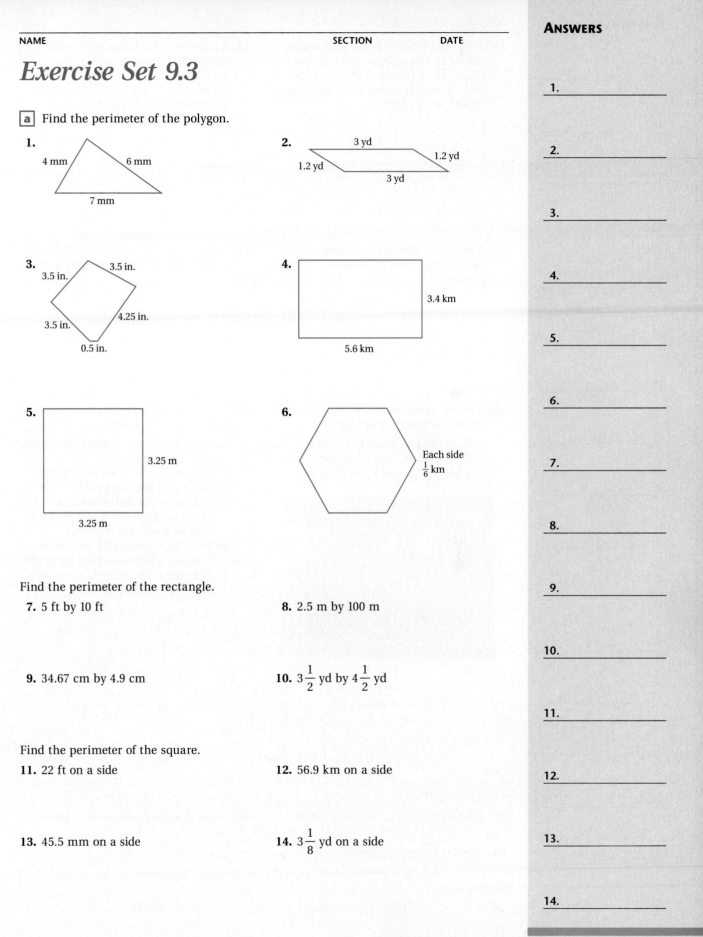

1.

4 mm   6 mm

7 mm

2.

3 yd

1.2 yd   1.2 yd

3 yd

3.

3.5 in.   3.5 in.

3.5 in.

4.25 in.

3.5 in.

0.5 in.

4.

3.4 km

5.6 km

5.

3.25 m

3.25 m

6.

Each side
$\frac{1}{6}$ km

Find the perimeter of the rectangle.

7. 5 ft by 10 ft

8. 2.5 m by 100 m

9. 34.67 cm by 4.9 cm

10. $3\frac{1}{2}$ yd by $4\frac{1}{2}$ yd

Find the perimeter of the square.

11. 22 ft on a side

12. 56.9 km on a side

13. 45.5 mm on a side

14. $3\frac{1}{8}$ yd on a side

15. _____

16. _____

17. _____

18. _____

19. a) _____

b) _____

20. a) _____

b) _____

c) _____

d) _____

e) _____

21. _____

22. _____

23. _____

24. _____

25. _____

26. _____

b Solve.

15. A security fence is to be built around a 173-m by 240-m field. What is the perimeter of the field? If fence wire costs $1.45 per meter, what will wire for the fence cost?

16. A standard-sized slow-pitch softball diamond is a square whose sides have length 65 ft. What is the perimeter of a softball diamond? (This is how far you would have to run if you hit a home run.)

$P = 65 \times 4$

17. A piece of flooring tile is a square 30.5 cm on a side. What is its perimeter?

18. A posterboard is 61.8 cm by 87.9 cm. What is the perimeter of the board?

19. A rain gutter is to be installed around the house shown in the figure.

a) Find the perimeter of the house.
b) If the gutter costs $4.59 per foot, what is the total cost of the gutter?

20. A carpenter is to build a fence around a 9-m by 12-m garden.

a) The posts are 3 m apart. How many posts will be needed?
b) The posts cost $2.40 each. How much will the posts cost?
c) The fence will surround all but 3 m of the garden, which will be a gate. How long will the fence be?
d) The fence costs $2.85 per meter. What will the cost of the fence be?
e) The gate costs $9.95. What is the total cost of the materials?

$P = 2(9) + 2(12) = 42$
A) $42 \div 3 = 14$

SKILL MAINTENANCE

21. Convert to decimal notation: 56.1%.

22. Convert to percent notation: 0.6734.

23. Convert to percent notation: $\dfrac{9}{8}$.

Evaluate.

24. 5^2

25. 10^2

26. 31^2

9.5 *Areas of Parallelograms, Triangles, and Trapezoids*

OBJECTIVES

After finishing Section 9.5, you should be able to:

a Find areas of parallelograms, triangles, and trapezoids.

b Solve problems involving areas of parallelograms, triangles, and trapezoids.

FOR EXTRA HELP

TAPE 16   TAPE 14A   MAC: 9
                                    IBM: 9

a FINDING OTHER AREAS

PARALLELOGRAMS

A **parallelogram** is a four-sided figure with two pairs of parallel sides, as shown below.

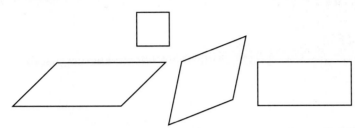

To find the area of a parallelogram, consider the one below.

If we cut off a piece and move it to the other end, we get a rectangle.

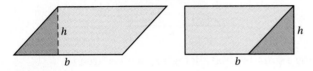

We can find the area by multiplying the length *b*, called a **base**, by *h*, called the **height**.

The *area of a parallelogram* is the product of the length of a base *b* and the height *h*:

$$A = b \cdot h.$$

EXAMPLE 1 Find the area of this parallelogram.

$A = b \cdot h$
$\quad = 7 \text{ km} \cdot 5 \text{ km}$
$\quad = 35 \text{ km}^2$

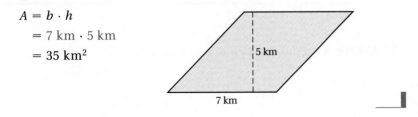

Find the area.

1.

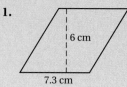

6 cm

7.3 cm

2.

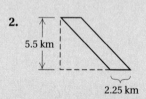

5.5 km

2.25 km

EXAMPLE 2   Find the area of this parallelogram.

$A = b \cdot h$

$= (1.2 \text{ m}) \times (6 \text{ m})$

$= 7.2 \text{ m}^2$

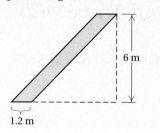

6 m

1.2 m

Do Exercises 1 and 2.

TRIANGLES

To find the area of a triangle, think of cutting out another just like it.

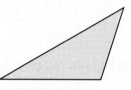

Then place the second one like this.

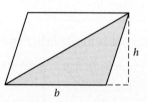

h

b

The resulting figure is a parallelogram whose area is

$b \cdot h.$

The triangle we started with has half the area of the parallelogram, or

$\frac{1}{2} \cdot b \cdot h.$

The *area of a triangle* is half the length of the base times the height:

$A = \frac{1}{2} \cdot b \cdot h.$

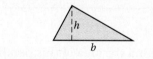

h

b

EXAMPLE 3   Find the area of this triangle.

$A = \frac{1}{2} \cdot b \cdot h$

$= \frac{1}{2} \cdot 9 \text{ m} \cdot 6 \text{ m}$

$= \frac{9 \cdot 6}{2} \text{ m}^2$

$= 27 \text{ m}^2$

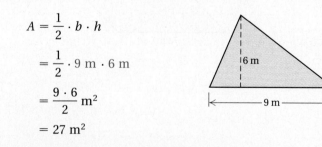

6 m

9 m

EXAMPLE 4 Find the area of this triangle.

$$A = \frac{1}{2} \cdot b \cdot h$$

$$= \frac{1}{2} \times 6.25 \text{ cm} \times 5.5 \text{ cm}$$

$$= 0.5 \times 6.25 \times 5.5 \text{ cm}^2$$

$$= 17.1875 \text{ cm}^2$$

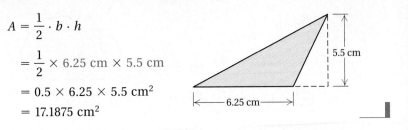

Do Exercises 3 and 4.

TRAPEZOIDS

A **trapezoid** is a four-sided figure with at least one pair of parallel sides, as shown below.

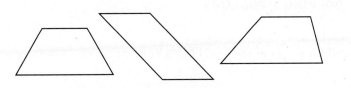

To find the area of a trapezoid, think of cutting out another just like it.

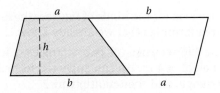

Then place the second one like this.

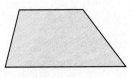

The resulting figure is a parallelogram whose area is

$$h \cdot (a + b). \qquad \text{The base is } a + b.$$

The trapezoid we started with has half the area of the parallelogram, or

$$\frac{1}{2} \cdot h \cdot (a + b).$$

The *area of a trapezoid* is half the product of the height and the sum of the lengths of the parallel sides, or the product of the height and the average length of the bases:

$$A = \frac{1}{2} \cdot h \cdot (a + b) = h \cdot \frac{a + b}{2}.$$

Find the area.

3.

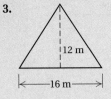

4.

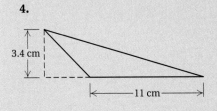

Answers on page A-5

Find the area.

5.

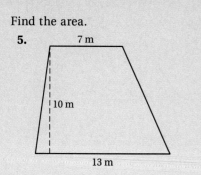
7 m

10 m

13 m

6.

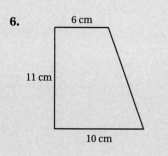
6 cm

11 cm

10 cm

7. Find the area.

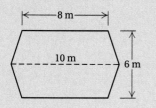
8 m

10 m

6 m

Answers on page A-5

EXAMPLE 5 Find the area of this trapezoid.

$$A = \frac{1}{2} \cdot h \cdot (a + b)$$

$$= \frac{1}{2} \cdot 7 \text{ cm} \cdot (12 + 18) \text{ cm}$$

$$= \frac{7 \cdot 30}{2} \cdot \text{cm}^2 = \frac{7 \cdot 15 \cdot 2}{1 \cdot 2} \text{ cm}^2$$

$$= \frac{7 \cdot 15}{1} \cdot \frac{2}{2} \text{ cm}^2$$

$$= 105 \text{ cm}^2$$

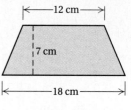
12 cm

7 cm

18 cm

Do Exercises 5 and 6.

b **SOLVING PROBLEMS**

EXAMPLE 6 Find the area of this kite.

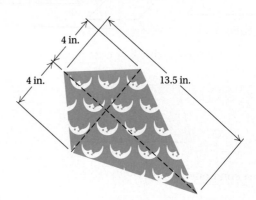
4 in.

4 in.

13.5 in.

1. ***Familiarize.*** We look for the kinds of figures whose areas we can calculate using area formulas that we already know.

2. ***Translate.*** The kite consists of two triangles, each with a base of 13.5 in. and a height of 4 in. We can apply the formula $A = \frac{1}{2} \cdot b \cdot h$ for the area of a triangle and then multiply by 2.

3. ***Solve.*** We have

$$A = \frac{1}{2} \cdot (13.5 \text{ in.}) \cdot (4 \text{ in.}) = 27 \text{ in}^2.$$

Then we multiply by 2:

$$2 \cdot 27 \text{ in}^2 = 54 \text{ in}^2.$$

4. ***Check.*** We can check by repeating the calculations.

5. ***State.*** The area of the kite is 54 in^2.

Do Exercise 7.

Exercise Set 9.5

a Find the area.

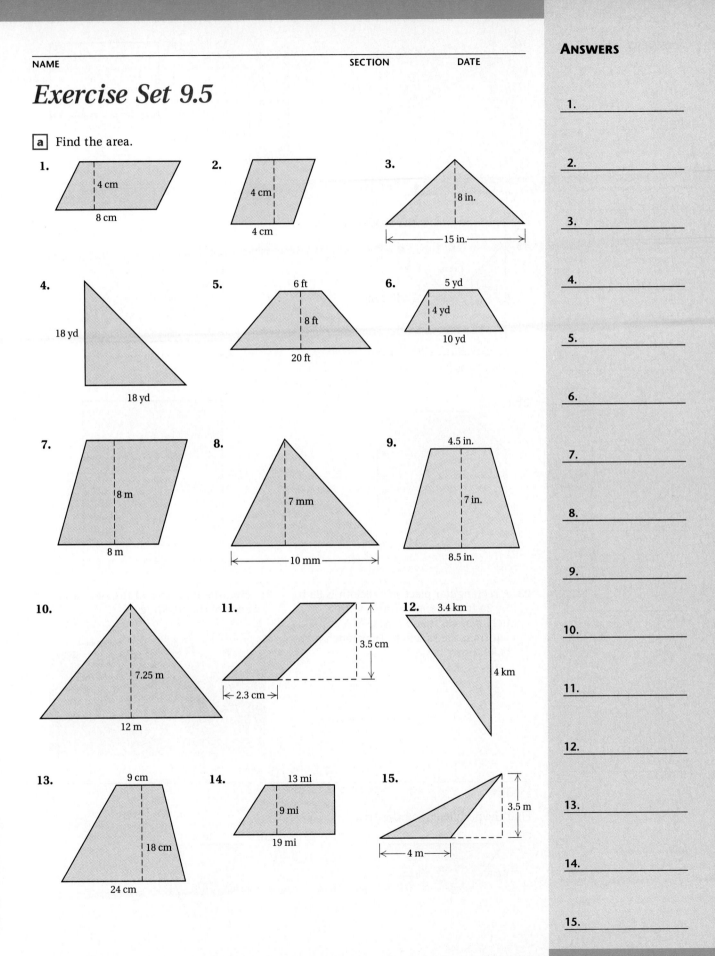

1.

2.

3.

4.

5.

6.

7.

8.

9.

10.

11.

12.

13.

14.

15.

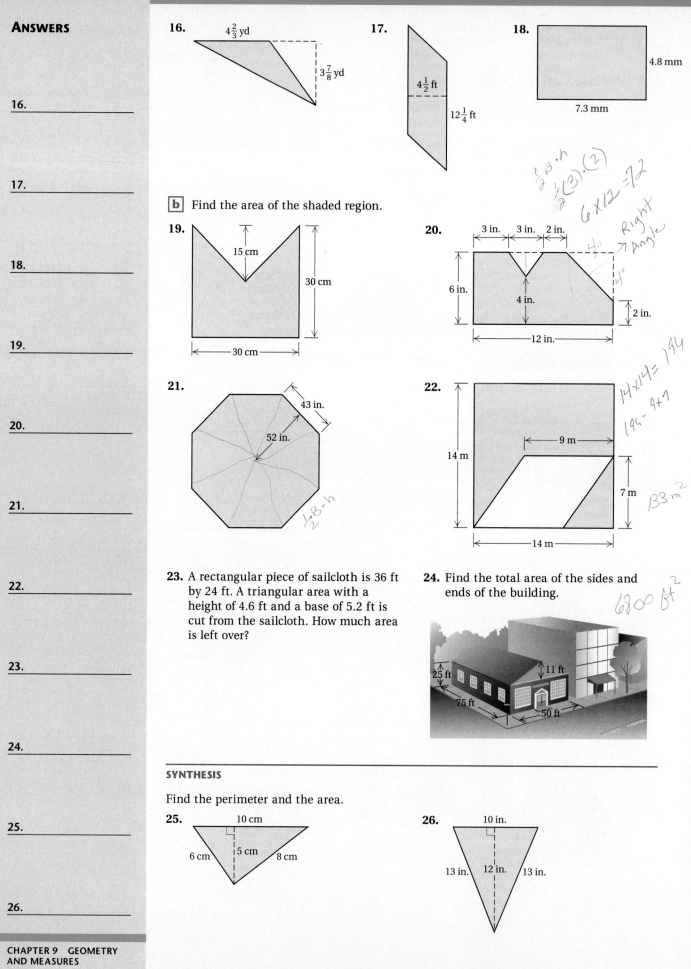

ANSWERS

16.

17.

18.

19.

20.

21.

22.

23.

24.

25.

26.

16. $4\frac{2}{3}$ yd, $3\frac{7}{8}$ yd

17. $4\frac{1}{2}$ ft, $12\frac{1}{4}$ ft

18. 4.8 mm, 7.3 mm

b Find the area of the shaded region.

19. 15 cm, 30 cm, 30 cm

20. 3 in., 3 in., 2 in., 6 in., 4 in., 2 in., 12 in.

21. 43 in., 52 in.

22. 9 m, 14 m, 7 m, 14 m

23. A rectangular piece of sailcloth is 36 ft by 24 ft. A triangular area with a height of 4.6 ft and a base of 5.2 ft is cut from the sailcloth. How much area is left over?

24. Find the total area of the sides and ends of the building.

25 ft, 11 ft, 75 ft, 50 ft

SYNTHESIS

Find the perimeter and the area.

25. 10 cm, 6 cm, 5 cm, 8 cm

26. 10 in., 13 in., 12 in., 13 in.

9.6 Circles

a RADIUS AND DIAMETER

At the right is a circle with center O. Segment $\overline{AC}$ is a *diameter*. A **diameter** is a segment that passes through the center of the circle and has endpoints on the circle. Segment $\overline{OB}$ is called a *radius*. A **radius** is a segment with one endpoint on the center and the other endpoint on the circle.

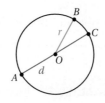

Suppose that d is the diameter of a circle and r is the radius. Then

$$d = 2 \cdot r \quad \text{and} \quad r = \frac{d}{2}.$$

EXAMPLE 1 Find the length of a radius of this circle.

$$r = \frac{d}{2}$$

$$= \frac{12 \text{ m}}{2}$$

$$= 6 \text{ m}$$

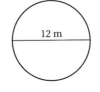

12 m

The radius is 6 m.

EXAMPLE 2 Find the length of a diameter of this circle.

$$d = 2 \cdot r$$

$$= 2 \cdot \frac{1}{4} \text{ ft}$$

$$= \frac{1}{2} \text{ ft}$$

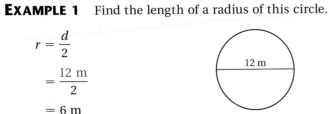

$\frac{1}{4}$ ft

The diameter is $\frac{1}{2}$ ft.

Do Exercises 1 and 2.

b CIRCUMFERENCE

The **circumference** of a circle is the distance around it. Calculating circumference is similar to finding the perimeter of a polygon.

Take a 12-oz soda can and measure the circumference C of the lid with a tape measure. Then measure the diameter d. Then find the ratio C/d.

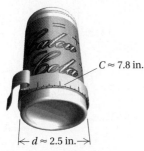

$C \approx 7.8$ in.

$\leftarrow d \approx 2.5$ in. $\rightarrow$

$$\frac{C}{d} \approx \frac{7.8 \text{ in.}}{2.5 \text{ in.}} \approx 3.1$$

1. Find the length of a radius.

24 km

2. Find the length of a diameter.

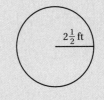

$2\frac{1}{2}$ ft

Answers on page A-5

3. Find the circumference of this circle. Use 3.14 for π.

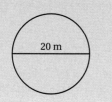
20 m

4. Find the circumference of this circle. Use $\frac{22}{7}$ for π.

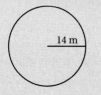
14 m

5. Find the circumference of this circle. Use 3.14 for π.

3.2 yd

Answers on page A-5

Suppose we did this with cans and circles of several sizes. We would get a number close to 3.1. For any circle, if we divide the circumference C by the diameter d, we get the same number. We call this number π (pi).

$$\frac{C}{d} = \pi \quad \text{or} \quad C = \pi \cdot d. \quad \text{The number } \pi \text{ is about 3.14, or about } \frac{22}{7}.$$

EXAMPLE 3   Find the circumference of this circle. Use 3.14 for π.

$C = \pi \cdot d$
$ \approx 3.14 \times 6 \text{ cm}$
$ \approx 18.84 \text{ cm}$

6 cm

The circumference is about 18.84 cm.

Do Exercise 3.

Since $d = 2 \cdot r$, where r is the length of a radius, it follows that

$C = \pi \cdot d = \pi \cdot (2 \cdot r).$

$$C = 2 \cdot \pi \cdot r$$

EXAMPLE 4   Find the circumference of this circle. Use $\frac{22}{7}$ for π.

$C = 2 \cdot \pi \cdot r$
$ \approx 2 \cdot \frac{22}{7} \cdot 70 \text{ in.}$
$ \approx 2 \cdot 22 \cdot \frac{70}{7} \text{ in.}$
$ \approx 44 \cdot 10 \text{ in.}$
$ \approx 440 \text{ in.}$

70 in.

The circumference is about 440 in.

EXAMPLE 5   Find the perimeter of this figure. Use 3.14 for π.

We let P = the perimeter. We see that we have half a circle attached to a square. Thus we add half the circumference to the lengths of the three line segments.

$P = 3 \times 9.4 \text{ km} + \frac{1}{2} \times 2 \times \pi \times 4.7 \text{ km}$
$ \approx 28.2 \text{ km} + 3.14 \times 4.7 \text{ km}$
$ \approx 28.2 \text{ km} + 14.758 \text{ km}$
$ \approx 42.958 \text{ km}$

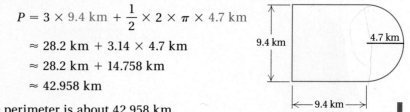
9.4 km   4.7 km   9.4 km

The perimeter is about 42.958 km.

Do Exercises 4 and 5.

C | AREA

Below is a circle of radius *r*.

Think of cutting half the circular region into small pieces and arranging them as shown below.

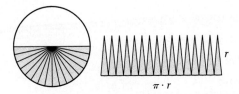

Then imagine cutting the other half of the circular region and arranging the pieces in with the others as shown below.

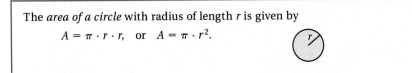

This is almost a parallelogram. The base has length $\frac{1}{2} \cdot 2 \cdot \pi \cdot r$, or $\pi \cdot r$ (half the circumference) and the height is *r*. Thus the area is

$(\pi \cdot r) \cdot r.$

This is the area of a circle.

The *area of a circle* with radius of length *r* is given by

$A = \pi \cdot r \cdot r, \quad \text{or} \quad A = \pi \cdot r^2.$

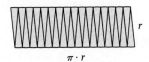

EXAMPLE 6   Find the area of this circle. Use $\frac{22}{7}$ for π.

$A = \pi \cdot r \cdot r$

$\approx \frac{22}{7} \cdot 14 \text{ cm} \cdot 14 \text{ cm}$

$\approx \frac{22}{7} \cdot 196 \text{ cm}^2$

$\approx 616 \text{ cm}^2$

The area is about 616 cm².

Do Exercise 6.

6. Find the area of this circle. Use $\frac{22}{7}$ for π.

5 km

31.41

Answer on page A-5

7. Find the area of this circle. Use 3.14 for π.

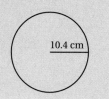

10.4 cm

8. Which is larger and by how much: a 10-ft square flower bed or a 12-ft diameter flower bed?

Caution!

Circumference $= \pi \cdot d = \pi \cdot (r + r) = \pi \cdot (2 \cdot r)$,

Area $= \pi \cdot r^2 = \pi \cdot (r \cdot r)$,

and

$r^2 \neq 2 \cdot r$.

EXAMPLE 7   Find the area of this circle. Use 3.14 for π. Round to the nearest hundredth.

$A = \pi \cdot r \cdot r$

$\approx 3.14 \times 2.1 \text{ m} \times 2.1 \text{ m}$

$\approx 3.14 \times 4.41 \text{ m}^2$

$\approx 13.8474 \text{ m}^2$

$\approx 13.85 \text{ m}^2$

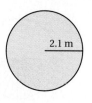

2.1 m

The area is about 13.85 m^2.

Do Exercise 7.

d   SOLVING PROBLEMS

EXAMPLE 8   Which makes a larger pizza and by how much: a 16-in. square pizza pan or a 16-in. diameter circular pizza pan?

First, we draw a picture of each.

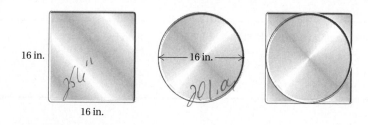

16 in.

16 in.

16 in.

Then we compute areas.

The area of the square is

$A = s \cdot s$

$= 16 \text{ in.} \times 16 \text{ in.} = 256 \text{ in}^2$.

The diameter of the circle is 16 in., so the radius is 16 in./2, or 8 in. The area of the circle is

$A = \pi \cdot r \cdot r$

$\approx 3.14 \times 8 \text{ in.} \times 8 \text{ in.} \approx 200.96 \text{ in}^2$.

We see that the square pizza pan is larger by about

$256 \text{ in}^2 - 200.96 \text{ in}^2$,   or 55.04 in^2.

Thus the square pan makes the larger pizza, by about 55.04 in^2.

Do Exercise 8.

Exercise Set 9.6

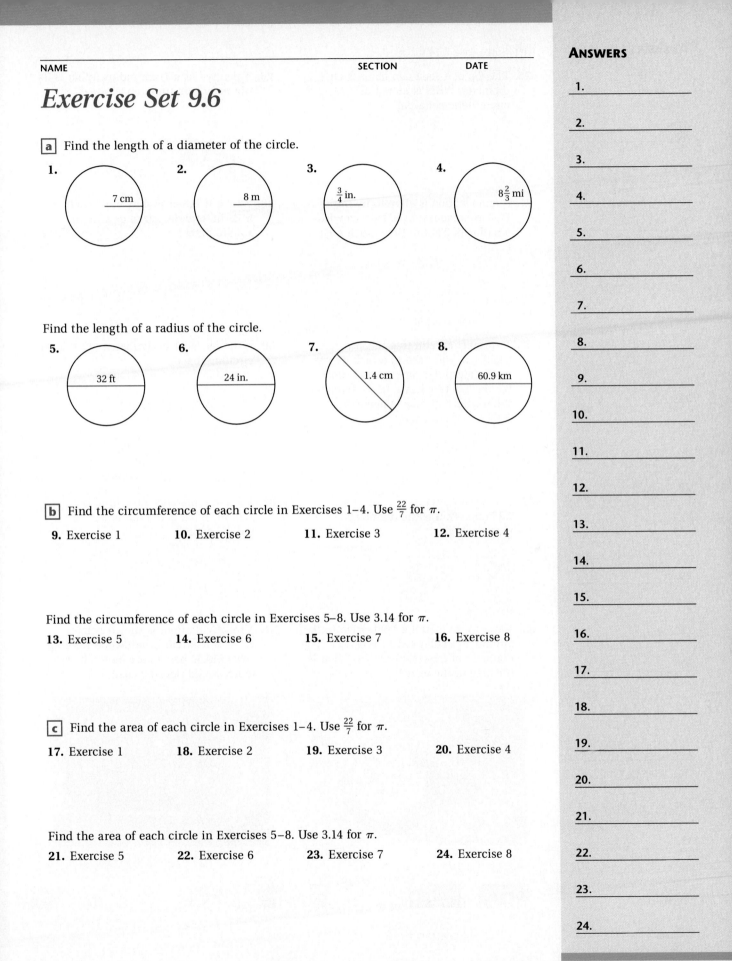

a   Find the length of a diameter of the circle.

1.   7 cm

2.   8 m

3. $\frac{3}{4}$ in.

4. $8\frac{2}{3}$ mi

Find the length of a radius of the circle.

5.   32 ft

6.   24 in.

7.   1.4 cm

8.   60.9 km

b   Find the circumference of each circle in Exercises 1–4. Use $\frac{22}{7}$ for π.

9. Exercise 1   **10.** Exercise 2   **11.** Exercise 3   **12.** Exercise 4

Find the circumference of each circle in Exercises 5–8. Use 3.14 for π.

13. Exercise 5   **14.** Exercise 6   **15.** Exercise 7   **16.** Exercise 8

c   Find the area of each circle in Exercises 1–4. Use $\frac{22}{7}$ for π.

17. Exercise 1   **18.** Exercise 2   **19.** Exercise 3   **20.** Exercise 4

Find the area of each circle in Exercises 5–8. Use 3.14 for π.

21. Exercise 5   **22.** Exercise 6   **23.** Exercise 7   **24.** Exercise 8

ANSWERS

1. _____
2. _____
3. _____
4. _____
5. _____
6. _____
7. _____
8. _____
9. _____
10. _____
11. _____
12. _____
13. _____
14. _____
15. _____
16. _____
17. _____
18. _____
19. _____
20. _____
21. _____
22. _____
23. _____
24. _____

d Solve. Use 3.14 for π.

25. The top of a soda can has a 6-cm diameter. What is its radius? circumference? area?

26. A penny has a 1-cm radius. What is its diameter? circumference? area?

27. A radio station is allowed by the FCC to broadcast over an area with a radius of 220 mi. How much area is this?

28. Which is larger and by how much: a 12-in. circular pizza or a 12-in. square pizza?

29. To protect an elm tree in your backyard, you need to attach gypsy moth caterpillar tape around the trunk. The tree has a 1.1-ft diameter. What length of tape is needed?

30. A silo has a 10-m diameter. What is its circumference?

31. The circumference of a quarter is 7.85 cm. What is the diameter? radius? area?

32. The circumference of a dime is 2.23 in. What is the diameter? radius? area?

33. You want to install a 1-yd–wide walk around a circular swimming pool. The diameter of the pool is 20 yd. What is the area of the walk?

34. A roller rink floor is shown below. What is its area? If hardwood flooring costs $10.50 per square meter, how much would flooring cost?

Find the perimeter. Use 3.14 for π.

35.

8 ft

8 ft

$8 + \frac{3}{2} \cdot \pi \cdot 8$

$8 + 1.5(3.14)(8)$

45.48

36.

4 cm   4 cm

4 cm

37.

4 yd

38.

8 in.   8 in.   8 in.   8 in.

39.

10 yd

10 yd

40.

12.8 cm

10.2 cm

Find the area of the shaded region. Use 3.14 for π.

41.

8 m

$A = r^2 \cdot \pi$

$A = \pi \cdot 4^2$

50.24

100.53

$A = \pi \cdot 8^2$

201.04

100.53

42.

10 yd

10 yd

ANSWERS

35. _____

36. _____

37. _____

38. _____

39. _____

40. _____

41. _____

42. _____

43.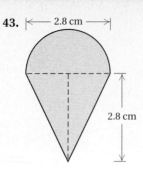

2.8 cm

2.8 cm

44.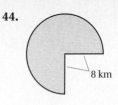

8 km

45.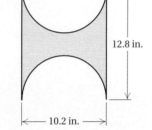

12.8 in.

10.2 in.

46.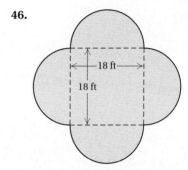

18 ft

18 ft

SKILL MAINTENANCE

Convert to percent notation.

47. 0.875   **48.** 0.58   **49.** $0.\overline{6}$ **50.** 0.4361

51. $\dfrac{3}{8}$ **52.** $\dfrac{5}{8}$ **53.** $\dfrac{2}{3}$ **54.** $\dfrac{1}{5}$

SYNTHESIS

55. ▦ $\pi \approx \dfrac{3927}{1250}$ is another approximation for π. Find decimal notation using your calculator. Round to the nearest thousandth.

56. ▦ The distance from Kansas City to Indianapolis is 500 mi. A car was driven this distance using tires with a radius of 14 in. How many revolutions of each tire occurred on the trip? Use $\dfrac{22}{7}$ for π.

57. Tennis balls are usually packed vertically three in a can, one on top of another. Suppose the diameter of a tennis ball is d. Find the height of the stack of balls. Find the circumference of one ball. Which is greater? Explain.

9.7 *Square Roots and the Pythagorean Theorem*

a SQUARE ROOTS

If a number is a product of two identical factors, then either factor is called a *square root* of the number. (If $a = c^2$, then c is a square root of a.) The symbol $\sqrt{}$ (called a *radical* sign) is used in naming square roots.

For example, $\sqrt{36}$ is the square root of 36. It follows that

$$\sqrt{36} = \sqrt{6 \cdot 6} = 6 \qquad \text{The square root of 36 is 6.}$$

because $6^2 = 36$.

EXAMPLE 1   Simplify: $\sqrt{25}$.

$$\sqrt{25} = \sqrt{5 \cdot 5} = 5 \qquad \text{The square root of 25 is 5 because } 5^2 = 25.$$

EXAMPLE 2   Simplify: $\sqrt{144}$.

$$\sqrt{144} = \sqrt{12 \cdot 12} = 12 \qquad \text{The square root of 144 is 12 because } 12^2 = 144.$$

Caution!   It is common to confuse squares and square roots. A number squared is that number multiplied by itself. For example, $16^2 = 16 \cdot 16 = 256$. A square root of a number is a number that when multiplied by itself gives the original number. For example, $\sqrt{16} = 4$, because $4 \cdot 4 = 16$.

EXAMPLES   Simplify.

3. $\sqrt{4} = 2$ **4.** $\sqrt{256} = 16$ **5.** $\sqrt{361} = 19$

Do Exercises 1–24.

b APPROXIMATING SQUARE ROOTS

Square roots of some numbers are not whole numbers or ordinary fractions. For example,

$$\sqrt{2}, \qquad \sqrt{3}, \qquad \sqrt{39}, \quad \text{and} \quad \sqrt{70}$$

are not whole numbers or ordinary fractions. We can approximate these square roots. For example, consider the following decimal approximations for $\sqrt{2}$. Each gives a closer approximation.

$$\sqrt{2} \approx 1.4 \qquad \text{because} \quad (1.4)^2 = 1.96,$$
$$\sqrt{2} \approx 1.41 \qquad \text{because} \quad (1.41)^2 = 1.9881$$
$$\sqrt{2} \approx 1.414 \qquad \text{because} \quad (1.414)^2 = 1.999396$$
$$\sqrt{2} \approx 1.4142 \qquad \text{because} \quad (1.4142)^2 = 1.99996164.$$

How do we find such approximations? The most common way is to use a calculator, but we can also use a table such as Table 1 at the back of the book.

OBJECTIVES

After finishing Section 9.7, you should be able to:

a   Simplify square roots of squares such as
$$\sqrt{25}.$$

b   Approximate square roots.

c   Given the lengths of any two sides of a right triangle, find the length of the third side.

FOR EXTRA HELP

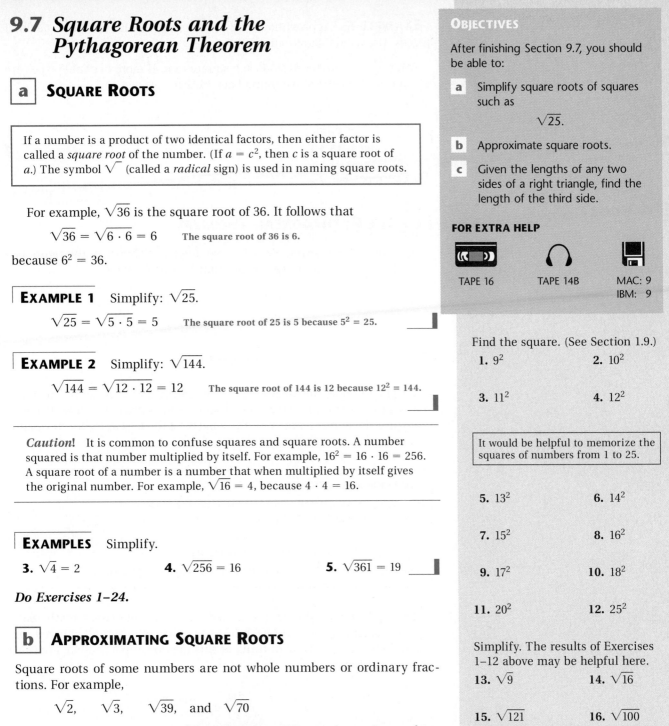

TAPE 16   TAPE 14B   MAC: 9
                               IBM: 9

Find the square. (See Section 1.9.)

1. 9^2 **2.** 10^2

3. 11^2 **4.** 12^2

It would be helpful to memorize the squares of numbers from 1 to 25.

5. 13^2 **6.** 14^2

7. 15^2 **8.** 16^2

9. 17^2 **10.** 18^2

11. 20^2 **12.** 25^2

Simplify. The results of Exercises 1–12 above may be helpful here.

13. $\sqrt{9}$ **14.** $\sqrt{16}$

15. $\sqrt{121}$ **16.** $\sqrt{100}$

17. $\sqrt{81}$ **18.** $\sqrt{64}$

19. $\sqrt{324}$ **20.** $\sqrt{400}$

21. $\sqrt{225}$ **22.** $\sqrt{169}$

23. $\sqrt{1}$ **24.** $\sqrt{0}$

Answers on page A-5

25. $\sqrt{5}$

EXAMPLE 6 Approximate $\sqrt{3}$, $\sqrt{27}$, and $\sqrt{180}$ to three decimal places. Use a calculator or Table 1.

We use a calculator to find each square root. If more decimal places are given than we ask for, we round back to three places.

$$\sqrt{3} \approx 1.732,$$
$$\sqrt{27} \approx 5.196,$$
$$\sqrt{180} \approx 13.416$$

Do Exercises 25–27.

26. $\sqrt{78}$

c THE PYTHAGOREAN THEOREM

A **right triangle** is a triangle with a 90° angle, as shown in the figure below. The small square in the corner indicates a 90° angle.

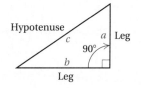

In a right triangle, the longest side is called the **hypotenuse**. It is also the side opposite the right angle. The other two sides are called **legs**. We generally use the letters a and b for the lengths of the legs and c for the length of the hypotenuse. They are related as follows.

THE PYTHAGOREAN THEOREM

In any right triangle, if a and b are the lengths of the legs and c is the length of the hypotenuse, then

$$a^2 + b^2 = c^2, \quad \text{or}$$
$$(\text{Leg})^2 + (\text{Other leg})^2 = (\text{Hypotenuse})^2.$$

The equation $a^2 + b^2 = c^2$ is called the *Pythagorean equation*.

---

27. $\sqrt{168}$

The Pythagorean theorem is named for the ancient Greek mathematician Pythagoras (569?–500? B.C.). It is uncertain who actually proved this result the first time. We can think of this relationship as adding areas, as illustrated below:

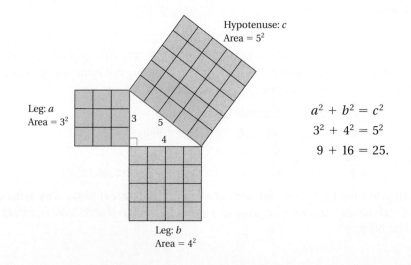

$$a^2 + b^2 = c^2$$
$$3^2 + 4^2 = 5^2$$
$$9 + 16 = 25.$$

Answers on page A-5

If we know the lengths of any two sides of a right triangle, we can find the length of the third side.

EXAMPLE 7 Find the length of the hypotenuse of this right triangle. Give an exact answer and an approximation to three decimal places.

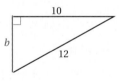

We substitute in the Pythagorean equation:

$$a^2 + b^2 = c^2$$
$$4^2 + 7^2 = c^2 \qquad \text{Substituting}$$
$$16 + 49 = c^2$$
$$65 = c^2.$$

The solution of this equation is the square root of c. We approximate the square root using a calculator or Table 1.

Exact answer: $\qquad c = \sqrt{65}$
Approximate answer: $\quad c \approx 8.062$

Do Exercise 28.

EXAMPLE 8 Find the length of the leg of this right triangle. Give an exact answer and an approximation to three decimal places.

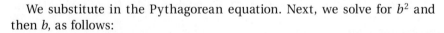

We substitute in the Pythagorean equation. Next, we solve for b^2 and then b, as follows:

$$a^2 + b^2 = c^2$$
$$10^2 + b^2 = 12^2 \qquad \text{Substituting}$$
$$100 + b^2 = 144$$
$$100 + b^2 - 100 = 144 - 100$$
$$b^2 = 144 - 100$$
$$b^2 = 44$$

Exact answer: $\quad b = \sqrt{44}$
Approximation: $\quad b \approx 6.633.$ $\qquad$ Using a calculator or Table 1

Do Exercises 29–31.

Note, in Example 8, that the approximate result is reasonable, because $\sqrt{36} = 6$ and $\sqrt{49} = 7$, so it would seem that $\sqrt{44}$ is somewhere between 6 and 7.

28. Find the length of the hypotenuse of this right triangle. Give an exact answer and an approximation to three decimal places.

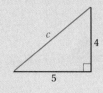

Find the length of the leg of the right triangle. Give an exact answer and an approximation to three decimal places.

29.

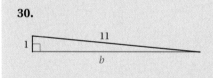

30.

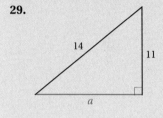

31.

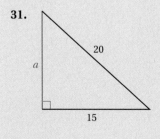

Answers on page A-5

32. How long is a guy wire reaching from the top of an 18-ft pole to a point on the ground 10 ft from the pole? Give an exact answer and an approximation to three decimal places.

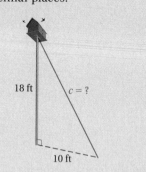

18 ft $c = ?$

10 ft

EXAMPLE 9 A 12-ft ladder leans against a building. The bottom of the ladder is 7 ft from the building. How high is the top of the ladder? Give an exact answer and an approximation to three decimal places.

1. Familiarize. We first make a drawing. In it we see a right triangle. We let $h =$ the unknown height.

2. Translate. We substitute 7 for a, h for b, and 12 for c in the Pythagorean equation:

$$a^2 + b^2 = c^2 \quad \text{Pythagorean equation}$$
$$7^2 + h^2 = 12^2.$$

3. Solve. We solve for h^2 and then h:

$$49 + h^2 = 144$$
$$49 + h^2 - 49 = 144 - 49$$
$$h^2 = 144 - 49$$
$$h^2 = 95$$

Exact answer: $h = \sqrt{95}$

Approximation: $h \approx 9.747$ ft.

4. Check. $7^2 + (\sqrt{95})^2 = 49 + 95 = 144 = 12^2.$

5. State. The top of the ladder is $\sqrt{95}$, or about 9.747 ft from the ground.

Do Exercise 32.

Answer on page A-5

Exercise Set 9.7

a Simplify.

1. $\sqrt{100}$

2. $\sqrt{25}$

3. $\sqrt{441}$

4. $\sqrt{225}$

5. $\sqrt{625}$

6. $\sqrt{576}$

7. $\sqrt{361}$

8. $\sqrt{484}$

9. $\sqrt{529}$

10. $\sqrt{169}$

11. $\sqrt{10,000}$

12. $\sqrt{4,000,000}$

b Approximate to three decimal places.

13. $\sqrt{48}$

14. $\sqrt{17}$

15. $\sqrt{8}$

16. $\sqrt{3}$

17. $\sqrt{18}$

18. $\sqrt{7}$

19. $\sqrt{6}$

20. $\sqrt{61}$

ANSWERS

1. _____

2. _____

3. _____

4. _____

5. _____

6. _____

7. _____

8. _____

9. _____

10. _____

11. _____

12. _____

13. _____

14. _____

15. _____

16. _____

17. _____

18. _____

19. _____

20. _____

21. $\sqrt{10}$ **22.** $\sqrt{21}$ **23.** $\sqrt{75}$ **24.** $\sqrt{220}$

25. $\sqrt{196}$ **26.** $\sqrt{123}$ **27.** $\sqrt{183}$ **28.** $\sqrt{300}$

21. _____

22. _____

23. _____

24. _____

25. _____

26. _____

27. _____

28. _____

29. _____

30. _____

31. _____

32. _____

33. _____

34. _____

c Find the length of the third side of the right triangle. Give an exact answer and an approximation to three decimal places.

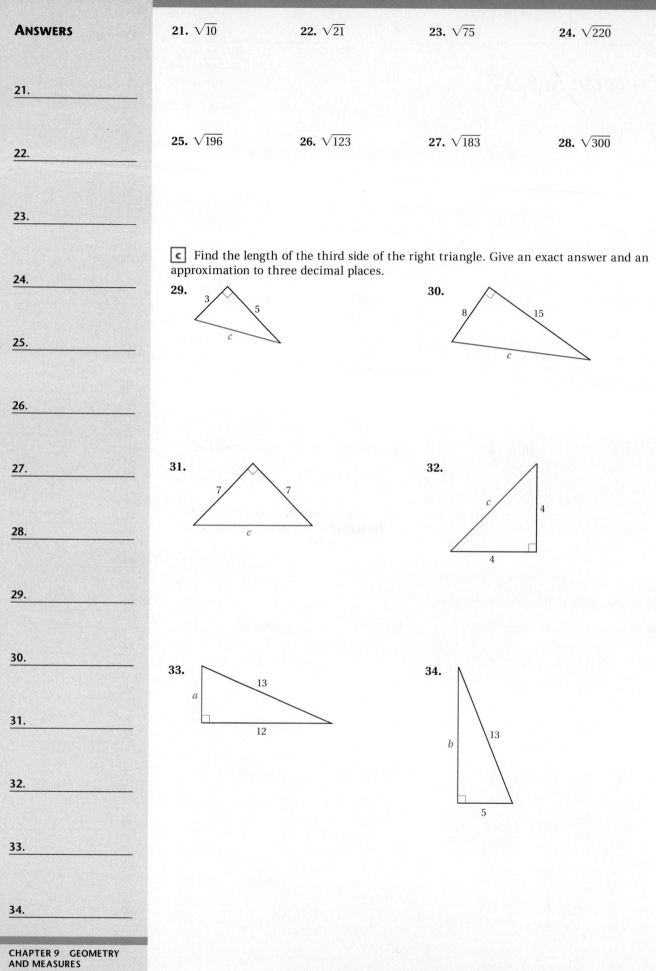

29.

3
5
c

30.

8
15
c

31.

7   7
c

32.

c
4
4

33.

13
a
12

34.

13
b
5

35.

10

b

6

36.

a

8

14

In a right triangle, find the length of the side not given. Give an exact answer and an approximation to three decimal places.

37. $a = 5$, $b = 12$

38. $a = 10$, $b = 24$

39. $a = 18$, $c = 30$

40. $a = 9$, $c = 15$

41. $b = 1$, $c = 20$

42. $a = 1$, $c = 32$

43. $a = 1$, $c = 15$

44. $a = 3$, $b = 4$

45. How long must a wire be to reach from the top of a 13-m telephone pole to a point on the ground 9 m from the base of the pole?

46. How long is a light cord reaching from the top of a 12-ft pole to a point 8 ft from the pole?

47. A slow-pitch softball diamond is actually a square 65 ft on a side. How far is it from home plate to second base?

48. A baseball diamond is actually a square 90 ft on a side. How far is it from home plate to second base?

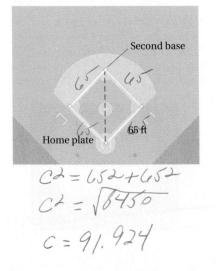

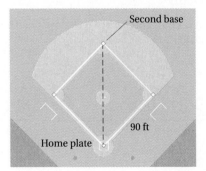

$C^2 = 65^2 + 65^2$

$C^2 = \sqrt{8450}$

$C = 91.924$

ANSWERS

35. _____

36. _____

37. _____

38. _____

39. _____

40. _____

41. _____

42. _____

43. _____

44. _____

45. _____

46. _____

47. _____

48. _____

49. How tall is this tree?

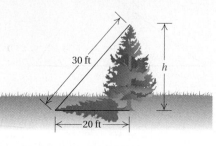

30 ft

h

20 ft

50. How far is the base of the fence post from point *A*?

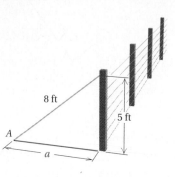

8 ft

5 ft

A

a

49. _____

50. _____

51. _____

52. _____

53. _____

54. _____

55. _____

56. _____

57. _____

58. _____

59. _____

60. _____

51. An airplane is flying at an altitude of 4100 ft. The slanted distance directly to the airport is 15,100 ft. How far is the airplane horizontally from the airport?

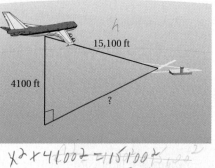

15,100 ft

4100 ft

?

$x^2 + 4100^2 = 15100^2$

$x^2 + 14810000 = 228010000$

$x^2 = 211,200,000 = 14532.7$

52. A surveyor had poles located at points *P*, *Q*, and *R* around a lake. The distances that the surveyor was able to measure are marked on the drawing. What is the approximate distance from *P* to *R* across the lake?

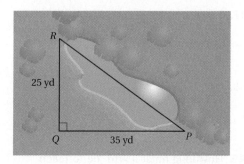

R

25 yd

Q   35 yd   *P*

SKILL MAINTENANCE

Convert to decimal notation.

53. 45.6%   **54.** 16.34%   **55.** 123%

56. 99%   **57.** 0.41%   **58.** 3%

SYNTHESIS

59. Which of the triangles below has the larger area?

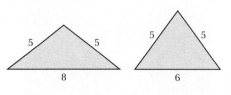

5   5

8

5   5

6

60. A 19-in. television set has a rectangular screen whose diagonal is 19 in. The ratio of length to width in any television set is 4 to 3. Find the length and the width of the screen.

19 in.

CALCULATOR CONNECTION

1. On certain calculators, there is a key with the Greek letter pi (π) printed on it. If such a key is on your calculator, to how many places does this key give the value of π?

If you have a $\boxed{\pi}$ key on your calculator, you can handle most computations without stopping to round the value of π. Rounding, if necessary, is done at the end. Find each of the following to three decimal places.

2. Find the circumference of a circle with a radius of 225.68 in.

3. Find the area of the circle in Exercise 2.

4. Find the area of a circle with a diameter of $46\frac{12}{13}$ in.

5. Find the area of a large irrigation circle with a diameter of 400 ft.

6. The area of a circular dart board is 254.469 in². Find the diameter, the radius, and the circumference.

7. The circumference of the face of a clock is $38\frac{1}{4}$ in. Find the radius, the diameter, and the area.

EXTENDED SYNTHESIS EXERCISES

1. According to specifications passed by the Federal Housing Authority (FHA), a room in a house should have a total window area of at least 10% of the total floor area of the room. A particular family room is 16 ft wide and 22 ft long.

   a) What should the minimum window area of the room be in order to meet FHA standards?

   b) Windows will be installed in the family room, all of which have dimensions 32 in. by 42 in. How many such windows will be needed to meet the minimum FHA standards?

2. A student finds the area of a circle using the formula $A = \pi r^2$, but substitutes the diameter instead of the radius. By what number and what operation could the wrong answer be adjusted in order to get the correct answer?

3. Mattie is training to run the mile. The first week of her training, she accomplishes the run in 6 min, 58 sec. In the second week, she improves her time by 12 sec. In the third week, she improves the preceding time by 10 sec, and in the fourth week, she improves her preceding time by 8 sec. If this pattern of improvement continues, what will her times be in the fifth, sixth, and seventh weeks? Make a line graph of Mattie's

times for the mile. Look for a pattern. Will Mattie eventually "bottom out" to a best time that she cannot improve? If so, when?

4. Typically, fingernails grow at the rate of 2 in. per year. How many inches do they grow in one month? Abdul usually trims his nails after they grow $\frac{1}{16}$ in. How many times will he have to trim his nails in one month? one year?

5. A square is cut in half so that the perimeter of the resulting rectangle is 30 ft. Find the area of the original square.

Find the area of the shaded region. Give the answer in square feet.

6.

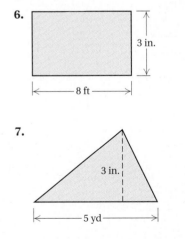

3 in.

8 ft

7.

3 in.

5 yd

(continued)

8.

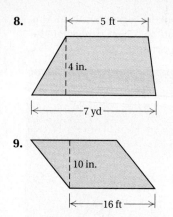

9.

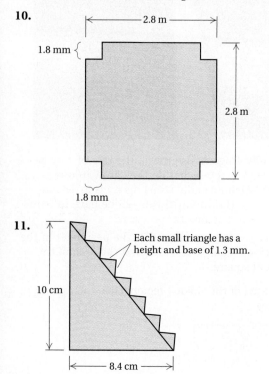

Find the area of the shaded region.

10.

11.

Each small triangle has a
height and base of 1.3 mm.

12. A factory is tooling a large sheet of metal for a machine. It is to be a 12-ft by 12-ft rectangle with 20 squares punched out, each of which is 3 in. on a side. Find the area of the resulting sheet. What percent of the original sheet of metal remains?

13. A 30-ft by 60-ft dance floor is to be turned into Jackson Eight's Dance Studio by placing an 18-ft by 42-ft dance floor in the middle and carpeting the rest of the room. The new dance floor is laid in pieces that are squares 8 in. by 8 in. How many such tiles are needed? What percent of the floor is the dance area?

EXERCISES FOR THINKING AND WRITING

1. List as many reasons as you can for using the metric system exclusively.

2. List as many reasons as you can for continuing our use of the American system.

3. Napoleon is credited with influencing the use of the metric system. Research this possibility and make a report.

Summary and Review: Chapter 9

IMPORTANT PROPERTIES AND FORMULAS

American Units of Length:	12 in. = 1 ft; 3 ft = 1 yd; 36 in. = 1 yd; 5280 ft = 1 mi
Metric Units of Length:	1 km = 1000 m; 1 hm = 100 m; 1 dam = 10 m;
	1 dm = 0.1 m; 1 cm = 0.01 m; 1 mm = 0.001 m
American–Metric Conversion:	1 m = 39.37 in.; 1 m = 3.3 ft; 2.54 cm = 1 in.;
	1 km = 0.621 mi; 1.609 km = 1 mi
Perimeter of a Rectangle:	$P = 2 \cdot (l + w)$, or $P = 2 \cdot l + 2 \cdot w$
Perimeter of a Square:	$P = 4 \cdot s$
Area of a Rectangle:	$A = l \cdot w$
Area of a Square:	$A = s \cdot s$, or $A = s^2$
Area of a Parallelogram:	$A = b \cdot h$
Area of a Triangle:	$A = \dfrac{1}{2} b \cdot h$
Area of a Trapezoid:	$A = \dfrac{1}{2} h \cdot (a + b)$
Radius and Diameter of a Circle:	$d = 2 \cdot r$, or $r = \dfrac{d}{2}$
Circumference of a Circle:	$C = \pi \cdot d$, or $C = 2 \cdot \pi \cdot r$
Area of a Circle:	$A = \pi \cdot r \cdot r$, or $A = \pi \cdot r^2$
Pythagorean Equation:	$a^2 + b^2 = c^2$

Review Exercises

The review objectives to be tested in addition to the material in this chapter are [7.1b, c] and [7.2a, b].

Complete.

1. 8 ft = _2.6_ yd

2. $\dfrac{5}{6}$ yd = _30_ in.

3. 0.3 mm = _.03_ cm

4. 4 m = _4000_ km   .004 ✓

5. 2 yd = _72_ in.

6. 4 km = _400000_ cm

7. 14 in. = _1.16_ ft

8. 15 cm = _.15_ m

9. 200 m = _218.72_ yd ✓ 20

10. 20 mi = _12.42_ km   32.18

Find the perimeter.

11. 23m

5 m, 3 m, 7 m, 4 m, 4 m

12. 4.4

0.5 m, 1.9 m, 0.8 m, 1.2 m

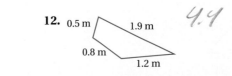

13. The dimensions of a standard-sized tennis court are 78 ft by 36 ft. Find the perimeter and the area of the tennis court.

228 / 2808

14. Find the length of a diagonal from one corner to another of the tennis court in Exercise 13.

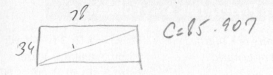

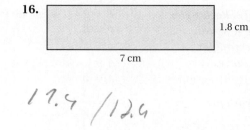

C=85.907

Find the perimeter and the area.

15.

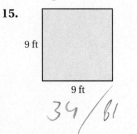

9 ft

9 ft

34 / 81

16.

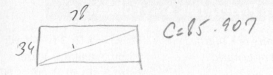

1.8 cm

7 cm

17.4 / 12.4

Find the area.

17.

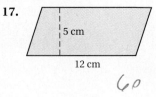

5 cm

12 cm

60

18.

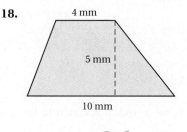

4 mm

5 mm

10 mm

35

19.

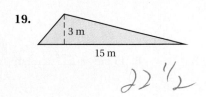

3 m

15 m

22 ½

20.

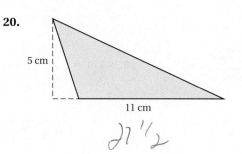

5 cm

11 cm

22 ½

21.

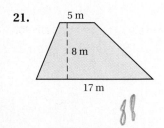

5 m

8 m

17 m

88

22.

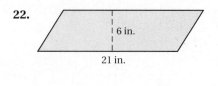

6 in.

21 in.

124

23. A grassy area is to be seeded around three sides of a building and has equal width on the three sides, as shown at right. What is the seeded area?

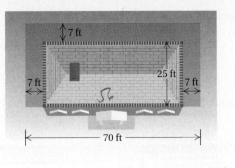

7 ft

25 ft   7 ft

7 ft

70 ft

$224r - 1400 = 640$

Find the length of a radius of the circle.

24.

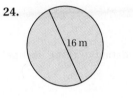

16 m

8

25.

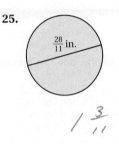

$\frac{28}{11}$ in.

$1\frac{3}{11}$

Find the length of a diameter of the circle.

26.

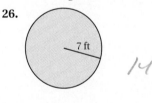

7 ft

14

27.

10 cm

20

28. Find the circumference of the circle in Exercise 24. Use 3.14 for π.

450.24

29. Find the circumference of the circle in Exercise 25. Use $\frac{22}{7}$ for π.

8

30. Find the area of the circle in Exercise 24. Use 3.14 for π.

200.94

31. Find the area of the circle in Exercise 25. Use $\frac{22}{7}$ for π.

32. Find the area of the shaded region. Use 3.14 for π.

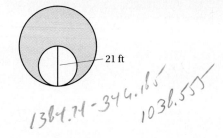

— 21 ft

1384.74 - 344.165
1036.555

33. Simplify: $\sqrt{64}$.

8

34. Approximate to three decimal places: $\sqrt{83}$.

9.110

In a right triangle, find the length of the side not given. Give an exact answer and an approximation to three decimal places.

35. $a = 15$, $b = 25$

$\sqrt{850}$ *25   29.10* $\sqrt{850}$

36. $a = 7$, $c = 10$

7.141

37.

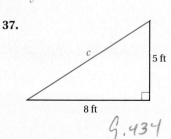

c 5 ft   8 ft

9.434

38.

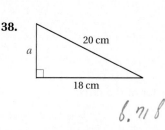

a 20 cm   18 cm

6.718

39. How long is a wire reaching from the top of a 24-ft pole to a point 16 ft from the pole?

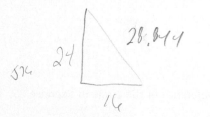

$\sqrt{24}$ 24   16   *28.844*

40. How tall is this tree?

360ρ -
1600

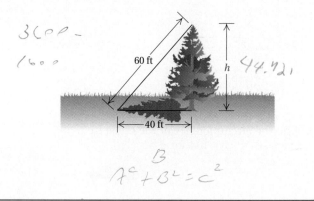

60 ft h *44.721*   40 ft

$a^2 + b^2 = c^2$

SKILL MAINTENANCE

41. Convert to percent notation: 0.47.   *47%*

42. Convert to percent notation: $\frac{23}{25}$.   *92%*

43. Convert to decimal notation: 56.7%.   *.567*

44. Convert to fractional notation: 73%.   *$\frac{73}{100}$*

Test: Chapter 9

Complete.

1. 4 ft = _____ in.

2. 4 in. = _____ ft

3. 6 km = _____ m

4. 8.7 mm = _____ cm

5. 200 yd = _____ m

6. 2400 km = _____ mi

Find the perimeter and the area.

7.

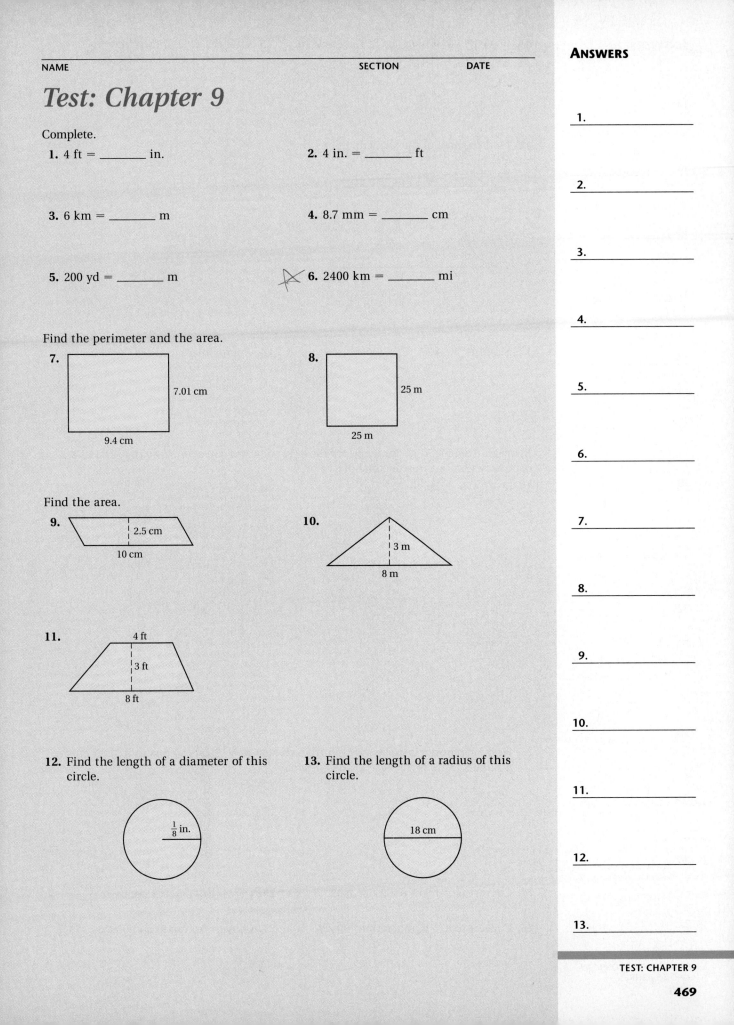

7.01 cm

9.4 cm

8.

25 m

25 m

Find the area.

9.

2.5 cm

10 cm

10.

3 m

8 m

11.

4 ft

3 ft

8 ft

12. Find the length of a diameter of this circle.

$\frac{1}{8}$ in.

13. Find the length of a radius of this circle.

18 cm

ANSWERS

1. _____

2. _____

3. _____

4. _____

5. _____

6. _____

7. _____

8. _____

9. _____

10. _____

11. _____

12. _____

13. _____

14. Find the circumference of the circle in Question 12. Use $\frac{22}{7}$ for π.

15. Find the area of the circle in Question 13. Use 3.14 for π.

16. Find the area of the shaded region.

18.6 km

9.0 km

17. Simplify: $\sqrt{225}$.

18. Approximate to three decimal places: $\sqrt{87}$.

In a right triangle, find the length of the side not given. Give an exact answer and an approximation to three decimal places.

19. $a = 24$, $b = 32$

20. $a = 2$, $c = 8$

21.

c 1

1

22.

7   10

b

23. How long must a wire be to reach from the top of a 13-m antenna to a point on the ground 9 m from the base of the antenna?

SKILL MAINTENANCE

24. Convert to percent notation: 0.93.

25. Convert to percent notation: $\frac{13}{16}$.

26. Convert to decimal notation: 93.2%.

27. Convert to fractional notation: $33\frac{1}{3}\%$.

Cumulative Review: Chapters 1–9

Perform the indicated operation and simplify.

1. $46{,}231 \times 1100$

2. $\dfrac{1}{10} \cdot \dfrac{5}{6}$

3. $14.5 + \dfrac{4}{5} - 0.1$

4. $2\dfrac{3}{5} \div 3\dfrac{9}{10}$

5. $0.1\overline{)3.56}$

6. $3\dfrac{1}{2} - 2\dfrac{2}{3}$

7. Determine whether 1,298,032 is divisible by 8.

8. Determine whether 5,024,120 is divisible by 3.

9. Find the prime factorization of 99.

10. Find the LCM of 35 and 49.

11. Round $35.\overline{7}$ to the nearest tenth.

12. Write a word name for 103.064.

13. Find the average and the median of this set of numbers:

   9, 13, 17, 18, 21, 29.

Find percent notation.

14. 0.08

15. $\dfrac{3}{5}$

16. Simplify: $\sqrt{121}$.

17. Approximate to two decimal places: $\sqrt{29}$.

18. Complete: 2 yd = _____ ft.

19. Find the perimeter.

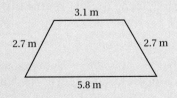

20. Find the area.

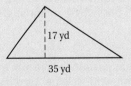

Solve.

21. $0.07 \cdot x = 10.535$

22. $x + 12{,}843 = 32{,}091$

23. $\dfrac{2}{3} \cdot y = 5$

24. $\dfrac{4}{5} + y = \dfrac{6}{7}$

This table shows typical sleep requirements in childhood.

AGE	HOURS OF DAYTIME SLEEP	HOURS OF NIGHTTIME SLEEP
1 week	8.0	8.5
1 month	7.0	8.5
3 months	5.5	9.5
6 months	3.3	11.0
9 months	2.5	11.5
12 months	2.3	11.5
18 months	2.0	11.5
2 years	1.5	11.5

25. How many hours of daytime sleep does a 3-month-old child need?

26. How many total hours of sleep does a 1-year-old child need?

27. How many more hours will a 6-month-old child sleep at night than a 1-week-old child?

28. How many hours would you expect a 2-month-old child to sleep at night?

Solve.

29. A microwave oven marked $220 was discounted to a sale price of $194. What was the rate of discount?

30. There are 11 million milk cows in America, each producing, on the average, 15,000 lb of milk per year. How many pounds of milk are produced each year in America?

31. The Schwartz family has a rectangular kitchen table measuring 52 in. by 30 in. They replace it with a circular table with a 48-in. diameter. How much bigger is their new table? Use 3.14 for π.

32. A person on a diet loses $3\frac{1}{2}$ lb in 2 weeks. At this rate, how many pounds will he lose in 5 weeks?

33. The U.S. Department of Agriculture requires that 80% of the seeds that a company produces must sprout. To find out about the quality of the seeds it has produced, a company takes 500 seeds and plants them. It finds that 417 of the seeds sprout. Did the seeds pass government standards?

34. A mechanic spent $\frac{1}{3}$ hr changing a car's oil, $\frac{1}{2}$ hr rotating the tires, $\frac{1}{10}$ hr changing the air filter, $\frac{1}{4}$ hr adjusting the idle speed, and $\frac{1}{15}$ hr checking the brake and transmission fluids. How many hours did the mechanic spend working on the car?

35. A driver bought gasoline when the odometer read 86,897.2. At the next gasoline purchase, the odometer read 87,153.0. How many miles had been driven?

36. A family has an annual income of $26,400. Of this, $\frac{1}{4}$ is spent for food. How much does the family spend for food?

SYNTHESIS

37. A homeowner is having a one-story addition built on an existing house. The addition measures 25 ft by 32 ft. The existing house measures 30 ft by 32 ft and has two stories. What is the percent increase in living area provided by the addition?

10

More on Measures

INTRODUCTION

In this chapter, we consider several other measures: volume, capacity, weight, mass, time, temperature, and units of area. Any of these topics can be omitted without loss of continuity to the remaining two chapters of the text.

• • • • • • • • • • • • • • • • • • • •

AN APPLICATION

How "big" is one million dollars? This photo shows one million one-dollar bills assembled by the Bureau of Engraving. The width of a dollar is 2.3125 in., the length is 6.0625 in., and the thickness is 0.0041 in. Find the volume occupied by one million one-dollar bills.

THE MATHEMATICS

To find the volume of a single one-dollar bill, we multiply the length times the width and then times the height, or thickness:

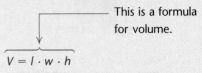

This is a formula for volume.

$V = l \cdot w \cdot h$

$= 6.0625 \times 2.3125 \times 0.0041.$

Then we multiply the result by one million.

3. Find the volume of the sphere. Use $\frac{22}{7}$ for π.

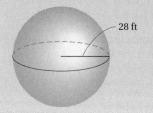
28 ft

4. The radius of a standard-sized golf ball is 2.1 cm. Find its volume. Use 3.14 for π.

5. Find the volume of this cone. Use 3.14 for π.

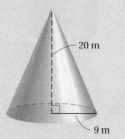
20 m

9 m

6. Find the volume of this cone. Use $\frac{22}{7}$ for π.

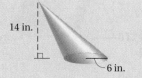
14 in.

6 in.

Answers on page A-5

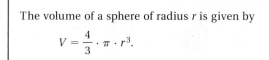

We find the volume of a sphere as follows.

> The volume of a sphere of radius r is given by
> $$V = \frac{4}{3} \cdot \pi \cdot r^3.$$

EXAMPLE 2   The radius of a standard-sized bowling ball is 4.2915 in. Find the volume of a standard-sized bowling ball. Round to the nearest hundredth of a cubic inch. Use 3.14 for π.

$$V = \frac{4}{3} \cdot \pi \cdot r^3 \approx \frac{4}{3} \times 3.14 \times (4.2915 \text{ in.})^3$$

$$\approx \frac{4 \times 3.14 \times 79.0364 \text{ in}^3}{3} \approx 330.90 \text{ in}^3$$

Do Exercises 3 and 4.

c | CONES

Consider a circle in a plane and choose any point not in the plane. The circular region, together with the set of all segments connecting P to a point on the circle, is called a **circular cone.**

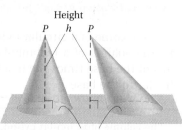
Height
P h P
Base

We find the volume of a cone as follows.

> The volume of a circular cone with base radius r is one-third the product of the base area and the height:
> $$V = \frac{1}{3} \cdot B \cdot h = \frac{1}{3} \pi \cdot r^2 \cdot h.$$

EXAMPLE 3   Find the volume of this circular cone. Use $\frac{22}{7}$ for π.

$$V = \frac{1}{3}\pi \cdot r^2 \cdot h$$
$$\approx \frac{1}{3} \times \frac{22}{7} \times 3 \text{ cm} \times 3 \text{ cm} \times 7 \text{ cm}$$
$$\approx 66 \text{ cm}^3$$

7 cm

3 cm

Do Exercises 5 and 6.

ANSWERS

Exercise Set 10.2

a Find the volume of the circular cylinder. Use 3.14 for π in Exercises 1–4. Use $\frac{22}{7}$ for π in Exercises 5 and 6.

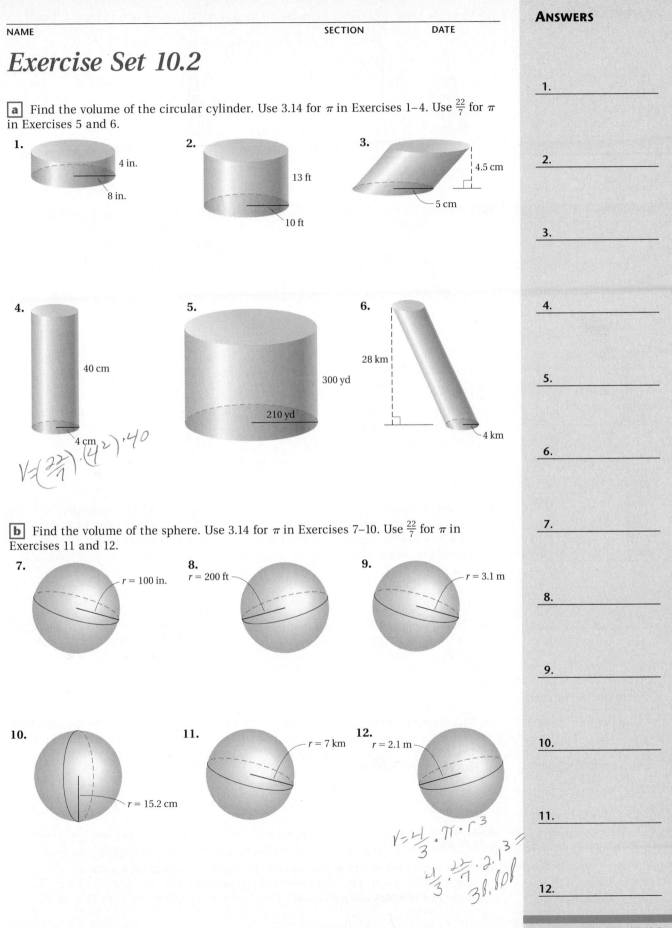

1.

4 in.

8 in.

2.

13 ft

10 ft

3.

4.5 cm

5 cm

4.

40 cm

4 cm

$V = \left(\frac{22}{7}\right) \cdot (4^2) \cdot 40$

5.

300 yd

210 yd

6.

28 km

4 km

b Find the volume of the sphere. Use 3.14 for π in Exercises 7–10. Use $\frac{22}{7}$ for π in Exercises 11 and 12.

7.

$r = 100$ in.

8.

$r = 200$ ft

9.

$r = 3.1$ m

10.

$r = 15.2$ cm

11.

$r = 7$ km

12.

$r = 2.1$ m

$V = \frac{4}{3} \cdot \pi \cdot r^3$

$\frac{4}{3} \cdot \frac{22}{7} \cdot 2.1^3 =$

38.808

1. _____

2. _____

3. _____

4. _____

5. _____

6. _____

7. _____

8. _____

9. _____

10. _____

11. _____

12. _____

13. _____

14. _____

15. _____

16. _____

17. _____

18. _____

19. _____

20. _____

21. _____

22. _____

23. _____

24. _____

25. _____

c Find the volume of the circular cone. Use 3.14 for π in Exercises 13 and 14. Use $\frac{22}{7}$ for π in Exercises 15 and 16.

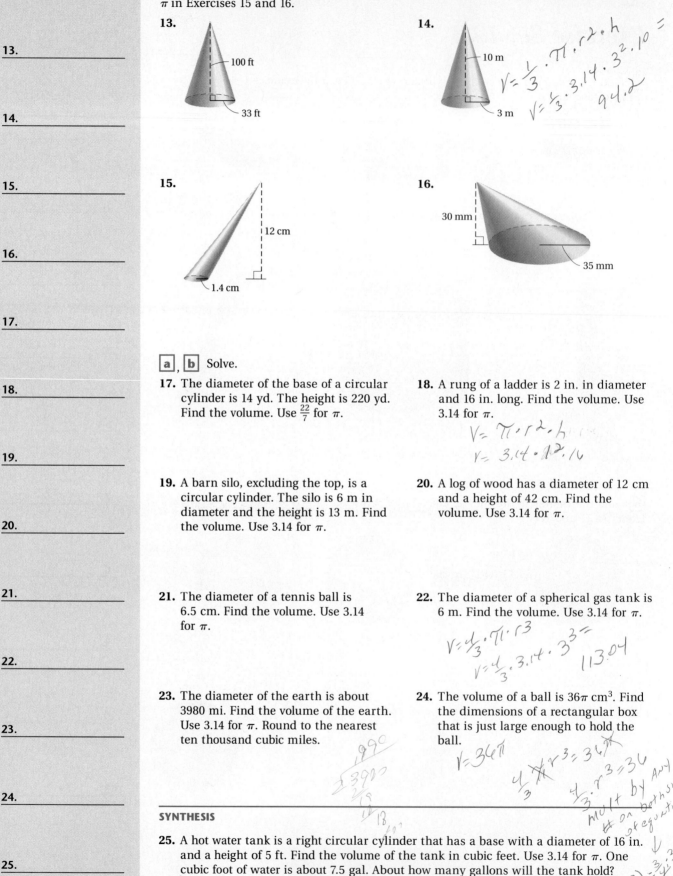

13.

100 ft

33 ft

14.

10 m

3 m

$V = \frac{1}{3} \cdot \pi \cdot r^2 \cdot h$

$V = \frac{1}{3} \cdot 3.14 \cdot 3^2 \cdot 10 =$

94.2

15.

12 cm

1.4 cm

16.

30 mm

35 mm

a , b Solve.

17. The diameter of the base of a circular cylinder is 14 yd. The height is 220 yd. Find the volume. Use $\frac{22}{7}$ for π.

18. A rung of a ladder is 2 in. in diameter and 16 in. long. Find the volume. Use 3.14 for π.

$V = \pi \cdot r^2 \cdot h$

$V = 3.14 \cdot 1^2 \cdot 16$

19. A barn silo, excluding the top, is a circular cylinder. The silo is 6 m in diameter and the height is 13 m. Find the volume. Use 3.14 for π.

20. A log of wood has a diameter of 12 cm and a height of 42 cm. Find the volume. Use 3.14 for π.

21. The diameter of a tennis ball is 6.5 cm. Find the volume. Use 3.14 for π.

22. The diameter of a spherical gas tank is 6 m. Find the volume. Use 3.14 for π.

$V = \frac{4}{3} \cdot \pi \cdot r^3$

$V = \frac{4}{3} \cdot 3.14 \cdot 3^3 = 113.04$

23. The diameter of the earth is about 3980 mi. Find the volume of the earth. Use 3.14 for π. Round to the nearest ten thousand cubic miles.

24. The volume of a ball is 36π cm^3. Find the dimensions of a rectangular box that is just large enough to hold the ball.

$V = 36\pi$

$\frac{4}{3} \cdot \pi \cdot r^3 = 36\pi$

$\frac{4}{3} \cdot r^3 = 36$

Mult by $\frac{3}{4}$ on both sides of equation

$r^3 = 27$

$r = 3$

SYNTHESIS

25. A hot water tank is a right circular cylinder that has a base with a diameter of 16 in. and a height of 5 ft. Find the volume of the tank in cubic feet. Use 3.14 for π. One cubic foot of water is about 7.5 gal. About how many gallons will the tank hold?

10.3 *Weight, Mass, and Time*

a ┃ WEIGHT: THE AMERICAN SYSTEM

The American units of weight are as follows.

> **AMERICAN UNITS OF WEIGHT**
>
> 1 ton (T) = 2000 pounds (lb)
>
> 1 lb = 16 ounces (oz)

EXAMPLE 1 A well-known hamburger is called a "quarter-pounder." Find its name in ounces: a "_____ ouncer."

$$\frac{1}{4} \text{ lb} = \frac{1}{4} \cdot 1 \text{ lb}$$

$$= \frac{1}{4} \cdot 16 \text{ oz} \qquad \text{Substituting 16 oz for 1 lb}$$

$$= 4 \text{ oz}$$

A "quarter-pounder" can also be called a "four-ouncer."

EXAMPLE 2 Complete: 15,360 lb = _____ T.

$$15{,}360 \text{ lb} = 15{,}360 \text{ lb} \times \frac{1 \text{ T}}{2000 \text{ lb}} \qquad \text{Multiplying by 1}$$

$$= \frac{15{,}360}{2000} \text{ T}$$

$$= 7.68 \text{ T}$$

Do Exercises 1–3.

b ┃ MASS: THE METRIC SYSTEM

There is a difference between **mass** and **weight**, but the terms are often used interchangeably. People sometimes use the word "weight" instead of "mass." Weight is related to the force of gravity. The farther you are from the center of the earth, the less you weigh. Your mass stays the same no matter where you are.

The basic unit of mass is the **gram** (g), which is the mass of 1 cubic centimeter (1 cm^3 or 1 mL) of water. Since a cubic centimeter is small, a gram is a small unit of mass.

$$1 \text{ g} = 1 \text{ gram} = \text{the mass of 1 cm}^3 \text{ (1 mL) of water}$$

OBJECTIVES

After finishing Section 10.3, you should be able to:

a Convert from one American unit of weight to another.

b Convert from one metric unit of mass to another.

c Convert from one unit of time to another.

FOR EXTRA HELP

| TAPE 17 | TAPE 15B | MAC: 10
IBM: 10 |

Complete.

1. 5 lb = _____ oz

2. 8640 lb = _____ T

3. 1 T = _____ oz

Answers on page A-5

The following table shows the metric units of mass. The prefixes are the same as those for length.

METRIC UNITS OF MASS

1 metric ton (t) = 1000 kilograms (kg)

1 *kilo*gram (kg) = 1000 grams (g)

1 *hecto*gram (hg) = 100 grams (g)

1 *deka*gram (dag) = 10 grams (g)

1 gram (g)

1 *deci*gram (dg) = $\frac{1}{10}$ gram (g)

1 *centi*gram (cg) = $\frac{1}{100}$ gram (g)

1 *milli*gram (mg) = $\frac{1}{1000}$ gram (g)

THINKING METRIC

One gram is about the mass of 1 raisin or 1 paperclip. Since 1 metric ton is 1000 kg and 1 kg is about 2.2 lb, it follows that 1 metric ton (t) is about 2200 lb, which is just a little more than 1 American ton (T).

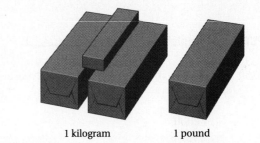

| 1 gram | 1 kilogram | 1 pound |

Small masses, such as dosages of medicine and vitamins, may be mea-sured in milligrams (mg). Grams (g) are used for objects ordinarily given in ounces, such as the mass of a letter, a piece of candy, a coin, or a small package of food.

Each 2.5 mg   15 g   2 g

Kilograms (kg) are used for larger food packages, such as meat, or for human body mass.

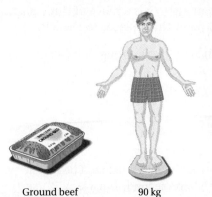

Ground beef
2 lb (0.9 kg)

90 kg

The metric ton (t) is used for very large masses, such as the mass of an automobile, a truckload of gravel, or an airliner.

Do Exercises 4–8.

CHANGING UNITS MENTALLY

As before, changing from one metric unit to another amounts to only the movement of a decimal point. We use this table.

1000	100	10	1	0.1	0.01	0.001
kg	hg	dag	g	dg	cg	mg

| **EXAMPLE 3**   Complete: 8 kg = _____ g.

Think: To go from kg to g in the table is a move of three places to the right. Thus we move the decimal point three places to the right.

   8.0   8.000.   8 kg = 8000 g

| **EXAMPLE 4**   Complete: 4235 g = _____ kg.

Think: To go from g to kg in the table is a move of three places to the left. Thus we move the decimal point three places to the left.

   4235.0   4.235.0   4235 g = 4.235 kg

Do Exercises 9 and 10.

| **EXAMPLE 5**   Complete: 6.98 cg = _____ mg.

Think: To go from cg to mg is a move of one place to the right. Thus we move the decimal point one place to the right.

   6.98   6.9.8   6.98 cg = 69.8 mg

The most commonly used metric units of mass are kg, g, cg, and mg. We have purposely used those more often than the others in the exercises.

Complete with mg, g, kg, or t.

4. A laptop computer has a mass of 6 _____.

5. That person has a body mass of 85.4 _____.

6. This is a 3-_____ vitamin.

7. A pen has a mass of 12 _____.

8. A minivan has a mass of 3 _____.

Complete.

9. 6.2 kg = _____ g

10. 304.8 cg = _____ g

Answers on page A-5

Complete.

11. 7.7 cg = _____ mg

12. 2344 mg = _____ cg

13. 67 dg = _____ mg

Complete.

14. 2 hr = _____ min

15. 4 yr = _____ days

16. 1 day = _____ min

17. 168 hr = _____ wk

Answers on page A-5

EXAMPLE 6   Complete: 89.21 mg = _____ g.

Think: To go from mg to g is a move of three places to the left. Thus we move the decimal point three places to the left.

$$89.21 \qquad 0.089.21 \qquad 89.21 \text{ mg} = 0.08921 \text{ g}$$

Do Exercises 11–13.

c   TIME

A table of units of time is shown below. The metric system sometimes uses "h" for hour and "s" for second, but we will use the more familiar "hr" and "sec."

UNITS OF TIME

1 day = 24 hours (hr)	1 year (yr) = $365\frac{1}{4}$ days
1 hr = 60 minutes (min)	1 week (wk) = 7 days
1 min = 60 seconds (sec)	

Since we cannot have $\frac{1}{4}$ day on the calendar, we give each year 365 days and every fourth year 366 days (a leap year), unless it is a year at the beginning of a century not divisible by 400.

EXAMPLE 7   Complete: 1 hr = _____ sec.

$$\begin{aligned} 1 \text{ hr} &= 60 \text{ min} \\ &= 60 \cdot 1 \text{ min} \\ &= 60 \cdot 60 \text{ sec} \qquad \text{Substituting 60 sec for 1 min} \\ &= 3600 \text{ sec} \end{aligned}$$

EXAMPLE 8   Complete: 5 yr = _____ days.

$$\begin{aligned} 5 \text{ yr} &= 5 \cdot 1 \text{ yr} \\ &= 5 \cdot 365\frac{1}{4} \text{ days} \qquad \text{Substituting } 365\frac{1}{4} \text{ days for 1 yr} \\ &= 1826\frac{1}{4} \text{ days} \end{aligned}$$

EXAMPLE 9   Complete: 4320 min = _____ days.

$$\begin{aligned} 4320 \text{ min} &= 4320 \text{ min} \cdot \frac{1 \text{ hr}}{60 \text{ min}} \cdot \frac{1 \text{ day}}{24 \text{ hr}} \\ &= \frac{4320}{60 \cdot 24} \text{ days} \\ &= 3 \text{ days} \end{aligned}$$

Do Exercises 14–17.

Exercise Set 10.3

ANSWERS

1. _____

2. _____

3. _____

4. _____

5. _____

6. _____

7. _____

8. _____

9. _____

10. _____

11. _____

12. _____

13. _____

14. _____

15. _____

16. _____

17. _____

18. _____

19. _____

20. _____

21. _____

22. _____

23. _____

24. _____

25. _____

26. _____

27. _____

a Complete.

1. 1 T = _2000_ lb

2. 1 lb = _____ oz

3. 6000 lb = _3_ T

4. 8 T = _16000_ lb

5. 4 lb = _____ oz

6. 10 lb = _160_ oz

7. 6.32 T = _____ lb

8. 8.07 T = _____ lb

9. 3200 oz = _0.1_ T

10. 6400 oz = _0.2_ T

$$\frac{6400}{16 \cdot 2000}$$

11. 80 oz = _5_ lb

12. 960 oz = _____ lb

b Complete.

13. 1 kg = _____ g

14. 1 hg = _____ g

15. 1 dag = _____ g

16. 1 dg = _____ g

17. 1 cg = _____ g

18. 1 mg = _____ g

19. 1 g = _1000_ mg

20. 1 g = _____ cg

21. 1 g = _____ dg

22. 25 kg = _25,000_ g

23. 234 kg = _234000_ g

24. 9403 g = _____ kg

25. 5200 g = _____ kg

26. 1.506 kg = _____ g

27. 67 hg = _____ kg

28. _____

29. _____

30. _____

31. _____

32. _____

33. _____

34. _____

35. _____

36. _____

37. _____

38. _____

39. _____

40. _____

41. _____

42. _____

43. _____

44. _____

45. _____

46. _____

47. _____

48. _____

49. _____

50. _____

28. 45 cg = _____ g

29. 0.502 dg = _____ g

30. 0.0025 cg = _____ mg

31. 8492 g = _____ kg

32. 9466 g = _9.464_ kg

33. 585 mg = _58.5_ cg

34. 96.1 mg = _____ cg

35. 8 kg = _800,000_ cg

36. 0.06 kg = _____ mg

37. 1 t = _1000_ kg

38. 2 t = _____ kg

39. 3.4 cg = _.0034_ dag

40. 115 mg = _____ g

c Complete.

41. 1 day = _____ hr

42. 1 hr = _____ min

43. 1 min = _____ sec

44. 1 wk = _____ days

45. 1 yr = _____ days

46. 2 yr = _____ days

47. 180 sec = _____ hr

48. 60 sec = _1/60 or .016_ hr

49. 492 sec = _____ min
(the amount of time it takes for the rays of the sun to reach the earth)

50. 18,000 sec = _____ hr

51. 156 hr = _____ days

52. 444 hr = _____ days

53. 645 min = _____ hr

54. 375 min = 6.25 hr

55. 2 wk = _____ hr

56. 4 hr = _____ sec

57. 756 hr = 4.5 wk

58. 166,320 min = _____ wk

59. 2922 wk = _____ yr

60. 623 days = _____ wk

SKILL MAINTENANCE

Evaluate.

61. 2^4

62. 17^2

63. 5^3

64. 8^2

Complete.

65. 5.43 m = _____ cm

66. 5.43 m = _____ km

SYNTHESIS

Complete. Use 1 kg = 2.205 lb and 453.5 g = 1 lb. Round to four decimal places.

67. 🖩 1 lb = _____ kg

68. 🖩 1 g = _____ lb

69. At $0.90 a dozen, the cost of eggs is $0.60 per pound. How much does an egg weigh?

70. Estimate the number of years in one million seconds.

71. Estimate the number of years in one billion seconds.

72. Estimate the number of years in one trillion seconds.

51. _____

52. _____

53. _____

54. _____

55. _____

56. _____

57. _____

58. _____

59. _____

60. _____

61. _____

62. _____

63. _____

64. _____

65. _____

66. _____

67. _____

68. _____

69. _____

70. _____

71. _____

72. _____

Medical applications. Another metric unit used in medicine is the microgram (μg). It is defined as follows.

$$1 \text{ microgram} = 1 \ \mu\text{g} = \frac{1}{1,000,000} \text{ g}; \quad 1,000,000 \ \mu\text{g} = 1 \text{ g}$$

Thus a microgram is one millionth of a gram, and one million micrograms is one gram.

Complete.

73. 1 mg = _____ μg

74. 1 μg = _____ mg

75. A physician orders 125 μg of digoxin. How many milligrams is the prescription?

76. A physician orders 0.25 mg of reserpine. How many micrograms is the prescription?

77. A medicine called sulfisoxazole usually comes in tablets that are 500 mg each. A standard dosage is 2 g. How many tablets would have to be taken in order to achieve this dosage?

78. Quinidine is a liquid mixture, part medicine and part water. There is 80 mg of Quinidine for every milliliter of liquid. A standard dosage is 200 mg. How much of the liquid mixture would be required in order to achieve the dosage?

79. A medicine called cephalexin is obtainable in a liquid mixture, part medicine and part water. There is 250 mg of cephalexin in 5 mL of liquid. A standard dosage is 400 mg. How much of the liquid would be required in order to achieve the dosage?

80. A medicine called Albuterol is used for the treatment of asthma. It typically comes in an inhaler that contains 18 g. One actuation, or spray, is 90 mg.

a) How many actuations are in one inhaler?

b) A student is going away for 4 months of college and wants to take enough Albuterol to last for that time. Assuming that the student will need 4 actuations per day, estimate about how many inhalers the student will need for the 4-month period.

10.4 *Temperature*

a | ESTIMATED CONVERSIONS

Below are two temperature scales: **Fahrenheit** for American measure and **Celsius** for metric measure.

By laying a ruler or a piece of paper horizontally between the scales, we can make an approximate conversion from one measure of temperature to another.

| EXAMPLES | Convert to Celsius. Approximate to the nearest ten degrees.

1. 212°F (Boiling point of water)   100°C   This is exact.
2. 32°F (Freezing point of water)   0°C   This is exact.

Do Exercises 1–3.

| EXAMPLES | Make an approximate conversion to Fahrenheit.

3. 44°C (Hot bath)   110°F   This is approximate.
4. 20°C (Room temperature)   68°F   This is approximate.

Do Exercises 4–6.

OBJECTIVES

After finishing Section 10.4, you should be able to:

a   Make an approximate conversion from Celsius temperature to Fahrenheit, and from Fahrenheit temperature to Celsius.

b   Convert from Celsius temperature to Fahrenheit and from Fahrenheit temperature to Celsius using the formulas $F = \frac{9}{5} \cdot C + 32$ and $C = \frac{5}{9} \cdot (F - 32)$.

FOR EXTRA HELP

TAPE 17   TAPE 15B   MAC: 10
                         IBM: 10

Convert to Celsius. Approximate to the nearest ten degrees.

1. 180°F (Brewing coffee)

2. 25°F (Cold day)

3. −10°F (Miserably cold day)

Convert to Fahrenheit. Approximate to the nearest ten degrees.

4. 25°C (Warm day at the beach)

5. 40°C (Temperature of a patient with a high fever)

6. 10°C (A cold bath)

Answers on page A-5

17.

18.

19.

20.

21.

22.

23.

24.

25.

26.

27.

28.

29.

30.

31.

32.

33.

34.

35.

b Convert to Fahrenheit. Use the formula $F = \dfrac{9}{5} \cdot C + 32$.

17. 25°C   **18.** 85°C   **19.** 40°C   **20.** 90°C

21. 3000°C
(melting point of iron)

22. 1000°C
(melting point of gold)

Convert to Celsius. Use the formula $C = \dfrac{5}{9} \cdot (F - 32)$.

23. 86°F   **24.** 59°F   **25.** 131°F   **26.** 140°F

27. 98.6°F
(normal body temperature)

28. 104°F
(high-fevered body temperature)

SKILL MAINTENANCE

Complete.

29. 23.4 cm = _____ mm

30. 0.23 km = _____ m

31. 28 ft = _____ in.

32. 72 ft = _____ yd

33. 72.4 m = _____ cm

34. 72.4 m = _____ km

SYNTHESIS

35. Another temperature scale often used is the **Kelvin** scale. Conversions from Celsius to Kelvin can be carried out using the formula

$K = C + 273$.

A chemistry textbook describes an experiment in which a reaction takes place at a temperature of 400° Kelvin. A student wishes to perform the experiment, but has only a Fahrenheit thermometer. At what Fahrenheit temperature will the reaction take place?

10.5 *Converting Units of Area*

a | AMERICAN UNITS

Let's do some conversions from one American unit of area to another.

EXAMPLE 1 Complete: $1 \text{ ft}^2 = $ _____ in^2.

$$1 \text{ ft}^2 = 1 \cdot (12 \text{ in.})^2 \quad \text{Substituting 12 in. for 1 ft}$$
$$= 12 \text{ in.} \cdot 12 \text{ in.}$$
$$= 144 \text{ in}^2$$

EXAMPLE 2 Complete: $8 \text{ yd}^2 = $ _____ ft^2.

$$8 \text{ yd}^2 = 8 \cdot (3 \text{ ft})^2 \quad \text{Substituting 3 ft for 1 yd}$$
$$= 8 \cdot 3 \text{ ft} \cdot 3 \text{ ft}$$
$$= 8 \cdot 3 \cdot 3 \cdot \text{ft} \cdot \text{ft}$$
$$= 72 \text{ ft}^2$$

Do Exercises 1–3.

American units are related as follows.

1 square yard (yd^2) = 9 square feet (ft^2)
1 square foot (ft^2) = 144 square inches (in^2)
1 square mile (mi^2) = 640 acres
1 acre = 43,560 ft^2

EXAMPLE 3 Complete: $36 \text{ ft}^2 = $ _____ yd^2.

We are converting from "ft^2" to "yd^2". Thus we choose a symbol for 1 with yd^2 on top and ft^2 on the bottom.

$$36 \text{ ft}^2 = 36 \text{ ft}^2 \times \frac{1 \text{ yd}^2}{9 \text{ ft}^2} \quad \text{Multiplying by 1 using } \frac{1 \text{ yd}^2}{9 \text{ ft}^2}$$
$$= \frac{36}{9} \times \frac{\text{ft}^2}{\text{ft}^2} \times 1 \text{ yd}^2$$
$$= 4 \text{ yd}^2$$

EXAMPLE 4 Complete: $7 \text{ mi}^2 = $ _____ acres.

$$7 \text{ mi}^2 = 7 \cdot 1 \text{ mi}^2$$
$$= 7 \cdot 640 \text{ acres} \quad \text{Substituting 640 acres for 1 mi}^2$$
$$= 4480 \text{ acres}$$

Do Exercises 4 and 5.

OBJECTIVES

After finishing Section 10.5, you should be able to:

a Convert from one American unit of area to another.

b Convert from one metric unit of area to another.

FOR EXTRA HELP

TAPE 17   TAPE 16A   MAC: 10
IBM: 10

Complete.
1. $1 \text{ yd}^2 = $ _____ ft^2

2. $5 \text{ yd}^2 = $ _____ ft^2

3. $20 \text{ ft}^2 = $ _____ in^2

Complete.
4. $360 \text{ in}^2 = $ _____ ft^2

5. $5 \text{ mi}^2 = $ _____ acres

Answers on page A-5

Complete.

6. $1 \text{ m}^2 = $ _____ mm^2

7. $1 \text{ cm}^2 = $ _____ mm^2

Complete.

8. $2.88 \text{ m}^2 = $ _____ cm^2

9. $4.3 \text{ mm}^2 = $ _____ cm^2

10. $678,000 \text{ m}^2 = $ _____ km^2

Answers on page A-5

b   METRIC UNITS

Let us convert from one metric unit of area to another.

EXAMPLE 5   Complete: $1 \text{ km}^2 = $ _____ m^2.

$$1 \text{ km}^2 = 1 \cdot (1000 \text{ m})^2 \qquad \text{Substituting 1000 m for 1 km}$$
$$= 1000 \text{ m} \cdot 1000 \text{ m}$$
$$= 1,000,000 \text{ m}^2$$

EXAMPLE 6   Complete: $1 \text{ m}^2 = $ _____ cm^2.

$$1 \text{ m}^2 = 1 \cdot (100 \text{ cm})^2 \qquad \text{Substituting 100 cm for 1 m}$$
$$= 100 \text{ cm} \cdot 100 \text{ cm}$$
$$= 10,000 \text{ cm}^2$$

Do Exercises 6 and 7.

MENTAL CONVERSION

To convert mentally, we use the table as before and multiply the number of moves by 2 to determine the number of moves of the decimal point.

1000	100	10	1	0.1	0.01	0.001
km	hm	dam	m	dm	cm	mm

EXAMPLE 7   Complete: $3.48 \text{ km}^2 = $ _____ m^2.

Think: To go from km to m in the table is a move of 3 places to the right. So we move the decimal point $2 \cdot 3$, or 6 places to the right.

$$3.48 \qquad 3.480000. \qquad 3.48 \text{ km}^2 = 3,480,000 \text{ m}^2$$

EXAMPLE 8   Complete: $586.78 \text{ cm}^2 = $ _____ m^2.

Think: To go from cm to m in the table is a move of 2 places to the left. So we move the decimal point $2 \cdot 2$, or 4 places to the left.

$$586.78 \qquad 0.0586.78 \qquad 586.78 \text{ cm}^2 = 0.058678 \text{ m}^2$$

Do Exercises 8–10.

Exercise Set 10.5

a Complete.

1. $1 \text{ ft}^2 = $ _____ in^2

2. $1 \text{ yd}^2 = $ _____ ft^2

3. $1 \text{ mi}^2 = $ _____ acres

4. $1 \text{ acre} = $ _____ ft^2

5. $1 \text{ in}^2 = $ _____ ft^2

6. $1 \text{ ft}^2 = $ _____ yd^2

7. $22 \text{ yd}^2 = $ _____ ft^2

8. $40 \text{ ft}^2 = $ _____ in^2

9. $44 \text{ yd}^2 = $ _____ ft^2

10. $144 \text{ ft}^2 = $ _____ yd^2

11. $20 \text{ mi}^2 = $ _____ acres

12. $576 \text{ in}^2 = $ _____ ft^2

13. $1 \text{ mi}^2 = $ _____ ft^2

14. $1 \text{ mi}^2 = $ _____ yd^2

15. $720 \text{ in}^2 = $ _____ ft^2

16. $27 \text{ ft}^2 = $ _____ yd^2

17. $144 \text{ in}^2 = $ _____ ft^2

18. $72 \text{ in}^2 = $ _____ ft^2

19. $1 \text{ acre} = $ _____ mi^2

20. $4 \text{ acres} = $ _____ ft^2

1. _____
2. _____
3. _____
4. _____
5. _____
6. _____
7. _____
8. _____
9. _____
10. _____
11. _____
12. _____
13. _____
14. _____
15. _____
16. _____
17. _____
18. _____
19. _____
20. _____

ANSWERS

21. _____

22. _____

23. _____

24. _____

25. _____

26. _____

27. _____

28. _____

29. _____

30. _____

31. _____

32. _____

33. _____

34. _____

35. _____

36. _____

37. _____

38. _____

39. _____

40. _____

41. _____

b Complete.

21. $5.21 \text{ km}^2 = \underline{\hspace{1cm}} \text{ m}^2$

22. $65 \text{ km}^2 = \underline{\hspace{1cm}} \text{ m}^2$

23. $0.014 \text{ m}^2 = \underline{\hspace{1cm}} \text{ cm}^2$

24. $0.028 \text{ m}^2 = \underline{\hspace{1cm}} \text{ mm}^2$

25. $2345.6 \text{ mm}^2 = \underline{\hspace{1cm}} \text{ cm}^2$

26. $8.38 \text{ cm}^2 = \underline{\hspace{1cm}} \text{ mm}^2$

27. $852.14 \text{ cm}^2 = \underline{\hspace{1cm}} \text{ m}^2$

28. $125 \text{ mm}^2 = \underline{\hspace{1cm}} \text{ m}^2$

29. $250,000 \text{ mm}^2 = \underline{\hspace{1cm}} \text{ cm}^2$

30. $2400 \text{ mm}^2 = \underline{\hspace{1cm}} \text{ cm}^2$

31. $472,800 \text{ m}^2 = \underline{\hspace{1cm}} \text{ km}^2$

32. $1.37 \text{ cm}^2 = \underline{\hspace{1cm}} \text{ mm}^2$

SKILL MAINTENANCE

33. Find the simple interest on $2000 at an interest rate of 8% for 1.5 yr.

34. Find the simple interest on $2000 at an interest rate of 5.3% for 2 yr.

Complete.

35. $22,176 \text{ ft} = \underline{\hspace{1cm}} \text{ mi}$

36. $22,176 \text{ mm} = \underline{\hspace{1cm}} \text{ km}$

SYNTHESIS

Complete.

37. $1 \text{ m}^2 = \underline{\hspace{1cm}} \text{ ft}^2$

38. $1 \text{ in}^2 = \underline{\hspace{1cm}} \text{ cm}^2$

39. $2 \text{ yd}^2 = \underline{\hspace{1cm}} \text{ m}^2$

40. $1 \text{ acre} = \underline{\hspace{1cm}} \text{ m}^2$

41. The president's family has about $20,175 \text{ ft}^2$ of living area in the White House. Estimate the living area in square meters.

CALCULATOR CONNECTION

1. *Measuring the mass of a cloud.* Using a calculator with a $\boxed{\pi}$ key, find the volume of a round cloud with a 1000-m diameter. Clouds consist of liquid water droplets. It is estimated that each cubic meter of a cloud has a mass of 0.6 g. What is the mass, in metric tons, of a cloud with a diameter of 1000 m?

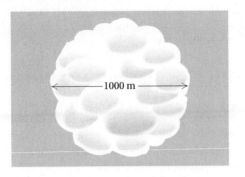

$\longleftarrow$ 1000 m $\longrightarrow$

EXTENDED SYNTHESIS EXERCISES

Find the volume of the solid. Give the answer in cubic feet. (Note that the solids are not drawn in perfect proportion.)

1.
12 ft 2.6 in. 3 in.

2.
14 ft 3 in. 4.6 in.

3.
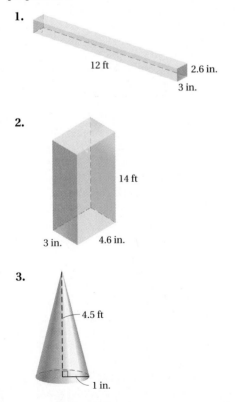
4.5 ft 1 in.

4.
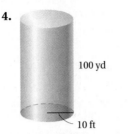
100 yd 10 ft

Find the volume.

5.
18 ft $\frac{3}{4}$ in.

6.
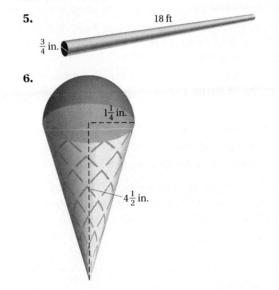
$1\frac{1}{4}$ in. $4\frac{1}{2}$ in.

7. Rafael is constructing concrete forms in order to install a new concrete driveway. The driveway will be a rectangle 20 ft wide and 58 ft long. The forms will allow the concrete to be 4 in. deep. How many cubic feet of concrete will Rafael need to order from the concrete company? When Rafael calls to order the concrete, he discovers that concrete is delivered only in cubic yards. How many cubic yards of concrete will he need?

8. J. C. Payne, a 71-year-old rancher at the time, recently asked the *Guinness Book of World Records* to accept a world record for constructing a ball of string. The ball was 13.2 ft in diameter. What was the volume of the ball? Assuming the diameter of the string was 0.1 in., how long was the string in feet? in miles?

(continued)

9. At the end of a recent year, it was estimated that there is 3,880,151 mi of roadway in the United States.

   a) Suppose that the average width of a road is 50 ft. How many square miles of land is covered by roads?

   b) Find the state in which this land would best fit.

   c) Assuming the pavement on the highways is 6 in. thick, what is the volume of the highways in cubic feet?

10. It is known that 1 gal of water weighs 8.3453 lb. Which weighs more, an ounce of pennies or an ounce (as capacity) of water?

11. A *board foot* is the amount of wood in a piece 12 in. by 12 in. by 1 in. A carpenter places the following order for a certain kind of lumber:

   25 pieces: 2 in. by 4 in. by 8 ft;

   32 pieces: 2 in. by 6 in. by 10 ft;

   24 pieces: 2 in. by 8 in. by 12 ft.

   The price of this type of lumber is $225 per thousand board feet. What is the total cost of the carpenter's order?

12. Lumber that starts out at a certain measure must be trimmed to take out warps and get boards that are straight. Because of trimming, a "two-by-four" is trimmed to an actual size of $1\frac{1}{2}$ in. by $3\frac{1}{2}$ in. What percent of the wood in a 10-ft board is lost by trimming?

13. A community center has a rectangular swimming pool that is 50 ft wide, 100 ft long, and 10 ft deep. The center decides to fill the pool with water to a line that is 1 ft from the top. Water costs $2.25 per 1000 ft^3. How much does it cost to fill the pool the first time?

EXERCISES FOR THINKING AND WRITING

1. Do a report on where the words *Fahrenheit* and *Celsius* originated. How is the word *Centigrade* related to temperature?

2. List and describe all the volume formulas that you have learned in this chapter.

Summary and Review: Chapter 10

pg. 4,5
503

IMPORTANT PROPERTIES AND FORMULAS

Volume of a Rectangular Solid:	$V = l \cdot w \cdot h$
American Units of Capacity:	1 gal = 4 qt; 1 qt = 2 pt; 1 pt = 16 oz; 1 pt = 2 cups; 1 cup = 8 oz
Metric Units of Capacity:	1 L = 1000 mL = 1000 cm^3
Volume of a Circular Cylinder:	$V = \pi \cdot r^2 \cdot h$
Volume of a Sphere:	$V = \frac{4}{3} \cdot \pi \cdot r^3$
Volume of a Cone:	$V = \frac{1}{3} \cdot \pi \cdot r^2 \cdot h$
American System of Weights:	1 T = 2000 lb; 1 lb = 16 oz
Metric System of Mass:	1 t = 1000 kg; 1 kg = 1000 g; 1 hg = 100 g;
	1 dag = 10 g; 1 dg = 0.1 g; 1 cg = 0.01 g; 1 mg = 0.001 g
Units of Time:	1 min = 60 sec; 1 hr = 60 min; 1 day = 24 hr; 1 wk = 7 days;
	1 yr = 365.25 days
Temperature Conversion:	$F = \frac{9}{5} \cdot C + 32;$ $C = \frac{5}{9} \cdot (F - 32)$

Review Exercises

The review objectives to be tested in addition to the material in this chapter are [1.9b], [7.8a], [9.1a], and [9.2a].

Find the volume.

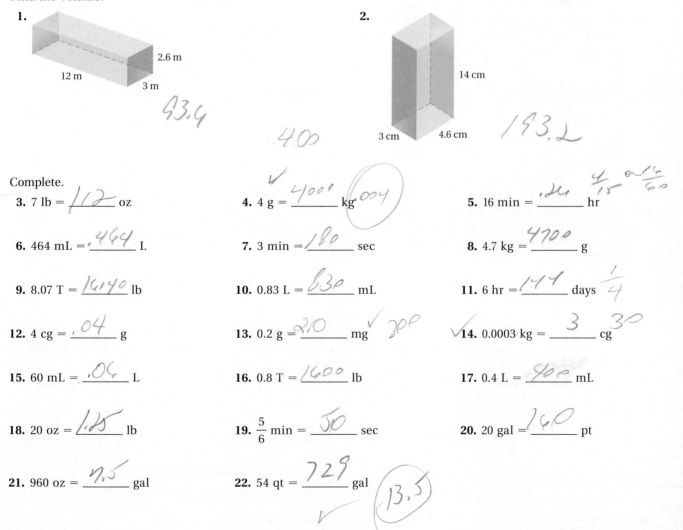

1.   2.6 m   12 m   3 m   *93.4*

2.   14 cm   3 cm   4.6 cm   *193.2*

400

Complete.

3. 7 lb = _*112*_ oz

4. 4 g = _*4000*_ kg   *004*

5. 16 min = _*.26*_ hr   *4/15 or 1/60*

6. 464 mL = _*.464*_ L

7. 3 min = _*180*_ sec

8. 4.7 kg = _*4700*_ g

9. 8.07 T = _*16140*_ lb

10. 0.83 L = _*830*_ mL

11. 6 hr = _*144*_ days   *1/4*

12. 4 cg = _*.04*_ g

13. 0.2 g = _*2,0*_ mg   *200*

14. 0.0003 kg = _*3*_ cg   *30*

15. 60 mL = _*.06*_ L

16. 0.8 T = _*1600*_ lb

17. 0.4 L = _*400*_ mL

18. 20 oz = _*1.25*_ lb

19. $\frac{5}{6}$ min = _*50*_ sec

20. 20 gal = _*160*_ pt

21. 960 oz = _*7,5*_ gal

22. 54 qt = _*729*_ gal   *13.5*

23. Convert 27 °C to Fahrenheit.

80.4   80.$\frac{3}{5}$

24. Convert 68°F to Celsius.

20

Find the volume. Use 3.14 for π.

25.

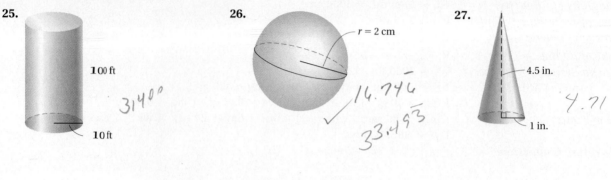

100 ft

10 ft

31400

26.

$r = 2$ cm

14.746

33.493 ✓

27.

4.5 in.

1 in.

4.71

Complete.

28. 4 yd^2 = ___ ft^2 *48   ✓34*

29. 0.3 km^2 = ___ m^2 *300,000*

30. 2070 in^2 = ___ ft^2 *14.375*

31. 600 cm^2 = ___ m^2 *.0600*

Solve.

32. A physician prescribed 780 mL per hour of a certain intravenous fluid for a patient. How many liters of fluid did this patient receive in one day?

18720 ML =

18.72

---

33. Find the simple interest on $5000 at 9.5% for 30 days.

Evaluate.

34. 3^3

35. $(4.7)^2$

36. 4.7^3

37. $\left(\dfrac{1}{2}\right)^4$

Complete.

38. 2.5 mi = ___ ft

39. 144 in. = ___ yd

40. 4568 cm = ___ m

41. 4568 cm = ___ mm

Test: Chapter 10

Find the volume.

1.

10.5 cm

4 cm

2 cm

Complete.

2. 3080 mL = _____ L

3. 0.24 L = _____ mL

4. 4 lb = _____ oz

5. 4.11 T = _____ lb

6. 3.8 kg = _____ g

7. 4.325 mg = _____ cg

8. 2200 mg = _____ g

9. 5 hr = _____ min

10. 15 days = _____ hr

11. 64 pt = _____ qt

12. 10 gal = _____ oz

13. 5 cups = _____ oz

14. Convert 95°F to Celsius.

15. Convert 59°C to Fahrenheit.

ANSWERS

1. _____
2. _____
3. _____
4. _____
5. _____
6. _____
7. _____
8. _____
9. _____
10. _____
11. _____
12. _____
13. _____
14. _____
15. _____

Complete.

16. $12 \text{ ft}^2 = $ _____ in^2

17. $3 \text{ cm}^2 = $ _____ m^2

16. _____

17. _____

18. _____

19. _____

20. _____

21. _____

22. _____

23. _____

24. _____

25. _____

26. _____

27. _____

28. _____

29. _____

30. _____

18. A twelve-box carton of 12-oz juice boxes comes in a rectangular box $10\frac{1}{2}$ in. by 8 in. by 5 in. What is its volume?

Find the volume. Use 3.14 for π.

19.

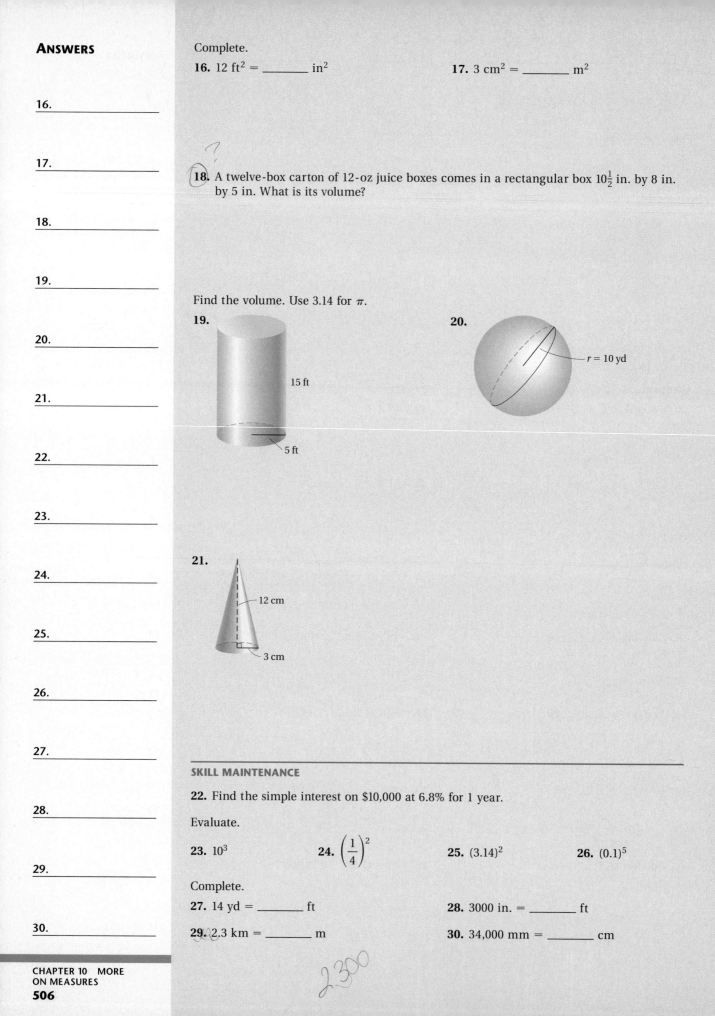

15 ft

5 ft

20.

$r = 10$ yd

21.

12 cm

3 cm

SKILL MAINTENANCE

22. Find the simple interest on $10,000 at 6.8% for 1 year.

Evaluate.

23. 10^3

24. $\left(\frac{1}{4}\right)^2$

25. $(3.14)^2$

26. $(0.1)^5$

Complete.

27. 14 yd = _____ ft

28. 3000 in. = _____ ft

29. 2.3 km = _____ m

30. 34,000 mm = _____ cm

Cumulative Review: Chapters 1–10

1. $1\frac{1}{2} + 2\frac{2}{3}$

2. $\left(\frac{1}{4}\right)^2 \div \left(\frac{1}{2}\right)^3 \times 2^4 + (10.3)(4)$

3. $120.5 - 32.98$

4. $22\overline{)27{,}148}$

5. $14 \div [33 \div 11 + 8 \times 2 - (15 - 3)]$

6. $8^3 + 45 \cdot 24 - 9^2 \div 3$

Find fractional notation.

7. 1.209

8. 17%

Use $<$, $>$, or $=$ for ▓ to write a true sentence.

9. $\frac{5}{6}$ ▓ $\frac{7}{8}$

10. $\frac{15}{18}$ ▓ $\frac{10}{12}$

Complete.

11. 6 oz = _____ lb

12. 15°C = _____ °F

13. 0.087 L = _____ mL

14. 9 sec = _____ min

15. 3 yd^2 = _____ ft^2

16. 17 cm = _____ m

Find the perimeter and the area.

17.

50 cm 80 cm 110 cm

18.

5.3 ft 6.8 ft 6.5 ft 8.1 ft 12.1 ft

The line graph shows the number of cars sold by McQuirk Motors over a 5-month period.

19. How many cars did the company sell in June?

20. In what month did the number of car purchases decrease?

Solve.

21. $\dfrac{12}{15} = \dfrac{x}{18}$

22. $\dfrac{3}{x} = \dfrac{7}{10}$

23. $25 \cdot x = 2835$

24. $x + \dfrac{3}{4} = \dfrac{7}{8}$

25. To get an A in math, a student must score an average of 90 on five tests. On the first four tests, the scores were 85, 92, 79, and 95. What is the lowest score that the student can get on the last test and still get an A?

26. Americans own 52 million dogs, 56 million cats, 45 million birds, 250 million fish, and 125 million other creatures as house pets. How many pets do Americans own?

27. The diameter of a basketball is 20 cm. What is its volume? Use 3.14 for π.

28. What is the simple interest on $800 at 12% for $\frac{1}{4}$ year?

29. How long must a rope be in order to reach from the top of an 8-m tree to a point on the ground 15 m from the bottom of the tree?

30. The sales tax on an office supply purchase of $5.50 is $0.33. What is the sales tax rate?

31. A bolt of fabric in a fabric store has $10\frac{3}{4}$ yd on it. A customer purchases $8\frac{5}{8}$ yd. How many yards remain on the bolt?

32. What is the cost, in dollars, of 15.6 gal of gasoline at 139.9¢ per gallon? Round to the nearest cent.

33. A box of powdered milk that makes 20 qt costs $4.99. A box that makes 8 qt costs $1.99. Which size has the lower unit price?

34. It is $\frac{7}{10}$ km from a student's dormitory to the library. Maria starts to walk there, changes her mind after going $\frac{1}{4}$ of the distance, and returns home. How far did she walk?

SYNTHESIS

35. Your house sits on a lot measuring 75 ft by 200 ft. The lot is at the intersection of two streets, so there are sidewalks on two sides of the lot. In the winter, you have to shovel the snow off the sidewalks. If the sidewalks are 3 ft wide and the snow is 4 in. deep, what volume of snow must you shovel?

11

The Real-Number System

INTRODUCTION

This chapter is the first of two that form an introduction to algebra. We expand on the arithmetic numbers to include new numbers called *real numbers*. We consider both the rational numbers and the irrational numbers, which make up the real numbers. We will learn to add, subtract, multiply, and divide real numbers.

AN APPLICATION

It is 3 seconds before liftoff of the space shuttle. Tell what integer corresponds to this situation.

THE MATHEMATICS

Three seconds before liftoff corresponds to

−3.

↑

This is a negative number.

The review objectives to be tested in addition to the material in this chapter are as follows.

[2.1d]   Find the prime factorization of a composite number.

[3.1a]   Find the LCM of two or more numbers.

[7.3b]   Solve basic percent problems.

[9.4a]   Find the area of a rectangle or square.

Pretest: Chapter 11

Use either < or > for ▨ to write a true sentence.

1. 0 ▨ −5   **2.** 10 ▨ −5   **3.** −35 ▨ −45   **4.** $-\dfrac{2}{3}$ ▨ $\dfrac{4}{5}$

Find decimal notation.

5. $-\dfrac{5}{8}$ **6.** $-\dfrac{2}{3}$ **7.** $-\dfrac{10}{11}$

Find the absolute value.

8. $|-12|$ **9.** $|2.3|$ **10.** $|0|$

Find the opposite, or additive inverse.

11. 5.4   **12.** $-\dfrac{2}{3}$

Compute and simplify.

13. $-9 + (-8)$ **14.** $20.2 - (-18.4)$ **15.** $-\dfrac{5}{6} - \dfrac{3}{10}$ **16.** $-11.5 + 6.5$

17. $-9(-7)$ **18.** $\dfrac{5}{8}\left(-\dfrac{2}{3}\right)$ **19.** $-19.6 \div 0.2$ **20.** $-56 \div (-7)$

21. $12 - (-6) + 14 - 8$ **22.** $20 - 10 \div 5 + 2^3$

11.1 The Real Numbers

EXAMPLE 4

The number
and 3.

EXAMPLE 5

The graph o

EXAMPLE 6

The numbe
way from 1 to

Do Exercises

c | **NOTAT**

Each rational

EXAMPLE 7

We first fin

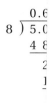

Thus, $\frac{5}{8} = 0.6$

Decimal n
nating decim

In this section, we introduce the *real numbers*. We begin with numbers called *integers* and build up to the real numbers.

We create the integers by starting with the whole numbers, 0, 1, 2, 3, and so on. For each number 1, 2, 3, and so on, we obtain a new number to the left of zero on the number line:

> For the number 1, there will be an *opposite* number −1 (negative 1).
>
> For the number 2, there will be an *opposite* number −2 (negative 2).
>
> For the number 3, there will be an *opposite* number −3 (negative 3), and so on.

The **integers** consist of the whole numbers and these new numbers. We picture them on a number line as follows.

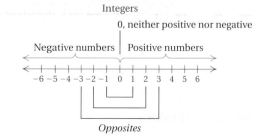

We call the newly obtained negative numbers **negative integers.** The natural numbers, or positive numbers, are called **positive integers.** Zero is neither positive nor negative. We call −1 and 1 opposites of each other. Similarly, −2 and 2 are opposites, −3 and 3 are opposites, −100 and 100 are opposites, and 0 is its own opposite. This gives us the integers, which extend infinitely to the left and right of zero.

> The integers: $\dots, -5, -4, -3, -2, -1, 0, 1, 2, 3, 4, 5, \dots$

a | INTEGERS AND THE REAL WORLD

Integers correspond to many real-world problems and situations. The following examples will help you get ready to translate problem situations to mathematical language.

EXAMPLE 1 Tell which integer corresponds to this situation: The temperature is 3 degrees below zero.

3° below zero is −3°

OBJECTIVES

After finishing Section 11.1, you should be able to:

a Tell which integers correspond to a real-world situation.

b Graph rational numbers on a number line.

c Convert from fractional notation for a rational number to decimal notation.

d Determine which of two real numbers is greater and indicate which, using < or >.

e Find the absolute value of a real number.

FOR EXTRA HELP

TAPE 18   TAPE 16A   MAC: 11A
IBM: 11A

Tell which intege
the situation.

1. The halfback
first down. T
was sacked fo
second down

2. The highest
recorded in t
was 134° in [
July 10, 1913.
temperature
the United S
below zero ir
in January of

3. At 10 sec bef
occurs. At 14
the first stag
the rocket.

4. A student ov
bookstore. T
$289 in a sa

Answers on pag

16. _____

17. _____

18. _____

19. _____

20. _____

21. _____

22. _____

23. _____

24. _____

25. _____

26. _____

27. _____

28. _____

29. _____

30. _____

31. _____

32. _____

Compute and simplify.

16. $3.1 - (-4.7)$

1.4

17. $-8 + 4 + (-7) + 3$

-4

18. $-\frac{1}{5} + \frac{3}{8}$

7/40

19. $2 - (-8)$

10

20. $3.2 - 5.7$

-2.5

21. $\frac{1}{8} - \left(-\frac{3}{4}\right)$

7/8

22. $4 \cdot (-12)$

-48

23. $-\frac{1}{2} \cdot \left(-\frac{3}{8}\right)$

3/16

24. $-45 \div 5$

-9

25. $-\frac{3}{5} \div \left(-\frac{4}{5}\right)$

3/4

26. $4.864 \div (-0.5)$

-9.728

27. $-2(16) - [2(-8) - 5^3]$

-32 - [(-16) - 125]
-141
101
109

SKILL MAINTENANCE

28. Find the area of a rectangle of length 12.4 ft and width 4.5 ft.

29. 24 is what percent of 50?

30. Find the prime factorization of 280.

31. Find the LCM of 16, 20, and 30.

SYNTHESIS

32. Simplify: $|-27 - 3(4)| - |-36| + |-12|$.

Cumulative Review: Chapters 1–11

Find decimal notation.

1. 26.3%

2. $-\dfrac{5}{11}$

Complete.

3. 83.4 cg = _____ mg

4. 2.75 mm^2 = _____ cm^2

5. Find the absolute value: $|-4.5|$.

6. Subtract: $2 - 13$.

7. What is the rate in meters per second?

   150 meters, 12 seconds

8. Find the radius, the circumference, and the area of this circle. Use $\frac{22}{7}$ for π.

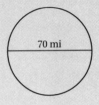

70 mi

9. Simplify: $\sqrt{225}$.

10. Approximate to two decimal places: $\sqrt{69}$.

11. Multiply: $(-2)(5)$.

12. Divide: $\dfrac{-48}{-16}$.

13. Add: $-2 + 10$.

14. Draw a pictograph representing the number of hours that each type of farmer works each week using the information given below. Use a clock symbol to represent 10 hours. Be sure to put in all of the appropriate labels.

Dairy	70
Cash grain	40
Tobacco/cotton	35
Beef/hog/sheep	30

Compute and simplify.

15. 14.85×0.001

16. $36 - (-3) + (-42)$

17. $\dfrac{5}{22} - \dfrac{4}{11}$

18. $\dfrac{2}{27} \cdot \left(-\dfrac{9}{16}\right)$

19. $4\dfrac{2}{9} - 2\dfrac{7}{18}$

20. $-\dfrac{3}{14} \div \dfrac{6}{7}$

21. $3(-4.5) + (2^2 - 3 \cdot 4^2)$

22. $12,854 \cdot 750,000$

23. $35.1 + (-2.61)$

24. $32 \div [(-2)(-8) - (15 - (-1))]$

Solve.

25. 7 is what percent of 8?

26. 4 is $12\frac{1}{2}\%$ of what number?

27. Kerry had $324.98 in a checking account. He wrote a check for $12.76, deposited $35.95, and wrote another check for $213.09. The bank paid $0.97 in interest and deducted a service charge of $3.00. How much is now in his checking account?

28. A can of fruit has a diameter of 7 cm and a height of 8 cm. Find the volume. Use 3.14 for π.

29. The following temperatures were recorded every four hours on a certain day in Seattle: 42°, 40°, 45°, 52°, 50°, 40°. What was the average temperature for the day?

30. Thirteen percent of a student body of 600 received all A's on their grade reports. How many students received all A's?

31. A lot is 125.5 m by 75 m. A house 60 m by 40.5 m and a rectangular swimming pool 10 m by 8 m are built on the lot. How much area is left?

32. A recipe for a pie crust calls for $1\frac{1}{4}$ cups of flour, and a recipe for a cake calls for $1\frac{2}{3}$ cups of flour. How many cups of flour are needed to make both recipes?

33. The four top television game show winners in a recent year won $74,834, $58,253, $57,200, and $49,154. How much did these four win in all? What were the average earnings?

34. A power walker circled a block 6.5 times. If the distance around the block is 0.7 km, how far did the walker go?

12

Algebra: Solving Equations and Problems

INTRODUCTION

In this chapter, we continue our introduction to algebra. We consider the manipulation of algebraic expressions and then use the manipulations to solve equations and problems.

AN APPLICATION

The top of the John Hancock Building in Chicago is a rectangle whose length is 60 ft more than the width. The perimeter is 520 ft. Find the width and the length of the rectangle.

THE MATHEMATICS

We let w = the width of the top of the building. We can then translate the problem to this *equation*:

$$2(w + 60) + 2w = 520.$$

The review objectives to be tested in addition to the material in this chapter are as follows.

[9.6a, b, c]   For circles, find the length of a radius, given the length of a diameter,
                  and conversely; the circumference, given the length of a diameter or
                  radius; and the area given the length of a radius.
[10.1a]   Find the volume of a rectangular solid using the formula $V = l \cdot w \cdot h$.
[11.2a]   Add real numbers.
[11.4a]   Multiply real numbers.

Pretest: Chapter 12

1. Evaluate $\dfrac{x}{2y}$ when $x = 5$ and $y = 8$.

2. Write an algebraic expression:

Seventy-eight percent of some number.

Multiply.

3. $9(z - 2)$

4. $-2(2a + b - 5c)$

Factor.

5. $4x - 12$

6. $6y - 9z - 18$

Collect like terms.

7. $5x - 8x$

8. $6x - 9y - 4x + 11y + 18$

Solve.

9. $-7x = 49$

10. $4y + 9 = 2y + 7$

11. $6a - 2 = 10$

12. $x + (x + 1) + (x + 2) = 12$

Solve.

13. The perimeter of a rectangular peach orchard is 146 m. The width is 5 m less than the length. Find the dimensions.

14. Money is invested in a savings account at 6% simple interest. After one year, there is $826.80 in the account. How much was originally invested?

12.1 *Introduction to Algebra*

Many types of problems require the use of equations in order to be solved effectively. The study of algebra involves the use of equations to solve problems. Equations are constructed from algebraic expressions.

ALGEBRAIC EXPRESSIONS

In arithmetic, you have worked with expressions such as

$$37 + 86, \quad 7 \times 8, \quad 19 - 7, \quad \text{and} \quad \frac{3}{8}.$$

In algebra, we use certain letters for numbers and work with *algebraic expressions* such as

$$x + 86, \quad 7 \times t, \quad 19 - y, \quad \text{and} \quad \frac{a}{b}.$$

Expressions like these should be familiar from the equation and problem solving that we have already done.

Sometimes a letter can stand for various numbers. In that case, we call the letter a **variable**. Let $a =$ your age. Then a is a variable since a changes from year to year. Sometimes a letter can stand for just one number. In that case, we call the letter a **constant**. Let $b =$ your date of birth. Then b is a constant.

An **algebraic expression** consists of variables, numerals, and operation signs. When we replace a variable by a number, we say that we are **substituting** for the variable. This process is called **evaluating the expression.**

EXAMPLE 1 Evaluate $x + y$ when $x = 37$ and $y = 29$.

We substitute 37 for x and 29 for y and carry out the addition:

$$x + y = 37 + 29 = 66.$$

The number 66 is called the **value** of the expression.

Algebraic expressions involving multiplication can be written in several ways. For example, "8 times a" can be written as $8 \times a$, $8 \cdot a$, $8(a)$, or simply $8a$. Two letters written together without a symbol, such as ab, also indicates a multiplication.

EXAMPLE 2 Evaluate $3y$ when $y = 14$.

$$3y = 3(14) = 42$$

Do Exercises 1–3.

1. Evaluate $a + b$ when $a = 38$ and $b = 26$.

2. Evaluate $x - y$ when $x = 57$ and $y = 29$.

3. Evaluate $4t$ when $t = 15$.

Answers on page A-6

4. Evaluate $\dfrac{a}{b}$ when $a = 200$ and $b = 8$.

5. Evaluate $\dfrac{10p}{q}$ when $p = 40$ and $q = 25$.

Complete the table by evaluating each expression for the given values.

6.

	$1 \cdot x$	x
$x = 3$	3	3
$x = -6$	-6	-6
$x = 4.8$	4.8	4.8

7.

	$2x$	$5x$
$x = 2$	4	10
$x = -6$	-12	-30
$x = 4.8$		

Answers on page A-6

Algebraic expressions involving division can also be written in several ways. For example, "8 divided by t" can be written as $8 \div t$, $8/t$, or $\dfrac{8}{t}$, where the fraction bar is a division symbol.

EXAMPLE 3   Evaluate $\dfrac{a}{b}$ when $a = 63$ and $b = 9$.

We substitute 63 for a and 9 for b and carry out the division:

$$\frac{a}{b} = \frac{63}{9} = 7.$$

$$\frac{63}{9} = 7$$

EXAMPLE 4   Evaluate $\dfrac{12m}{n}$ when $m = 8$ and $n = 16$.

$$\frac{12m}{n} = \frac{12 \cdot 8}{16} = \frac{96}{16} = 6$$

Do Exercises 4 and 5.

b   EQUIVALENT EXPRESSIONS AND THE DISTRIBUTIVE LAWS

In solving equations and doing other kinds of work in algebra, we manipulate expressions in various ways. To see how to do this, we consider some examples in which we evaluate expressions.

EXAMPLE 5   Evaluate $1 \cdot x$ when $x = 5$ and $x = -8$ and compare the results to x.

We substitute 5 for x:

$$1 \cdot x = 1 \cdot 5 = 5.$$

Then we substitute -8 for x:

$$1 \cdot x = 1 \cdot (-8) = -8.$$

We see that $1 \cdot x$ and x represent the same number.

Do Exercises 6 and 7.

We see in Example 5 and Margin Exercise 6 that the expressions represent the same number for any replacement of x that is meaningful. In that sense, the expressions $1 \cdot x$ and x are **equivalent**.

> Two expressions that have the same value for all meaningful replacements are called *equivalent*.

We see in Margin Exercise 7 that the expressions $2x$ and $5x$ are *not* equivalent.

The fact that $1 \cdot x$ and x are equivalent is a law of real numbers. It is called the **identity property of 1.** We often refer to the use of the identity property of 1 as "multiplying by 1." We have used multiplying by 1 for understanding many times in this text.

THE IDENTITY PROPERTY OF 1

For any real number a,

$$a \cdot 1 = 1 \cdot a = a.$$

(The number 1 is the *multiplicative identity*.)

We now consider two other laws of real numbers called the **distributive laws.** They are the basis of many procedures in both arithmetic and algebra and are probably the most important laws that we use to manipulate algebraic expressions. The first distributive law involves two operations: addition and multiplication.

Let us begin by considering a multiplication problem from arithmetic:

$$
\begin{array}{r}
4\ 5 \\
\times \quad 7 \\
\hline
3\ 5 \\
2\ 8\ 0 \\
3\ 1\ 5 \\
\end{array}
$$

$3\ 5 \leftarrow$ This is $7 \cdot 5$.
$2\ 8\ 0 \leftarrow$ This is $7 \cdot 40$.
$3\ 1\ 5 \leftarrow$ This is the sum $7 \cdot 40 + 7 \cdot 5$.

To carry out the multiplication, we actually added two products. That is,

$$7 \cdot 45 = 7(40 + 5) = 7 \cdot 40 + 7 \cdot 5.$$

Let us examine this further. If we wish to multiply a sum of several numbers by a factor, we can either add and then multiply, or multiply and then add.

EXAMPLE 6   Evaluate $5(x + y)$ and $5x + 5y$ when $x = 2$ and $y = 8$ and compare the results.

We substitute 2 for x and 8 for y in each expression. Then we use our rules for order of operations to calculate.

a) $5(x + y) = 5(2 + 8)$
$\qquad = 5(10)$ **Adding within parentheses first, and then multiplying**
$\qquad = 50$

b) $5x + 5y = 5 \cdot 2 + 5 \cdot 8$
$\qquad = 10 + 40$ **Multiplying first and then adding**
$\qquad = 50$

We see that the expressions $5(x + y)$ and $5x + 5y$ are equivalent.

Do Exercises 8–10.

8. Evaluate $3(x + y)$ and $3x + 3y$ when $x = 5$ and $y = 7$.

9. Evaluate $6x + 6y$ and $6(x + y)$ when $x = 10$ and $y = 5$.

10. Evaluate $4(x + y)$ and $4x + 4y$ when $x = 11$ and $y = 5$.

Answers on page A-6

11. Evaluate $7(x - y)$ and $7x - 7y$ when $x = 9$ and $y = 7$.

12. Evaluate $6x - 6y$ and $6(x - y)$ when $x = 10$ and $y = 5$.

13. Evaluate $2(x - y)$ and $2x - 2y$ when $x = 11$ and $y = 5$.

What are the terms of the expression?

14. $5x - 4y + 3$

15. $-4y - 2x + 3z$

> **THE DISTRIBUTIVE LAW OF MULTIPLICATION OVER ADDITION**
>
> For any numbers a, b, and c,
>
> $$a(b + c) = ab + ac.$$

In the statement of the distributive law, we know that in an expression such as $ab + ac$, the multiplications are to be done first according to the rules for order of operations. So, instead of writing $(4 \cdot 5) + (4 \cdot 7)$, we can write $4 \cdot 5 + 4 \cdot 7$. However, in $a(b + c)$, we cannot omit the parentheses. If we did we would have $ab + c$, which means $(ab) + c$. For example, $3(4 + 2) = 18$, but $3 \cdot 4 + 2 = 14$.

The second distributive law relates multiplication and subtraction. This law says that to multiply by a difference, we can either subtract and then multiply or multiply and then subtract.

> **THE DISTRIBUTIVE LAW OF MULTIPLICATION OVER SUBTRACTION**
>
> For any numbers a, b, and c,
>
> $$a(b - c) = ab - ac.$$

We often refer to "*the* distributive law" when we mean *either* of these laws.

Do Exercises 11–13.

What do we mean by the *terms* of an expression? **Terms** are separated by addition signs. If there are subtraction signs, we can find an equivalent expression that uses addition signs.

EXAMPLE 7   What are the terms of $3x - 4y + 2z$?

$$3x - 4y + 2z = 3x + (-4y) + 2z \qquad \text{Separating parts with + signs}$$

The terms are $3x$, $-4y$, and $2z$.

Do Exercises 14 and 15.

The distributive laws are the basis for a procedure in algebra called **multiplying**. In an expression such as $8(a + 2b - 7)$, we multiply each term inside the parentheses by 8:

$$8(a + 2b - 7) = 8 \cdot a + 8 \cdot 2b - 8 \cdot 7 = 8a + 16b - 56.$$

EXAMPLES   Multiply.

8. $9(x - 5) = 9x - 9(5)$ Using the distributive law of multiplication over subtraction

$$= 9x - 45$$

9. $2(w + 1) = 2 \cdot w + 2 \cdot 1$ Using the distributive law of multiplication over addition

$$= 2w + 2$$

EXAMPLE 10 Multiply: $-4(x - 2y + 3z)$.

$$-4(x - 2y + 3z) = -4 \cdot x - (-4)(2y) + (-4)(3z)$$

Using both distributive laws

$$= -4x - (-8y) + (-12z)$$
$$= -4x + 8y - 12z$$

We can also do this problem by first finding an equivalent expression with all plus signs and then multiplying:

$$-4(x - 2y + 3z) = -4[x + (-2y) + 3z]$$
$$= -4 \cdot x + (-4)(-2y) + (-4)(3z) = -4x + 8y - 12z.$$

Do Exercises 16–20.

c FACTORING

Factoring is the reverse of multiplying. To factor, we can use the distributive laws in reverse:

$$ab + ac = a(b + c) \quad \text{and} \quad ab - ac = a(b - c).$$

> To *factor* an expression is to find an equivalent expression that is a product.

Look at Example 8. To *factor* $9x - 45$, we find an equivalent expression that is a product, $9(x - 5)$. When all the terms of an expression have a factor in common, we can "factor it out" using the distributive laws. Note the following.

$9x$ has the factors $9, -9, 3, -3, 1, -1, x, -x, 3x, -3x, 9x, -9x$;
-45 has the factors $1, -1, 3, -3, 5, -5, 9, -9, 15, -15, 45, -45$.

We usually remove the largest common factor. In this case, that factor is 9. Thus,

$$9x - 45 = 9 \cdot x - 9 \cdot 5$$
$$= 9(x - 5).$$

Remember that an expression is factored when we find an equivalent expression that is a product.

EXAMPLES Factor.

11. $5x - 10 = 5 \cdot x - 5 \cdot 2$ Try to do this step mentally.
$$= 5(x - 2) \quad \longleftarrow \text{You can check by multiplying.}$$

12. $9x + 27y - 9 = 9 \cdot x + 9 \cdot 3y - 9 \cdot 1$
$$= 9(x + 3y - 1)$$

Caution! Note that although $3(3x + 9y - 3)$ is also equivalent to $9x + 27y - 9$, it is *not* the desired form. However, we can complete the process by factoring out another factor of 3:

$$9x + 27y - 9 = 3(3x + 9y - 3) = 3 \cdot 3(x + 3y - 1) = 9(x + 3y - 1).$$

Remember to factor out the *largest common factor.*

Multiply.

16. $3(x - 5)$

17. $5(x + 1)$

18. $5(x - y + 4)$

19. $-2(x - 3)$

20. $-5(x - 2y + 4z)$

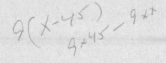

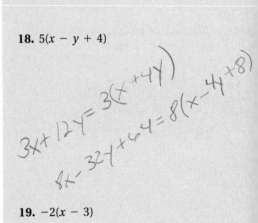

Answers on page A-6

Factor.

21. $6z - 12$

$6(z - 2)$

22. $3x - 6y + 9$

$3(x - 2y + 3)$

23. $16a - 36b + 42$

$16(a - 2.25b + 2.63)$

24. $-12x + 32y - 16z$

Collect like terms.

25. $6x - 3x$

26. $7x - x$

27. $x - 9x$

28. $x - 0.41x$

29. $5x + 4y - 2x - y$

30. $3x - 7x - 11 + 8y + 4 - 13y$

Answers on page A-6

EXAMPLES Factor. Try to write just the answer, if you can.

13. $5x - 5y = 5(x - y)$

14. $-3x + 6y - 9z = -3(x - 2y + 3z)$

We usually factor out a negative when the first term is negative. The way we factor can depend on the situation in which we are working. We might also factor the expression in Example 14 as follows:

$$-3x + 6y - 9z = 3(-x + 2y - 3z).$$

15. $18z - 12x - 24 = 6(3z - 2x - 4)$

> *Remember:* An expression is factored when it is written as a product.

Do Exercises 21–24.

d COLLECTING LIKE TERMS

Terms such as $5x$ and $-4x$, whose variable factors are exactly the same, are called **like terms.** Similarly, numbers, such as -7 and 13, are like terms. Also, $3y^2$ and $9y^2$ are like terms because the variables are raised to the same power. Terms such as $4y$ and $5y^2$ are not like terms, and $7x$ and $2y$ are not like terms.

The process of **collecting like terms** is based on the distributive laws. We can also apply the distributive law when a factor is on the right.

EXAMPLES Collect like terms. Try to write just the answer, if you can.

16. $4x + 2x = (4 + 2)x = 6x$ Factoring out the x using a distributive law

17. $2x + 3y - 5x - 2y = 2x - 5x + 3y - 2y$

$$= (2 - 5)x + (3 - 2)y = -3x + y$$

18. $3x - x = (3 - 1)x = 2x$

19. $x - 0.24x = 1 \cdot x - 0.24x = (1 - 0.24)x = 0.76x$

20. $x - 6x = 1 \cdot x - 6 \cdot x = (1 - 6)x = -5x$

21. $4x - 7y + 9x - 5 + 3y - 8 = 13x - 4y - 13$

Do Exercises 25–30.

Exercise Set 12.1

[a] Evaluate.

1. $6x$ when $x = 7$

2. $9t$ when $t = 8$

3. $\dfrac{x}{y}$ when $x = 9$ and $y = 3$

4. $\dfrac{m}{n}$ when $m = 18$ and $n = 3$

5. $\dfrac{3p}{q}$ when $p = 2$ and $q = 6$

6. $\dfrac{5y}{z}$ when $y = 15$ and $z = 25$

7. $\dfrac{x + y}{5}$ when $x = 10$ and $y = 20$

8. $\dfrac{p - q}{2}$ when $p = 17$ and $q = 3$

[b] Evaluate.

9. $10(x + y)$ and $10x + 10y$ when $x = 20$ and $y = 4$

10. $5(a + b)$ and $5a + 5b$ when $a = 16$ and $b = 6$

11. $10(x - y)$ and $10x - 10y$ when $x = 20$ and $y = 4$

12. $5(a - b)$ and $5a - 5b$ when $a = 16$ and $b = 6$

Multiply.

13. $2(b + 5)$

14. $4(x + 3)$

15. $7(1 - t)$

16. $4(1 - y)$

17. $6(5x + 2)$

18. $9(6m + 7)$

19. $7(x + 4 + 6y)$

20. $4(5x + 8 + 3p)$

21. $-7(y - 2)$

22. $-9(y - 7)$

23. $-9(-5x - 6y + 8)$

24. $-7(-2x - 5y + 9)$

25. $-4(x - 3y - 2z)$

26. $8(2x - 5y - 8z)$

27. $3.1(-1.2x + 3.2y - 1.1)$

28. $-2.1(-4.2x - 4.3y - 2.2)$

[c] Factor. Check by multiplying.

29. $2x + 4$

30. $5y + 20$

31. $30 + 5y$

32. $7x + 28$

ANSWERS

1. 42
2. 72
3. $\frac{9}{3}$ or 3
4. 6
5. 1
6.
7. 30/5 a 4
8.
9. 240
10.
11. 160
12.
13. $2B + 10$
14.
15. $7 - 7t$
16.
17. $30x + 12$
18.
19. $7x + 28 + 42y$
20.
21. $-7y + 14$
22.
23. $-45x + 54y - 72$
24.
25. $-4x + 12y + 8z$
26.
27. $-3.72x + 9.92y - 3.41$
28.
29. $2(x + 2)$
30.
31. $5(y + 6)$
32.

33. _7(2x+3y)_

34. _____

35. _5(x+2+3y)_

36. _____

37. _8(x-3)_

38. _____

39. _4(8-y)_

40. _____

41. _8(x+80y-22)_ ✓

42. _____

43. _6(-3x-2y+6)_

44. _____

45. _19a_

46. _____

47. _10a_ ✓ _9a_

48. _____

49. _____

50. _____

51. _-19a+88_

52. _____

53. _-2+5+iT+7y+y_ ✓

54. _____

55. _18x_

56. _____

57. _5N_

58. _____

59. _____

60. _____

61. _____

62. _____

63. _____

64. _____

65. _____

66. _____

67. _____

68. _____

33. $14x + 21y$ 34. $18a + 24b$ 35. $5x + 10 + 15y^3$

36. $9a + 27b + 81$ 37. $8x - 24$ 38. $10x - 50$

39. $32 - 4y$ 40. $24 - 6m$ 41. $8x + 10y - 22$

42. $9a + 6b - 15$ 43. $-18x - 12y + 6$ 44. $-14x + 21y + 7$

d Collect like terms.

45. $9a + 10a$ 46. $14x + 3x$ 47. $10a - a$

48. $-10x + x$ 49. $2x + 9z + 6x$ 50. $3a - 5b + 4a$

51. $41a + 90 - 60a - 2$ 52. $42x - 6 - 4x + 20$

53. $23 + 5t + 7y - t - y - 27$ 54. $95 - 90d - 87 - 9d + 3 + 7d$

55. $11x - 3x$ 56. $9t - 13t$

57. $6n - n$ 58. $10t - t$

59. $y - 17y$ 60. $5m - 8m + 4$

61. $-8 + 11a - 5b + 6a - 7b + 7$ 62. $8x - 5x + 6 + 3y - 2y - 4$

63. $9x + 2y - 5x$ 64. $8y - 3z + 4y$

65. $11x + 2y - 4x - y$ 66. $13a + 9b - 2a - 4b$

67. $2.7x + 2.3y - 1.9x - 1.8y$ 68. $6.7a + 4.3b - 4.1a - 2.9b$

12.2 Solving Equations: The Addition Principle

OBJECTIVE

After finishing Section 12.2, you should be able to:

a Solve equations using the addition principle.

FOR EXTRA HELP

TAPE 20 TAPE 17B MAC: 12
IBM: 12

a USING THE ADDITION PRINCIPLE

Consider the equation

$x = 7$.

We can easily "see" that the solution of this equation is 7. If we replace x by 7, we get

$7 = 7$, which is true.

Now consider the equation

$x + 6 = 13$.

The solution of this equation is also 7, but the fact that 7 is the solution is not as obvious. We now begin to consider principles that allow us to start with an equation and end up with an equation like $x = 7$, in which the variable is alone on one side and for which the solution is easy to find. The equations $x + 6 = 13$ and $x = 7$ are **equivalent**.

> Equations with the same solutions are called *equivalent equations*.

One principle that we use to solve equations concerns the addition principle, which we have used throughout this text.

> **THE ADDITION PRINCIPLE**
> If an equation $a = b$ is true, then
> $$a + c = b + c$$
> is true for any number c.

Let us again solve $x + 6 = 13$ using the addition principle. We want to get x alone on one side. To do so, we use the addition principle, adding -6 on both sides:

$$x + 6 = 13$$
$$x + 6 + (-6) = 13 + (-6) \qquad \text{Using the addition principle; adding } -6 \text{ on both sides}$$
$$x + 0 = 7 \qquad \text{Simplifying}$$
$$x = 7 \qquad \text{Identity property of 0}$$

Do Exercise 1.

1. Solve $x + 2 = 11$ using the addition principle.

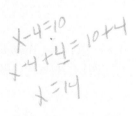

Answer on page A-6

2. Solve using the addition principle:

$$x + 7 = 2.$$

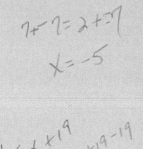

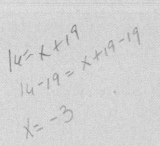

Solve.

3. $8.7 = n - 4.5$

4. $y + 17.4 = 10.9$

When we use the addition principle, we sometimes say that we "add the same number on both sides of the equation." This is also true for subtraction, since we can express every subtraction as an addition. That is, since

$$a - c = b - c \quad \text{means} \quad a + (-c) = b + (-c),$$

the addition principle tells us that we can "subtract the same number on both sides of an equation."

EXAMPLE 1   Solve: $x + 5 = -7$.

We have

$$x + 5 = -7$$
$$x + 5 - 5 = -7 - 5 \qquad \text{Using the addition principle: adding } -5 \text{ on both sides or subtracting 5 on both sides}$$
$$x + 0 = -12 \qquad \text{Simplifying}$$
$$x = -12. \qquad \text{Identity property of 0}$$

We can see that the solution of $x = -12$ is the number -12. To check the answer, we substitute -12 in the original equation.

CHECK:

$$\begin{array}{c|c} x + 5 = -7 \\ \hline -12 + 5 & -7 \\ -7 & \text{TRUE} \end{array}$$

The solution of the original equation is -12.

In Example 1, to get x alone, we used the addition principle and subtracted 5 on both sides. This eliminated the 5 on the left. We started with $x + 5 = -7$, and using the addition principle we found a simpler equation $x = -12$, for which it was easy to "see" the solution. The equations $x + 5 = -7$ and $x = -12$ are equivalent.

Do Exercise 2.

Now we solve an equation with a subtraction using the addition principle.

EXAMPLE 2   Solve: $-6.5 = y - 8.4$.

We have

$$-6.5 = y - 8.4$$
$$-6.5 + 8.4 = y - 8.4 + 8.4 \qquad \text{Using the addition principle: adding 8.4 to eliminate } -8.4 \text{ on the right}$$
$$1.9 = y.$$

CHECK:

$$\begin{array}{c|c} -6.5 = y - 8.4 \\ \hline -6.5 & 1.9 - 8.4 \\ & -6.5 \qquad \text{TRUE} \end{array}$$

The solution is 1.9.

Note that equations are reversible. That is, if $a = b$ is true, then $b = a$ is true. Thus, when we solve $-6.5 = y - 8.4$, we can reverse it and solve $y - 8.4 = -6.5$ if we wish.

Do Exercises 3 and 4.

Exercise Set 12.2

a Solve using the addition principle. Don't forget to check!

1. $x + 5 = 12$

$x + 5 - 5 = 12 - 5$
$x = 7$

2. $x + 3 = 7$

3. $x + 15 = -5$

$x + 15 - 15 = -5 - 15$
$x = -20$

4. $y + 8 = 37$

5. $x + 6 = -8$

6. $t + 8 = -14$

7. $x + 16 = -2$

8. $y + 34 = -8$

9. $x - 9 = 6$

10. $x - 9 = 2$

11. $x - 7 = -21$

$x = -21 + 7 \quad -14$
$x = -14$

12. $x - 5 = -16$

13. $5 + t = 7$

14. $6 + y = 22$

15. $-7 + y = 13$

16. $-8 + z = 16$

17. $-3 + t = -9$

18. $-8 + y = -23$

19. $r + \dfrac{1}{3} = \dfrac{8}{3}$

20. $t + \dfrac{3}{8} = \dfrac{5}{8}$

21. $m + \dfrac{5}{6} = -\dfrac{11}{12}$

$+ -\dfrac{5}{6}\,\dfrac{10}{12}$

22. $x + \dfrac{2}{3} = -\dfrac{5}{6}$

$x + \dfrac{2}{3} - \dfrac{2}{3} = -\dfrac{5}{6} - \dfrac{2}{3}$

$x = -\dfrac{9}{6} = -\dfrac{3}{2}$

23. $x - \dfrac{5}{6} = \dfrac{7}{8}$

24. $y - \dfrac{3}{4} = \dfrac{5}{6}$

1. 7
2.
3.
4.
5.
6.
7.
8.
9.
10.
11.
12.
13.
14.
15.
16.
17.
18.
19.
20.
21.
22.
23.
24.

25. $-\dfrac{1}{5} + z = -\dfrac{1}{4}$

26. $-\dfrac{1}{8} + y = -\dfrac{3}{4}$

27. $7.4 = x + 2.3$

$-\frac{1}{8} + y + \frac{1}{8} = -\frac{3}{4} + \frac{1}{8}$

$y = \frac{4}{8} + \frac{1}{8}$

$y = -\frac{5}{8}$

28. $9.3 = 4.6 + x$

29. $7.6 = x - 4.8$

30. $9.5 = y - 8.3$

31. $-9.7 = -4.7 + y$

32. $-7.8 = 2.8 + x$

33. $5\dfrac{1}{6} + x = 7$

34. $5\dfrac{1}{4} = 4\dfrac{2}{3} + x$

35. $q + \dfrac{1}{3} = -\dfrac{1}{7}$

36. $47\dfrac{1}{8} = -76 + z$

SKILL MAINTENANCE

37. Add: $-3 + (-8)$.

38. Subtract: $-3 - (-8)$.

39. Multiply: $-\dfrac{2}{3} \cdot \dfrac{5}{8}$.

40. Divide: $-\dfrac{3}{7} \div \left(-\dfrac{9}{7}\right)$.

41. Multiply: $(-3.2) \cdot (-4.1)$.

42. Add: $-\dfrac{2}{3} + \dfrac{5}{8}$.

SYNTHESIS

Solve.

43. ▦ $-356.788 = -699.034 + t$

44. $-\dfrac{4}{5} + \dfrac{7}{10} = x - \dfrac{3}{4}$

45. $x + \dfrac{4}{5} = -\dfrac{2}{3} - \dfrac{4}{15}$

46. $8 - 25 = 8 + x - 21$

47. $16 + x - 22 = -16$

48. $x + x = x$

49. $x + 3 = 3 + x$

50. $x + 4 = 5 + x$

51. $-\dfrac{3}{2} + x = -\dfrac{5}{17} - \dfrac{3}{2}$

52. $|x| = 5$

12.3 Solving Equations: The Multiplication Principle

a USING THE MULTIPLICATION PRINCIPLE

Suppose that $a = b$ is true and we multiply a by some number c. We get the same answer if we multiply b by c, because a and b are the same number.

THE MULTIPLICATION PRINCIPLE

If an equation $a = b$ is true, then

$$a \cdot c = b \cdot c$$

is true for any number c.

When using the multiplication principle, we sometimes say that we "multiply on both sides by the same number."

EXAMPLE 1 Solve: $\frac{2}{3}x = 18$.

To get x alone, we multiply by the *multiplicative inverse*, or *reciprocal*, of $\frac{2}{3}$. Then we get the *multiplicative identity*, 1, times x, or $1 \cdot x$, which simplifies to x. This allows us to eliminate the $\frac{2}{3}$ on the left.

$$\frac{2}{3}x = 18$$

$$\frac{3}{2} \cdot \frac{2}{3}x = \frac{3}{2} \cdot 18 \qquad \text{Using the multiplication principle: multiplying by } \frac{3}{2} \text{ to eliminate } \frac{2}{3} \text{ on the left}$$

$$1 \cdot x = 27 \qquad \text{Simplifying}$$

$$x = 27$$

CHECK:

$$\frac{\frac{2}{3}x = 18}{\frac{2}{3} \cdot 27 \; \Big| \; 18}$$
$$18 \; \Big| \qquad \text{TRUE}$$

The solution is 27.

Do Exercises 1 and 2.

OBJECTIVE

After finishing Section 12.3, you should be able to:

a Solve equations using the multiplication principle.

FOR EXTRA HELP

TAPE 20   TAPE 18A   MAC: 12
                                   IBM: 12

Solve.

1. $\frac{4}{5}x = 24$

2. $4x = -7$

Answers on page A-6

3. Solve: $5x = 40$.

The multiplication principle also tells us that we can "divide on both sides by a nonzero number." This is because division is the same as multiplying by a reciprocal. That is,

$$\frac{a}{c} = \frac{b}{c} \quad \text{means} \quad a \cdot \frac{1}{c} = b \cdot \frac{1}{c}, \quad \text{when } c \neq 0.$$

In an expression like $3x$, the number 3 is called the **coefficient**. In practice it is usually more convenient to "divide" on both sides of the equation if the coefficient of the variable is in decimal notation or is an integer. When the coefficient is in fractional notation, it is more convenient to "multiply" by a reciprocal.

EXAMPLE 2 Solve: $3x = 9$.

We have

$$3x = 9$$

$$\frac{3x}{3} = \frac{9}{3} \qquad \text{Using the multiplication principle: multiplying by } \tfrac{1}{3} \text{ on both sides or dividing by 3 on both sides}$$

$$1 \cdot x = 3 \qquad \text{Simplifying}$$

$$x = 3. \qquad \text{Identity property of 1}$$

It is now easy to see that the solution is 3.

CHECK:
$$\begin{array}{c|c} 3x = 9 \\ \hline 3 \cdot 3 & 9 \\ 9 & \end{array} \quad \text{TRUE}$$

The solution of the original equation is 3.

Do Exercise 3.

4. Solve: $108 = -6x$.

EXAMPLE 3 Solve: $92 = -4x$.

$$92 = -4x$$

$$\frac{92}{-4} = \frac{-4x}{-4} \qquad \text{Using the multiplication principle. Dividing on both sides by } -4 \text{ is the same as multiplying by } -\tfrac{1}{4}.$$

$$-23 = 1 \cdot x \qquad \text{Simplifying}$$

$$-23 = x \qquad \text{Identity property of 1}$$

CHECK:
$$\begin{array}{c|c} 92 = -4x \\ \hline 92 & -4(-23) \\ & 92 \end{array} \quad \text{TRUE}$$

The solution is -23.

Note that equations are reversible. That is, if $a = b$ is true, then $b = a$ is true. Thus, when we solve $92 = -4x$, we can reverse it and solve $-4x = 92$.

Do Exercise 4.

Answers on page A-6

Exercise Set 12.3

ANSWERS

1. _____

2. _____

3. _____

4. _____

5. _____

6. _____

7. _____

8. _____

9. _____

10. _____

11. _____

12. _____

13. _____

14. _____

15. _____

16. _____

17. _____

18. _____

19. _____

20. _____

21. _____

22. _____

23. _____

24. _____

a Solve using the multiplication principle. Don't forget to check!

1. $6x = 36$ **2.** $4x = 52$ **3.** $5x = 45$ **4.** $8x = 56$

5. $84 = 7x$ **6.** $63 = 7x$ **7.** $-x = 40$ **8.** $50 = -x$

$$-1(-x) = 40(-1)$$
$$x = -40$$

9. $-2x = -10$ **10.** $-78 = -39p$ **11.** $7x = -49$ **12.** $9x = -54$

$$-2x = \frac{-10}{-2}$$

13. $-12x = 72$ **14.** $-15x = 105$ **15.** $-21x = -126$

$$\frac{-12}{-12} = \frac{12}{-12}$$

16. $-13x = -104$ **17.** $\dfrac{1}{7}t = -9$ **18.** $-\dfrac{1}{8}y = 11$

$$\frac{1}{7}t = -9$$
$$\frac{1}{7} \times \frac{7}{1} = -\frac{9}{1} \times \frac{7}{1} = 63$$

19. $\dfrac{3}{4}x = 27$ **20.** $\dfrac{4}{5}x = 16$ **21.** $-\dfrac{1}{3}t = 7$

22. $-\dfrac{1}{6}x = 9$ **23.** $-\dfrac{1}{3}m = \dfrac{1}{5}$ **24.** $\dfrac{1}{5} = -\dfrac{1}{8}z$

25. _____

26. _____

27. _____

28. _____

29. _____

30. _____

31. _____

32. _____

33. _____

34. _____

35. _____

36. _____

37. _____

38. _____

39. _____

40. _____

41. _____

42. _____

43. _____

44. _____

45. _____

46. _____

25. $-\dfrac{3}{5}r = \dfrac{9}{10}$

26. $\dfrac{2}{5}y = -\dfrac{4}{15}$

27. $-\dfrac{3}{2}r = -\dfrac{27}{4}$

28. $-\dfrac{5}{7}x = -\dfrac{10}{14}$

29. $6.3x = 44.1$

$\dfrac{44.1}{6.3} = 7$

$x = 7$

30. $2.7y = 54$

31. $-3.1y = 21.7$

32. $-3.3y = 6.6$

33. $38.7m = 309.6$

34. $29.4m = 235.2$

35. $-\dfrac{2}{3}y = -10.6$

$y = -10.6 \times \dfrac{3}{2}$

15.9

36. $-\dfrac{9}{7}y = 12.06$

SKILL MAINTENANCE

37. Find the circumference, the diameter, and the area of a circle painted on a driveway. Its radius is 10 ft. Use 3.14 for π.

38. Find the circumference, the radius, and the area of a circle whose diameter is 24 cm. Use 3.14 for π.

39. Find the volume of a rectangular block of granite of length 25 ft, width 10 ft, and height 32 ft.

40. Find the volume of a rectangular solid of length 1.3 cm, width 10 cm, and height 2.4 cm.

SYNTHESIS

Solve.

41. ▦ $-0.2344m = 2028.732$

42. $0 \cdot x = 0$

43. $0 \cdot x = 9$

44. $4|x| = 48$

45. $2|x| = -12$

46. A student makes a calculation and gets an answer of 22.5. On the last step, the student multiplies by 0.3 when a division by 0.3 should have been done. What should the correct answer be?

12.4 *Using the Principles Together*

OBJECTIVES

After finishing Section 12.4, you should be able to:

a Solve equations using both the addition and the multiplication principles.

b Solve equations in which like terms may need to be collected.

c Solve equations by first removing parentheses and collecting like terms.

FOR EXTRA HELP

TAPE 20   TAPE 18   MAC: 12
                      IBM: 12

a | APPLYING BOTH PRINCIPLES

Consider the equation $3x + 4 = 13$. It is more complicated than those in the preceding two sections. In order to solve such an equation, we first isolate the x-term, $3x$, using the addition principle. Then we apply the multiplication principle to get x by itself.

EXAMPLE 1 Solve: $3x + 4 = 13$.

$$3x + 4 = 13$$

$$3x + 4 - 4 = 13 - 4 \quad \text{Using the addition principle: adding } -4 \text{ or subtracting 4 on both sides}$$

$$3x = 9 \quad \text{Simplifying}$$

$$\frac{3x}{3} = \frac{9}{3} \quad \text{Using the multiplication principle: dividing on both sides by 3}$$

$$x = 3 \quad \text{Simplifying}$$

CHECK:
$$\frac{3x + 4 = 13}{\begin{array}{c|c} 3 \cdot 3 + 4 & 13 \\ 9 + 4 & \\ 13 & \end{array}} \quad \text{TRUE}$$

The solution is 3.

Do Exercise 1.

EXAMPLE 2 Solve: $-5x - 6 = 16$.

$$-5x - 6 = 16$$

$$-5x - 6 + 6 = 16 + 6 \quad \text{Adding 6 on both sides}$$

$$-5x = 22$$

$$\frac{-5x}{-5} = \frac{22}{-5} \quad \text{Dividing on both sides by } -5$$

$$x = -\frac{22}{5}, \text{ or } -4\frac{2}{5} \quad \text{Simplifying}$$

CHECK:
$$\frac{-5x - 6 = 16}{\begin{array}{c|c} -5\left(-\frac{22}{5}\right) - 6 & 16 \\ 22 - 6 & \\ 16 & \end{array}} \quad \text{TRUE}$$

The solution is $-\frac{22}{5}$.

Do Exercises 2 and 3.

1. Solve: $9x + 6 = 51$.

Solve.

2. $8x - 4 = 28$

3. $-\dfrac{1}{2}x + 3 = 1$

Answers on page A-6

4. Solve: $-18 - x = -57$.

Solve.

5. $-4 - 8x = 8$

6. $41.68 = 4.7 - 8.6y$

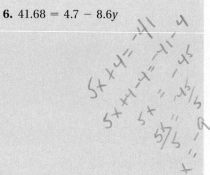

Solve.

7. $4x + 3x = -21$

8. $x - 0.09x = 728$

Answers on page A-6

EXAMPLE 3   Solve: $45 - x = 13$.

$$45 - x = 13$$
$$-45 + 45 - x = -45 + 13 \qquad \text{Adding } -45 \text{ on both sides}$$
$$-x = -32$$
$$-1 \cdot x = -32 \qquad -x = -1 \cdot x$$
$$\frac{-1 \cdot x}{-1} = \frac{-32}{-1} \qquad$$

Dividing on both sides by -1. (We could have multiplied on both sides by -1 instead. That would also change the sign on both sides.)

$$x = 32$$

CHECK:

$$\begin{array}{c|c} 45 - x = 13 \\ \hline 45 - 32 & 13 \\ 13 & \text{TRUE} \end{array}$$

The solution is 32.

Do Exercise 4.

EXAMPLE 4   Solve: $16.3 - 7.2y = -8.18$.

$$16.3 - 7.2y = -8.18$$
$$-16.3 + 16.3 - 7.2y = -16.3 + (-8.18) \qquad \text{Adding } -16.3 \text{ on both sides}$$
$$-7.2y = -24.48$$
$$\frac{-7.2y}{-7.2} = \frac{-24.48}{-7.2} \qquad \text{Dividing by } -7.2 \text{ on both sides}$$
$$y = 3.4$$

CHECK:

$$\begin{array}{c|c} 16.3 - 7.2y = -8.18 \\ \hline 16.3 - 7.2(3.4) & -8.18 \\ 16.3 - 24.48 & \\ -8.18 & \text{TRUE} \end{array}$$

The solution is 3.4.

Do Exercises 5 and 6.

b COLLECTING LIKE TERMS

If there are like terms on one side of the equation, we collect them before using the addition or multiplication principles.

EXAMPLE 5   Solve: $3x + 4x = -14$.

$$3x + 4x = -14$$
$$7x = -14 \qquad \text{Collecting like terms}$$
$$\frac{7x}{7} = \frac{-14}{7} \qquad \text{Dividing by 7 on both sides}$$
$$x = -2.$$

The number -2 checks, so the solution is -2.

Do Exercises 7 and 8.

If there are like terms on opposite sides of the equation, we get them on the same side by using the addition principle. Then we collect them. In other words, we get all terms with a variable on one side and all numbers on the other.

9. Solve: $7y + 5 = 2y + 10$.

| **EXAMPLE 6**   Solve: $2x - 2 = -3x + 3$.

$$2x - 2 = -3x + 3$$

$$2x - 2 + 2 = -3x + 3 + 2 \qquad \text{Adding 2}$$

$$2x = -3x + 5 \qquad \text{Collecting like terms}$$

$$2x + 3x = -3x + 3x + 5 \qquad \text{Adding } 3x$$

$$5x = 5 \qquad \text{Simplifying}$$

$$\frac{5x}{5} = \frac{5}{5} \qquad \text{Dividing by 5}$$

$$x = 1 \qquad \text{Simplifying}$$

Solve.
10. $5 - 2y = 3y - 5$

CHECK:

$$\begin{array}{c|c} 2x - 2 = -3x + 3 \\ \hline 2 \cdot 1 - 2 & -3 \cdot 1 + 3 \\ 2 - 2 & -3 + 3 \\ 0 & 0 \qquad \text{TRUE} \end{array}$$

The solution is 1.

Do Exercise 9.

In Example 6, we used the addition principle to get all terms with a variable on one side and all numbers on the other side. Then we collected like terms and proceeded as before. If there are like terms on one side at the outset, they should be collected first.

11. $7x - 17 + 2x = 2 - 8x + 15$

| **EXAMPLE 7**   Solve: $6x + 5 - 7x = 10 - 4x + 3$.

$$6x + 5 - 7x = 10 - 4x + 3$$

$$-x + 5 = 13 - 4x \qquad \text{Collecting like terms}$$

$$4x - x + 5 = 13 - 4x + 4x \qquad \text{Adding } 4x$$

$$3x + 5 = 13 \qquad \text{Simplifying}$$

$$3x + 5 - 5 = 13 - 5 \qquad \text{Subtracting 5}$$

$$3x = 8 \qquad \text{Simplifying}$$

$$\frac{3x}{3} = \frac{8}{3} \qquad \text{Dividing by 3}$$

$$x = \frac{8}{3}. \qquad \text{Simplifying}$$

12. $3x - 15 = 5x + 2 - 4x$

The number $\frac{8}{3}$ checks, so $\frac{8}{3}$ is the solution.

Do Exercises 10–12.

Answers on page A-6

13. Solve: $\dfrac{7}{8}x - \dfrac{1}{4} + \dfrac{1}{2}x = \dfrac{3}{4} + x.$

CLEARING FRACTIONS AND DECIMALS

For the equations considered thus far, we generally use the addition principle first. There are, however, some situations in which it is to our advantage to use the multiplication principle first. Consider, for example,

$$\frac{1}{2}x = \frac{3}{4}.$$

If we multiply by 4 on both sides, we get $2x = 3$, which has no fractions. We have "cleared fractions." Consider

$$2.3x = 5.$$

If we multiply by 10 on both sides, we get $23x = 50$, which has no decimal points. We have "cleared decimals." The equations are then easier to solve. It is your choice whether to clear fractions or decimals, but doing so often eases computations.

In what follows, we use the multiplication principle first to "clear," or "eliminate," fractions or decimals. For fractions, the number by which we multiply is the **least common multiple of all the denominators.**

EXAMPLE 8   Solve:

$$\frac{2}{3}x - \frac{1}{6} + \frac{1}{2}x = \frac{7}{6} + 2x.$$

The number 6 is the least common multiple of all the denominators. We multiply on both sides by 6:

$$6\left(\frac{2}{3}x - \frac{1}{6} + \frac{1}{2}x\right) = 6\left(\frac{7}{6} + 2x\right) \qquad \text{Multiplying by 6 on both sides}$$

$$6 \cdot \frac{2}{3}x - 6 \cdot \frac{1}{6} + 6 \cdot \frac{1}{2}x = 6 \cdot \frac{7}{6} + 6 \cdot 2x$$

Using the distributive laws. (*Caution!* Be sure to multiply all the terms by 6.)

$$4x - 1 + 3x = 7 + 12x \qquad \text{Simplifying. Note that the fractions are cleared.}$$

$$7x - 1 = 7 + 12x \qquad \text{Collecting like terms}$$

$$7x - 1 - 12x = 7 + 12x - 12x \qquad \text{Subtracting } 12x$$

$$-5x - 1 = 7 \qquad \text{Simplifying}$$

$$-5x - 1 + 1 = 7 + 1 \qquad \text{Adding 1}$$

$$-5x = 8 \qquad \text{Collecting like terms}$$

$$x = -\frac{8}{5}. \qquad \text{Multiplying by } -\frac{1}{5} \text{ or dividing by } -5$$

The number $-\frac{8}{5}$ checks and is the solution.

Do Exercise 13.

Answer on page A-6

To illustrate clearing decimals, we repeat Example 4, but this time we clear the decimals first.

EXAMPLE 9 Solve: $16.3 - 7.2y = -8.18$.

The greatest number of decimal places in any one number is *two*. Multiplying by 100, which has *two* 0's, will clear the decimals.

$$100(16.3 - 7.2y) = 100(-8.18)$$ Multiplying by 100 on both sides

$$100(16.3) - 100(7.2y) = 100(-8.18)$$ Using a distributive law

$$1630 - 720y = -818$$ Simplifying

$$1630 - 720y - 1630 = -818 - 1630$$ Subtracting 1630 on both sides

$$-720y = -2448$$ Collecting like terms

$$\frac{-720y}{-720} = \frac{-2448}{-720}$$ Dividing by -720 on both sides

$$y = 3.4$$

The number 3.4 checks and is the solution.

Do Exercise 14.

c | EQUATIONS CONTAINING PARENTHESES

To solve certain kinds of equations that contain parentheses, we use the distributive laws to first remove the parentheses. Then we proceed as before.

EXAMPLE 10 Solve: $4x = 2(12 - 2x)$.

$$4x = 2(12 - 2x)$$

$$4x = 24 - 4x$$ Using a distributive law to multiply and remove parentheses

$$4x + 4x = 24 - 4x + 4x$$ Adding $4x$ to get all x-terms on one side

$$8x = 24$$ Collecting like terms

$$\frac{8x}{8} = \frac{24}{8}$$ Dividing by 8

$$x = 3$$

CHECK:

$$
\begin{array}{c|c}
4x & 2(12 - 2x) \\
\hline
4 \cdot 3 & 2(12 - 2 \cdot 3) \\
12 & 2(12 - 6) \\
 & 2 \cdot 6 \\
 & 12 \quad \text{TRUE}
\end{array}
$$

We use the rules for order of operations to carry out the calculations on each side of the equation.

The solution is 3.

Do Exercises 15 and 16.

14. Solve: $41.68 = 4.7 - 8.6y$.

Solve.

15. $2(2y + 3) = 14$

16. $5(3x - 2) = 35$

Answers on page A-6

Solve.

17. $3(7 + 2x) = 30 + 7(x - 1)$

Here is a procedure for solving the types of equation discussed in this section.

AN EQUATION-SOLVING PROCEDURE

1. Multiply on both sides to clear the equation of fractions or decimals. (This is optional, but it can ease computations.)

2. If parentheses occur, multiply using the *distributive laws* to remove them.

3. Collect like terms on each side, if necessary.

4. Get all terms with variables on one side and all constant terms on the other side, using the *addition principle*.

5. Collect like terms again, if necessary.

6. Multiply or divide to solve for the variable, using the *multiplication principle*.

7. Check all possible solutions in the original equation.

---

18. $4(3 + 5x) - 4 = 3 + 2(x - 2)$

EXAMPLE 11 Solve: $2 - 5(x + 5) = 3(x - 2) - 1$.

$$2 - 5(x + 5) = 3(x - 2) - 1$$

$$2 - 5x - 25 = 3x - 6 - 1 \qquad \text{Using the distributive laws to multiply and remove parentheses}$$

$$-5x - 23 = 3x - 7 \qquad \text{Collecting like terms}$$

$$-5x - 23 + 5x = 3x - 7 + 5x \qquad \text{Adding } 5x$$

$$-23 = 8x - 7 \qquad \text{Collecting like terms}$$

$$-23 + 7 = 8x - 7 + 7 \qquad \text{Adding } 7$$

$$-16 = 8x \qquad \text{Collecting like terms}$$

$$\frac{-16}{8} = \frac{8x}{8} \qquad \text{Dividing by 8}$$

$$-2 = x$$

CHECK:

$2 - 5(x + 5) = 3(x - 2) - 1$	
$2 - 5(-2 + 5)$	$3(-2 - 2) - 1$
$2 - 5(3)$	$3(-4) - 1$
$2 - 15$	$-12 - 1$
-13	-13 TRUE

The solution is -2.

Note that the solution of $-2 = x$ is -2, which is also the solution of $x = -2$.

Do Exercises 17 and 18.

Exercise Set 12.4

a Solve. Don't forget to check!

1. $5x + 6 = 31$ **2.** $8x + 6 = 30$ **3.** $8x + 4 = 68$ **4.** $8z + 7 = 79$

5. $4x - 6 = 34$ **6.** $4x - 11 = 21$ **7.** $3x - 9 = 33$ **8.** $6x - 9 = 57$

9. $7x + 2 = -54$ **10.** $5x + 4 = -41$ **11.** $-45 = 3 + 6y$

$$5x + 4 - 4 = -41 - 4$$
$$5x = -45$$
$$\frac{5x}{5} = \frac{-45}{5} \quad x = -9$$

12. $-91 = 9t + 8$ **13.** $-4x + 7 = 35$ **14.** $-5x - 7 = 108$

15. $-7x - 24 = -129$ **16.** $-6z - 18 = -132$

b Solve.

17. $5x + 7x = 72$ **18.** $4x + 5x = 45$ **19.** $8x + 7x = 60$

$$9x = 45 \qquad\qquad 15x = 60$$
$$\frac{9x}{9} = \frac{45}{9} = 5 \qquad \frac{15x}{15} = \frac{60}{15} = 4$$

20. $3x + 9x = 96$ **21.** $4x + 3x = 42$ **22.** $6x + 19x = 100$

23. $-6y - 3y = 27$ **24.** $-4y - 8y = 48$ **25.** $-7y - 8y = -15$

$$-12y = 48$$
$$\frac{-12y}{-12} = \frac{48}{-12}$$
$$y = -4$$

ANSWERS

1.
2.
3.
4.
5.
6.
7.
8.
9.
10.
11.
12.
13.
14.
15.
16.
17.
18.
19.   4
20.
21.
22.
23.
24.
25.

26. _____

27. _____

28. _____

29. _____

30. _____

31. _____

32. _____

33. _____

34. _____

35. _____

36. _____

37. _____

38. _____

39. _____

40. _____

41. _____

42. _____

43. _____

44. _____

45. _____

46. _____

47. _____

48. _____

49. _____

50. _____

26. $-10y - 3y = -39$

27. $10.2y - 7.3y = -58$

28. $6.8y - 2.4y = -88$

29. $x + \dfrac{1}{3}x = 8$

(handwritten)
$3x + 1x = 24$
$4t = 24; \ 4 = 6$

30. $x + \dfrac{1}{4}x = 10$

31. $8y - 35 = 3y$

32. $4x - 6 = 6x$

33. $8x - 1 = 23 - 4x$

(handwritten)
$12x - 1 = 23$
$12x = 24$
$x = 2$

34. $5y - 2 = 28 - y$

35. $2x - 1 = 4 + x$

(handwritten)
$x - 1 = 4$
$x = 5$

36. $5x - 2 = 6 + x$

37. $6x + 3 = 2x + 11$

38. $5y + 3 = 2y + 15$

39. $5 - 2x = 3x - 7x + 25$

(handwritten)
$5 - 2x = -4x + 25$
$5 - 2x + 4x = -4x + 4x + 25$
$5 + 2x = 25$
$2x = 20 \quad x = 10$

40. $10 - 3x = 2x - 8x + 40$

41. $4 + 3x - 6 = 3x + 2 - x$

42. $5 + 4x - 7 = 4x - 2 - x$

43. $4y - 4 + y + 24 = 6y + 20 - 4y$

44. $5y - 7 + y = 7y + 21 - 5y$

(handwritten)
$6y - 7 = 2y + 21$
$6y - 7 - 2y = 2y + 21 - 2y$
$4y - 7 = 21$
$4y - 7 + 7 = 21 + 7 = 28 \qquad \dfrac{4y}{4} = \dfrac{28}{4} = 7$
$y = \boxed{7}$

Solve. Clear fractions or decimals first.

45. $\dfrac{7}{2}x + \dfrac{1}{2}x = 3x + \dfrac{3}{2} + \dfrac{5}{2}x$

(handwritten)
$7x + 1x = 4x + 3 + 5x$
$8x = 11x + 3$
$-3x = 3 \quad \dfrac{3}{-3} = -1$

46. $\dfrac{7}{8}x - \dfrac{1}{4} + \dfrac{3}{4}x = \dfrac{1}{16} + x$

(handwritten)
✷ MULT EA TERM BY
LCD TO GET RID OF
FRACTIONS

47. $\dfrac{2}{3} + \dfrac{1}{4}t = \dfrac{1}{3}$

48. $-\dfrac{3}{2} + x = -\dfrac{5}{6} - \dfrac{4}{3}$

49. $\dfrac{2}{3} + 3y = 5y - \dfrac{2}{15}$

50. $\dfrac{1}{2} + 4m = 3m - \dfrac{5}{2}$

(handwritten)
$16\left(\dfrac{7}{8}x - \dfrac{1}{4} + \dfrac{3}{4}x\right) = \left(\dfrac{1}{16} + x\right)16$
$14x - 4 + 12x = 1 + 16x$
$24x - 4 = 1 + 16x$
$26x - 4 - 14x = 1 + 16x - 16x$
$10x - 4 = 1$
$10x = 5$
$x = \dfrac{5}{10} = \boxed{+\dfrac{1}{2}}$

51. $\dfrac{5}{3} + \dfrac{2}{3}x = \dfrac{25}{12} + \dfrac{5}{4}x + \dfrac{3}{4}$

52. $1 - \dfrac{2}{3}y = \dfrac{9}{5} - \dfrac{y}{5} + \dfrac{3}{5}$

53. $2.1x + 45.2 = 3.2 - 8.4x$

54. $0.96y - 0.79 = 0.21y + 0.46$

subtract .21y

Amult by 100 to get rid of decimals

$96y - 79 = 21y + 46$

$75y - 79 = 46$

$75y = 46 + 79 = 125$

$\dfrac{125}{75} = \dfrac{5}{3}$

$x = \dfrac{5}{3}$

55. $1.03 - 0.62x = 0.71 - 0.22x$

56. $1.7t + 8 - 1.62t = 0.4t - 0.32 + 8$

$14\left(\dfrac{}{}\right)$

57. $\dfrac{2}{7}x - \dfrac{1}{2}x = \dfrac{3}{4}x + 1$

$\dfrac{4x - 7x = 3x + 14}{14}$

58. $\dfrac{5}{16}y + \dfrac{3}{8}y = 2 + \dfrac{1}{4}y$

c Solve.

59. $3(2y - 3) = 27$

60. $4(2y - 3) = 28$

61. $40 = 5(3x + 2)$

62. $9 = 3(5x - 2)$

63. $2(3 + 4m) - 9 = 45$

64. $3(5 + 3m) - 8 = 88$

65. $5r - (2r + 8) = 16$

$5r - 2r - 8 = 16$

$3r - 8 = 14$

$3r = 24$

$24/3 = 8$

66. $6b - (3b + 8) = 16$

67. $6 - 2(3x - 1) = 2$

68. $10 - 3(2x - 1) = 1$

69. $5(d + 4) = 7(d - 2)$

$5D + 20 = 7D - 14$

$-2D + 20 = -14$

$-2D = -34 \quad (17)$

70. $3(t - 2) = 9(t + 2)$

71. $8(2t + 1) = 4(7t + 7)$

72. $7(5x - 2) = 6(6x - 1)$

ANSWERS

51. _____
52. _____
53. _____
54. _____
55. _____
56. _____
57. _____
58. _____
59. _____
60. _____
61. _____
62. _____
63. _____
64. _____
65. _____
66. _____
67. _____
68. _____
69. _____
70. _____
71. _____
72. _____

ANSWERS

73. _____

74. _____

75. _____

76. _____

77. _____

78. _____

79. _____

80. _____

81. _____

82. _____

83. _____

84. _____

85. _____

86. _____

87. _____

88. _____

89. _____

90. _____

91. _____

92. _____

73. $3(r - 6) + 2 = 4(r + 2) - 21$

74. $5(t + 3) + 9 = 3(t - 2) + 6$

$5t + 15 + 9 = 3t - 6 + 6$

$5t + 24 = 3T$

$5T - 3T + 24 = 3T - 3T$

$2T + 24 = \emptyset \quad 2T = -24$

$\dfrac{-24}{2}$

$T = 12$

75. $19 - (2x + 3) = 2(x + 3) + x$

76. $13 - (2c + 2) = 2(c + 2) + 3c$

77. $2[4 - 2(3 - x)] - 1 = 4[2(4x - 3) + 7] - 25$

$2[4 - 6 - 2x] - 1 = 4[8x - 6 + 7] - 25 \qquad -5 = 28x - 21$

$2[-2 + 2x] = 4[8x + 1] - 25 \qquad 16 = 28x$

$-4 + 4x - 1 = 32x + 4 - 25$

$4x - 5 = 32x - 21 \qquad \dfrac{16}{28} = \dfrac{4}{7}$

78. $5[3(7 - t) - 4(8 + 2t)] - 20 = -6[2(6 + 3t) - 4]$

79. $0.7(3x + 6) = 1.1 - (x + 2)$

80. $0.9(2x + 8) = 20 - (x + 5)$

81. $a + (a - 3) = (a + 2) - (a + 1)$

$a + a - 3 = a + 2 - a - 1$

$2a - 3 = 2 - 1$

$2c - 3 = 1$

$2a = 4 \quad a = 2$

82. $0.8 - 4(b - 1) = 0.2 + 3(4 - b)$

SKILL MAINTENANCE

83. Divide: $-22.1 \div 3.4$.

84. Factor: $7x - 21 - 14y$.

85. Use $<$ or $>$ for ▦ to write a true sentence:

-15 ▦ -13.

86. Find $-(-x)$ when $x = -14$.

SYNTHESIS

Solve.

87. $\dfrac{y - 2}{3} = \dfrac{2 - y}{5}$

88. $3x = 4x$

89. $\dfrac{5 + 2y}{3} = \dfrac{25}{12} + \dfrac{5y + 3}{4}$

90. ▦ $0.05y - 1.82 = 0.708y - 0.504$

91. $\dfrac{2}{3}(2x - 1) = 10$

92. $\dfrac{2}{3}\left(\dfrac{7}{8} - 4x\right) - \dfrac{5}{8} = \dfrac{3}{8}$

12.5 *Solving Problems*

OBJECTIVES

After finishing Section 12.5, you should be able to:

a Translate phrases to algebraic expressions.

b Solve problems by translating to equations.

FOR EXTRA HELP

TAPE 20   TAPE 18B   MAC: 12
IBM: 12

a | TRANSLATING TO ALGEBRAIC EXPRESSIONS

In algebra, we translate problems to equations. The different parts of an equation are translations of word phrases to algebraic expressions. To translate, it helps to learn which words translate to certain operation symbols.

KEY WORDS			
ADDITION (+)	**SUBTRACTION (−)**	**MULTIPLICATION (·)**	**DIVISION (÷)**
add	subtract	multiply	divide
sum	difference	product	quotient
plus	minus	times	divided by
more than	less than	twice	
increased by	decreased by	of	
	take from		

EXAMPLE 1 Translate to an algebraic expression:

Twice (or two times) some number.

Think of some number, say, 8. What number is twice 8? It is 16. How did you get 16? You multiplied by 2. Do the same thing using a variable. We can use any variable we wish, such as x, y, m, or n. Let's use y to stand for some number. If we multiply by 2, we get an expression

$$y \times 2, \quad 2 \times y, \quad 2 \cdot y, \quad \text{or} \quad 2y.$$

EXAMPLE 2 Translate to an algebraic expression:

Seven less than some number.

We let

x = the number.

Now if the number were 23, then the translation would be $23 - 7$. If we knew the number to be 345, then the translation would be $345 - 7$. The translation is found as follows:

$$\underbrace{\text{Seven}} \quad \underbrace{\text{less than}} \quad \underbrace{\text{some number}}$$
$$x - 7.$$

Note that $7 - x$ is *not* a correct translation of the expression in Example 2. The expression $7 - x$ is a translation of "seven minus some number" or "some number less than seven."

EXAMPLE 3 Translate to an algebraic expression:

Eighteen more than a number.

We let

t = the number.

Translate to an algebraic expression.

1. Twelve less than some number

2. Twelve more than some number

3. Four less than some number

4. Half of some number

5. Six more than eight times some number

6. The difference of two numbers

7. Fifty-nine percent of some number

8. Two hundred less than the product of two numbers

9. The sum of two numbers

Answers on page A-6

Now if the number were 26, then the translation would be 18 + 26. If we knew the number to be 174, then the translation would be 18 + 174. The translation is

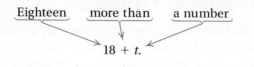

EXAMPLE 4   Translate to an algebraic expression:

A number divided by 5.

We let

m = the number.

Now if the number were 76, then the translation would be 76 ÷ 5, or 76/5, or $\frac{76}{5}$. If the number were 213, then the translation would be 213 ÷ 5, or 213/5, or $\frac{213}{5}$. The translation is found as follows:

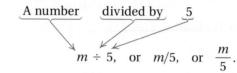

EXAMPLE 5   Translate to an algebraic expression.

PHRASE	ALGEBRAIC EXPRESSION
Five more than some number	$5 + n$, or $n + 5$
Half of a number	$\frac{1}{2}t$, or $\frac{t}{2}$
Five more than three times some number	$5 + 3p$, or $3p + 5$
The difference of two numbers	$x - y$
Six less than the product of two numbers	$mn - 6$
Seventy-six percent of some number	$76\%z$, or $0.76z$

Do Exercises 1–9.

b   **FIVE STEPS FOR SOLVING PROBLEMS**

We have studied many new equation-solving tools in this chapter. We now apply them to problem solving. We have purposely used the following strategy to introduce you to algebra.

FIVE STEPS FOR PROBLEM SOLVING IN ALGEBRA

1. *Familiarize* yourself with the problem situation.
2. *Translate* to an equation.
3. *Solve* the equation.
4. *Check* your possible answer in the original problem.
5. *State* your answer clearly.

Of the five steps, probably the most important is the first one: becoming familiar with the problem situation. Here are some hints for familiarization.

To familiarize yourself with the problem situation:

1. If a problem is given in words, read it carefully.

2. Reread the problem, perhaps aloud. Try to verbalize the problem to yourself.

3. List the information given and the questions to be answered. Choose a variable (or variables) to represent the unknown and clearly state what the variable represents. Be descriptive! For example, let L = length, d = distance, and so on.

4. Make a drawing and label it with known information. Also, indicate unknown information, using specific units if given.

5. Find further information. Look up a formula on the inside back cover of this book or in a reference book. Talk to a reference librarian or an expert in the field.

6. Make a table of the given information and the information you have collected. Look for patterns that may help in the translation to an equation.

7. Guess or estimate the answer.

EXAMPLE 6 A 72-in. board is cut into two pieces, and one piece is twice as long as the other. How long are the pieces?

1. Familiarize. We first draw a picture. We let

$$x = \text{the length of the shorter piece.}$$

Then $2x$ = the length of the longer piece.

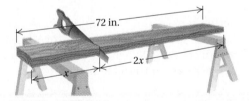

We can further familiarize ourselves with the problem by making some guesses. Let us suppose that $x = 31$ in. Then $2x = 62$ in., and $x + 2x = 93$ in. This is not correct but the procedure does help us to become familiar with the problem.

2. Translate. From the figure, we can see that the lengths of the two pieces add up to 72 in. That gives us our translation.

Length of one piece	plus	Length of other	is	72
↓	↓	↓	↓	↓
x	$+$	$2x$	$=$	72

3. Solve. We solve the equation:

$$x + 2x = 72$$

$$3x = 72 \qquad \text{Collecting like terms}$$

$$\frac{3x}{3} = \frac{72}{3} \qquad \text{Dividing by 3}$$

$$x = 24.$$

10. A 58-in. board is cut into two pieces. One piece is 2 in. longer than the other. How long are the pieces?

4. Check. Do we have an answer to the *problem?* If one piece is 24 in. long, the other, to be twice as long, must be 48 in. long. The lengths of the pieces add up to 72 in. This checks.

5. State. One piece is 24 in. long, and the other is 48 in. long.

Do Exercise 10.

EXAMPLE 7 Five plus three more than a number is nineteen. What is the number?

1. Familiarize. Let x = the number. Then "three more than a number" translates to $x + 3$, and "five plus three more than a number" translates to $5 + (x + 3)$.

2. Translate. The familiarization leads us to the following translation:

Five	plus	Three more than a number	is	Nineteen.
↓	↓	↓	↓	↓
5	+	$(x + 3)$	=	19

3. Solve. We solve the equation:

$$5 + (x + 3) = 19$$
$$x + 8 = 19 \quad \text{Collecting like terms}$$
$$x + 8 - 8 = 19 - 8 \quad \text{Subtracting 8}$$
$$x = 11.$$

4. Check. Three more than 11 is 14. Adding 5 to 14, we get 19. This checks.

5. State. The number is 11.

Do Exercise 11.

11. If 5 is subtracted from three times a certain number, the result is 10. What is the number?

EXAMPLE 8 Better Rent-A-Car rents an intermediate-sized car (such as a Chevrolet, Ford, or Honda) at a daily rate of $44.95 plus 29 cents a mile. A salesperson can spend $100 per day on car rental. How many miles can the person drive per day on the $100 budget?

1. Familiarize. Suppose the salesperson drives 75 mi. Then the cost is

	Daily charge	plus			Mileage charge	
or	↓	↓			↓	
	$44.95	plus	Cost per mile	times	Number of miles driven	
	$44.95	+	$0.29	·	75,	

which is $44.95 + $21.75, or $66.70. This familiarizes us with the way in which a calculation is made. Note that we convert 29 cents to $0.29 so that we have the same units, dollars, throughout the equation. Otherwise, we will not get a correct answer.

Let m = the number of miles that can be driven on a $100 budget.

2. Translate. We reword the problem and translate as follows.

Daily rate ⟶ plus ⟶ Cost per mile ⟶ times ⟶ Number of miles driven ⟶ is ⟶ Cost

$44.95 + $0.29 · m = $100

3. Solve. We solve the equation:

$$44.95 + 0.29m = 100$$

$$100(44.95 + 0.29m) = 100(100)$$ **Multiplying by 100 on both sides to clear the decimals**

$$100(44.95) + 100(0.29m) = 10,000$$ **Using the distributive law**

$$4495 + 29m = 10,000$$ **Simplifying**

$$4495 + 29m - 4495 = 10,000 - 4495$$ **Subtracting 4495**

$$29m = 5505$$

$$\frac{29m}{29} = \frac{5505}{29}$$ **Dividing by 29**

$$m \approx 189.8.$$ **Rounding to the nearest tenth. "≈" means "is approximately equal to."**

4. Check. We check in the original problem. We multiply 189.8 by $0.29, obtaining $55.042. Then we add $55.042 to $44.95 and get $99.992, which is just about the $100 allotted.

5. State. The person can drive about 189.8 mi on the car rental allotment of $100.

Do Exercise 12.

EXAMPLE 9 The state of Colorado is roughly a rectangle whose perimeter is 1300 mi. The length is 110 mi more than the width. Find the dimensions.

1. Familiarize. We first draw a picture. We let

$$w = \text{the width of the rectangle.}$$

Then $w + 110$ = the length.

(We can also let l = the length and $l - 110$ = the width.)
The perimeter P of a rectangle is the distance around it and is given by the formula $2l + 2w = P$, where l = the length and w = the width.

2. Translate. To translate the problem, we substitute $w + 110$ for l and 1300 for P, as follows:

$$2l + 2w = P$$

$$2(w + 110) + 2w = 1300.$$

3. Solve. We solve the equation:

$$2(w + 110) + 2w = 1300$$

$$2w + 220 + 2w = 1300$$ **Multiplying using a distributive law**

$$4w + 220 = 1300$$

$$4w = 1080$$

$$w = 270.$$

Possible dimensions are $w = 270$ mi and $w + 110 = 380$ mi.

12. Better also rents compact cars at a rate of $34.95 plus 27 cents per mile. What mileage will allow the salesperson to stay within a budget of $100?

$w + 110$

$w + 110$

Answer on page A-6

13. A hooked rug has a perimeter of 42 ft. The length is 3 ft more than the width. Find the dimensions of the rug.

4. Check. If the width is 270 mi and the length is 110 mi + 270 mi, or 380 mi, the perimeter is 2(380 mi) + 2(270 mi), or 1300 mi. This checks.

5. State. The width is 270 mi, and the length is 380 mi.

Do Exercise 13.

Answer on page A-6

SIDELIGHTS

STUDY TIPS: EXTRA TIPS ON PROBLEM SOLVING

We have often presented some tips and guidelines to enhance your learning abilities. The following tips, which are focused on problem solving, summarize some points already considered and propose some new ones.

• *The following are the five steps for problem solving:*

1. *Familiarize* **yourself with the problem situation.**

2. *Translate* **the problem to an equation.** As you study more mathematics, you will find that the translation may be to some other kind of mathematical language, such as an inequality.

3. *Solve* **the equation.** If the translation is to some other kind of mathematical language, you would carry out some kind of mathematical manipulation—in the case of an inequality, you would solve it.

4. *Check* **the answer in the original problem.** This does not mean to check in the translated equation. It means to go back to the original worded problem.

5. *State* **the answer to the problem clearly.**

For Step 4 on checking, some further comment is appropriate. *You may be able to translate to an equation and to solve the equation, but you may find that none of the solutions of the equation is the solution of the original problem.* To see how this can happen, consider this example.

EXAMPLE   The sum of two consecutive even integers is 537. Find the integers.

1. *Familiarize.* Suppose we let x = the first number. Then $x + 2$ = the second number.

2. *Translate.* The problem can be translated to the following equation: $x + (x + 2) = 537$.

3. *Solve.* We solve the equation as follows:

$$2x + 2 = 537$$
$$2x = 535$$
$$x = \frac{535}{2}, \text{ or } 267.5.$$

4. *Check.* Then $x + 2 = 269.5$. However, the numbers are not only not even, but they are not integers.

5. *State.* The problem has no solution.

The following are some other tips.

• *To be good at problem solving, do lots of problems.* The situation is similar to learning a skill, such as playing golf. At first you may not be successful, but the more you practice and work at improving your skills, the more successful you will become. For problem solving, do more than just two or three odd-numbered assigned problems. Do them all, and if you have time, do the even-numbered problems as well. Then find another book on the same subject and do problems in that book.

• *Look for patterns when solving problems.* You will eventually see patterns in similar kinds of problems. For example, there is a pattern in the way that you solve problems involving consecutive integers.

• *When translating to an equation, or some other mathematical language, consider the dimensions of the variables and the constants in the equation.* The variables that represent length should all be in the same unit, those that represent money should all be in dollars or in cents, and so on.

Exercise Set 12.5

ANSWERS

1. _____
2. _____
3. _____
4. _____
5. _____
6. _____
7. _____
8. _____
9. _____
10. _____
11. _____
12. _____
13. _____
14. _____
15. _____
16. _____
17. _____
18. _____
19. _____
20. _____
21. _____
22. _____
23. _____
24. _____
25. _____
26. _____
27. _____
28. _____

a Translate to an algebraic expression.

1. 6 more than b

2. 8 more than t

3. 9 less than c

4. 4 less than d

5. 6 increased by q

6. 11 increased by z

7. b more than a

8. c more than d

9. x less than y

10. c less than h

11. x added to w

12. s added to t

13. The sum of r and s

14. The sum of d and f

15. Twice x

16. Three times p

17. 5 multiplied by t

18. The product of 3 and b

19. The product of 97% and some number

20. 43% of some number

b Solve.

21. What number added to 85 is 117?

22. Eight times what number is 2552?

23. An energy saving consultant charges $80 an hour. How many hours did the consultant work in order to make $53,400?

24. A game board has 64 squares. If you win 35 squares and your opponent wins the rest, how many does your opponent win?

25. In a recent year, the cost of four 12-oz boxes of Post® Oat Flakes was $9.56. How much did one box cost?

26. The total amount spent on women's blouses in a recent year was $6.5 billion. This was $0.2 billion more than was spent on women's dresses. How much was spent on women's dresses?

27. When 17 is subtracted from four times a certain number, the result is 211. What is the number?

28. When 36 is subtracted from five times a certain number, the result is 374. What is the number?

Solve.

22. A color television sold for $629 in May. This was $38 more than the January cost. Find the January cost.

23. Selma gets a $4 commission for each appliance that she sells. One week she received $108 in commissions. How many appliances did she sell?

24. An 8-m length of rope is cut into two pieces. One piece is 2 m longer than the other. How long are the pieces?

25. If 14 is added to three times a certain number, the result is 41. Find the number.

26. The perimeter of a rectangular pad is 56 cm. The width is 6 cm less than the length. Find the width and the length.

27. A car rental agency rents compact cars at $41.95 plus 12 cents per mile. A businessperson has a car rental allotment of $258 per day. How many miles can the businessperson travel on the $258 budget?

SKILL MAINTENANCE

28. Find the diameter, the circumference, and the area of a circle when $r = 20$ ft. Use 3.14 for π.

29. Find the volume of a rectangular solid when the length is 20 cm, the width is 18.5 cm, and the height is 4.6 cm.

30. Add: $-12 + 10 + (-19) + (-24)$.

31. Multiply: $(-2) \cdot (-3) \cdot (-5) \cdot (-2) \cdot (-1)$.

SYNTHESIS

32. The total length of the Nile and Amazon Rivers is 13,108 km. If the Amazon were 234 km longer, it would be as long as the Nile. Find the length of each river.

Solve.

33. $2|n| + 4 = 50$

34. $|3n| = 60$

Test: Chapter 12

Solve.

1. $x + 7 = 15$

$x + 7 - 7 = 15 - 7$
8

2. $t - 9 = 17$

$17 + 9 = 26$

3. $3x = -18$

$\dfrac{3x}{3} = \dfrac{-18}{3} \; -6$

4. $-\dfrac{4}{7}x = -28$

$-\dfrac{4}{7}x \cdot \dfrac{7}{4}x = -28 \cdot \dfrac{7}{4} = +49$

Get all T's on left

5. $3t + 7 = 2t - 5$

$3T - 2T + 7 = -5 -$
-12

must by 10 (LCD)

6. $\dfrac{1}{2}x - \dfrac{3}{5} = \dfrac{2}{5}$

$5x - 6 = 4$
$5x - 6 + 6 = 4 + 6$
$5x = 10 \div 5 = 2$

7. $8 - y = 16$

$8 - y - 8 = 16 - 8$
$-1(-y) = 8(-1)$
$y = -8$

8. $-\dfrac{2}{5} + x = -\dfrac{3}{4}$ $(\times 20)$

$-8 + 20x = -15$
$-8 + 8 + 20x = -15 + 8$
$20x = -7$
Divide by $\dfrac{20}{5}$ $\dfrac{-7}{20}$

9. $0.4p + 0.2 = 4.2p - 7.8 - 0.6p$

$\times$ by 10 dec
gets 1 digit to right

$4p + 2 = 42p - 78 - 6p$
$4p + 2 = 36p - 78 -$ subtract 36p from ea side
$-32p + 2 = -78$
$-32p = -80$ subt 2
$p = \dfrac{80}{32} = \dfrac{5}{2}$

10. $3(x + 2) = 27$

$3x + 6 = 27$
$3x = 21 \quad x = 7$

11. $-3x - 6(x - 4) = 9$

$-3x - 6x + 24 = 9$
$-9x + 24 = 9$
$-9x = -15$
$x = \dfrac{15}{9} = \dfrac{5}{3}$

ANSWERS

1. 8
2. 26
3. -6
4.
5.
6.
7.
8.
9.
10.
11.

Solve.

12. The perimeter of a rectangular piece of cardboard is 36 cm. The length is 4 cm greater than the width. Find the width and the length.

13. If you triple a number and then subtract 14, you get $\frac{2}{3}$ of the original number. What is the original number?

12. _____

13. _____

14. _____

14. Translate to an algebraic expression:

Nine less than some number.

15. _____

16. _____

SKILL MAINTENANCE

17. _____

15. Multiply:

$(-9) \cdot (-2) \cdot (-2) \cdot (-5)$.

16. Add:

$$\frac{2}{3} + \left(-\frac{8}{9}\right).$$

17. Find the diameter, the circumference, and the area of a circle when the radius is 70 yd. Use 3.14 for π.

18. _____

18. Find the volume of a rectangular solid when the length is 22 ft, the width is 10 ft, and the height is 6 ft.

SYNTHESIS

19. Solve: $3|w| - 8 = 37$.

19. _____

20. A movie theater had a certain number of tickets to give away. Five people got the tickets. The first got $\frac{1}{3}$ of the tickets, the second got $\frac{1}{4}$ of the tickets, and the third got $\frac{1}{5}$ of the tickets. The fourth person got eight tickets, and there were five tickets left for the fifth person. Find the total number of tickets given away.

20. _____

Cumulative Review: Chapters 1–12

This is a review of the entire textbook. A question that may occur at this point is what notation to use for a particular problem or exercise. Although there is no hard-and-fast rule, especially as you use mathematics outside the classroom, here is the guideline that we follow: Use the notation given in the problem. That is, if the problem is given using mixed numerals, give the answer in mixed numerals. If the problem is given in decimal notation, give the answer in decimal notation.

1. In 47201, what digit tells the number of thousands?

2. Write expanded notation for 7405.

Add and simplify, if appropriate.

3.
$$\begin{array}{r} 7\ 4\ 1 \\ +\ 2\ 7\ 1 \\ \hline \end{array}$$

4.
$$\begin{array}{r} 4\ 9\ 0\ 3 \\ 5\ 2\ 7\ 8 \\ 6\ 3\ 9\ 1 \\ +\ 4\ 5\ 1\ 3 \\ \hline \end{array}$$

5. $\dfrac{2}{13} + \dfrac{1}{26}$

6.
$$\begin{array}{r} 2\dfrac{4}{9} \\ +\ 3\dfrac{1}{3} \\ \hline \end{array}$$

7.
$$\begin{array}{r} 2.0\ 4\ 8 \\ 6\ 3.9\ 1\ 4 \\ +\ 4\ 2\ 8.0\ 0\ 9 \\ \hline \end{array}$$

8. $34.56 + 2.783 + 0.433 + 765.1$

Subtract and simplify, if possible.

9.
$$\begin{array}{r} 6\ 7\ 4 \\ -\ 5\ 2\ 2 \\ \hline \end{array}$$

10.
$$\begin{array}{r} 9\ 4\ 6\ 5 \\ -\ 8\ 7\ 9\ 1 \\ \hline \end{array}$$

11. $\dfrac{7}{8} - \dfrac{2}{3}$

12.
$$\begin{array}{r} 4\dfrac{1}{3} \\ -\ 1\dfrac{5}{8} \\ \hline \end{array}$$

13.
$$\begin{array}{r} 2\ 0.0 \\ -\ \ \ \ 0.0\ 0\ 2\ 7 \\ \hline \end{array}$$

14. $40.03 - 5.789$

Multiply and simplify, if possible.

15.
$$\begin{array}{r} 2\ 9\ 7 \\ \times\ \ \ 1\ 6 \\ \hline \end{array}$$

16.
$$\begin{array}{r} 3\ 4\ 9 \\ \times\ 7\ 6\ 3 \\ \hline \end{array}$$

17. $1\dfrac{3}{4} \cdot 2\dfrac{1}{3}$

18. $\dfrac{9}{7} \cdot \dfrac{14}{15}$

19. $12 \cdot \dfrac{5}{6}$

20.
$$\begin{array}{r} 3\ 4.0\ 9 \\ \times\ \ \ \ \ 7.6 \\ \hline \end{array}$$

Divide and simplify. State the answer using a whole-number quotient and remainder.

21. $6\ )\overline{\ 3\ 4\ 3\ 8}$

22. $3\ 4\ )\overline{\ 1\ 9\ 1\ 4}$

23. Give a mixed numeral for the quotient in Exercise 22.

24. $\dfrac{4}{5} \div \dfrac{8}{15}$

25. $2\dfrac{1}{3} \div 30$

26. $2.7\ )\overline{\ 1\ 0\ 5.3}$

27. Round 68,489 to the nearest thousand.

28. Round 0.4275 to the nearest thousandth.

29. Round $21.8\overline{3}$ to the nearest hundredth.

30. Determine whether 1368 is divisible by 8.

31. Find all the factors of 15.

32. Find the LCM of 16, 25, and 32.

Simplify.

33. $\dfrac{21}{30}$

34. $\dfrac{275}{5}$

35. Convert to a mixed numeral: $\dfrac{18}{5}$.

36. Use $=$ or $\neq$ for ▨ to write a true sentence:

$\dfrac{4}{7}$ ▨ $\dfrac{3}{5}$.

37. Use $<$ or $>$ for ▨ to write a true sentence:

$\dfrac{4}{7}$ ▨ $\dfrac{3}{5}$.

38. Which number is greater, 1.001 or 0.9976?

39. Use $<$ or $>$ for ▨ to write a true sentence:

987 ▨ 879.

40. What part is shaded?

Convert to decimal notation.

41. $\dfrac{37}{1000}$

42. $\dfrac{13}{25}$

43. $\dfrac{8}{9}$

44. 7%

Convert to fractional notation.

45. 4.63

46. $7\dfrac{1}{4}$

47. 40%

Convert to percent notation.

48. $\dfrac{17}{20}$

49. 1.5

Solve.

50. $234 + y = 789$

51. $3.9 \times y = 249.6$

52. $\dfrac{2}{3} \cdot t = \dfrac{5}{6}$

53. $\dfrac{8}{17} = \dfrac{36}{x}$

Solve.

54. Lorenzo makes donations of $627 and $48. What was the total donation?

55. A machine wraps 134 candy bars per minute. How long does it take this machine to wrap 8710 bars?

56. A share of stock bought for $29\frac{5}{8}$ dropped $3\frac{7}{8}$ before it was resold. What was the price when it was resold?

57. At the start of a trip, a car's odometer read 27,428.6 mi, and at the end of the trip, the reading was 27,914.5 mi. How long was the trip?

58. From an income of $12,000, amounts of $2300 and $1600 are paid for federal and state taxes. How much remains after these taxes have been paid?

59. A substitute teacher is paid $47 a day for 9 days. How much was received?

60. A person walks $\frac{3}{5}$ km per hour. At this rate, how far would the person walk in $\frac{1}{2}$ hr?

61. Eight identical dresses cost a total of $679.68. What is the cost of each dress?

62. Eight gallons of exterior paint cover 400 ft². How much paint is needed to cover 650 ft²?

63. Eighteen ounces of a fruit drink costs $3.06. Find the unit price in cents per ounce.

64. What is the simple interest on $4000 principal at 5% for $\frac{3}{4}$ year?

65. A real estate agent received $5880 commission on the sale of an $84,000 home. What was the rate of commission?

66. The population of a city is 29,000 this year and is increasing at 4% per year. What will the population be next year?

67. The ages of students in a math class at a community college are as follows:

18, 21, 26, 31, 32, 18, 50.

Find the average, the median, and the mode of their ages.

Evaluate.

68. 18^2

69. 20^2

Simplify.

70. $\sqrt{9}$

71. $\sqrt{121}$

72. Approximate to three decimal places: $\sqrt{20}$.

73. Find the length of the third side of this right triangle. Give an exact answer and an approximation to three decimal places.

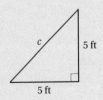

Complete.

74. $\dfrac{1}{3}$ yd = _____ in.

75. 4280 mm = _____ cm

76. 3 days = _____ hr

77. 20,000 g = _____ kg

78. 5 lb = _____ oz

79. 0.008 cg = _____ mg

80. 8190 mL = _____ L

81. 20 qt = _____ gal

82. Find the perimeter and the area.

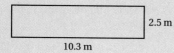

2.5 m

10.3 m

Find the area.

83.

5 in.

10 in.

84.

10.8 yd

8.3 yd

20.2 yd

85.

4 cm

15.4 cm

86. Find the diameter, the circumference, and the area of this circle. Use 3.14 for π.

10.4 in.

87. Find the volume.

2.3 m

2.3 m

10 m

Find the volume. Use 3.14 for π.

88.

16 ft

4 ft

89.

16 cm

4 cm

90.

4 mi

Simplify.

91. $12 \times 20 - 10 \div 5$

92. $4^3 - 5^2 + (16 \cdot 4 + 23 \cdot 3)$

93. $|(-1) \cdot 3|$

94. Add: $17 + (-3)$.

95. Subtract: $\left(-\dfrac{1}{3}\right) - \left(-\dfrac{2}{3}\right)$.

96. Multiply: $(-6) \cdot (-5)$.

97. Multiply: $-\dfrac{5}{7} \cdot \dfrac{14}{35}$.

98. Divide: $\dfrac{48}{-6}$.

Solve.

99. $7 - x = 12$

100. $-4.3x = -17.2$

101. $5x + 7 = 3x - 9$

102. $5(x - 2) - 8(x - 4) = 20$

Translate to an algebraic expression.

103. 17 more than some number

104. 38 percent of some number

105. A game board has 64 squares. If you win 25 squares and your opponent wins the rest, how many does your opponent get?

106. If you add one-third of a number to the number itself, you get 48. What is the number?

Final Examination

1. Write expanded notation for 8345.

2. In 3784, what digit tells the number of hundreds?

Add and simplify, if possible.

3.
$$\begin{array}{r} 4\ 1.3\ 8 \\ 2.0\ 1\ 3 \\ +\ 1\ 7\ 2.2\ 2\ 4\ 7 \\ \hline \end{array}$$

4.
$$\begin{array}{r} 3\dfrac{1}{4} \\ +\ 5\dfrac{1}{2} \\ \hline \end{array}$$

5. $\dfrac{7}{5} + \dfrac{4}{15}$

6.
$$\begin{array}{r} 4\ 3\ 2 \\ +\ 3\ 2\ 7 \\ \hline \end{array}$$

7.
$$\begin{array}{r} 6\ 2\ 0\ 9 \\ 2\ 1\ 3\ 4 \\ 9\ 1\ 8\ 7 \\ +\ 4\ 0\ 3\ 2 \\ \hline \end{array}$$

8. $0.456 + 34.5 + 0.94 + 122.9877$

Subtract and simplify, if possible.

9.
$$\begin{array}{r} 8\ 9\ 8\ 7 \\ -\ 3\ 4\ 2\ 6 \\ \hline \end{array}$$

10.
$$\begin{array}{r} 9\ 0\ 0\ 6 \\ -\ 3\ 0\ 6\ 9 \\ \hline \end{array}$$

11.
$$\begin{array}{r} 3\ 1.2 \\ -\ \ \ 0.8\ 0\ 8 \\ \hline \end{array}$$

12.
$$\begin{array}{r} 3\dfrac{1}{2} \\ -\ 2\dfrac{7}{8} \\ \hline \end{array}$$

13. $\dfrac{3}{4} - \dfrac{2}{3}$

14. $123.04 - 23.88$

Multiply and simplify, if possible.

15. $3\dfrac{1}{4} \cdot 7\dfrac{1}{2}$

16. $\dfrac{8}{9} \cdot \dfrac{3}{4}$

17. $\dfrac{2}{5} \cdot 15$

18.
$$\begin{array}{r} 3\ 4\ 2 \\ \times\ \ \ 1\ 7 \\ \hline \end{array}$$

19.
$$\begin{array}{r} 9\ 8\ 7 \\ \times\ 2\ 3\ 8 \\ \hline \end{array}$$

20.
$$\begin{array}{r} 2\ 5.4\ 3 \\ \times\ \ \ \ \ 8.9 \\ \hline \end{array}$$

Divide and simplify, if possible.

21. $8\,\overline{)\,4\ 1\ 3\ 7}$
State the answer using a whole-number quotient and remainder.

22. Give a mixed numeral for the quotient in Exercise 21.

23. $2\ 1\,\overline{)\,4\ 1\ 3\ 7}$

24. $\dfrac{3}{5} \div \dfrac{9}{10}$

25. $5\dfrac{2}{3} \div 4\dfrac{2}{5}$

26. $1.6\,\overline{)\,7\ 6.8}$

27. Round 42,574 to the nearest thousand.

28. Round 6.7892 to the nearest hundredth.

29. Round $7.3\overline{8}$ to the nearest thousandth.

ANSWERS

1. _____
2. _____
3. _____
4. _____
5. _____
6. _____
7. _____
8. _____
9. _____
10. _____
11. _____
12. _____
13. _____
14. _____
15. _____
16. _____
17. _____
18. _____
19. _____
20. _____
21. _____
22. _____
23. _____
24. _____
25. _____
26. _____
27. _____
28. _____
29. _____

ANSWERS

30. _____

31. _____

32. _____

33. _____

34. _____

35. _____

36. _____

37. _____

38. _____

39. _____

40. _____

41. _____

42. _____

43. _____

44. _____

45. _____

46. _____

47. _____

48. _____

49. _____

50. _____

51. _____

52. _____

53. _____

54. _____

55. _____

56. _____

30. Determine whether 3312 is divisible by 9.

31. Find all the factors of 8.

32. Find the LCM of 23, 46, and 10.

Simplify.

33. $\dfrac{63}{42}$

34. $\dfrac{100}{10}$

35. Convert to a mixed numeral: $\dfrac{23}{3}$.

36. Use = or ≠ for ▧ to write a true sentence:

$\dfrac{3}{5}$ ▧ $\dfrac{6}{10}$.

37. Use < or > for ▧ to write a true sentence:

$\dfrac{6}{11}$ ▧ $\dfrac{5}{9}$.

38. Which is greater, 0.089 or 0.9?

39. Use < or > for ▧ to write a true sentence:

456 ▧ 546.

40. What part is shaded?

Convert to decimal notation.

41. 49.9%

42. $\dfrac{6}{25}$

43. $\dfrac{3}{11}$

44. $\dfrac{786}{100}$

Convert to fractional notation.

45. $5\dfrac{3}{4}$

46. 37%

47. 0.897

Convert to percent notation.

48. 0.77

49. $\dfrac{24}{25}$

Solve.

50. $\dfrac{25}{12} = \dfrac{8}{x}$

51. $y + \dfrac{2}{5} = \dfrac{11}{25}$

52. $78 \cdot t = 1950$

53. $3.9 + y = 249.6$

Solve.

54. The enrollment in a college increased from 3000 to 3150. Find the percent of increase.

55. A consumer spent $83 for groceries, $204.89 for clothes, and $24.71 for gasoline. How much was spent in all?

56. How many $\frac{1}{4}$-lb boxes of chocolate can be filled with 20 lb of chocolates?

57. A part-time worker is paid $58 a day for 6 days. How much was received?

58. A $5\frac{1}{2}$-m flagpole was set $1\frac{3}{4}$ m into the ground. How much was above the ground?

59. A person received checks of $324 and $987. What was the total?

60. A student has $75 in a checking account. Checks of $17 and $19 are written. How much is left in the account?

61. A driver travels 325 mi on 25 gal of gasoline. How many miles per gallon did the driver obtain?

62. A lab technician paid $149.88 for 6 identical lab coats. How much did each lab coat cost?

63. A driver traveled 216 km in 6 hr. At this rate, how far would the driver travel in 15 hr?

64. A 3-lb package of meat costs $11.95. Find the unit cost in dollars per pound.

65. A student got 78% of the questions correct on a test. There were 50 questions. How many of the questions were correct?

66. What is the simple interest on $2000 principal at 6% for $\frac{1}{2}$ year?

67. Find the average, the median, and the mode of this set of numbers:

$11, $12, $12, $12, $19, $25.

68. The circle graph shows color preference for a new car.

a) Which is the favorite color?
b) The survey considered 5000 people. How many preferred red?

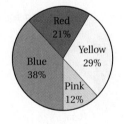

Evaluate.

69. 25^2

70. 16^2

Simplify.

71. $\sqrt{49}$

72. $\sqrt{625}$

73. Approximate to three decimal places: $\sqrt{24}$.

74. Find the length of the third side of this right triangle. Give an exact answer and an approximation to three decimal places.

11 ft
6 ft
a

Complete.

75. 15 ft = _____ yd

76. 2371 m = _____ km

77. 5 L = _____ mL

78. 7 T = _____ lb

79. 24 hr = _____ min

80. 5.34 kg = _____ g

81. 75.4 mg = _____ cg

82. 80 oz = _____ pt

ANSWERS

83. _____

84. _____

85. _____

86. _____

87. _____

88. _____

89. _____

90. _____

91. _____

92. _____

93. _____

94. _____

95. _____

96. _____

97. _____

98. _____

99. _____

100. _____

101. _____

102. _____

103. _____

104. _____

83. Find the perimeter and the area.

2.8 m

9.6 m

Find the area.

84.

3.9 ft

4.7 ft

12.6 ft

85.

17 m

18 m

86.

24 cm

9 cm

87. Find the radius, the circumference, and the area of this circle. Use 3.14 for π.

8.6 yd

88. Find the volume.

4.1 ft

4.1 ft

4.1 ft

Find the volume. Use 3.14 for π.

89.

1000 mi

10 mi

90.

10 m

91.

1000 in.

10 in.

Simplify.

92. $200 \div 25 + 125 \cdot 3$

93. $(2 + 3)^3 - 4^3 + 19 \cdot 2$

94. $|-32|$

95. Subtract: $-7 - 15$.

96. Add: $-7 + (-15)$.

97. Multiply: $-5 \cdot (-6)$.

98. Multiply: $-\dfrac{2}{3} \cdot \dfrac{5}{6}$.

99. Divide: $\dfrac{42}{-7}$.

Solve.

100. $x + 25 = -51.4$

101. $\dfrac{2}{3}x = 18$

102. $0.5m - 13 = 17 + 2.5m$

103. A consultant charges $80 an hour. How many hours did the consultant work in order to make $42,600?

104. If you add two-fifths of a number to the number itself, you get 56. What is the number?

Developmental Units

INTRODUCTION

These developmental units are meant to provide extra instruction for students who have difficulty with any of Sections 1.2, 1.3, 1.5, or 1.6.

· ·

A *Addition*

S *Subtraction*

M *Multiplication*

D *Division*

After finishing Section A, you should be able to:

a Add any two of the numbers 0, 1, 2, 3, 4, 5, 6, 7, 8, 9.

b Find certain sums of three numbers such as 1 + 7 + 9.

c Add two whole numbers when carrying is not necessary.

d Add two whole numbers when carrying is necessary.

Add; think of joining sets of objects.

1. 4 + 5   **2.** 3 + 4

3.   9   **4.**   8
       + 5   + 8

5.   9   **6.**   7
       + 7   + 9

The first printed use of the + symbol was in a book by a German, Johann Widmann, in 1498.

Answers on page A-6

A Addition

a BASIC ADDITION

Basic addition can be explained by counting. The sum

$$3 + 4$$

can be found by counting out a set of 3 objects and a separate set of 4 objects, putting them together, and counting all the objects.

A set of 3   +   A set of 4   =   A set of 7

The numbers to be added are called **addends**. The result is the **sum**.

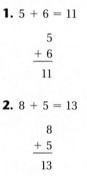

$$3 \quad + \quad 4 \quad = \quad 7$$

Addend   Addend   Sum

EXAMPLES   Add. Think of putting sets of objects together.

1. 5 + 6 = 11

$$\begin{array}{r} 5 \\ + 6 \\ \hline 11 \end{array}$$

2. 8 + 5 = 13

$$\begin{array}{r} 8 \\ + 5 \\ \hline 13 \end{array}$$

We can also do these problems by counting up from one of the numbers. For example, in Example 1, we start at 5 and count up 6 times: 6, 7, 8, 9, 10, 11.

Do Exercises 1–6.

What happens when we add 0? Think of a set of 5 objects. If we add 0 objects to it, we still have 5 objects. Similarly, if we have a set with 0 objects in it and add 5 objects to it, we have a set with 5 objects. Thus,

$$5 + 0 = 5 \quad \text{and} \quad 0 + 5 = 5.$$

> Adding 0 to a number does not change the number:
> $a + 0 = 0 + a = a.$

EXAMPLES Add.

3. $0 + 9 = 9$

$$\begin{array}{r} 0 \\ +\,9 \\ \hline 9 \end{array}$$

4. $0 + 0 = 0$

$$\begin{array}{r} 0 \\ +\,0 \\ \hline 0 \end{array}$$

5. $97 + 0 = 97$

$$\begin{array}{r} 97 \\ +\,0 \\ \hline 97 \end{array}$$

Do Exercises 7–12.

Your objective for this part of the section is to be able to add any of the numbers 0, 1, 2, 3, 4, 5, 6, 7, 8, 9. Adding 0 is easy. The rest of the sums are listed in this table. Memorize the table by saying it to yourself over and over or by using flash cards.

+	1	2	3	4	5	6	7	8	9
1	2	3	4	5	6	7	8	9	10
2	3	4	5	6	7	8	9	10	11
3	4	5	6	7	8	9	10	11	12
4	5	6	7	8	9	10	11	12	13
5	6	7	8	9	10	11	12	13	14
6	7	8	9	10	11	12	13	14	15
7	8	9	10	11	12	13	14	15	16
8	9	10	11	12	13	14	15	16	17
9	10	11	12	13	14	15	16	17	18

$6 + 7 = 13$
Find 6 at the left, and 7 at the top.

$7 + 6 = 13$
Find 7 at the left, and 6 at the top.

It is very important that you *memorize* the basic addition facts! If you do not, you will always have trouble with addition.

Note the following.

$3 + 4 = 7$	$7 + 6 = 13$	$7 + 2 = 9$
$4 + 3 = 7$	$6 + 7 = 13$	$2 + 7 = 9$

We can add whole numbers in any order. This is the *commutative law of addition.* Because of this law, you need to learn only about half the table above, as shown by the shading.

Do Exercises 13 and 14.

b CERTAIN SUMS OF THREE NUMBERS

To add $3 + 5 + 4$, we can add 3 and 5, then 4:

$$\begin{array}{c} 3 + 5 + 4 \\ \downarrow\!\swarrow \\ 8 + 4 \\ \downarrow\!\swarrow \\ 12. \end{array}$$

We can also add 5 and 4, then 3:

$$\begin{array}{c} 3 + 5 + 4 \\ \downarrow\!\swarrow \\ 3 + 9 \\ \downarrow\!\swarrow \\ 12. \end{array}$$

Either way we get 12.

Add.

7. $8 + 0$

8. $0 + 8$

9.
$$\begin{array}{r} 7 \\ +\,0 \\ \hline \end{array}$$

10.
$$\begin{array}{r} 46 \\ +\,0 \\ \hline \end{array}$$

11. $0 + 13$

12. $58 + 0$

Complete the table.

13.

+	1	2	3	4	5
1			4		
2					
3			7		
4					
5					

14.

+	6	5	7	4	9
7			14		
9					
5			9		
8					
4					

Answers on page A-6

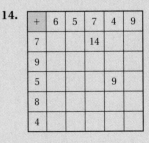

Add from the top mentally.

15.
```
   1
   6
 + 9
```

16.
```
   2
   3
 + 4
```

17.
```
   6
   1
 + 4
```

18.
```
   5
   2
 + 8
```

Add.

19.
```
   2 4
 + 3 5
```

20.
```
   3 4 6
 + 2 0 3
```

21.
```
   8 3 2 7
 + 1 6 5 2
```

22.
```
   3 4 6 1
 + 2 0 3 5
```

EXAMPLE 6   Add from the top mentally.

```
   1       We first add 1 and 7,        1
   7       getting 8. Then we add       7 ⟶ 8
 + 9       8 and 9, getting 17.       + 9   9 ⟶ 17
```

EXAMPLE 7   Add from the top mentally.

```
   2
   4 ⟶ 6
 + 8   8 ⟶ 14
```

Do Exercises 15–18.

c ADDITION (NO CARRYING)

To add larger numbers, we can add the ones first, then the tens, then the hundreds, and so on.

EXAMPLE 8   Add: 5722 + 3234.

```
   5 7 2 2     Add ones.
 + 3 2 3 4
         6
```

```
   5 7 2 2     Add tens.
 + 3 2 3 4
       5 6
```

```
   5 7 2 2     Add hundreds.
 + 3 2 3 4
     9 5 6
```

```
   5 7 2 2     Add thousands.
 + 3 2 3 4
   8 9 5 6
```

This is for explanation.

```
   5 7 2 2
 + 3 2 3 4
   8 9 5 6
```

You should write only this.

Do Exercises 19–22.

d | ADDITION (WITH CARRYING)

CARRYING TENS

| EXAMPLE 9 | Add: 18 + 27.

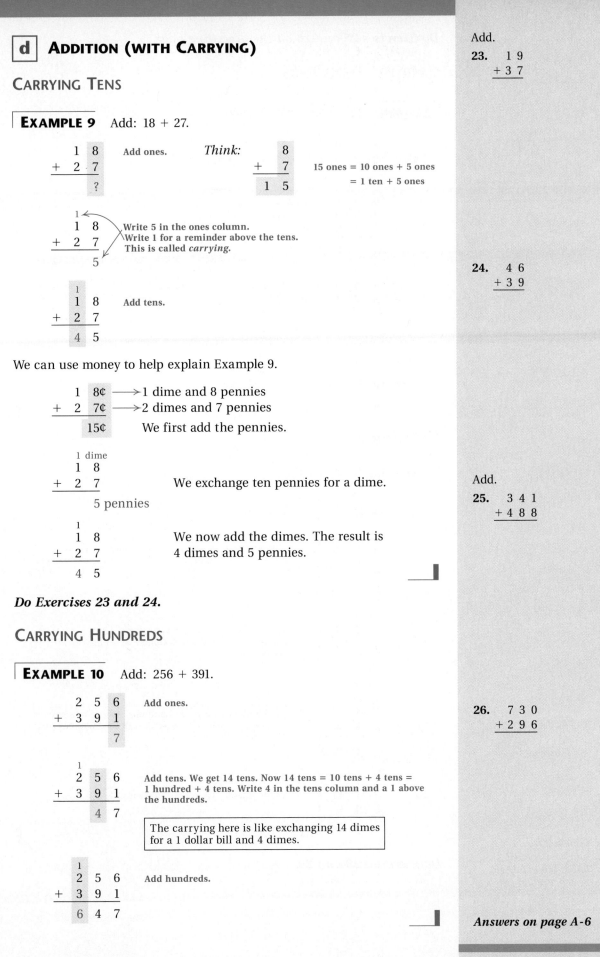

```
  1  8      Add ones.        Think:        8
+ 2  7                                   +  7      15 ones = 10 ones + 5 ones
     ?                                     1 5             = 1 ten + 5 ones
```

```
   1
   1  8    Write 5 in the ones column.
+  2  7    Write 1 for a reminder above the tens.
   ───     This is called carrying.
      5
```

```
   1
   1  8    Add tens.
+  2  7
   ───
   4  5
```

We can use money to help explain Example 9.

```
  1  8¢ ──→ 1 dime and 8 pennies
+ 2  7¢ ──→ 2 dimes and 7 pennies
     15¢    We first add the pennies.
```

```
   1 dime
   1  8
+  2  7    We exchange ten pennies for a dime.
   ─────
   5 pennies
```

```
   1
   1  8    We now add the dimes. The result is
+  2  7    4 dimes and 5 pennies.
   ───
   4  5
```

Do Exercises 23 and 24.

CARRYING HUNDREDS

| EXAMPLE 10 | Add: 256 + 391.

```
  2  5  6     Add ones.
+ 3  9  1
  ───────
        7
```

```
   1
  2  5  6     Add tens. We get 14 tens. Now 14 tens = 10 tens + 4 tens =
+ 3  9  1     1 hundred + 4 tens. Write 4 in the tens column and a 1 above
  ───────     the hundreds.
     4  7
```

┌───┐
│ The carrying here is like exchanging 14 dimes   │
│ for a 1 dollar bill and 4 dimes.   │
└───┘

```
   1
  2  5  6     Add hundreds.
+ 3  9  1
  ───────
  6  4  7
```

Add.

23.   1 9
        + 3 7

24.   4 6
        + 3 9

Add.

25.   3 4 1
        + 4 8 8

26.   7 3 0
        + 2 9 6

Answers on page A-6

A   ADDITION

601

27. Add.

$$\begin{array}{r} 7\ 8\ 5\ 0 \\ +\ 4\ 8\ 4\ 8 \\ \hline \end{array}$$

Add.

28.
$$\begin{array}{r} 7\ 9\ 8\ 9 \\ +\ 5\ 6\ 7\ 2 \\ \hline \end{array}$$

29.
$$\begin{array}{r} 5\ 6,7\ 8\ 9 \\ +\ 1\ 4,5\ 3\ 9 \\ \hline \end{array}$$

Answers on page A-6

Do Exercises 25 and 26 on the preceding page.

CARRYING THOUSANDS

EXAMPLE 11 Add: 4803 + 3792.

$$\begin{array}{r} 4\ 8\ 0\ 3 \\ +\ 3\ 7\ 9\ 2 \\ \hline 5 \end{array}$$ Add ones.

$$\begin{array}{r} 4\ 8\ 0\ 3 \\ +\ 3\ 7\ 9\ 2 \\ \hline 9\ 5 \end{array}$$ Add tens.

$$\begin{array}{r} ^{1} \\ 4\ 8\ 0\ 3 \\ +\ 3\ 7\ 9\ 2 \\ \hline 5\ 9\ 5 \end{array}$$ Add hundreds. We get 15 hundreds. Now 15 hundreds = 10 hundreds + 5 hundreds = 1 thousand + 5 hundreds. Write 5 in the hundreds column and 1 above the thousands.

$$\begin{array}{r} ^{1} \\ 4\ 8\ 0\ 3 \\ +\ 3\ 7\ 9\ 2 \\ \hline 8\ 5\ 9\ 5 \end{array}$$ Add thousands.

Do Exercise 27.

COMBINED CARRYING

EXAMPLE 12 Add: 5767 + 4993.

$$\begin{array}{r} ^{1} \\ 5\ 7\ 6\ 7 \\ +\ 4\ 9\ 9\ 3 \\ \hline 0 \end{array}$$ Add ones. We get 10 ones. Now 10 ones = 1 ten + 0 ones. Write 0 in the ones column and 1 above the tens.

$$\begin{array}{r} ^{1\ 1} \\ 5\ 7\ 6\ 7 \\ +\ 4\ 9\ 9\ 3 \\ \hline 6\ 0 \end{array}$$ Add tens. We get 16 tens. Now 16 tens = 1 hundred + 6 tens. Write 6 in the tens column and 1 above the hundreds.

$$\begin{array}{r} ^{1\ 1\ 1} \\ 5\ 7\ 6\ 7 \\ +\ 4\ 9\ 9\ 3 \\ \hline 7\ 6\ 0 \end{array}$$ Add hundreds. We get 17 hundreds. Now 17 hundreds = 1 thousand + 7 hundreds. Write 7 in the hundreds column and 1 above the thousands.

$$\begin{array}{r} ^{1\ 1\ 1} \\ 5\ 7\ 6\ 7 \\ +\ 4\ 9\ 9\ 3 \\ \hline 1\ 0\ 7\ 6\ 0 \end{array}$$ Add thousands. We get 10 thousands.

Do Exercises 28 and 29.

Exercise Set A

ANSWERS

1. _____
2. _____
3. _____
4. _____
5. _____
6. _____
7. _____
8. _____
9. _____
10. _____
11. _____
12. _____
13. _____
14. _____
15. _____
16. _____
17. _____
18. _____
19. _____
20. _____
21. _____
22. _____
23. _____
24. _____
25. _____
26. _____
27. _____
28. _____
29. _____
30. _____
31. _____
32. _____
33. _____
34. _____
35. _____
36. _____
37. _____
38. _____
39. _____
40. _____
41. _____
42. _____

a Add. Try to do these mentally. If you have trouble, think of putting objects together.

1. $\begin{array}{r} 9 \\ + 2 \\ \hline \end{array}$

2. $\begin{array}{r} 8 \\ + 9 \\ \hline \end{array}$

3. $\begin{array}{r} 8 \\ + 7 \\ \hline \end{array}$

4. $\begin{array}{r} 6 \\ + 7 \\ \hline \end{array}$

5. $\begin{array}{r} 7 \\ + 8 \\ \hline \end{array}$

6. $\begin{array}{r} 9 \\ + 5 \\ \hline \end{array}$

7. $\begin{array}{r} 5 \\ + 7 \\ \hline \end{array}$

8. $\begin{array}{r} 7 \\ + 5 \\ \hline \end{array}$

9. $\begin{array}{r} 7 \\ + 6 \\ \hline \end{array}$

10. $\begin{array}{r} 5 \\ + 6 \\ \hline \end{array}$

11. $\begin{array}{r} 7 \\ + 9 \\ \hline \end{array}$

12. $\begin{array}{r} 9 \\ + 8 \\ \hline \end{array}$

13. $\begin{array}{r} 9 \\ + 7 \\ \hline \end{array}$

14. $\begin{array}{r} 8 \\ + 4 \\ \hline \end{array}$

15. $\begin{array}{r} 9 \\ + 1 \\ \hline \end{array}$

16. $\begin{array}{r} 8 \\ + 2 \\ \hline \end{array}$

17. $\begin{array}{r} 3 \\ + 8 \\ \hline \end{array}$

18. $\begin{array}{r} 0 \\ + 7 \\ \hline \end{array}$

19. $\begin{array}{r} 4 \\ + 3 \\ \hline \end{array}$

20. $\begin{array}{r} 2 \\ + 9 \\ \hline \end{array}$

21. $\begin{array}{r} 4 \\ + 4 \\ \hline \end{array}$

22. $\begin{array}{r} 0 \\ + 0 \\ \hline \end{array}$

23. $\begin{array}{r} 3 \\ + 0 \\ \hline \end{array}$

24. $\begin{array}{r} 9 \\ + 9 \\ \hline \end{array}$

25. $\begin{array}{r} 8 \\ + 6 \\ \hline \end{array}$

26. $\begin{array}{r} 0 \\ + 9 \\ \hline \end{array}$

27. $\begin{array}{r} 3 \\ + 7 \\ \hline \end{array}$

28. $\begin{array}{r} 6 \\ + 8 \\ \hline \end{array}$

29. $\begin{array}{r} 2 \\ + 2 \\ \hline \end{array}$

30. $\begin{array}{r} 7 \\ + 7 \\ \hline \end{array}$

31. $\begin{array}{r} 6 \\ + 5 \\ \hline \end{array}$

32. $\begin{array}{r} 2 \\ + 0 \\ \hline \end{array}$

33. $\begin{array}{r} 7 \\ + 8 \\ \hline \end{array}$

34. $\begin{array}{r} 7 \\ + 9 \\ \hline \end{array}$

35. $\begin{array}{r} 8 \\ + 8 \\ \hline \end{array}$

36. $\begin{array}{r} 8 \\ + 1 \\ \hline \end{array}$

37. $\begin{array}{r} 8 \\ + 3 \\ \hline \end{array}$

38. $\begin{array}{r} 5 \\ + 8 \\ \hline \end{array}$

39. $\begin{array}{r} 5 \\ + 9 \\ \hline \end{array}$

40. $\begin{array}{r} 4 \\ + 1 \\ \hline \end{array}$

41. $\begin{array}{r} 4 \\ + 7 \\ \hline \end{array}$

42. $\begin{array}{r} 6 \\ + 1 \\ \hline \end{array}$

43. $6 + 7$ **44.** $7 + 7$ **45.** $3 + 9$ **46.** $6 + 0$ **47.** $6 + 4$

48. $9 + 3$ **49.** $5 + 5$ **50.** $5 + 3$ **51.** $1 + 1$ **52.** $4 + 5$

53. $9 + 4$ **54.** $0 + 8$ **55.** $4 + 6$ **56.** $2 + 7$ **57.** $3 + 7$

58. $3 + 3$ **59.** $5 + 8$ **60.** $3 + 6$ **61.** $4 + 4$ **62.** $4 + 7$

63. $8 + 8$ **64.** $5 + 2$ **65.** $4 + 8$ **66.** $6 + 6$ **67.** $3 + 5$

68. $0 + 4$ **69.** $3 + 4$ **70.** $2 + 8$ **71.** $6 + 9$ **72.** $4 + 9$

b Add from the top mentally.

73.
$$\begin{array}{r} 1 \\ 1 \\ + 2 \\ \hline \end{array}$$
74.
$$\begin{array}{r} 1 \\ 2 \\ + 3 \\ \hline \end{array}$$
75.
$$\begin{array}{r} 1 \\ 4 \\ + 3 \\ \hline \end{array}$$
76.
$$\begin{array}{r} 1 \\ 3 \\ + 4 \\ \hline \end{array}$$
77.
$$\begin{array}{r} 1 \\ 6 \\ + 8 \\ \hline \end{array}$$

78.
$$\begin{array}{r} 1 \\ 8 \\ + 3 \\ \hline \end{array}$$
79.
$$\begin{array}{r} 1 \\ 7 \\ + 5 \\ \hline \end{array}$$
80.
$$\begin{array}{r} 3 \\ 2 \\ + 5 \\ \hline \end{array}$$
81.
$$\begin{array}{r} 4 \\ 3 \\ + 5 \\ \hline \end{array}$$
82.
$$\begin{array}{r} 1 \\ 7 \\ + 9 \\ \hline \end{array}$$

83.
$$\begin{array}{r} 5 \\ 2 \\ + 6 \\ \hline \end{array}$$
84.
$$\begin{array}{r} 4 \\ 5 \\ + 1 \\ \hline \end{array}$$
85.
$$\begin{array}{r} 1 \\ 9 \\ + 6 \\ \hline \end{array}$$
86.
$$\begin{array}{r} 1 \\ 8 \\ + 7 \\ \hline \end{array}$$

c Add.

87.
```
  1 4
+   5
```

88.
```
  2 3
+ 1 6
```

89.
```
  3 4
+ 6 1
```

90.
```
  5 4
+ 3 5
```

91.
```
  6 7
+ 2 0
```

92.
```
  7 8
+   1
```

93.
```
  3 0 8
+ 5 4 1
```

94.
```
  4 9 6
+ 5 0 3
```

95.
```
  7 0 0
+ 2 0 0
```

96.
```
  8 0 1
+   6 7
```

97.
```
  7 6 5
+ 1 1 0
```

98.
```
  6 6 6
+ 3 3 3
```

99.
```
  5 2 3
+ 3 2 5
```

100.
```
  7 4 7
+ 1 3 0
```

101.
```
  7 8 9
+ 9 0 0
```

102.
```
  8 2 5 0
+ 9 4 3 0
```

103.
```
  6 5 5 2
+ 4 3 2 1
```

104.
```
  3 4 0 6
+ 1 2 9 3
```

105.
```
  7 2 2 5
+ 2 5 2 2
```

106.
```
  7 3 4 0
+ 3 5 2 7
```

107.
```
  4 8 2 5
+ 5 0 7 0
```

108.
```
  2 0 7 3
+ 1 9 2 5
```

109.
```
  7 3 4 0
+ 2 6 5 8
```

110.
```
  9 1 1 1
+ 9 1 1 1
```

111.
```
  2 3 4 5
+ 5 4 3 2
```

112.
```
  7 8 8 9
+ 9 0 0 0
```

113.
```
  9 9 8 6
+ 9 0 1 3
```

114.
```
  5 2,4 3 3
+ 1 2,0 5 6
```

115.
```
  4 3,7 2 3
+ 5 6,2 7 6
```

116.
```
  5 1,6 7 0
+ 2 6,1 0 7
```

ANSWERS

87. _____

88. _____

89. _____

90. _____

91. _____

92. _____

93. _____

94. _____

95. _____

96. _____

97. _____

98. _____

99. _____

100. _____

101. _____

102. _____

103. _____

104. _____

105. _____

106. _____

107. _____

108. _____

109. _____

110. _____

111. _____

112. _____

113. _____

114. _____

115. _____

116. _____

d Add.

117.
```
   3 8
 +   8
```

118.
```
   1 7
 +   9
```

119.
```
   1 7
 + 3 8
```

120.
```
   9 5
 +   6
```

121.
```
   8 6 2
 + 7 8 1
```

122.
```
   9 9 9
 +   1 1
```

123.
```
   3 5 5
 + 4 9 1
```

124.
```
   2 8 0
 + 3 4 8
```

125.
```
   8 1 4
 + 3 9 0
```

126.
```
   2 7 4
 + 3 3 3
```

127.
```
   9 9 9 0
 +     1 0
```

128.
```
   9 9 9
 +   1 1
```

129.
```
   9 9 9
 + 1 1 1
```

130.
```
   8 3 9
 + 3 8 8
```

131.
```
   9 0 9
 + 2 0 2
```

132.
```
   8 0 8
 + 9 0 9
```

133.
```
   8 7 1 8
 + 1 4 2 0
```

134.
```
   3 8 5 4
 + 2 7 0 0
```

135.
```
   4 8 2 8
 + 1 2 8 3
```

136.
```
   6 9 9 5
 + 1 4 3 2
```

137.
```
   9 8 8 9
 +       1
```

138.
```
   6 8 8 9
 + 4 7 2 3
```

139.
```
   9 1 2 8
 + 1 9 9 7
```

140.
```
   8 8 9 8
 + 6 6 4 5
```

141.
```
   9 9 8 9
 + 6 7 8 5
```

142.
```
   4 6,8 8 9
 + 2 1,7 8 6
```

143.
```
   2 3,4 4 8
 + 1 0,9 8 9
```

144.
```
   6 7,6 5 8
 + 9 8,7 8 6
```

145.
```
   7 7,5 4 8
 + 2 3,7 6 7
```

146.
```
   4 4,6 8 4
 +   4,7 6 5
```

SYNTHESIS

147. Find the next five terms of this sequence.

1, 13, 25, 37, _____, _____, _____, _____, _____

148. Look for a pattern and complete two more rows of this number "triangle."

```
        1
       1 1
      1 2 1
     1 3 3 1
    1 4 6 4 1
  1 5 10 10 5 1
```

S Subtraction

a BASIC SUBTRACTION

Subtraction can be explained by taking away part of a set.

| **EXAMPLE 1** Subtract: $7 - 3$.

We can do this by counting out 7 objects and then taking away 3 of them. Then we count the number that remain: $7 - 3 = 4$.

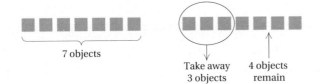

7 objects

Take away
3 objects

4 objects
remain

We could also do this mentally by starting at 7 and counting down 3 times: 6, 5, 4.

| **EXAMPLES** Subtract. Think of "take away."

2. $11 - 6 = 5$ *Take away: "11 take away 6 is 5."*

$$\begin{array}{r} 11 \\ -\ 6 \\ \hline 5 \end{array}$$

3. $17 - 9 = 8$

$$\begin{array}{r} 17 \\ -\ 9 \\ \hline 8 \end{array}$$

Do Exercises 1–4.

In Developmental Unit A, you memorized an addition table. That table will enable you to subtract also. First, let's recall how addition and subtraction are related.

An addition:

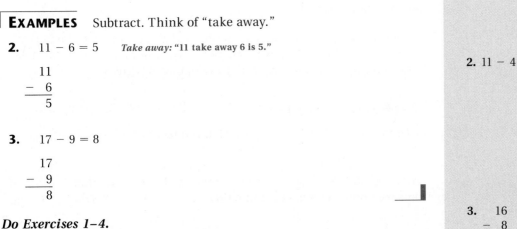

4   +   3   =   7

Two related subtractions.

A.

$7 - 3$ ⟵   =   4

B.

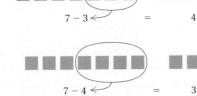

$7 - 4$ ⟵   =   3

Answers on page A-6

OBJECTIVES

After finishing Section S, you should be able to:

a Find basic differences such as $5 - 3$, $13 - 8$, and so on.

b Subtract one whole number from another when borrowing is not necessary.

c Subtract one whole number from another when borrowing is necessary.

Subtract.

1. $10 - 6$

2. $11 - 4$

3. $\begin{array}{r} 16 \\ -\ 8 \\ \hline \end{array}$

4. $\begin{array}{r} 10 \\ -\ 7 \\ \hline \end{array}$

For each addition fact, write two subtraction facts.

5. $8 + 4 = 12$

6. $6 + 7 = 13$

Subtract. Try to do these mentally.

7. $14 - 6$

8. $12 - 5$

9. $\begin{array}{r} 1\ 3 \\ -\quad 4 \\ \hline \end{array}$

10. $\begin{array}{r} 1\ 1 \\ -\quad 7 \\ \hline \end{array}$

Answers on page A-6

Since we know that

$$4 + 3 = 7, \qquad \text{A basic addition fact}$$

we also know the two subtraction facts

$$7 - 3 = 4 \quad \text{and} \quad 7 - 4 = 3.$$

EXAMPLE 4 From $8 + 9 = 17$, write two subtraction facts.

a) We first move the 8.

$$8 + 9 = 17$$

The related sentence is

$$17 - 8 = 9.$$

b) We next move the 9.

$$8 + 9 = 17$$

The related sentence is

$$17 - 9 = 8.$$

Do Exercises 5 and 6.

We can use the idea that subtraction is defined in terms of addition to think of subtraction as "how much more." We also make use of the table of basic addition facts considered in Developmental Unit A.

EXAMPLE 5 Find: $15 - 7$.

To find $15 - 7$, we ask, "7 plus what number is 15?"

$$7 + \blacksquare = 15$$

From the addition table that we memorized, we know that $7 + 8 = 15$. Thus we know the related subtraction

$$15 - 7 = 8.$$

Do Exercises 7–10.

b SUBTRACTION (NO BORROWING)

To subtract larger numbers, we can subtract the ones first, then the tens, then the hundreds, and so on.

EXAMPLE 6   Subtract: 5787 − 3214.

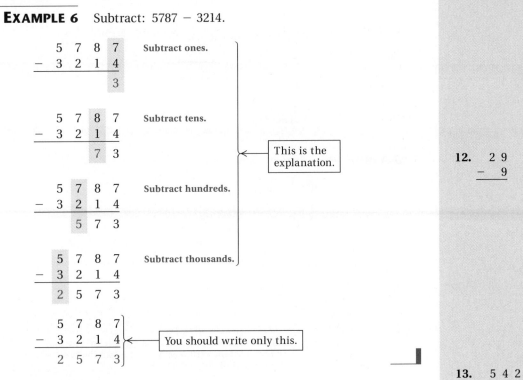

Do Exercises 11–14.

c SUBTRACTION (WITH BORROWING)

We now consider subtraction when borrowing is necessary.

BORROWING FROM THE TENS PLACE

EXAMPLE 7   Subtract: 37 − 18.

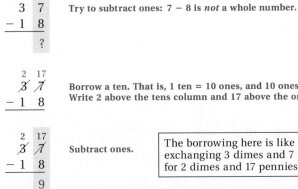

$$\begin{array}{r} 3\ 7 \\ -\ 1\ 8 \\ \hline ? \end{array}$$ Try to subtract ones: 7 − 8 is *not* a whole number.

Borrow a ten. That is, 1 ten = 10 ones, and 10 ones + 7 ones = 17 ones. Write 2 above the tens column and 17 above the ones.

Subtract ones.

The borrowing here is like exchanging 3 dimes and 7 pennies for 2 dimes and 17 pennies.

---

11. $\begin{array}{r} 7\ 8 \\ -\ 6\ 4 \\ \hline \end{array}$

12. $\begin{array}{r} 2\ 9 \\ -\ \ \ 9 \\ \hline \end{array}$

13. $\begin{array}{r} 5\ 4\ 2 \\ -\ 3\ 0\ 1 \\ \hline \end{array}$

14. $\begin{array}{r} 6\ 8\ 9\ 6 \\ -\ 4\ 8\ 7\ 1 \\ \hline \end{array}$

Answers on page A-6

Subtract.

15.
```
   4 6
 − 2 9
```

16.
```
   7 4
 − 3 8
```

Subtract.

17.
```
   6 4 6
 − 1 9 2
```

18.
```
   7 3 3
 − 4 8 3
```

```
   2  17
   3̶  7̶
 −    1  8
      1  9
```
Subtract tens.

```
   2  17
   3̶  7̶
 −    1  8
      1  9
```
You should write only this.

Do Exercises 15 and 16.

BORROWING HUNDREDS

EXAMPLE 8   Subtract: 538 − 275.

```
   5  3  8
 − 2  7  5
         3
```
Subtract ones.

```
   5  3  8
 − 2  7  5
      ?  3
```
Try to subtract tens: 30 − 70 is not a whole number.

```
   4  13
   5̶  3̶  8
 − 2     7  5
            3
```
Borrow a hundred. That is, 1 hundred = 10 tens, and 10 tens + 3 tens = 13 tens. Write 4 above the hundreds column and 13 above the tens.

> The borrowing is like exchanging 5 dollars, 3 dimes, and 8 pennies for 4 dollars, 13 dimes, and 8 pennies.

```
   4  13
   5̶  3̶  8
 − 2  7  5
      6  3
```
Subtract tens.

```
   4  13
   5̶  3̶  8
 − 2  7  5
   2  6  3
```
Subtract hundreds.

```
   4  13
   5̶  3̶  8
 − 2  7  5
   2  6  3
```
You should write only this.

Do Exercises 17 and 18.

BORROWING MORE THAN ONCE

Sometimes we must borrow more than once.

EXAMPLE 9   Subtract: 672 − 394.

$$
\begin{array}{r}
{\scriptstyle 6\ 12} \\
6\,\not7\,\not2 \\
-\ 3\ 9\ 4 \\
\hline
8
\end{array}
$$
  Borrowing a ten to subtract ones.

$$
\begin{array}{r}
{\scriptstyle\quad 16} \\
{\scriptstyle 5\ 6\ 12} \\
\not6\,\not7\,\not2 \\
-\ 3\ 9\ 4 \\
\hline
2\ 7\ 8
\end{array}
$$
  Borrowing a hundred to subtract tens.

Do Exercises 19 and 20.

EXAMPLE 10   Subtract: 6357 − 1769.

$$
\begin{array}{r}
{\scriptstyle\quad\ \ 4\ 17} \\
6\ 3\,\not5\,\not7 \\
-\ 1\ 7\ 6\ 9 \\
\hline
8
\end{array}
$$
  We cannot subtract 9 from 7.
We borrow a ten.

$$
\begin{array}{r}
{\scriptstyle\quad\ 14} \\
{\scriptstyle\ \ 2\ \not4\ 17} \\
6\,\not3\,\not5\,\not7 \\
-\ 1\ 7\ 6\ 9 \\
\hline
8\ 8
\end{array}
$$
  We cannot subtract 6 tens from 4 tens.
We borrow a hundred.

$$
\begin{array}{r}
{\scriptstyle\ 12\ 14} \\
{\scriptstyle 5\ \not2\ \not4\ 17} \\
\not6\,\not3\,\not5\,\not7 \\
-\ 1\ 7\ 6\ 9 \\
\hline
4\ 5\ 8\ 8
\end{array}
$$
  We cannot subtract 7 hundreds from 2 hundreds.
We borrow a thousand.

We can always check by adding the answer to the number being subtracted.

EXAMPLE 11   Subtract: 8341 − 2673. Check by adding.

We check by adding 5668 and 2673.

$$
\begin{array}{r}
{\scriptstyle 12\ 13} \\
{\scriptstyle 7\ \not2\ \not3\ 11} \\
\not8\,\not3\,\not4\,\not1 \\
-\ 2\ 6\ 7\ 3 \\
\hline
5\ 6\ 6\ 8
\end{array}
\qquad
\begin{array}{r}
Check: \\
\\
\end{array}
\qquad
\begin{array}{r}
{\scriptstyle 1\ 1\ 1} \\
5\ 6\ 6\ 8 \\
+\ 2\ 6\ 7\ 3 \\
\hline
8\ 3\ 4\ 1
\end{array}
$$

Do Exercises 21 and 22.

ZEROS IN SUBTRACTION

Before subtracting note the following:

50 is 5 tens;

70 is 7 tens.

Subtract.

19.
$$
\begin{array}{r}
5\ 6\ 3 \\
-\ 1\ 8\ 7 \\
\hline
\end{array}
$$

20.
$$
\begin{array}{r}
7\ 3\ 3 \\
-\ 4\ 8\ 8 \\
\hline
\end{array}
$$

Subtract. Check by adding.

21.
$$
\begin{array}{r}
4\ 2\ 3\ 6 \\
-\ 1\ 6\ 7\ 9 \\
\hline
\end{array}
$$

22.
$$
\begin{array}{r}
7\ 5\ 4\ 1 \\
-\ 3\ 8\ 6\ 7 \\
\hline
\end{array}
$$

Complete.

23. 80 = _____ tens

24. 60 = _____ tens

25. 300 = _____ tens

26. 900 = _____ tens

Answers on page A-6

Complete.

27. 5000 = _____ tens

28. 9000 = _____ tens

29. 5380 = _____ tens

30. 6770 = _____ tens

Subtract.

31.   6 0
        − 1 8

32.   4 8 0
        − 2 5 6

Subtract.

33.   6 0 2
        − 4 6 4

34.   4 0 8
        − 3 6 4

Subtract.

35.   4 0 0 6
        − 1 2 3 8

36.   9 0 0 1
        − 7 8 0 4

Subtract.

37.   3 0 0 0
        − 1 7 5 4

38.   8 0 1 7
        − 3 2 8 9

Answers on page A-6

Then

100 is 10 tens;

200 is 20 tens.

Do Exercises 23–26 on the preceding page.

Also,

230 is 2 hundreds + 3 tens

or 20 tens + 3 tens

or 23 tens.

Similarly,

1000 is 100 tens;

2000 is 200 tens;

4670 is 467 tens.

Do Exercises 27–30.

EXAMPLE 12 Subtract: 50 − 37.

$$\begin{array}{r}\overset{4\ 10}{\cancel{5}\ \cancel{0}} \\ -\ 3\ 7 \\ \hline 1\ 3\end{array}$$

We have 5 tens.

We keep 4 of them in the tens column.

We put 1 ten, or 10 ones, with the ones.

Do Exercises 31 and 32.

EXAMPLE 13 Subtract: 803 − 547.

$$\begin{array}{r}\overset{7\ 9\ 13}{8\ 0\ \cancel{3}} \\ -\ 5\ 4\ 7 \\ \hline 2\ 5\ 6\end{array}$$

We have 8 hundreds, or 80 tens.

We keep 79 tens.

We put 1 ten, or 10 ones, with the ones.

Do Exercises 33 and 34.

EXAMPLE 14 Subtract: 9003 − 2789.

$$\begin{array}{r}\overset{8\ 9\ 9\ 13}{9\ 0\ 0\ \cancel{3}} \\ -\ 2\ 7\ 8\ 9 \\ \hline 6\ 2\ 1\ 4\end{array}$$

We have 9 thousands, or 900 tens.

We keep 899 tens.

We put 1 ten, or 10 ones, with the ones.

Do Exercises 35 and 36.

EXAMPLES Subtract.

15.
$$\begin{array}{r}\overset{4\ 9\ 9\ 10}{5\ 0\ 0\ 0} \\ -\ 2\ 8\ 6\ 1 \\ \hline 2\ 1\ 3\ 9\end{array}$$

16.
$$\begin{array}{r}\overset{\quad\quad 10}{\overset{4\ 9\ 0\ 13}{5\ 0\ 1\ 3}} \\ -\ 1\ 8\ 5\ 7 \\ \hline 3\ 1\ 5\ 6\end{array}$$

Do Exercises 37 and 38.

Exercise Set S

a Subtract. Try to do these mentally.

1. $7 - 3$ **2.** $3 - 2$ **3.** $4 - 1$ **4.** $2 - 0$

5. $3 - 3$ **6.** $8 - 8$ **7.** $5 - 0$ **8.** $6 - 3$

9. $7 - 6$ **10.** $9 - 8$ **11.** $10 - 3$ **12.** $10 - 5$

13. $\begin{array}{r} 7 \\ -\ 0 \\ \hline \end{array}$ **14.** $\begin{array}{r} 9 \\ -\ 0 \\ \hline \end{array}$ **15.** $\begin{array}{r} 8 \\ -\ 8 \\ \hline \end{array}$ **16.** $\begin{array}{r} 7 \\ -\ 7 \\ \hline \end{array}$

17. $\begin{array}{r} 8 \\ -\ 3 \\ \hline \end{array}$ **18.** $\begin{array}{r} 5 \\ -\ 2 \\ \hline \end{array}$ **19.** $\begin{array}{r} 16 \\ -\ 8 \\ \hline \end{array}$ **20.** $\begin{array}{r} 17 \\ -\ 9 \\ \hline \end{array}$

21. $\begin{array}{r} 12 \\ -\ 6 \\ \hline \end{array}$ **22.** $\begin{array}{r} 13 \\ -\ 8 \\ \hline \end{array}$ **23.** $\begin{array}{r} 11 \\ -\ 4 \\ \hline \end{array}$ **24.** $\begin{array}{r} 12 \\ -\ 9 \\ \hline \end{array}$

25. $\begin{array}{r} 14 \\ -\ 7 \\ \hline \end{array}$ **26.** $\begin{array}{r} 18 \\ -\ 9 \\ \hline \end{array}$ **27.** $\begin{array}{r} 13 \\ -\ 7 \\ \hline \end{array}$ **28.** $\begin{array}{r} 15 \\ -\ 9 \\ \hline \end{array}$

29. $\begin{array}{r} 8 \\ -\ 6 \\ \hline \end{array}$ **30.** $\begin{array}{r} 9 \\ -\ 7 \\ \hline \end{array}$

ANSWERS

1. _____
2. _____
3. _____
4. _____
5. _____
6. _____
7. _____
8. _____
9. _____
10. _____
11. _____
12. _____
13. _____
14. _____
15. _____
16. _____
17. _____
18. _____
19. _____
20. _____
21. _____
22. _____
23. _____
24. _____
25. _____
26. _____
27. _____
28. _____
29. _____
30. _____

31. _____
32. _____
33. _____
34. _____
35. _____
36. _____
37. _____
38. _____
39. _____
40. _____
41. _____
42. _____
43. _____
44. _____
45. _____
46. _____
47. _____
48. _____
49. _____
50. _____
51. _____
52. _____
53. _____
54. _____
55. _____
56. _____
57. _____
58. _____
59. _____
60. _____
61. _____
62. _____
63. _____
64. _____
65. _____
66. _____

Subtract.

31. $6 - 6$ **32.** $7 - 7$ **33.** $11 - 7$ **34.** $12 - 8$

35. $5 - 0$ **36.** $4 - 0$ **37.** $13 - 9$ **38.** $14 - 9$

39. $11 - 2$ **40.** $12 - 3$ **41.** $16 - 9$ **42.** $18 - 9$

43. $11 - 6$ **44.** $11 - 5$ **45.** $10 - 4$ **46.** $10 - 8$

47. $14 - 8$ **48.** $15 - 8$ **49.** $9 - 8$ **50.** $10 - 2$

b Subtract.

51.
$$\begin{array}{r} 15 \\ -\ 4 \\ \hline \end{array}$$
52.
$$\begin{array}{r} 26 \\ -13 \\ \hline \end{array}$$
53.
$$\begin{array}{r} 64 \\ -31 \\ \hline \end{array}$$
54.
$$\begin{array}{r} 55 \\ -34 \\ \hline \end{array}$$

55.
$$\begin{array}{r} 67 \\ -20 \\ \hline \end{array}$$
56.
$$\begin{array}{r} 78 \\ -11 \\ \hline \end{array}$$
57.
$$\begin{array}{r} 548 \\ -301 \\ \hline \end{array}$$
58.
$$\begin{array}{r} 596 \\ -403 \\ \hline \end{array}$$

59.
$$\begin{array}{r} 700 \\ -200 \\ \hline \end{array}$$
60.
$$\begin{array}{r} 867 \\ -101 \\ \hline \end{array}$$
61.
$$\begin{array}{r} 765 \\ -111 \\ \hline \end{array}$$
62.
$$\begin{array}{r} 666 \\ -333 \\ \hline \end{array}$$

63.
$$\begin{array}{r} 525 \\ -323 \\ \hline \end{array}$$
64.
$$\begin{array}{r} 747 \\ -130 \\ \hline \end{array}$$
65.
$$\begin{array}{r} 988 \\ -700 \\ \hline \end{array}$$
66.
$$\begin{array}{r} 9450 \\ -8230 \\ \hline \end{array}$$

67.	6552 −4321	68.	3496 −1235	69.	7525 −2522	70.	7547 −3421

71.	5875 −2111	72.	16,843 − 4,321	73.	38,695 −37,004	74.	23,707 −11,607

75.	67,899 −66,673	76.	99,999 − 1	77.	56,780 −56,770	78.	42,111 −32,010

79.	77,654 −66,611	80.	23,456 −12,345

c Subtract.

81.	34 −16	82.	86 −47	83.	93 −28	84.	42 −13

85.	86 −78	86.	98 −89	87.	625 −317	88.	726 −409

89.	735 −609	90.	853 −236	91.	961 −747	92.	787 −698

93.	6769 −2367	94.	6431 −2876	95.	3982 −2479	96.	7654 −1765

97.	5246 −2859	98.	6328 −2679	99.	7641 −3809	100.	8743 − 599

ANSWERS

67. _____
68. _____
69. _____
70. _____
71. _____
72. _____
73. _____
74. _____
75. _____
76. _____
77. _____
78. _____
79. _____
80. _____
81. _____
82. _____
83. _____
84. _____
85. _____
86. _____
87. _____
88. _____
89. _____
90. _____
91. _____
92. _____
93. _____
94. _____
95. _____
96. _____
97. _____
98. _____
99. _____
100. _____

101. _____

102. _____

103. _____

104. _____

105. _____

106. _____

107. _____

108. _____

109. _____

110. _____

111. _____

112. _____

113. _____

114. _____

115. _____

116. _____

117. _____

118. _____

119. _____

120. _____

121. _____

122. _____

123. _____

124. _____

125. _____

101.
$$\begin{array}{r} 12{,}647 \\ -\ \ 4{,}897 \\ \hline \end{array}$$

102.
$$\begin{array}{r} 16{,}222 \\ -\ \ 5{,}777 \\ \hline \end{array}$$

103.
$$\begin{array}{r} 46{,}781 \\ -12{,}988 \\ \hline \end{array}$$

104.
$$\begin{array}{r} 75{,}654 \\ -48{,}785 \\ \hline \end{array}$$

105.
$$\begin{array}{r} 40 \\ -24 \\ \hline \end{array}$$

106.
$$\begin{array}{r} 50 \\ -37 \\ \hline \end{array}$$

107.
$$\begin{array}{r} 70 \\ -54 \\ \hline \end{array}$$

108.
$$\begin{array}{r} 90 \\ -73 \\ \hline \end{array}$$

109.
$$\begin{array}{r} 140 \\ -\ 54 \\ \hline \end{array}$$

110.
$$\begin{array}{r} 470 \\ -189 \\ \hline \end{array}$$

111.
$$\begin{array}{r} 690 \\ -235 \\ \hline \end{array}$$

112.
$$\begin{array}{r} 803 \\ -414 \\ \hline \end{array}$$

113.
$$\begin{array}{r} 703 \\ -132 \\ \hline \end{array}$$

114.
$$\begin{array}{r} 6406 \\ -\ 258 \\ \hline \end{array}$$

115.
$$\begin{array}{r} 2309 \\ -\ 109 \\ \hline \end{array}$$

116.
$$\begin{array}{r} 3406 \\ -1293 \\ \hline \end{array}$$

117.
$$\begin{array}{r} 6807 \\ -3059 \\ \hline \end{array}$$

118.
$$\begin{array}{r} 7340 \\ -3027 \\ \hline \end{array}$$

119.
$$\begin{array}{r} 4037 \\ -2974 \\ \hline \end{array}$$

120.
$$\begin{array}{r} 4007 \\ -1589 \\ \hline \end{array}$$

121.
$$\begin{array}{r} 8000 \\ -2794 \\ \hline \end{array}$$

122.
$$\begin{array}{r} 8002 \\ -6543 \\ \hline \end{array}$$

123.
$$\begin{array}{r} 38{,}000 \\ -37{,}695 \\ \hline \end{array}$$

124.
$$\begin{array}{r} 16{,}043 \\ -11{,}588 \\ \hline \end{array}$$

SYNTHESIS

125. A parent sends a child to a river with two buckets. One bucket measures 3 gal and the other 5 gal. How can the child return with exactly 1 gal?

M Multiplication

a BASIC MULTIPLICATION

To multiply, we begin with two numbers, called **factors**, and get a third number, called a **product**. Multiplication can be explained by counting. The product 3×5 can be found by counting out 3 sets of 5 objects each, joining them (in a rectangular array if desired), and counting all the objects.

$$3 \times 5 = 15$$

Factor   Factor   Product

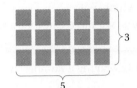

We can also think of multiplication as repeated addition.

$$3 \times 5 = \underbrace{5 + 5 + 5}_{\text{3 addends of 5}} = 15$$

EXAMPLES Multiply. If you have trouble, think either of putting sets of objects together in a rectangular array or of repeated addition.

1. $5 \times 6 = 30$

$$\begin{array}{r} 6 \\ \times\ 5 \\ \hline 30 \end{array}$$

2. $8 \times 4 = 32$

$$\begin{array}{r} 4 \\ \times\ 8 \\ \hline 32 \end{array}$$

Do Exercises 1–4.

MULTIPLYING BY 0

How do we multiply by 0? Consider $4 \cdot 0$. Using repeated addition, we see that

$$4 \cdot 0 = \underbrace{0 + 0 + 0 + 0}_{\text{4 addends of 0}} = 0.$$

We can also think of this using sets. That is, $4 \cdot 0$ is 4 sets with 0 objects in each set, so the total is 0.

Consider $0 \cdot 4$. Using repeated addition, we say that this is 0 addends of 4, which is 0. Using sets, we say that this is 0 sets with 4 objects in each set, which is 0. Thus, we have the following.

> Multiplying by 0 gives 0.

EXAMPLES Multiply.

3. $13 \times 0 = 0$

$$\begin{array}{r} 0 \\ \times\ 13 \\ \hline 0 \end{array}$$

4. $0 \cdot 11 = 0$

$$\begin{array}{r} 11 \\ \times\ 0 \\ \hline 0 \end{array}$$

5. $0 \cdot 0 = 0$

$$\begin{array}{r} 0 \\ \times\ 0 \\ \hline 0 \end{array}$$

Do Exercises 5 and 6.

OBJECTIVES

After finishing Section M, you should be able to:

a Multiply any two of the numbers 0, 1, 2, 3, 4, 5, 6, 7, 8, 9.

b Multiply multiples of 10, 100, and 1000.

c Multiply larger numbers by 0, 1, 2, 3, 4, 5, 6, 7, 8, and 9.

d Multiply by multiples of 10, 100, and 1000.

Multiply. Think of joining sets in a rectangular array or of repeated addition.

1. $7 \cdot 8$ (The dot " $\cdot$ " means the same as " $\times$ ".)

2. $\begin{array}{r} 9 \\ \times\ 4 \\ \hline \end{array}$

3. $4 \cdot 7$

4. $\begin{array}{r} 7 \\ \times\ 6 \\ \hline \end{array}$

Multiply.

5. $8 \cdot 0$

6. $\begin{array}{r} 17 \\ \times\ 0 \\ \hline \end{array}$

Answers on page A-6

Multiply.

7. $8 \cdot 1$

8.
$$\begin{array}{r} 2\,3 \\ \times \quad 1 \\ \hline \end{array}$$

9. Complete the table.

×	2	3	4	5
2				
3			12	
4				
5		15		
6				

10.

×	6	7	8	9
5				
6			48	
7				
8		56		
9				

MULTIPLYING BY 1

How do we multiply by 1? Consider $5 \cdot 1$. Using repeated addition, we see that

$$5 \cdot 1 = \underbrace{1 + 1 + 1 + 1 + 1}_{\text{5 addends of 1}} = 5.$$

We can also think of this using sets. That is, $5 \cdot 1$ is 5 sets with 1 object in each set, so the total is 5.

Consider $1 \cdot 5$. Using repeated addition, we say that this is 1 addend of 5, which is 5. Using sets, we say that this is 1 set of 5 objects, which is 5. Thus, we have the following.

> Multiplying a number by 1 does not change the number:
> $$a \cdot 1 = 1 \cdot a = a.$$

This is a very important property.

EXAMPLES Multiply.

6. $13 \cdot 1 = 13$

$$\begin{array}{r} 1 \\ \times\, 13 \\ \hline 13 \end{array}$$

7. $1 \cdot 7 = 7$

$$\begin{array}{r} 7 \\ \times\, 1 \\ \hline 7 \end{array}$$

8. $1 \cdot 1 = 1$

$$\begin{array}{r} 1 \\ \times\, 1 \\ \hline 1 \end{array}$$

Do Exercises 7 and 8.

You should be able to multiply any of the numbers 0, 1, 2, 3, 4, 5, 6, 7, 8, 9. Multiplying by 0 and 1 is easy. The rest of the products are listed in the table below.

×	2	3	4	5	6	7	8	9
2	4	6	8	10	12	14	16	18
3	6	9	12	15	18	21	24	27
4	8	12	16	20	24	28	32	36
5	10	15	20	25	30	35	40	45
6	12	18	24	30	36	42	48	54
7	14	21	28	35	42	49	56	63
8	16	24	32	40	48	56	64	72
9	18	27	36	45	54	63	72	81

$5 \times 7 = 35$
Find 5 at the left, and 7 at the top.

It is *very* important that you have the basic multiplication facts *memorized*. If you do not, you will always have trouble with multiplication.

The commutative law says that we can multiply numbers in any order. Thus you need to learn only about half the table.

Do Exercises 9 and 10.

Answers on page A-6

b MULTIPLYING MULTIPLES OF 10, 100, AND 1000

We now consider multiplication by multiples of 10, 100, and 1000. These are numbers such as 10, 20, 30, 100, 400, 1000, and 7000.

MULTIPLYING BY A MULTIPLE OF 10

We know that

$$50 = 5 \text{ tens} \qquad 340 = 34 \text{ tens} \quad \text{and} \quad 2340 = 234 \text{ tens}$$
$$= 5 \cdot 10, \qquad\qquad = 34 \cdot 10, \qquad\qquad = 234 \cdot 10.$$

Turning this around, we see that to multiply any number by 10, all we need do is write a 0 on the end of the number.

To multiply a number by 10, write 0 on the end of the number.

EXAMPLES Multiply.

9. $10 \cdot 6 = 60$

10. $10 \cdot 47 = 470$

11. $10 \cdot 583 = 5830$

Do Exercises 11–15.

Let us find $4 \cdot 90$. This is $4 \cdot (9 \text{ tens})$, or 36 tens. The procedure is the same as multiplying 4 and 9 and writing a 0 on the end. Thus, $4 \cdot 90 = 360$.

EXAMPLES Multiply.

12. $5 \cdot 70 = 350$
$\qquad\qquad$ 5 · 7, then write a 0

13. $8 \cdot 80 = 640$

14. $5 \cdot 60 = 300$

Do Exercises 16 and 17.

MULTIPLYING BY A MULTIPLE OF 100

Note the following:

$$300 = 3 \text{ hundreds} \qquad 4700 = 47 \text{ hundreds} \quad \text{and} \quad 56{,}800 = 568 \text{ hundreds}$$
$$= 3 \cdot 100, \qquad\qquad = 47 \cdot 100, \qquad\qquad = 568 \cdot 100.$$

Turning this around, we see that to multiply any number by 100, all we need do is write two 0's on the end of the number.

To multiply a number by 100, write two 0's on the end of the number.

Multiply.
11. $10 \cdot 7$

12. $10 \cdot 45$

13. $10 \cdot 273$

14. $10 \cdot 10$

15. $10 \cdot 100$

Multiply.
16. $\begin{array}{r} 7\,0 \\ \times \quad 8 \\ \hline \end{array}$

17. $\begin{array}{r} 6\,0 \\ \times \quad 6 \\ \hline \end{array}$

Answers on page A-7

Multiply.

18. $100 \cdot 7$ **19.** $100 \cdot 23$

20. $100 \cdot 723$ **21.** $100 \cdot 100$

22. $100 \cdot 1000$

Multiply.

23.
$$\begin{array}{r} 7\,0\,0 \\ \times\quad 8 \\ \hline \end{array}$$

24.
$$\begin{array}{r} 4\,0\,0 \\ \times\quad 4 \\ \hline \end{array}$$

Multiply.

25. $1000 \cdot 9$ **26.** $1000 \cdot 852$

27. $1000 \cdot 10$ **28.** $4000 \cdot 300$

29. $2000 \cdot 3000$

Answers on page A-7

EXAMPLES Multiply.

15. $100 \cdot 6 = 600$

16. $100 \cdot 39 = 3900$

17. $100 \cdot 448 = 44{,}800$

Do Exercises 18–22.

Let us find $4 \cdot 900$. This is $4 \cdot$ (9 hundreds). If we use addition, this is

9 hundreds + 9 hundreds + 9 hundreds + 9 hundreds,

or 36 hundreds,

which is the same as multiplying 4 and 9 and writing two 0's on the end. Thus, $4 \cdot 900 = 3600$.

EXAMPLES Multiply.

18. $6 \cdot 800 = 4800$
$6 \cdot 8$, then write 00

19. $9 \cdot 700 = 6300$

20. $5 \cdot 500 = 2500$

Do Exercises 23 and 24.

MULTIPLYING BY A MULTIPLE OF 1000

Note the following:

$$6000 = 6 \text{ thousands} \quad \text{and} \quad 19{,}000 = 19 \text{ thousands}$$
$$= 6 \cdot 1000 \qquad\qquad\qquad = 19 \cdot 1000.$$

Turning this around, we see that to multiply any number by 1000, all we need do is write three 0's on the end of the number.

To multiply a number by 1000, write three 0's on the end of the number.

EXAMPLES Multiply.

21. $1000 \cdot 8 = 8000$

22. $2000 \cdot 13 = 26{,}000$

23. $1000 \cdot 567 = 567{,}000$

Do Exercises 25–29.

MULTIPLYING MULTIPLES BY MULTIPLES

Let us multiply 50 and 30. This is $50 \cdot$ (3 tens), or 150 tens, or 1500. The procedure is the same as multiplying 5 and 3 and writing two 0's on the end.

To multiply multiples of tens, hundreds, thousands, and so on:

a) Multiply the one-digit numbers.

b) Count the number of zeros.

c) Write that many 0's on the end.

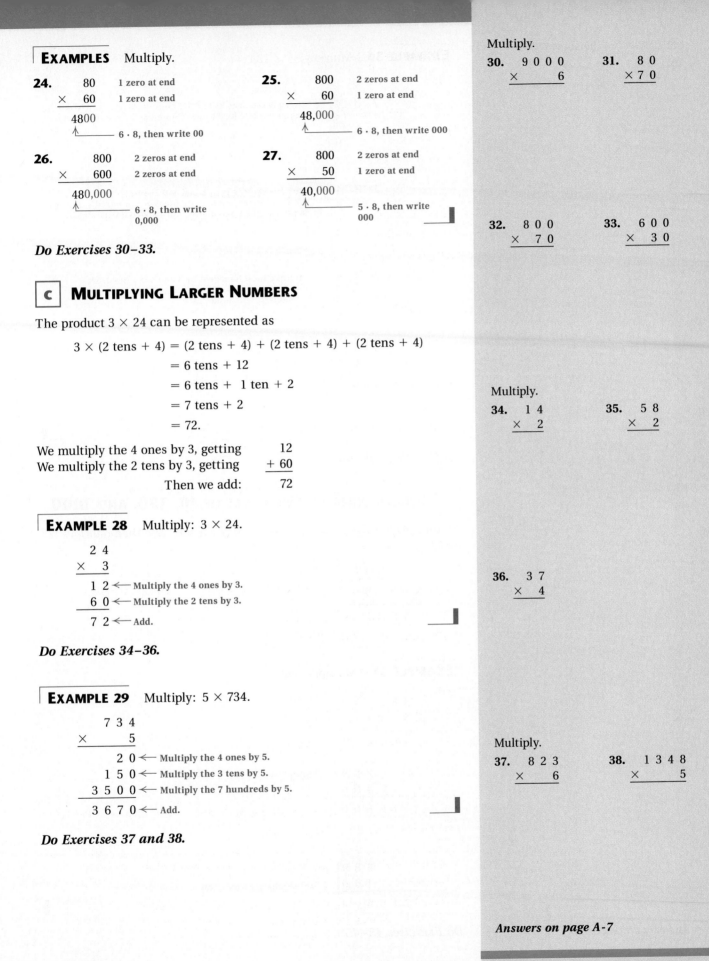

EXAMPLES Multiply.

24. $\begin{array}{r} 80 \\ \times\ 60 \\ \hline 4800 \end{array}$ 1 zero at end
   1 zero at end

   6 · 8, then write 00

25. $\begin{array}{r} 800 \\ \times\ 60 \\ \hline 48,000 \end{array}$ 2 zeros at end
   1 zero at end

   6 · 8, then write 000

26. $\begin{array}{r} 800 \\ \times\ 600 \\ \hline 480,000 \end{array}$ 2 zeros at end
   2 zeros at end

   6 · 8, then write 0,000

27. $\begin{array}{r} 800 \\ \times\ 50 \\ \hline 40,000 \end{array}$ 2 zeros at end
   1 zero at end

   5 · 8, then write 000

Do Exercises 30–33.

c | MULTIPLYING LARGER NUMBERS

The product 3×24 can be represented as

$$3 \times (2\ \text{tens} + 4) = (2\ \text{tens} + 4) + (2\ \text{tens} + 4) + (2\ \text{tens} + 4)$$
$$= 6\ \text{tens} + 12$$
$$= 6\ \text{tens} + 1\ \text{ten} + 2$$
$$= 7\ \text{tens} + 2$$
$$= 72.$$

We multiply the 4 ones by 3, getting 12
We multiply the 2 tens by 3, getting $+\ 60$
Then we add: 72

EXAMPLE 28 Multiply: 3×24.

$$\begin{array}{r} 2\ 4 \\ \times\quad 3 \\ \hline 1\ 2 \\ 6\ 0 \\ \hline 7\ 2 \end{array}$$

← Multiply the 4 ones by 3.
← Multiply the 2 tens by 3.
← Add.

Do Exercises 34–36.

EXAMPLE 29 Multiply: 5×734.

$$\begin{array}{r} 7\ 3\ 4 \\ \times\qquad 5 \\ \hline 2\ 0 \\ 1\ 5\ 0 \\ 3\ 5\ 0\ 0 \\ \hline 3\ 6\ 7\ 0 \end{array}$$

← Multiply the 4 ones by 5.
← Multiply the 3 tens by 5.
← Multiply the 7 hundreds by 5.
← Add.

Do Exercises 37 and 38.

Multiply.

30. $\begin{array}{r} 9\ 0\ 0\ 0 \\ \times\qquad 6 \\ \hline \end{array}$

31. $\begin{array}{r} 8\ 0 \\ \times 7\ 0 \\ \hline \end{array}$

32. $\begin{array}{r} 8\ 0\ 0 \\ \times\ \ 7\ 0 \\ \hline \end{array}$

33. $\begin{array}{r} 6\ 0\ 0 \\ \times\ \ 3\ 0 \\ \hline \end{array}$

Multiply.

34. $\begin{array}{r} 1\ 4 \\ \times\ \ 2 \\ \hline \end{array}$

35. $\begin{array}{r} 5\ 8 \\ \times\ \ 2 \\ \hline \end{array}$

36. $\begin{array}{r} 3\ 7 \\ \times\ \ 4 \\ \hline \end{array}$

Multiply.

37. $\begin{array}{r} 8\ 2\ 3 \\ \times\ \ \ 6 \\ \hline \end{array}$

38. $\begin{array}{r} 1\ 3\ 4\ 8 \\ \times\qquad 5 \\ \hline \end{array}$

Answers on page A-7

Multiply using the short form.

39. 5 8
      × 2

40. 3 7
      × 4

41. 8 2 3
       × 6

42. 1 3 4 8
       × 5

Multiply.

43. 7 4 6
       × 8

44. 7 4 6
       × 8 0

45. 7 4 6
       × 8 0 0

EXAMPLE 30 Multiply: 5×734.

$$
\begin{array}{r}
^2 \\
7\ 3\ 4 \\
\times\ 5 \\
\hline
0
\end{array}
$$

Multiply the ones by 5: $5 \cdot (4 \text{ ones}) = 20 \text{ ones} = 2 \text{ tens} + 0 \text{ ones}$. Write 0 in the ones column and 2 above the tens.

$$
\begin{array}{r}
^1\ ^2 \\
7\ 3\ 4 \\
\times\ 5 \\
\hline
7\ 0
\end{array}
$$

Multiply the 3 tens by 5 and add 2 tens: $5 \cdot (3 \text{ tens}) = 15 \text{ tens}$, $15 \text{ tens} + 2 \text{ tens} = 17 \text{ tens} = 1 \text{ hundred} + 7 \text{ tens}$. Write 7 in the tens column and 1 above the hundreds.

$$
\begin{array}{r}
^1\ ^2 \\
7\ 3\ 4 \\
\times\ 5 \\
\hline
3\ 6\ 7\ 0
\end{array}
$$

Multiply the 7 hundreds by 5 and add 1 hundred: $5 \cdot (7 \text{ hundreds}) = 35 \text{ hundreds}$, $35 \text{ hundreds} + 1 \text{ hundred} = 36 \text{ hundreds}$.

$$
\left.\begin{array}{r}
^1\ ^2 \\
7\ 3\ 4 \\
\times\ 5 \\
\hline
3\ 6\ 7\ 0
\end{array}\right\}
$$

You should write only this.

Avoid writing the reminders unless necessary.

Do Exercises 39–42.

d MULTIPLYING BY MULTIPLES OF 10, 100, AND 1000

To multiply 327 by 50, we multiply by 10 (write a 0), and then multiply 327 by 5.

$$
\begin{array}{r}
3\ 2\ 7 \\
\times\ 5\ \boxed{0} \\
\hline
1\ 6,3\ 5\ 0
\end{array}
$$

← Write a 0.

Multiply $5 \cdot 327$.

EXAMPLE 31 Multiply: 400×289.

$$
\begin{array}{r}
2\ 8\ 9 \\
\times\ 4\ \boxed{0\ 0} \\
\hline
0\ 0
\end{array}
$$

← Write two 0's.

$$
\begin{array}{r}
2\ 8\ 9 \\
\times\ 4\ 0\ 0 \\
\hline
1\ 1\ 5,6\ 0\ 0
\end{array}
$$

Multiply 4 and 289:

$$
\begin{array}{r}
^3\ ^3 \\
2\ 8\ 9 \\
\times\ 4 \\
\hline
1\ 1\ 5\ 6
\end{array}
$$

$$
\left.\begin{array}{r}
^3\ ^3 \\
2\ 8\ 9 \\
\times\ 4\ 0\ 0 \\
\hline
1\ 1\ 5,6\ 0\ 0
\end{array}\right\}
$$

You should write only this.

Do Exercises 43–45.

Exercise Set M

a Multiply. Try to do these mentally.

1.   3
   $\times\,4$

2.   1
   $\times\,3$

3.   6
   $\times\,4$

4.   6
   $\times\,0$

5.   7
   $\times\,1$

6.   0
   $\times\,2$

7.   4
   $\times\,6$

8.   2
   $\times\,2$

9.   10
   $\times\,1$

10.   6
   $\times\,5$

11.   1
   $\times\,10$

12.   5
   $\times\,2$

13.   2
   $\times\,5$

14.   9
   $\times\,7$

15.   3
   $\times\,7$

16.   9
   $\times\,6$

17.   2
   $\times\,6$

18.   7
   $\times\,0$

19.   9
   $\times\,8$

20.   18
   $\times\,1$

21.   8
   $\times\,9$

22.   1
   $\times\,8$

23.   0
   $\times\,4$

24.   8
   $\times\,0$

25.   4
   $\times\,7$

26.   3
   $\times\,8$

27.   1
   $\times\,7$

28.   5
   $\times\,9$

29.   0
   $\times\,5$

30.   2
   $\times\,9$

31.   6
   $\times\,7$

32.   7
   $\times\,1$

33.   0
   $\times\,7$

34.   5
   $\times\,7$

35.   1
   $\times\,9$

36.   9
   $\times\,5$

37.   5
   $\times\,8$

38.   0
   $\times\,0$

39.   8
   $\times\,5$

40.   2
   $\times\,8$

ANSWERS

1. _____
2. _____
3. _____
4. _____
5. _____
6. _____
7. _____
8. _____
9. _____
10. _____
11. _____
12. _____
13. _____
14. _____
15. _____
16. _____
17. _____
18. _____
19. _____
20. _____
21. _____
22. _____
23. _____
24. _____
25. _____
26. _____
27. _____
28. _____
29. _____
30. _____
31. _____
32. _____
33. _____
34. _____
35. _____
36. _____
37. _____
38. _____
39. _____
40. _____

41. _____
42. _____
43. _____
44. _____
45. _____
46. _____
47. _____
48. _____
49. _____
50. _____
51. _____
52. _____
53. _____
54. _____
55. _____
56. _____
57. _____
58. _____
59. _____
60. _____
61. _____
62. _____
63. _____
64. _____
65. _____
66. _____
67. _____
68. _____
69. _____
70. _____
71. _____
72. _____
73. _____
74. _____
75. _____
76. _____
77. _____
78. _____
79. _____
80. _____
81. _____
82. _____
83. _____
84. _____

41. $4 \cdot 4$ **42.** $5 \cdot 5$ **43.** $8 \cdot 8$ **44.** $9 \cdot 9$

45. $1 \cdot 1$ **46.** $0 \cdot 0$ **47.** $3 \cdot 3$ **48.** $2 \cdot 2$

49. $6 \cdot 6$ **50.** $1 \cdot 8$ **51.** $8 \cdot 6$ **52.** $0 \cdot 1$

53. $3 \cdot 9$ **54.** $2 \cdot 9$ **55.** $9 \cdot 5$ **56.** $6 \cdot 0$

57. $6 \cdot 7$ **58.** $10 \cdot 1$ **59.** $0 \cdot 10$ **60.** $6 \cdot 8$

61. $9 \cdot 6$ **62.** $8 \cdot 0$ **63.** $5 \cdot 5$ **64.** $9 \cdot 8$

65. $7 \cdot 7$ **66.** $3 \cdot 5$ **67.** $9 \cdot 3$ **68.** $0 \cdot 2$

69. $1 \cdot 8$ **70.** $1 \cdot 9$ **71.** $2 \cdot 1$ **72.** $5 \cdot 1$

73. $8 \cdot 4$ **74.** $3 \cdot 2$ **75.** $5 \cdot 3$ **76.** $1 \cdot 6$

77. $4 \cdot 2$ **78.** $2 \cdot 4$ **79.** $4 \cdot 5$ **80.** $5 \cdot 4$

81. $8 \cdot 8$ **82.** $4 \cdot 4$ **83.** $5 \cdot 2$ **84.** $8 \cdot 0$

DEVELOPMENTAL UNITS

624

b Multiply.

85.
```
    1 0
  ×   8
```

86.
```
    1 0
  ×   6
```

87.
```
    1 0
  ×   9
```

88.
```
      9
  × 1 0
```

89.
```
      7
  × 1 0
```

90.
```
    2 0
  ×   8
```

91.
```
    3 0
  ×   7
```

92.
```
    4 5
  × 1 0
```

93.
```
    7 8
  × 1 0
```

94.
```
    5 0
  ×   9
```

95.
```
    8 0
  ×   7
```

96.
```
    9 0
  ×   4
```

97.
```
    1 0 0
  ×     8
```

98.
```
    1 0 0
  ×     3
```

99.
```
    1 0 0
  ×     9
```

100.
```
    1 0 0
  ×   1 0
```

101.
```
      6 7
  × 1 0 0
```

102.
```
    3 2 1
  × 1 0 0
```

103.
```
    3 4 5 7
  ×   1 0 0
```

104.
```
    4 0 0
  ×     3
```

105.
```
    7 0 0
  ×     7
```

106.
```
    5 0 0
  ×     8
```

107.
```
    6 0 0
  ×     7
```

108.
```
    1 0 0
  × 1 0 0
```

109.
```
    1 0 0 0
  ×       7
```

110.
```
    1 0 0 0
  ×       9
```

111.
```
    1 0 0 0
  ×       2
```

112.
```
      4 5 7
  × 1 0 0 0
```

113.
```
    7 8 8 8
  × 1 0 0 0
```

114.
```
    6 7 6 9
  × 1 0 0 0
```

115.
```
    2 0 0 0
  ×       9
```

116.
```
    5 0 0 0
  ×       4
```

117.
```
    6 0 0 0
  ×       8
```

118.
```
    8 0 0 0
  ×       2
```

119.
```
    3 0 0 0
  ×       2
```

120.
```
    1 0 0 0
  × 1 0 0 0
```

121.
```
    4 0
  × 3 0
```

122.
```
    7 0
  × 7 0
```

123.
```
    2 0
  × 1 0
```

124.
```
    8 0
  × 5 0
```

ANSWERS

85. _____

86. _____

87. _____

88. _____

89. _____

90. _____

91. _____

92. _____

93. _____

94. _____

95. _____

96. _____

97. _____

98. _____

99. _____

100. _____

101. _____

102. _____

103. _____

104. _____

105. _____

106. _____

107. _____

108. _____

109. _____

110. _____

111. _____

112. _____

113. _____

114. _____

115. _____

116. _____

117. _____

118. _____

119. _____

120. _____

121. _____

122. _____

123. _____

124. _____

ANSWERS

125. _____
126. _____
127. _____
128. _____
129. _____
130. _____
131. _____
132. _____
133. _____
134. _____
135. _____
136. _____
137. _____
138. _____
139. _____
140. _____
141. _____
142. _____
143. _____
144. _____
145. _____
146. _____
147. _____
148. _____
149. _____
150. _____
151. _____
152. _____
153. _____
154. _____
155. _____
156. _____
157. _____
158. _____
159. _____
160. _____
161. _____
162. _____
163. _____
164. _____

125.
$$\begin{array}{r} 50 \\ \times\ 50 \\ \hline \end{array}$$

126.
$$\begin{array}{r} 400 \\ \times\ 30 \\ \hline \end{array}$$

127.
$$\begin{array}{r} 700 \\ \times\ 70 \\ \hline \end{array}$$

128.
$$\begin{array}{r} 200 \\ \times\ 30 \\ \hline \end{array}$$

129.
$$\begin{array}{r} 700 \\ \times\ 90 \\ \hline \end{array}$$

130.
$$\begin{array}{r} 400 \\ \times 300 \\ \hline \end{array}$$

131.
$$\begin{array}{r} 500 \\ \times 300 \\ \hline \end{array}$$

132.
$$\begin{array}{r} 4000 \\ \times\ 200 \\ \hline \end{array}$$

133.
$$\begin{array}{r} 6000 \\ \times\ 20 \\ \hline \end{array}$$

134.
$$\begin{array}{r} 6000 \\ \times 4000 \\ \hline \end{array}$$

135.
$$\begin{array}{r} 4000 \\ \times 4000 \\ \hline \end{array}$$

136.
$$\begin{array}{r} 8000 \\ \times\ 10 \\ \hline \end{array}$$

c Multiply.

137.
$$\begin{array}{r} 49 \\ \times\ 3 \\ \hline \end{array}$$

138.
$$\begin{array}{r} 74 \\ \times\ 6 \\ \hline \end{array}$$

139.
$$\begin{array}{r} 593 \\ \times\ 5 \\ \hline \end{array}$$

140.
$$\begin{array}{r} 609 \\ \times\ 8 \\ \hline \end{array}$$

141.
$$\begin{array}{r} 899 \\ \times\ 7 \\ \hline \end{array}$$

142.
$$\begin{array}{r} 865 \\ \times\ 4 \\ \hline \end{array}$$

143.
$$\begin{array}{r} 8118 \\ \times\ 2 \\ \hline \end{array}$$

144.
$$\begin{array}{r} 3264 \\ \times\ 9 \\ \hline \end{array}$$

145.
$$\begin{array}{r} 7731 \\ \times\ 4 \\ \hline \end{array}$$

146.
$$\begin{array}{r} 6754 \\ \times\ 2 \\ \hline \end{array}$$

147.
$$\begin{array}{r} 43,777 \\ \times\ 2 \\ \hline \end{array}$$

148.
$$\begin{array}{r} 32,564 \\ \times\ 6 \\ \hline \end{array}$$

d Multiply.

149.
$$\begin{array}{r} 58 \\ \times 60 \\ \hline \end{array}$$

150.
$$\begin{array}{r} 93 \\ \times 30 \\ \hline \end{array}$$

151.
$$\begin{array}{r} 42 \\ \times 80 \\ \hline \end{array}$$

152.
$$\begin{array}{r} 78 \\ \times 90 \\ \hline \end{array}$$

153.
$$\begin{array}{r} 346 \\ \times\ 60 \\ \hline \end{array}$$

154.
$$\begin{array}{r} 723 \\ \times\ 50 \\ \hline \end{array}$$

155.
$$\begin{array}{r} 342 \\ \times\ 20 \\ \hline \end{array}$$

156.
$$\begin{array}{r} 267 \\ \times\ 40 \\ \hline \end{array}$$

157.
$$\begin{array}{r} 897 \\ \times 400 \\ \hline \end{array}$$

158.
$$\begin{array}{r} 366 \\ \times 300 \\ \hline \end{array}$$

159.
$$\begin{array}{r} 834 \\ \times 700 \\ \hline \end{array}$$

160.
$$\begin{array}{r} 333 \\ \times 900 \\ \hline \end{array}$$

161.
$$\begin{array}{r} 5673 \\ \times 2000 \\ \hline \end{array}$$

162.
$$\begin{array}{r} 4678 \\ \times 5000 \\ \hline \end{array}$$

163.
$$\begin{array}{r} 6788 \\ \times 9000 \\ \hline \end{array}$$

164.
$$\begin{array}{r} 9129 \\ \times 8000 \\ \hline \end{array}$$

D Division

a BASIC DIVISION

Division can be explained by arranging a set of objects in a rectangular array. This can be done in two ways.

EXAMPLE 1 Divide: $18 \div 6$.

METHOD 1 We can do this division by taking 18 objects and determining how many rows, each with 6 objects, we can arrange the objects into.

3 rows of 6 objects

Since there are 3 rows of 6 objects, we have

$$18 \div 6 = 3.$$

METHOD 2 We can also arrange the objects into 6 rows and determine how many objects are in each row.

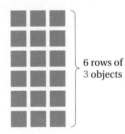

6 rows of 3 objects

Since there are 3 objects in each of the 6 rows, we have

$$18 \div 6 = 3.$$

We can also use fractional notation for division. That is,

$$18 \div 6 = 18/6 = \frac{18}{6}.$$

EXAMPLES Divide.

2. $36 \div 9 = 4$ *Think:* 36 objects: How many rows, each with 9 objects? or 36 objects: How many objects in each of 9 rows?

3. $\dfrac{42}{7} = 6$

4. $\dfrac{24}{3} = 8$

Do Exercises 1–4.

OBJECTIVES

After finishing Section D, you should be able to:

a Find basic quotients such as $20 \div 5$, $56 \div 7$, and so on.

b Divide using the "guess, multiply, and subtract" method.

c Divide by estimating multiples of thousands, hundreds, tens, and ones.

Divide.

1. $24 \div 6$

2. $64 \div 8$

3. $\dfrac{63}{7}$

4. $\dfrac{27}{9}$

Answers on page A-7

For each multiplication fact, write two division facts.

5. $6 \cdot 2 = 12$

6. $7 \times 6 = 42$

Answers on page A-7

In Developmental Unit M, you memorized a multiplication table. That table will enable you to divide as well. First, let us recall how multiplication and division are related.

A multiplication: $5 \cdot 4 = 20$.

Two related divisions:

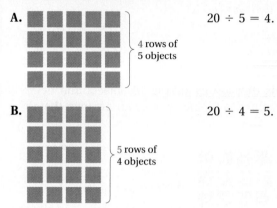

A. $\qquad$ 4 rows of 5 objects $\qquad$ $20 \div 5 = 4$.

B. $\qquad$ 5 rows of 4 objects $\qquad$ $20 \div 4 = 5$.

Since we know that

$$5 \cdot 4 = 20, \qquad \text{A basic multiplication fact}$$

we also know the two division facts

$$20 \div 5 = 4 \quad \text{and} \quad 20 \div 4 = 5.$$

EXAMPLE 5 From $7 \cdot 8 = 56$, write two division facts.

a) We first move the 7.

$$7 \cdot 8 = 56$$

The related sentence is

$$56 \div 7 = 8.$$

b) We next move the 8.

$$7 \cdot 8 = 56$$

The related sentence is

$$56 \div 8 = 7.$$

Do Exercises 5 and 6.

We can use the idea that division is defined in terms of multiplication to do basic divisions. We also make use of the table of basic multiplication facts considered in Developmental Unit M.

EXAMPLE 6   Find: $35 \div 7$.

To find $35 \div 7$, we ask, "7 times what number is 35?"

$$7 \cdot \blacksquare = 35$$

From the multiplication table that we memorized, we know that $7 \cdot 5 = 35$. Thus we know the related division

$$35 \div 7 = 5.$$

Do Exercises 7–10.

DIVISION BY 1

Note that

$$3 \div 1 = 3 \quad \text{because} \quad 3 = 3 \cdot 1; \qquad \frac{14}{1} = 14 \quad \text{because} \quad 14 = 14 \cdot 1.$$

Any number divided by 1 is that same number:

$$a \div 1 = \frac{a}{1} = a.$$

EXAMPLES   Divide.

7. $\dfrac{8}{1} = 8$

8. $6 \div 1 = 6$

9. $34 \div 1 = 34$

Do Exercises 11–13.

DIVISION BY 0

Why can't we divide by 0? Suppose the number 4 could be divided by 0. Then if $\square$ were the answer,

$$4 \div 0 = \square \quad \text{and this would mean} \quad 4 = \square \cdot 0 = 0. \quad \text{False!}$$

Suppose 12 could be divided by 0. If $\square$ were the answer,

$$12 \div 0 = \square \quad \text{and this would mean} \quad 12 = \square \cdot 0 = 0. \quad \text{False!}$$

Thus, $a \div 0$ would be some number $\square$ such that $a = \square \cdot 0 = 0$. So the only possible number that could be divided by 0 would be 0 itself.

Divide.

7. $28 \div 4$

8. $81 \div 9$

9. $\dfrac{16}{2}$

10. $\dfrac{54}{6}$

Divide.

11. $6 \div 1$

12. $\dfrac{13}{1}$

13. $1 \div 1$

Answers on page A-7

Divide, if possible. If not possible, write "not defined."

14. $\dfrac{8}{4}$ 15. $\dfrac{5}{0}$

16. $\dfrac{0}{5}$ 17. $\dfrac{0}{0}$

18. $12 \div 0$ 19. $100 \div 10$

20. $\dfrac{5}{3-3}$ 21. $\dfrac{8-8}{4}$

Divide.

22. $23 \div 23$ 23. $\dfrac{67}{67}$

24. $\dfrac{41}{41}$ 25. $17 \div 17$

26. $17 \div 1$ 27. $\dfrac{54}{54}$

Answers on page A-7

But such a division would give us any number we wish, for

$$0 \div 0 = 8 \quad \text{because} \quad 0 = 8 \cdot 0;$$
$$0 \div 0 = 3 \quad \text{because} \quad 0 = 3 \cdot 0;$$
$$0 \div 0 = 7 \quad \text{because} \quad 0 = 7 \cdot 0.$$

All true!

We avoid the preceding difficulties by agreeing to exclude division by 0.

> Division by 0 is not defined. (We agree not to divide by 0.)

DIVIDING 0 BY OTHER NUMBERS

Note that

$$0 \div 3 = 0 \quad \text{because} \quad 0 = 0 \cdot 3; \qquad \dfrac{0}{12} = 0 \quad \text{because } 0 = 0 \cdot 12.$$

> Zero divided by any number greater than 0 is 0:
>
> $$\dfrac{0}{a} = 0, \quad a > 0.$$

EXAMPLES   Divide.

10. $0 \div 8 = 0$

11. $0 \div 22 = 0$

12. $\dfrac{0}{9} = 0$

Do Exercises 14–21.

DIVISION OF A NUMBER BY ITSELF

Note that

$$3 \div 3 = 1 \quad \text{because} \quad 3 = 1 \cdot 3; \qquad \dfrac{34}{34} = 1 \quad \text{because} \quad 34 = 1 \cdot 34.$$

> Any number greater than 0 divided by itself is 1:
>
> $$\dfrac{a}{a} = 1, \quad a > 0.$$

EXAMPLES   Divide.

13. $8 \div 8 = 1$

14. $27 \div 27 = 1$

15. $\dfrac{32}{32} = 1$

Do Exercises 22–27.

b DIVIDING BY "GUESS, MULTIPLY, AND SUBTRACT"

To understand the process of division, we use a method known as "guess, multiply, and subtract."

EXAMPLE 16   Divide $275 \div 4$. Use "guess, multiply, and subtract."

We *guess* a partial quotient of 35. We could guess *any* number—say, 4, 16, or 30. We *multiply* and *subtract* as follows:

```
    3 5 ← Partial quotient
4 ) 2 7 5
    1 4 0 ← 35 · 4
    1 3 5 ← Remainder
```

Next, look at 135 and *guess* another partial quotient—say, 20. Then *multiply* and *subtract:*

```
      2 0 ← Second partial quotient
      3 5
4 ) 2 7 5
    1 4 0
    1 3 5
      8 0 ← 20 · 4
      5 5 ← Remainder
```

Next, look at 55 and *guess* another partial quotient—say, 13. Then *multiply* and *subtract:*

```
      1 3 ← Third partial quotient
      2 0
      3 5
4 ) 2 7 5
    1 4 0
    1 3 5
      8 0
      5 5
      5 2 ← 13 · 4
        3 ← Remainder is less than 4
```

Since we cannot subtract any more 4's, the division is finished. We add our partial quotients.

```
      6 8 ← Quotient (sum of guesses)
      1 3
      2 0
      3 5
4 ) 2 7 5
    1 4 0
    1 3 5
      8 0
      5 5
      5 2
        3
```

CHECK:
$$275 = (4 \times 68) + 3$$

$$275 \mid \begin{array}{l} 272 + 3 \\ 275 \end{array}$$

The answer is 68 R 3. This tells us that with 275 objects, we could make 68 rows of 4 and have 3 left over.

Divide using the "guess, multiply, and subtract" method.

28. $6 \overline{)454}$

29. $32 \overline{)747}$

Answers on page A-7

Divide using the "guess, multiply, and subtract" method.

30. 7) 6 7 8 9

The partial quotients (guesses) can be made in any manner so long as subtraction is possible.

Do Exercises 28 and 29 on the preceding page.

EXAMPLE 17 Divide: $1506 \div 32$.

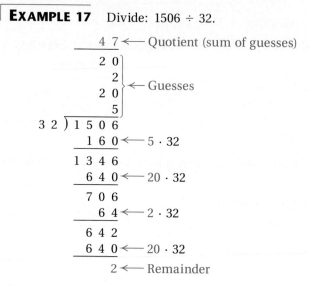

The answer is 47 R 2.

Remember, you can *guess any partial quotient* as long as subtraction is possible.

Do Exercises 30 and 31.

31. 6 4) 3 0 1 2

| c | **DIVIDING BY ESTIMATING MULTIPLES** |

Let us refine the guessing. We guess multiples of 10, 100, and 1000, and so on.

EXAMPLE 18 Divide: $7643 \div 3$.

a) Are there any thousands in the quotient? Yes, $3 \cdot 1000 = 3000$, which is less than 7643. To find how many thousands, we find products of 3 and multiples of 1000.

$$3 \cdot 1000 = 3000$$
$$3 \cdot 2000 = 6000$$
$$3 \cdot 3000 = 9000 \leftarrow$$ 7643 is here, so there are 2000 threes in the quotient.

```
        2 0 0 0
  3 ) 7 6 4 3
      6 0 0 0
      1 6 4 3
```

Answers on page A-7

b) Now go to the hundreds place. Are there any hundreds in the quotient?

$$3 \cdot 100 = 300$$
$$3 \cdot 200 = 600$$
$$3 \cdot 300 = 900$$
$$3 \cdot 400 = 1200$$
$$3 \cdot 500 = 1500$$
$$3 \cdot 600 = 1800$$

$\leftarrow$ 1643

```
          5 0 0
        2 0 0 0
    3 ) 7 6 4 3
        6 0 0 0
        1 6 4 3
        1 5 0 0
          1 4 3
```

c) Now go to the tens place. Are there any tens in the quotient?

$$3 \cdot 10 = 30$$
$$3 \cdot 20 = 60$$
$$3 \cdot 30 = 90$$
$$3 \cdot 40 = 120$$
$$3 \cdot 50 = 150$$

$\leftarrow$ 143

```
            4 0
          5 0 0
        2 0 0 0
    3 ) 7 6 4 3
        6 0 0 0
        1 6 4 3
        1 5 0 0
          1 4 3
          1 2 0
            2 3
```

d) Now go to the ones place. Are there any ones in the quotient?

$$3 \cdot 1 = 3$$
$$3 \cdot 2 = 6$$
$$3 \cdot 3 = 9$$
$$3 \cdot 4 = 12$$
$$3 \cdot 5 = 15$$
$$3 \cdot 6 = 18$$
$$3 \cdot 7 = 21$$
$$3 \cdot 8 = 24$$

$\leftarrow$ 23

```
        2 5 4 7
              7
            4 0
          5 0 0
        2 0 0 0
    3 ) 7 6 4 3
        6 0 0 0
        1 6 4 3
        1 5 0 0
          1 4 3
          1 2 0
            2 3
            2 1
              2
```

The answer is 2547 R 2.

Do Exercises 32 and 33.

Divide.

32. 4) 3 8 5

33. 7) 8 8 4 6

Divide using the short form.

34. 2) 6 4 8

35. 9) 3 7 5 8

Divide.

36. 1 1) 4 1 5

37. 4 6) 1 0 7 5

A SHORT FORM

Here is a shorter way to write Example 18.

Instead of this,

```
            2 5 4 7
                  7
                4 0
              5 0 0
            2 0 0 0
      3 ) 7 6 4 3
            6 0 0 0
            1 6 4 3
            1 5 0 0
              1 4 3
              1 2 0
                2 3
                2 1
                  2
```

Short form

we write this.

```
            2 5 4 7
      3 ) 7 6 4 3
            6 0 0 0
            1 6 4 3
            1 5 0 0
              1 4 3
              1 2 0
                2 3
                2 1
                  2
```

We write a 2 above the thousands digit in the dividend to record 2000.
We write a 5 to record 500.
We write a 4 to record 40.
We write a 7 to record 7.

Do Exercises 34 and 35.

EXAMPLE 19   Divide 2637 ÷ 41. Use the short form.

```
                6
      4 1 ) 2 6 3 7
            2 4 6 0
              1 7 7

              6 4
      4 1 ) 2 6 3 7
            2 4 6 0
              1 7 7
              1 6 4
                1 3
```

The answer is 64 R 13.

Do Exercises 36 and 37.

Exercise Set D

a Divide, if possible.

1. 24 ÷ 8   **2.** 72 ÷ 9   **3.** 28 ÷ 7   **4.** 48 ÷ 8

5. 22 ÷ 22   **6.** 32 ÷ 1   **7.** 45 ÷ 5   **8.** 15 ÷ 3

9. 14 ÷ 2   **10.** 40 ÷ 8   **11.** 37 ÷ 1   **12.** 29 ÷ 1

13. 10 ÷ 2   **14.** 36 ÷ 4   **15.** 12 ÷ 3   **16.** 35 ÷ 7

17. 54 ÷ 9   **18.** 18 ÷ 2   **19.** 20 ÷ 4   **20.** 9 ÷ 1

21. 16 ÷ 2   **22.** 72 ÷ 8   **23.** 42 ÷ 7   **24.** 63 ÷ 9

25. 12 ÷ 4   **26.** 8 ÷ 4   **27.** 54 ÷ 6   **28.** 18 ÷ 3

29. 18 ÷ 9   **30.** 9 ÷ 3   **31.** 28 ÷ 4   **32.** 6 ÷ 3

33. 56 ÷ 7   **34.** 24 ÷ 6   **35.** 14 ÷ 2   **36.** 27 ÷ 9

37. 14 ÷ 7   **38.** 21 ÷ 7   **39.** 36 ÷ 6   **40.** 12 ÷ 2

41. _____
42. _____
43. _____
44. _____
45. _____
46. _____
47. _____
48. _____
49. _____
50. _____
51. _____
52. _____
53. _____
54. _____
55. _____
56. _____
57. _____
58. _____
59. _____
60. _____
61. _____
62. _____
63. _____
64. _____
65. _____
66. _____
67. _____
68. _____
69. _____
70. _____
71. _____
72. _____
73. _____
74. _____
75. _____
76. _____
77. _____
78. _____
79. _____
80. _____

41. $8 \div 8$　　**42.** $32 \div 8$　　**43.** $30 \div 5$　　**44.** $18 \div 6$

45. $49 \div 7$　　**46.** $81 \div 9$　　**47.** $6 \div 6$　　**48.** $0 \div 7$

49. $9 \div 0$　　**50.** $16 \div 0$　　**51.** $42 \div 6$　　**52.** $63 \div 7$

53. $\dfrac{48}{6}$　　**54.** $\dfrac{35}{5}$　　**55.** $\dfrac{9}{9}$　　**56.** $\dfrac{45}{9}$

57. $\dfrac{0}{5}$　　**58.** $\dfrac{7}{7}$　　**59.** $\dfrac{0}{8}$　　**60.** $\dfrac{6}{2}$

61. $\dfrac{3}{3}$　　**62.** $\dfrac{8}{2}$　　**63.** $\dfrac{7}{1}$　　**64.** $\dfrac{5}{5}$

65. $\dfrac{6}{1}$　　**66.** $\dfrac{0}{3}$　　**67.** $\dfrac{2}{2}$　　**68.** $\dfrac{25}{5}$

69. $\dfrac{4}{2}$　　**70.** $\dfrac{24}{3}$　　**71.** $\dfrac{0}{9}$　　**72.** $\dfrac{16}{4}$

73. $\dfrac{0}{4}$　　**74.** $\dfrac{40}{5}$　　**75.** $\dfrac{3}{1}$　　**76.** $\dfrac{24}{3}$

77. $\dfrac{9}{0}$　　**78.** $\dfrac{16}{0}$　　**79.** $\dfrac{32}{8}$　　**80.** $\dfrac{9}{9}$

b Divide using the "guess, multiply, and subtract" method.

81. 4) 2 7 7

82. 2) 3 9 9

83. 8) 7 3 7

84. 6) 8 3 1

85. 5) 8 6 1 9

86. 3) 8 7 7 5

87. 9) 7 7 7 7

88. 8) 4 1 7 9

89. 7) 3 6 9 1

90. 2) 5 7 9 4

91. 2 0) 8 7 5

92. 3 0) 9 8 7

93. 2 1) 9 9 9

94. 2 3) 9 7 5

95. 8 5) 7 7 5 7

96. 5 4) 2 8 2 1

97. 1 1 1) 3 2 1 9

98. 1 0 2) 5 6 1 2

99. 3 4 6) 7 8,9 1 0

100. 7 8 1) 1 5,9 9 9

ANSWERS

81. _____

82. _____

83. _____

84. _____

85. _____

86. _____

87. _____

88. _____

89. _____

90. _____

91. _____

92. _____

93. _____

94. _____

95. _____

96. _____

97. _____

98. _____

99. _____

100. _____

[c] Divide.

101. $5 \overline{)\ 1\ 0\ 5}$ **102.** $6 \overline{)\ 7\ 0\ 8}$ **103.** $9 \overline{)\ 8\ 2\ 0}$

104. $3 \overline{)\ 9\ 6\ 5}$ **105.** $5 \overline{)\ 4\ 8\ 2\ 3}$ **106.** $8 \overline{)\ 5\ 4\ 3\ 7}$

107. $7 \overline{)\ 9\ 2\ 9\ 8}$ **108.** $4\ 1 \overline{)\ 1\ 1\ 1\ 5}$ **109.** $4\ 6 \overline{)\ 1\ 0\ 5\ 8}$

110. $2\ 4 \overline{)\ 7\ 7\ 2\ 2}$ **111.** $3\ 8 \overline{)\ 8\ 5\ 2\ 2}$ **112.** $8\ 1 \overline{)\ 2\ 2\ 4\ 7}$

113. $7\ 2 \overline{)\ 6\ 2\ 6\ 5}$ **114.** $7\ 4 \overline{)\ 5\ 5\ 5\ 0}$ **115.** $9\ 4 \overline{)\ 2\ 1\ 5\ 3}$

116. $8\ 2 \overline{)\ 4\ 0\ 6\ 4}$ **117.** $1\ 1\ 7 \overline{)\ 4\ 4,9\ 0\ 2}$ **118.** $7\ 4\ 0 \overline{)\ 5\ 5,2\ 0\ 0}$

TABLE 1

Square Roots

N	$\sqrt{N}$	N	$\sqrt{N}$	N	$\sqrt{N}$	N	$\sqrt{N}$
2	1.414	27	5.196	52	7.211	77	8.775
3	1.732	28	5.292	53	7.280	78	8.832
4	2	29	5.385	54	7.348	79	8.888
5	2.236	30	5.477	55	7.416	80	8.944
6	2.449	31	5.568	56	7.483	81	9
7	2.646	32	5.657	57	7.550	82	9.055
8	2.828	33	5.745	58	7.616	83	9.110
9	3	34	5.831	59	7.681	84	9.165
10	3.162	35	5.916	60	7.746	85	9.220
11	3.317	36	6	61	7.810	86	9.274
12	3.464	37	6.083	62	7.874	87	9.327
13	3.606	38	6.164	63	7.937	88	9.381
14	3.742	39	6.245	64	8	89	9.434
15	3.873	40	6.325	65	8.062	90	9.487
16	4	41	6.403	66	8.124	91	9.539
17	4.123	42	6.481	67	8.185	92	9.592
18	4.243	43	6.557	68	8.246	93	9.644
19	4.359	44	6.633	69	8.307	94	9.695
20	4.472	45	6.708	70	8.367	95	9.747
21	4.583	46	6.782	71	8.426	96	9.798
22	4.690	47	6.856	72	8.485	97	9.849
23	4.796	48	6.928	73	8.544	98	9.899
24	4.899	49	7	74	8.602	99	9.950
25	5	50	7.071	75	8.660	100	10
26	5.099	51	7.141	76	8.718		

TABLE 1

639

The following is a list of text exercises that appear on the "Math Makes a Difference" videotapes.

TEXT/VIDEO SECTION	EXERCISE NUMBERS
Section 1.1	3, 21, 37, 43, 45
Section 1.2	7, 9, 37, 43
Section 1.3	39
Section 1.4	23, 29
Section 1.5	21
Section 1.6	11
Section 1.7	1, 5, 7, 41, 45, 55, 57
Section 1.8	23
Section 1.9	5, 7, 9, 11, 35, 51
Section 2.1	13, 47, 59, 65
Section 2.2	1, 3, 5, 7, 9, 11, 15
Section 2.3	21
Section 2.4	1, 5, 27, 37
Section 2.5	3, 21, 33, 37, 39
Section 2.6	1, 7, 9, 19, 35, 45, 53
Section 2.7	9, 13, 19, 21, 33, 35, 45
Section 3.1	1, 3, 7, 9, 19, 23, 29, 39
Section 3.2	3, 5, 11, 15, 29, 33, 40
Section 3.3	9, 29, 43, 53
Section 3.4	7, 15, 23, 41, 43, 51
Section 3.5	1, 9, 15, 17, 19, 23, 35
Section 3.6	3, 7, 20, 26, 37, 43
Section 3.7	7, 11, 21, 27
Section 4.1	9, 19, 23, 33, 37
Section 4.2	3, 7, 11, 37, 53
Section 4.3	9, 11, 15, 31, 61, 63, 67
Section 4.4	6, 16, 23, 24, 25, 26
Section 5.1	3, 5, 7, 23, 33, 57
Section 5.2	33, 35, 59, 69
Section 5.3	1, 3, 15, 23, 32
Section 5.4	1, 5, 23
Section 5.5	7, 40
Section 6.1	3, 7, 13, 22, 23, 31, 35
Section 6.2	1, 3, 7, 9, 21
Section 6.3	1, 5, 13, 19, 21, 29, 35, 37
Section 6.4	3, 15, 17
Section 6.5	9

Section 7.1	1, 23, 47, 51
Section 7.2	11, 15, 19, 32
Section 7.3	3, 5, 23, 25, 31
Section 7.4	3, 5, 13, 17
Section 7.5	3, 15, 17, 19
Section 7.6	1, 5
Section 7.7	9
Section 7.8	none
Section 8.1	1, 7, 20
Section 8.2	none
Section 8.3	11, 13, 15, 25
Section 8.4	3
Section 9.1	8, 11, 17, 21, 27, 35
Section 9.2	7, 21, 31, 35, 41
Section 9.3	3, 5, 15
Section 9.4	13, 21
Section 9.5	1, 9, 19
Section 9.6	25, 31
Section 9.7	1, 3, 17, 19, 25, 29, 33, 45
Section 10.1	1, 19, 25
Section 10.2	15, 21, 23
Section 10.3	5, 51
Section 10.4	19, 21
Section 10.5	1, 15, 19, 27
Section 11.1	5, 15, 21, 22, 31, 37, 43, 45, 51
Section 11.2	11, 15, 31, 39, 47, 49, 63, 65, 67
Section 11.3	1, 3, 7, 9, 31, 39, 57, 67, 71
Section 11.4	7, 13, 22, 27
Section 11.5	9, 11, 15, 17, 21, 23, 25, 37, 75
Section 12.1	3, 9, 13, 23, 29, 33, 37, 43, 45, 47, 49, 59
Section 12.2	5, 15, 17, 19, 21
Section 12.3	1, 7, 9, 13, 23, 25
Section 12.4	7, 11, 17, 35, 41, 47, 53
Section 12.5	1, 3, 6, 15, 17, 24, 31, 37

Using a Scientific Calculator

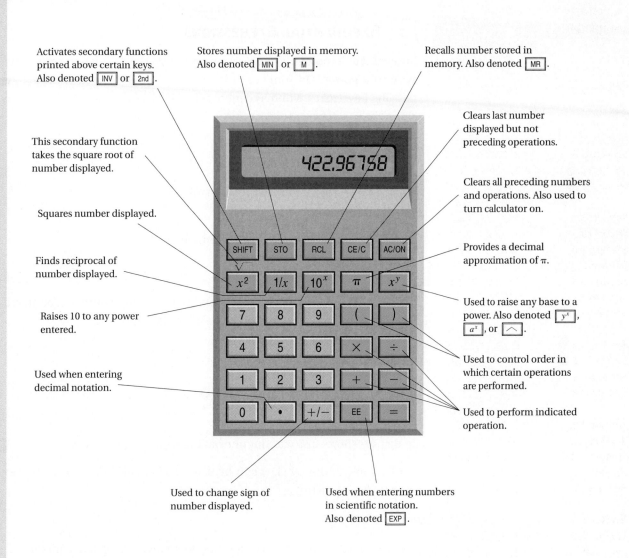

Activates secondary functions printed above certain keys. Also denoted INV or 2nd.

Stores number displayed in memory. Also denoted MIN or M.

Recalls number stored in memory. Also denoted MR.

This secondary function takes the square root of number displayed.

Clears last number displayed but not preceding operations.

Squares number displayed.

Clears all preceding numbers and operations. Also used to turn calculator on.

Finds reciprocal of number displayed.

Provides a decimal approximation of π.

Raises 10 to any power entered.

Used to raise any base to a power. Also denoted y^x, a^x, or $\wedge$.

Used when entering decimal notation.

Used to control order in which certain operations are performed.

Used to perform indicated operation.

Used to change sign of number displayed.

Used when entering numbers in scientific notation. Also denoted EXP.

After finishing this appendix, you should be able to use a calculator to:

a   Evaluate exponential expressions.

b   Evaluate expressions using the built-in order of operations.

c   Calculate with fractions and mixed numerals.

d   Find square roots.

e   Calculate with real numbers.

f   Check solutions of equations.

Evaluate.

1. 3^8

2. 15^3

3. 23^2

4. 2^{10}

5. 7^4

Evaluate.

6. $(1.8)^4$

7. $(0.3)^5$

Answers on page A-24

Using a Scientific Calculator

Calculator usage has become commonplace in business and daily living. For this reason, we have included a brief section on the basic operation of a scientific calculator.

Features and required procedures vary widely among scientific calculators. If you are unfamiliar with the use of your particular calculator, consult its manual or your instructor.

The calculator is a remarkable tool for both students and teachers, but remember, it is not a substitute for learning the concepts.

a   EXPONENTIAL EXPRESSIONS

To evaluate exponential expressions, we use the $\boxed{x^y}$ key to raise any base to a power. On some calculators, this key is denoted $\boxed{y^x}$, $\boxed{a^x}$, or $\boxed{\wedge}$. The keystrokes will make more sense if we say "raised to the" as we press the $\boxed{x^y}$ key.

EXAMPLE 1   Evaluate: 3^5.

We find 3^5 using the following keystrokes:

$\boxed{3}$ $\boxed{x^y}$ $\boxed{5}$ $\boxed{=}$.

The answer, 243, is displayed in the window:

         243 .

Since raising a number to the second power is frequently encountered in mathematics, most calculators have a key for this, $\boxed{x^2}$.

EXAMPLE 2   Evaluate: 15^2.

We find 15^2 using the following keystrokes:

$\boxed{1}$ $\boxed{5}$ $\boxed{x^2}$.

The display shows 225 without our having pressed the $\boxed{=}$ key.

In Example 2, the $\boxed{x^y}$ key can be used to evaluate 15^2, but the $\boxed{x^2}$ key requires fewer keystrokes.

Do Exercises 1–5.

We can easily raise decimal notation to a power using a calculator.

EXAMPLE 3   Find: $(1.2)^3$.

To find $(1.2)^3$, we press the following keys:

$\boxed{1}$ $\boxed{.}$ $\boxed{2}$ $\boxed{x^y}$ $\boxed{3}$ $\boxed{=}$.

The answer, 1.728, appears in the window.

Do Exercises 6 and 7.

b ORDER OF OPERATIONS

The order of operations is built into most scientific calculators.

EXAMPLE 4   Calculate: $36 \div 2 \cdot 3 - 4 \cdot 4$.

To calculate $36 \div 2 \cdot 3 - 4 \cdot 4$, we press the following keys:

$\boxed{3}\ \boxed{6}\ \boxed{\div}\ \boxed{2}\ \boxed{\times}\ \boxed{3}\ \boxed{-}\ \boxed{4}\ \boxed{\times}\ \boxed{4}\ \boxed{=}$.

The answer, 38, is displayed in the window:

$\boxed{38}$.

When parentheses appear in the problem, we must enter the operation preceding the parentheses.

EXAMPLE 5   Calculate: $36 \div (2 \cdot 3 - 4) \cdot 4$.

We press the following keys:

$\boxed{3}\ \boxed{6}\ \boxed{\div}\ \boxed{(}\ \boxed{2}\ \boxed{\times}\ \boxed{3}\ \boxed{-}\ \boxed{4}\ \boxed{)}\ \boxed{\times}\ \boxed{4}\ \boxed{=}$.

The answer, 72, is displayed in the window.

EXAMPLE 6   Calculate: $(15 + 3)^3 + 4(12 - 7)^2$.

We calculate using the following keystrokes:

$\boxed{(}\ \boxed{1}\ \boxed{5}\ \boxed{+}\ \boxed{3}\ \boxed{)}\ \boxed{x^y}\ \boxed{3}\ \boxed{+}\ \boxed{4}\ \boxed{\times}\ \boxed{(}\ \boxed{1}\ \boxed{2}\ \boxed{-}\ \boxed{7}\ \boxed{)}\ \boxed{x^2}\ \boxed{=}$.

The answer is 5932.

Do Exercises 8–12.

Even when the order of operations is built in, parentheses must be inserted at times.

EXAMPLE 7   Calculate: $\dfrac{80}{8 - 6}$.

We press

$\boxed{8}\ \boxed{0}\ \boxed{\div}\ \boxed{(}\ \boxed{8}\ \boxed{-}\ \boxed{6}\ \boxed{)}\ \boxed{=}$.

The display reads 40, the correct answer:

$\boxed{40}$.

In Example 7, if we had not inserted the parentheses, 80 would have been divided by 8 *before* the 6 had been subtracted, given an incorrect answer of 4.

$\boxed{8}\ \boxed{0}\ \boxed{\div}\ \boxed{8}\ \boxed{-}\ \boxed{6}\ \boxed{=}\qquad\boxed{4}$

↑   ↑ —————— Wrong!

Do Exercises 13 and 14.

Calculate.

8. $68 - 8 \div 4 + 3 \cdot 5$

9. $50 - 8 \cdot 3 + 4(5^2 - 2)$

10. $3 + 4[15 - 2(6 - 3)]$

11. $[3 + 2(10 - 4)^2] - 20 \div 5$

12. $\{(150 \cdot 5) \div [(3 \cdot 16) \div (8 \cdot 3)]\} + 25 \cdot (12 \div 4)$

Calculate.

13. $\dfrac{1200}{30 - 18}$

14. $\dfrac{50 - 5}{5 + 10}$

Answers on page A-24

Calculate.

15. $\dfrac{7}{8} - \dfrac{1}{3}$

16. $\dfrac{2}{3} + \dfrac{1}{4} - \dfrac{2}{5}$

17. $3\dfrac{1}{4} + 9\dfrac{5}{6}$

18. $\dfrac{2}{5} \cdot \dfrac{14}{11} - \dfrac{1}{2}$

19. $2\dfrac{1}{2} \div 1\dfrac{1}{4}$

Evaluate and express the answer in decimal notation.

20. $\dfrac{3}{5}\left(\dfrac{19}{4} - \dfrac{1}{8}\right)$

21. $\dfrac{3}{8} + \dfrac{3}{4} + \dfrac{1}{2}$

22. $\dfrac{1}{8}\left(\dfrac{1}{2} \div \dfrac{2}{3}\right) + \dfrac{7}{32}$

---

c | CALCULATING WITH FRACTIONAL NOTATION

The $\boxed{\text{a}^{\text{b}/\text{c}}}$ key allows us to compute with fractional notation and mixed numerals.

> **EXAMPLE 8** Calculate: $\dfrac{1}{2} + 3\dfrac{1}{5} - \dfrac{4}{9}$.

To enter the problem, we use the following keystrokes:

The display in the window is in the form

   $\boxed{3 \lrcorner 23 \lrcorner 90}$.

This means that the answer is $3\frac{23}{90}$.

If we press the $\boxed{\text{Shift}}$ key and the $\boxed{\text{d/c}}$ key, the answer is converted from a mixed numeral to fractional notation:

   $\boxed{293 \lrcorner 90}$,

which means $\frac{293}{90}$.

Do Exercises 15–19.

The $\boxed{\text{a}^{\text{b}/\text{c}}}$ key can also convert an answer in fractional notation to decimal notation.

> **EXAMPLE 9** Calculate $\frac{1}{2}\left(\frac{3}{5} - \frac{1}{8}\right)$ and express the answer in decimal notation.

We press the following keys:

$\boxed{1}\ \boxed{\text{a}^{\text{b}/\text{c}}}\ \boxed{2}\ \boxed{\times}\ \boxed{(}\ \boxed{3}\ \boxed{\text{a}^{\text{b}/\text{c}}}\ \boxed{5}\ \boxed{-}\ \boxed{1}\ \boxed{\text{a}^{\text{b}/\text{c}}}\ \boxed{8}\ \boxed{)}\ \boxed{=}$.

The display reads

   $\boxed{19 \lrcorner 80}$,

which means $\frac{19}{80}$. We now press $\boxed{\text{a}^{\text{b}/\text{c}}}$ to convert $\frac{19}{80}$ to decimal notation:

   $\boxed{0.2375}$.

The answer is 0.2375.

Do Exercises 20–22.

Answers on page A-24

d | FINDING SQUARE ROOTS

To find square roots, we use the $\boxed{\sqrt{}}$ key.

EXAMPLES   Simplify.

10. $\sqrt{64}$

 We press the following keys:

  $\boxed{6}\ \boxed{4}\ \boxed{\sqrt{}}$.

 The display shows 8 without our having used the $\boxed{=}$ key.

11. $\sqrt{0.000625}$

 We enter

  $\boxed{.}\ \boxed{0}\ \boxed{0}\ \boxed{0}\ \boxed{6}\ \boxed{2}\ \boxed{5}\ \boxed{\sqrt{}}$.

 The answer is 0.025.

Do Exercises 23–26.

 We can approximate some square roots as follows.

EXAMPLE 12   Approximate $\sqrt{31}$ to three decimal places.

 To find $\sqrt{31}$, we press

  $\boxed{3}\ \boxed{1}\ \boxed{\sqrt{}}$.

The display shows

  5.5677644 .

Thus, $\sqrt{31} \approx 5.568$, if we round to three decimal places. The answer seems reasonable since $5^2 < 31 < 6^2$.

Do Exercises 27–30.

e | OPERATIONS WITH REAL NUMBERS

To enter a negative number, we use the $\boxed{+/-}$ key. To enter -5, we press $\boxed{5}$ and then $\boxed{+/-}$. The display then reads

  -5 .

EXAMPLE 13   Evaluate: $-8 - (-2.3)$.

 We press the following keys:

  $\boxed{8}\ \boxed{+/-}\ \boxed{-}\ \boxed{2}\ \boxed{.}\ \boxed{3}\ \boxed{+/-}\ \boxed{=}$.

The answer is -5.7.

Do Exercises 31–34.

Simplify.

23. $\sqrt{0.0001}$

24. $\sqrt{169}$

25. $\sqrt{62,500}$

26. $\sqrt{0.001764}$

Approximate the square root to three decimal places.

27. $\sqrt{11}$

28. $\sqrt{270}$

29. $\sqrt{6.4}$

30. $\sqrt{0.009}$

Evaluate.

31. $-5 - (-13)$

32. $-13 + 72 + (-20) + (-23)$

33. $(-6)^2$

34. $35 - (-16) - (-21) + 9^2$

Answers on page A-24

Evaluate.

35. $-8 + 4(7 - 9) + 5$

36. $-3[2 + (-5)]$

37. $7[4 - (-3)] + 5[3^2 - (-4)]$

Check to see if the given number is a solution of the equation using a calculator.

38. $3(5x - 9) + 2(x + 3) =$
$-5(3x - 7) + 8; \quad x = 2$

39. $2(3x - 5) + 7x =$
$3x - (x + 8) + 20; \quad x = 3$

40. $20(x - 39) = 5x - 432;$
$x = 23\frac{1}{5}$

41. $-\frac{1}{2}x + 8 = 1\frac{1}{2}x - 6; \quad x = -28$

42. $-(x - 3) - (x - 4) = 2x + 3;$
$x = 1$

Answers on page A-24

EXAMPLE 14   Evaluate: $-7(2 - 9) - 20$.

We enter

$\boxed{7}\ \boxed{+/-}\ \boxed{\times}\ \boxed{(}\ \boxed{2}\ \boxed{-}\ \boxed{9}\ \boxed{)}\ \boxed{-}\ \boxed{2}\ \boxed{0}\ \boxed{=}.$

The answer is 29.

Do Exercises 35–37.

f CHECKING SOLUTIONS OF EQUATIONS

Calculators can be used to check solutions of equations. We can replace the variable with the solution and evaluate each side of the equation separately.

EXAMPLE 15   Check to see if 5 is a solution of the equation:

$$3(x - 5) + 4(x + 3) = 2(x - 6) + 34.$$

To check using a calculator, we first evaluate the left side:

$\boxed{3}\ \boxed{\times}\ \boxed{(}\ \boxed{5}\ \boxed{-}\ \boxed{5}\ \boxed{)}\ \boxed{+}\ \boxed{4}\ \boxed{\times}\ \boxed{(}\ \boxed{5}\ \boxed{+}\ \boxed{3}\ \boxed{)}\ \boxed{=}.$

The display reads 32:

$\qquad\qquad$ 32 .

We then evaluate the right side:

$\boxed{2}\ \boxed{\times}\ \boxed{(}\ \boxed{5}\ \boxed{-}\ \boxed{6}\ \boxed{)}\ \boxed{+}\ \boxed{3}\ \boxed{4}\ \boxed{=}.$

Again, the display reads 32:

$\qquad\qquad$ 32 .

Since the left side is the same as the right side, the solution checks.

Do Exercises 38–42.

Answers

Margin Exercise Answers

CHAPTER 1

Margin Exercises, Section 1.1, pp. 3–6

1. 1 thousand + 8 hundreds + 7 tens + 5 ones
2. 3 ten thousands + 6 thousands + 2 hundreds + 2 tens + 3 ones   **3.** 3 thousands + 2 tens + 1 one
4. 2 thousands + 9 ones   **5.** 5 thousands + 7 hundreds
6. 5689   **7.** 87,128   **8.** 9003   **9.** Fifty-seven
10. Twenty-nine   **11.** Eighty-eight   **12.** Two hundred four   **13.** Nineteen thousand, two hundred four
14. One million, seven hundred nineteen thousand, two hundred four   **15.** Twenty-two billion, three hundred one million, seven hundred nineteen thousand, two hundred four   **16.** 213,105,329   **17.** 2 ten thousands
18. 2 hundred thousands   **19.** 2 millions
20. 2 ten millions   **21.** 4   **22.** 7   **23.** 9   **24.** 0

Margin Exercises, Section 1.2, pp. 9–13

1. $4 + 6 = 10$ **2.** $15 + $13 = 28 **3.** 40 mi + 50 mi = 90 mi   **4.** 5 ft + 7 ft = 12 ft   **5.** 4 in. + 5 in. + 9 in. + 6 in. + 5 in. = 29 in.   **6.** 5 ft + 6 ft + 5 ft + 6 ft = 22 ft   **7.** $80 \text{ in}^2 + 90 \text{ in}^2 = 170 \text{ in}^2$
8. $100 \text{ mi}^2 + 200 \text{ mi}^2 = 300 \text{ mi}^2$ **9.** 10 gal + 18 gal = 28 gal   **10.** 3000 tons + 7000 tons = 10,000 tons
11. 9745   **12.** 13,465   **13.** 16,182   **14.** 27   **15.** 34
16. 27   **17.** 38   **18.** 47   **19.** 61   **20.** 27,474

Margin Exercises, Section 1.3, pp. 19–22

1. 16 oz − 5 oz = 11 oz   **2.** 400 acres − 100 acres = 300 acres   **3.** $7 = 2 + 5$, or $7 = 5 + 2$ **4.** $17 = 9 + 8$, or $17 = 8 + 9$ **5.** $5 = 13 − 8; 8 = 13 − 5$ **6.** $11 = 14 − 3; 3 = 14 − 11$ **7.** 15 + ■ = 32; ■ = 32 − 15
8. 10 + ■ = 23; ■ = 23 − 10   **9.** 3801   **10.** 6328
11. 4747   **12.** 56   **13.** 205   **14.** 658   **15.** 2851
16. 1546

Margin Exercises, Section 1.4, pp. 25–28

1. 40   **2.** 50   **3.** 70   **4.** 100   **5.** 40   **6.** 80   **7.** 90
8. 140   **9.** 470   **10.** 240   **11.** 290   **12.** 600   **13.** 800

14. 800   **15.** 9300   **16.** 8000   **17.** 8000   **18.** 19,000
19. 69,000   **20.** 200   **21.** 1800   **22.** 2600   **23.** 11,000
24. <   **25.** >   **26.** >   **27.** <   **28.** <   **29.** >

Margin Exercises, Section 1.5, pp. 32–36

1. $8 \cdot 4 = 32$ mi   **2.** $10 \cdot 75 = 750$ mL   **3.** $12 \cdot 20 = 240$
4. $4 \cdot 6 = 24 \text{ ft}^2$ **5.** 1035   **6.** 3024   **7.** 46,252
8. 205,065   **9.** 144,432   **10.** 287,232   **11.** 14,075,720
12. 391,760   **13.** 17,345,600   **14.** 56,200   **15.** 562,000
16. (a) 1081; (b) 1081; (c) same   **17.** 40   **18.** 15
19. 210,000; 160,000

Margin Exercises, Section 1.6, pp. 40–45

1. 112 ÷ 14 = ■   **2.** 112 ÷ 8 = ■   **3.** $15 = 5 \cdot 3$, or $15 = 3 \cdot 5$ **4.** $72 = 9 \cdot 8$, or $72 = 8 \cdot 9$ **5.** $6 = 12 ÷ 2$; $2 = 12 ÷ 6$ **6.** $6 = 42 ÷ 7; 7 = 42 ÷ 6$ **7.** 6; $6 \cdot 9 = 54$
8. 6 R 7; $6 \cdot 9 = 54, 54 + 7 = 61$ **9.** 4 R 5; $4 \cdot 12 = 48, 48 + 5 = 53$ **10.** 6 R 13; $6 \cdot 24 = 144, 144 + 13 = 157$
11. 59 R 3   **12.** 1475 R 5   **13.** 1015   **14.** 134
15. 63 R 12   **16.** 807 R 4   **17.** 1088   **18.** 360 R 4
19. 800 R 47

Margin Exercises, Section 1.7, pp. 49–52

1. 7   **2.** 5   **3.** No   **4.** Yes   **5.** 5   **6.** 10   **7.** 5
8. 22   **9.** 22,490   **10.** 9022   **11.** 570   **12.** 3661   **13.** 8
14. 45   **15.** 77   **16.** 3311   **17.** 6114   **18.** 8   **19.** 16
20. 644   **21.** 96   **22.** 94

Margin Exercises, Section 1.8, pp. 56–62

1. $1324   **2.** 98 gal   **3.** 218,000   **4.** 591,971   **5.** $521
6. 17,956; 1,507,359   **7.** 10,127   **8.** $10,843   **9.** 2800
10. $9462   **11.** $100,375 \text{ mi}^2$ **12.** 4320 min
13. 275 packages; 5 cans left over   **14.** 54 weeks; 1 episode left over   **15.** 45 gal   **16.** 70 min   **17.** 106

Margin Exercises, Section 1.9, pp. 67–70

1. 5^4 **2.** 5^5 **3.** 10^2 **4.** 10^4 **5.** 10,000   **6.** 100
7. 512   **8.** 32   **9.** 51   **10.** 30   **11.** 584   **12.** 84

13. 4; 1 **14.** 52; 52 **15.** 29 **16.** 1880 **17.** 305
18. 93 **19.** 100; 52 **20.** 1880 **21.** 305 **22.** 93
23. 70 **24.** 46 **25.** 4

CHAPTER 2

Margin Exercises, Section 2.1, pp. 81–85

1. 1, 2, 3, 6 **2.** 1, 2, 4, 8 **3.** 1, 2, 5, 10 **4.** 1, 2, 4, 8, 16,
32 **5.** $5 = 1 \cdot 5$, $45 = 9 \cdot 5$, $100 = 20 \cdot 5$ **6.** $10 = 1 \cdot 10$,
$60 = 6 \cdot 10$, $110 = 11 \cdot 10$ **7.** 5, 10, 15, 20, 25, 30, 35, 40,
45, 50 **8.** Yes **9.** Yes **10.** No **11.** 13, 19, 41 are
prime; 4, 6, 8 are composite; 1 is neither **12.** $2 \cdot 3$
13. $2 \cdot 2 \cdot 3$ **14.** $3 \cdot 3 \cdot 5$ **15.** $2 \cdot 7 \cdot 7$ **16.** $2 \cdot 3 \cdot 3 \cdot 7$
17. $2 \cdot 2 \cdot 2 \cdot 2 \cdot 3 \cdot 3$

Margin Exercises, Section 2.2, pp. 89–92

1. Yes **2.** No **3.** Yes **4.** No **5.** Yes **6.** No
7. Yes **8.** No **9.** Yes **10.** No **11.** No **12.** Yes
13. No **14.** Yes **15.** No **16.** Yes **17.** No **18.** Yes
19. No **20.** Yes **21.** Yes **22.** No **23.** No **24.** Yes
25. Yes **26.** No **27.** No **28.** Yes **29.** No **30.** Yes
31. Yes **32.** No

Margin Exercises, Section 2.3, pp. 95–98

1. 1 numerator; 6 denominator **2.** 5 numerator;
7 denominator **3.** 22 numerator; 3 denominator
4. $\frac{1}{2}$ **5.** $\frac{1}{3}$ **6.** $\frac{1}{3}$ **7.** $\frac{1}{6}$ **8.** $\frac{5}{8}$ **9.** $\frac{2}{3}$ **10.** $\frac{3}{4}$ **11.** $\frac{4}{6}$
12. $\frac{4}{3}$ **13.** $\frac{5}{5}$ **14.** $\frac{5}{4}$ **15.** $\frac{7}{4}$ **16.** $\frac{2}{5}$ **17.** $\frac{2}{3}$ **18.** $\frac{2}{6}, \frac{4}{6}$
19. 1 **20.** 1 **21.** 1 **22.** 1 **23.** 1 **24.** 1 **25.** 0
26. 0 **27.** 0 **28.** 0 **29.** Not defined
30. Not defined **31.** 8 **32.** 10 **33.** 346 **34.** 1

Margin Exercises, Section 2.4, pp. 101–104

1. $\frac{2}{3}$ **2.** $\frac{5}{8}$ **3.** $\frac{10}{3}$ **4.** $\frac{33}{8}$ **5.** $\frac{46}{5}$ **6.** $\frac{15}{56}$ **7.** $\frac{32}{15}$ **8.** $\frac{3}{100}$
9. $\frac{14}{3}$
10.

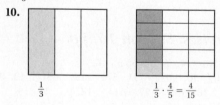

$$\frac{1}{3} \qquad \frac{1}{3} \cdot \frac{4}{5} = \frac{4}{15}$$

11. $\frac{3}{8}$ **12.** $\frac{63}{100}$ cm^2 **13.** $\frac{3}{40}$

Margin Exercises, Section 2.5, pp. 107–110

1. $\frac{8}{16}$ **2.** $\frac{30}{50}$ **3.** $\frac{52}{100}$ **4.** $\frac{200}{100}$ **5.** $\frac{12}{9}$ **6.** $\frac{18}{24}$ **7.** $\frac{90}{100}$
8. $\frac{9}{45}$ **9.** $\frac{56}{49}$ **10.** $\frac{1}{4}$ **11.** $\frac{5}{6}$ **12.** 5 **13.** $\frac{4}{3}$ **14.** $\frac{7}{8}$
15. $\frac{89}{78}$ **16.** $\frac{8}{7}$ **17.** $\frac{1}{4}$ **18.** $\frac{2}{100} = \frac{1}{50}$; $\frac{4}{100} = \frac{1}{25}$; $\frac{32}{100} = \frac{8}{25}$;
$\frac{44}{100} = \frac{11}{25}$; $\frac{18}{100} = \frac{9}{50}$ **19.** $=$ **20.** $\neq$

Margin Exercises, Section 2.6, p. 114

1. $\frac{7}{12}$ **2.** $\frac{1}{3}$ **3.** 6 **4.** $\frac{5}{2}$ **5.** 10 lb

Margin Exercises, Section 2.7, pp. 119–122

1. $\frac{5}{2}$ **2.** $\frac{7}{10}$ **3.** $\frac{1}{9}$ **4.** 5 **5.** $\frac{8}{7}$ **6.** $\frac{8}{3}$ **7.** $\frac{1}{10}$ **8.** 100
9. 1 **10.** $\frac{14}{15}$ **11.** $\frac{4}{5}$ **12.** 32 **13.** 320 **14.** 200 gal

CHAPTER 3

Margin Exercises, Section 3.1, pp. 135–139

1. 45 **2.** 40 **3.** 30 **4.** 24 **5.** 10 **6.** 80 **7.** 40
8. 360 **9.** 864 **10.** 2520 **11.** 18 **12.** 24 **13.** 36
14. 210 **15.** 2520 **16.** 3780

Margin Exercises, Section 3.2, pp. 143–146

1. $\frac{4}{5}$ **2.** 1 **3.** $\frac{1}{2}$ **4.** $\frac{3}{4}$ **5.** $\frac{5}{6}$ **6.** $\frac{29}{24}$ **7.** $\frac{5}{9}$ **8.** $\frac{413}{1000}$
9. $\frac{759}{1000}$ **10.** $\frac{197}{210}$ **11.** $\frac{11}{10}$ lb

Margin Exercises, Section 3.3, pp. 149–152

1. $\frac{1}{2}$ **2.** $\frac{3}{8}$ **3.** $\frac{1}{2}$ **4.** $\frac{1}{12}$ **5.** $\frac{13}{18}$ **6.** $\frac{1}{2}$ **7.** $\frac{9}{112}$ **8.** $<$
9. $>$ **10.** $>$ **11.** $>$ **12.** $<$ **13.** $\frac{1}{6}$ **14.** $\frac{11}{40}$
15. $\frac{11}{20}$ cup

Margin Exercises, Section 3.4, pp. 155–158

1. $1\frac{2}{3}$ **2.** $8\frac{3}{4}$ **3.** $12\frac{2}{3}$ **4.** $\frac{22}{5}$ **5.** $\frac{61}{10}$ **6.** $\frac{29}{6}$ **7.** $\frac{37}{4}$
8. $\frac{62}{3}$ **9.** $2\frac{1}{3}$ **10.** $1\frac{1}{10}$ **11.** $18\frac{1}{3}$ **12.** $807\frac{2}{3}$ **13.** $134\frac{23}{45}$

Margin Exercises, Section 3.5, pp. 161–164

1. $7\frac{2}{5}$ **2.** $12\frac{1}{10}$ **3.** $13\frac{7}{12}$ **4.** $1\frac{1}{2}$ **5.** $3\frac{1}{6}$ **6.** $3\frac{1}{3}$ **7.** $3\frac{2}{3}$
8. $17\frac{1}{12}$ yd **9.** $3\frac{3}{4}$ ft **10.** $23\frac{1}{4}$ gal

Margin Exercises, Section 3.6, pp. 169–172

1. 20 **2.** $1\frac{7}{8}$ **3.** $12\frac{4}{5}$ **4.** $8\frac{1}{3}$ **5.** 16 **6.** $7\frac{3}{7}$ **7.** $1\frac{7}{8}$
8. $\frac{7}{10}$ **9.** $227\frac{1}{2}$ mi **10.** 20 **11.** $240\frac{3}{4}$ ft^2

Margin Exercises, Section 3.7, pp. 177–178

1. $\frac{1}{2}$ **2.** $\frac{3}{10}$ **3.** $20\frac{2}{3}$ **4.** $\frac{5}{9}$ **5.** $\frac{31}{40}$ **6.** $\frac{27}{56}$

CHAPTER 4

Margin Exercises, Section 4.1, pp. 192–194

1. Twenty-seven and three tenths **2.** Two and four
thousand five hundred thirty-three ten-thousandths
3. Two hundred forty-five and eighty-nine hundredths
4. Thirty-one thousand, seventy-nine and seven hundred
sixty-four thousandths **5.** Four thousand, two hundred
seventeen and $\frac{56}{100}$ dollars **6.** Thirteen and $\frac{98}{100}$ dollars
7. $\frac{896}{1000}$ **8.** $\frac{2378}{100}$ **9.** $\frac{56,789}{10,000}$ **10.** $\frac{19}{10}$ **11.** 7.43 **12.** 0.406
13. 6.7089 **14.** 0.9 **15.** 0.057 **16.** 0.083

Margin Exercises, Section 4.2, pp. 197–198

1. 2.04　**2.** 0.06　**3.** 0.58　**4.** 1　**5.** 0.8989　**6.** 21.05
7. 2.8　**8.** 13.9　**9.** 234.4　**10.** 7.0　**11.** 0.64　**12.** 7.83
13. 34.68　**14.** 0.03　**15.** 0.943　**16.** 8.004　**17.** 43.112
18. 37.401　**19.** 7459.355　**20.** 7459.35　**21.** 7459.4
22. 7459　**23.** 7460　**24.** 7500　**25.** 7000

Margin Exercises, Section 4.3, pp. 201–204

1. 10.917　**2.** 34.2079　**3.** 4.969　**4.** 3.5617　**5.** 9.40544
6. 912.67　**7.** 2514.773　**8.** 10.754　**9.** 0.339
10. 0.5345　**11.** 0.5172　**12.** 7.36992　**13.** 1194.22
14. 4.9911　**15.** 38.534　**16.** 14.164　**17.** 2133.5

Margin Exercises, Section 4.4, pp. 209–210

1. 8.4°　**2.** 137.5 gal

CHAPTER 5

Margin Exercises, Section 5.1, pp. 226–229

1. 529.48　**2.** 5.0594　**3.** 34.2906　**4.** 0.348　**5.** 0.0348
6. 0.00348　**7.** 0.000348　**8.** 34.8　**9.** 348　**10.** 3480
11. 34,800　**12.** 17,000,000　**13.** 5,600,000,000
14. 1569¢　**15.** 17¢　**16.** $0.35　**17.** $5.77

Margin Exercises, Section 5.2, pp. 233–238

1. 0.6　**2.** 1.5　**3.** 0.47　**4.** 0.32　**5.** 3.75　**6.** 0.25
7. **(a)** 375; **(b)** 15　**8.** 4.9　**9.** 12.8　**10.** 15.625
11. 12.78　**12.** 0.001278　**13.** 0.09847　**14.** 67.832
15. 0.78314　**16.** 1105.6　**17.** 0.2426　**18.** 593.44
19. 13.65 million

Margin Exercises, Section 5.3, pp. 243–246

1. 0.8　**2.** 0.45　**3.** 0.275　**4.** 1.32　**5.** 0.4　**6.** 0.375
7. $0.1\overline{6}$　**8.** $0.\overline{6}$　**9.** $0.\overline{45}$　**10.** $1.\overline{09}$　**11.** $0.\overline{428571}$
12. 0.7; 0.67; 0.667　**13.** 0.8; 0.81; 0.808　**14.** 6.2; 6.25;
6.245　**15.** 0.72　**16.** 0.552　**17.** 9.6575

Margin Exercises, Section 5.4, pp. 249–251

1. (b)　**2.** (a)　**3.** (d)　**4.** (b)　**5.** (a)　**6.** (d)　**7.** (b)
8. (c)　**9.** (b)　**10.** (b)　**11.** (c)　**12.** (a)　**13.** (c)
14. (c)

Margin Exercises, Section 5.5, pp. 255–258

1. $37.28　**2.** $368.75　**3.** 96.52 cm²　**4.** $0.89
5. 28.6 miles per gallon

CHAPTER 6

Margin Exercises, Section 6.1, pp. 273–274

1. $\frac{5}{11}$　**2.** $\frac{57.3}{86.1}$　**3.** $\frac{6\frac{3}{4}}{7\frac{2}{5}}$　**4.** $\frac{21.1}{182.5}$　**5.** $\frac{107}{365}$　**6.** $\frac{4}{7\frac{2}{3}}$　**7.** $\frac{38.2}{56.1}$

8. 18 is to 27 as 2 is to 3.　**9.** 3.6 is to 12 as 3 is to 10.
10. 1.2 is to 1.5 as 4 is to 5.　**11.** $\frac{3}{4}$

Margin Exercises, Section 6.2, pp. 277–278

1. 5 mi/hr　**2.** 12 mi/hr　**3.** 0.3 mi/hr　**4.** 1100 ft/sec
5. 4 ft/sec　**6.** 14.5 ft/sec　**7.** 250 ft/sec　**8.** 2 gal/day
9. 20.64 ¢/oz　**10.** Jar A

Margin Exercises, Section 6.3, pp. 283–286

1. Yes　**2.** No　**3.** No　**4.** Yes　**5.** No　**6.** Yes　**7.** 14
8. $11\frac{1}{4}$　**9.** 10.5　**10.** 9　**11.** 10.8　**12.** $\frac{125}{42}$, or $2\frac{41}{42}$

Margin Exercises, Section 6.4, pp. 289–291

1. 3360 mi　**2.** 8　**3.** 3.24　**4.** 21 in.　**5.** 1620
6. 140 oz

Margin Exercises, Section 6.5, pp. 295–296

1. 15　**2.** 24.75 ft

CHAPTER 7

Margin Exercises, Section 7.1, pp. 310–312

1. $\frac{70}{100}$; $70 \times \frac{1}{100}$; 70×0.01　**2.** $\frac{23.4}{100}$; $23.4 \times \frac{1}{100}$; 23.4×0.01
3. $\frac{100}{100}$; $100 \times \frac{1}{100}$; 100×0.01　**4.** 0.34　**5.** 0.789
6. 0.1208　**7.** 0.021　**8.** 24%　**9.** 347%　**10.** 100%
11. 10%　**12.** 52%

Margin Exercises, Section 7.2, pp. 316–317

1. 25%　**2.** 87.5%, or $87\frac{1}{2}$%　**3.** $66.\overline{6}$%, or $66\frac{2}{3}$%
4. $83.\overline{3}$%, or $83\frac{1}{3}$%　**5.** 57%　**6.** 76%　**7.** $\frac{3}{5}$　**8.** $\frac{13}{400}$
9. $\frac{2}{3}$

10.

$\frac{1}{5}$	$\frac{5}{6}$	$\frac{3}{8}$
0.2	$0.83\overline{3}$	0.375
20%	$83.\overline{3}$%, or $83\frac{1}{3}$%	$37\frac{1}{2}$%

Margin Exercises, Section 7.3, pp. 323–326

1. $12\% \times 50 = a$　**2.** $a = 40\% \times 60$　**3.** $45 = 20\% \times t$
4. $120\% \times y = 60$　**5.** $16 = n \times 40$　**6.** $b \times 84 = 10.5$
7. 6　**8.** $35.20　**9.** 225　**10.** $50　**11.** 40%
12. 12.5%

Margin Exercises, Section 7.4, pp. 330–332

1. $\frac{12}{100} = \frac{a}{50}$　**2.** $\frac{40}{100} = \frac{a}{60}$　**3.** $\frac{20}{100} = \frac{45}{b}$　**4.** $\frac{120}{100} = \frac{60}{b}$
5. $\frac{n}{100} = \frac{16}{40}$　**6.** $\frac{n}{100} = \frac{10.5}{84}$　**7.** 6　**8.** 35.2　**9.** $225
10. 50　**11.** 40%　**12.** 12.5%

Margin Exercises, Section 7.5, pp. 335–340

1. 44% **2.** 5 lb **3.** 9% **4.** 4% **5.** $10,682

Margin Exercises, Section 7.6, pp. 345–346

1. $53.52; $722.47 **2.** 6% **3.** $420

Margin Exercises, Section 7.7, pp. 349–352

1. $5628 **2.** 25% **3.** $1675 **4.** $33.60; $106.40

Margin Exercises, Section 7.8, pp. 355–358

1. $344 **2.** $258 **3.** $56 **4.** $2205 **5.** $2101.25

CHAPTER 8

Margin Exercises, Section 8.1, pp. 371–374

1. 75 **2.** 54.9 **3.** 81 **4.** 19.4 **5.** 143.1 yd/game
6. 28 mpg **7.** 2.54 **8.** 94 **9.** 17 **10.** 17 **11.** 91
12. 17 **13.** 67.5 **14.** 45 **15.** 34, 67 **16.** 13, 24, 27,
28, 67, 89 **17. (a)** 82; **(b)** 81; **(c)** 74, 86, 96, 67, 82

Margin Exercises, Section 8.2, pp. 377–381

1. McGriff **2.** Morris **3.** 7.2 **4.** Clark, Grace,
McGriff **5.** Dallas, Denver, and San Francisco
6. $140–$190 **7.** None **8.** 600 **9.** Iraq and Israel
10. 8.625 × 50, or 431.25 **11.** Brazil, Germany, Japan, US
12. Brazil and Japan **13.** India and Germany
14.

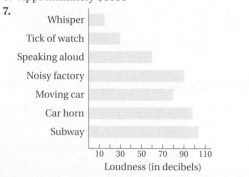

Margin Exercises, Section 8.3, pp. 387–392

1. Boredom **2.** Work or military service **3.** About 5%
4. Approximately $17,000 **5.** Approximately $14,000
6. Approximately $6000
7.

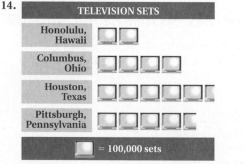

8. Ninth **9.** Third and fifth **10.** 250 **11.** $20,000
12. Age 4 **13.** $10,000
14.

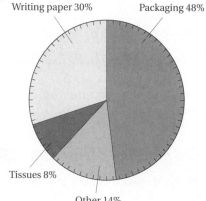

Margin Exercises, Section 8.4, pp. 397–400

1. $540 **2.** 38% **3.** 6 to 1 **4. (a)** 173°; **(b)** 108°;
(c) 29°; **(d)** 50°;
(e)

Writing paper 30%   Packaging 48%   Tissues 8%   Other 14%

CHAPTER 9

Margin Exercises, Section 9.1, pp. 419–421

1. 2 **2.** 3 **3.** $1\frac{1}{2}$ **4.** $2\frac{1}{2}$ **5.** 288 **6.** 43.5
7. 240,768 **8.** 6 **9.** $1\frac{5}{12}$ **10.** 8 **11.** $11\frac{2}{3}$, or $11.\overline{6}$
12. 5 **13.** 0.5 **14.** 288

Margin Exercises, Section 9.2, pp. 427–430

1. cm **2.** km **3.** mm **4.** m **5.** cm **6.** m
7. 23,000 **8.** 400 **9.** 178 **10.** 9040 **11.** 7.814
12. 781.4 **13.** 0.967 **14.** 8,900,000 **15.** 6.78 **16.** 97.4
17. 0.1 **18.** 8.451 **19.** 90.909 **20.** 804.5 **21.** 1479.843

Margin Exercises, Section 9.3, pp. 433–434

1. 26 cm **2.** 46 in. **3.** 12 cm **4.** 17.5 yd **5.** 32 km
6. 40 km **7.** 21 yd **8.** 31.2 km **9.** 70 ft; $346.50

Margin Exercises, Section 9.4, pp. 437–438

1. 8 cm² **2.** 56 km² **3.** $18\frac{3}{8}$ yd² **4.** 144 km²
5. 118.81 m² **6.** $12\frac{1}{4}$ yd² **7.** 659.75 m²

Margin Exercises, Section 9.5, pp. 442–444

1. 43.8 cm² **2.** 12.375 km² **3.** 96 m² **4.** 18.7 cm²
5. 100 m² **6.** 88 cm² **7.** 54 m²

Margin Exercises, Section 9.6, pp. 447–450

1. 12 km **2.** 5 ft **3.** 62.8 m **4.** 88 m **5.** 20.096 yd
6. $78\frac{4}{7}$ km² **7.** 339.62 cm² **8.** 12-ft diameter flower bed; by about 13.04 ft²

Margin Exercises, Section 9.7, pp. 455–458

1. 81 **2.** 100 **3.** 121 **4.** 144 **5.** 169 **6.** 196
7. 225 **8.** 256 **9.** 289 **10.** 324 **11.** 400 **12.** 625
13. 3 **14.** 4 **15.** 11 **16.** 10 **17.** 9 **18.** 8 **19.** 18
20. 20 **21.** 15 **22.** 13 **23.** 1 **24.** 0 **25.** 2.236
26. 8.832 **27.** 12.961 **28.** $c = \sqrt{41}$; $c \approx 6.403$
29. $a = \sqrt{75}$; $a \approx 8.660$ **30.** $b = \sqrt{120}$; $b \approx 10.954$
31. $a = \sqrt{175}$; $a \approx 13.229$ **32.** $\sqrt{424}$ ft ≈ 20.591 ft

CHAPTER 10

Margin Exercises, Section 10.1, pp. 475–478

1. 12 cm³ **2.** 20 ft³ **3.** 128 ft³ **4.** 40 **5.** 20
6. mL **7.** mL **8.** L **9.** L **10.** 970 **11.** 8.99
12. 4.8 L **13.** (a) 118.28 mL; (b) 0.11828 L **14.** $1)

Margin Exercises, Section 10.2, pp. 481–482

1. 785 ft³ **2.** 67,914 m³ **3.** $91,989\frac{1}{3}$ ft³
4. 38.77272 cm³ **5.** 1695.6 m³ **6.** 528 in³

Margin Exercises, Section 10.3, pp. 485–488

1. 80 **2.** 4.32 **3.** 32,000 **4.** kg **5.** kg **6.** mg
7. g **8.** t **9.** 6200 **10.** 3.048 **11.** 77 **12.** 234.4
13. 6700 **14.** 120 **15.** 1461 **16.** 1440 **17.** 1

Margin Exercises, Section 10.4, pp. 493–494

1. 80°C **2.** 0°C **3.** −20°C **4.** 80°F **5.** 100°F
6. 50°F **7.** 176°F **8.** 95°F **9.** 35°C **10.** 45°C

Margin Exercises, Section 10.5, pp. 497–498

1. 9 **2.** 45 **3.** 2880 **4.** 2.5 **5.** 3200 **6.** 1,000,000
7. 100 **8.** 28,800 **9.** 0.043 **10.** 0.678

CHAPTER 11

Margin Exercises, Section 11.1, pp. 512–516

1. 8; −5 **2.** 134; −76 **3.** −10; 148 **4.** −137; 289
5.

6.

7.

8. −0.375 **9.** $-0.\overline{54}$ **10.** $1.\overline{3}$ **11.** < **12.** < **13.** >
14. > **15.** > **16.** < **17.** < **18.** > **19.** 8 **20.** 0
21. 9 **22.** $\frac{2}{3}$ **23.** 5.6

Margin Exercises, Section 11.2, pp. 519–522

1. −8 **2.** −3 **3.** −8 **4.** 4 **5.** 0 **6.** −2 **7.** −11
8. −12 **9.** 2 **10.** −4 **11.** −2 **12.** 0 **13.** −22
14. 3 **15.** 0.53 **16.** 2.3 **17.** −7.7 **18.** −6.2 **19.** $-\frac{2}{9}$
20. $-\frac{19}{20}$ **21.** −58 **22.** −56 **23.** −14 **24.** −12
25. 4 **26.** −8.7 **27.** 7.74 **28.** $\frac{8}{9}$ **29.** 0 **30.** −12
31. −14; 14 **32.** −1; 1 **33.** 19; −19 **34.** 1.6; −1.6
35. $-\frac{2}{3}$; $\frac{2}{3}$ **36.** $\frac{9}{8}$; $-\frac{9}{8}$ **37.** 4 **38.** 13.4 **39.** 0
40. $-\frac{1}{4}$

Margin Exercises, Section 11.3, pp. 525–526

1. −10 **2.** 3 **3.** −5 **4.** −2 **5.** −11 **6.** 4 **7.** −2
8. −6 **9.** −16 **10.** 7.1 **11.** 3 **12.** 0 **13.** $\frac{3}{2}$ **14.** −8
15. 7 **16.** −3 **17.** −23.3 **18.** 0 **19.** −9 **20.** 17
21. 12.7

Margin Exercises, Section 11.4, pp. 529–530

1. $2 \cdot 10 = 20$; $1 \cdot 10 = 10$; $0 \cdot 10 = 0$; $-1 \cdot 10 = -10$;
$-2 \cdot 10 = -20$; $-3 \cdot 10 = -30$ **2.** −18 **3.** −100
4. −80 **5.** $-\frac{5}{9}$ **6.** −30.033 **7.** $-\frac{7}{10}$ **8.** $1 \cdot (-10) =$
-10; $0 \cdot (-10) = 0$; $-1 \cdot (-10) = 10$; $-2 \cdot (-10) = 20$;
$-3 \cdot (-10) = 30$ **9.** 12 **10.** 32 **11.** 35 **12.** $\frac{20}{63}$ **13.** $\frac{2}{3}$
14. 13.455 **15.** −30 **16.** 30 **17.** −32 **18.** $-\frac{8}{3}$
19. −30 **20.** −30.75 **21.** $-\frac{5}{3}$ **22.** 120 **23.** −120
24. 6

Margin Exercises, Section 11.5, pp. 533–536

1. −2 **2.** 5 **3.** −3 **4.** 9 **5.** −6 **6.** $-\frac{30}{7}$
7. Undefined **8.** 0 **9.** $\frac{3}{2}$ **10.** $-\frac{4}{5}$ **11.** $-\frac{1}{3}$ **12.** −5
13. $\frac{1}{5.78}$ **14.** $-\frac{7}{2}$

15.

NUMBER	OPPOSITE	RECIPROCAL
$\frac{2}{3}$	$-\frac{2}{3}$	$\frac{3}{2}$
$-\frac{5}{4}$	$\frac{5}{4}$	$-\frac{4}{5}$
0	0	Undefined
1	-1	1
4.5	4.5	$-\frac{1}{4.5}$

16. $\frac{4}{7} \cdot \left(-\frac{5}{3}\right)$ **17.** $5 \cdot \left(-\frac{1}{8}\right)$ **18.** $-10 \cdot \left(\frac{1}{7}\right)$ **19.** $-\frac{2}{3} \cdot \frac{7}{4}$
20. $-5 \cdot \left(\frac{1}{7}\right)$ **21.** $-\frac{20}{21}$ **22.** $-\frac{12}{5}$ **23.** $\frac{16}{7}$ **24.** -7
25. -1237 **26.** 8 **27.** 381 **28.** -12

Chapter 12

Margin Exercises, Section 12.1, pp. 549–554

1. 64 **2.** 28 **3.** 60 **4.** 25 **5.** 16
6.

	$1 \cdot x$	x
$x = 3$	3	3
$x = -6$	-6	-6
$x = 4.8$	4.8	4.8

7.

	$2x$	$5x$
$x = 2$	4	10
$x = -6$	-12	-30
$x = 4.8$	9.6	24

8. 36; 36 **9.** 90; 90 **10.** 64; 64 **11.** 14; 14 **12.** 30; 30
13. 12; 12 **14.** $5x$; $-4y$; 3 **15.** $-4y$; $-2x$; $3z$
16. $3x - 15$ **17.** $5x + 5$ **18.** $5x - 5y + 20$
19. $-2x + 6$ **20.** $-5x + 10y - 20z$
21. $6(z - 2)$ **22.** $3(x - 2y + 3)$ **23.** $2(8a - 18b + 21)$
24. $-4(3x - 8y + 4z)$ **25.** $3x$ **26.** $6x$ **27.** $-8x$
28. $0.59x$ **29.** $3x + 3y$ **30.** $-4x - 5y - 7$

Margin Exercises, Section 12.2, pp. 557–558

1. 9 **2.** -5 **3.** 13.2 **4.** -6.5

Margin Exercises, Section 12.3, pp. 561–562

1. 30 **2.** $-\frac{7}{4}$ **3.** 8 **4.** -18

Margin Exercises, Section 12.4, pp. 565–570

1. 5 **2.** 4 **3.** 4 **4.** 39 **5.** $-\frac{3}{2}$ **6.** -4.3 **7.** -3

8. 800 **9.** 1 **10.** 2 **11.** 2 **12.** $\frac{17}{2}$ **13.** $\frac{8}{3}$ **14.** -4.3
15. 2 **16.** 3 **17.** -2 **18.** $-\frac{1}{2}$

Margin Exercises, Section 12.5, pp. 576–580

1. $x - 12$ **2.** $y + 12$, or $12 + y$ **3.** $m - 4$ **4.** $\frac{1}{2} \cdot p$
5. $6 + 8x$, or $8x + 6$ **6.** $a - b$ **7.** $59\%x$, or $0.59x$
8. $xy - 200$ **9.** $p + q$ **10.** 28 in., 30 in. **11.** 5
12. 240.9 mi **13.** Width: 9 ft; length: 12 ft

Developmental Units

Margin Exercises, Section A, pp. 598–602

1. 9 **2.** 7 **3.** 14 **4.** 16 **5.** 16 **6.** 16 **7.** 8 **8.** 8
9. 7 **10.** 46 **11.** 13 **12.** 58
13.

+	1	2	3	4	5
1	2	3	4	5	6
2	3	4	5	6	7
3	4	5	6	7	8
4	5	6	7	8	9
5	6	7	8	9	10

14.

+	6	5	7	4	9
7	13	12	14	11	16
9	15	14	16	13	18
5	11	10	12	9	14
8	14	13	15	12	17
4	10	9	11	8	13

15. 16 **16.** 9 **17.** 11 **18.** 15 **19.** 59 **20.** 549
21. 9979 **22.** 5496 **23.** 56 **24.** 85 **25.** 829
26. 1026 **27.** 12,698 **28.** 13,661 **29.** 71,328

Margin Exercises, Section S, pp. 607–612

1. 4 **2.** 7 **3.** 8 **4.** 3 **5.** $12 - 8 = 4$; $12 - 4 = 8$
6. $13 - 6 = 7$; $13 - 7 = 6$ **7.** 8 **8.** 7 **9.** 9 **10.** 4
11. 14 **12.** 20 **13.** 241 **14.** 2025 **15.** 17 **16.** 36
17. 454 **18.** 250 **19.** 376 **20.** 245 **21.** 2557
22. 3674 **23.** 8 **24.** 6 **25.** 30 **26.** 90 **27.** 500
28. 900 **29.** 538 **30.** 677 **31.** 42 **32.** 224
33. 138 **34.** 44 **35.** 2768 **36.** 1197 **37.** 1246
38. 4728

Margin Exercises, Section M, pp. 617–622

1. 56 **2.** 36 **3.** 28 **4.** 42 **5.** 0 **6.** 0 **7.** 8
8. 23

9.

×	2	3	4	5
2	4	6	8	10
3	6	9	12	15
4	8	12	16	20
5	10	15	20	25
6	12	18	24	30

10.

×	6	7	8	9
5	30	35	40	45
6	36	42	48	54
7	42	49	56	63
8	48	56	64	72
9	54	63	72	81

11. 70 **12.** 450 **13.** 2730 **14.** 100 **15.** 1000
16. 560 **17.** 360 **18.** 700 **19.** 2300 **20.** 72,300
21. 10,000 **22.** 100,000 **23.** 5600 **24.** 1600
25. 9000 **26.** 852,000 **27.** 10,000 **28.** 1,200,000
29. 6,000,000 **30.** 54,000 **31.** 5600 **32.** 56,000
33. 18,000 **34.** 28 **35.** 116 **36.** 148 **37.** 4938
38. 6740 **39.** 116 **40.** 148 **41.** 4938 **42.** 6740
43. 5968 **44.** 59,680 **45.** 596,800

Margin Exercises, Section D, pp. 627–634

1. 4 **2.** 8 **3.** 9 **4.** 3 **5.** 12 ÷ 2 = 6; 12 ÷ 6 = 2
6. 42 ÷ 6 = 7; 42 ÷ 7 = 6 **7.** 7 **8.** 9 **9.** 8 **10.** 9
11. 6 **12.** 13 **13.** 1 **14.** 2 **15.** Not defined **16.** 0
17. Not defined **18.** Not defined **19.** 10 **20.** Not
defined **21.** 0 **22.** 1 **23.** 1 **24.** 1 **25.** 1 **26.** 17
27. 1 **28.** 75 R 4 **29.** 23 R 11 **30.** 969 R 6
31. 47 R 4 **32.** 96 R 1 **33.** 1263 R 5 **34.** 324
35. 417 R 5 **36.** 37 R 8 **37.** 23 R 17

Exercise Set and Test Answers

Diagnostic PRETEST, P. xxvii

1. [1.2b] 1807 **2.** [1.7b] 29 **3.** [1.5b] 15,087
4. [1.8a] 21 glasses, 2 oz left over **5.** [2.7c] $\frac{3}{10}$
6. [2.4a] $\frac{3}{2}$ **7.** [2.1d] 2 · 2 · 2 · 2 · 3 · 3 **8.** [2.6b] $\frac{1}{2}$ cup
9. [3.1a] 48 **10.** [3.2b] $\frac{13}{24}$ **11.** [3.6a] $15\frac{2}{5}$
12. [3.5c] $1\frac{3}{5}$ m **13.** [3.6c] 30 **14.** [4.2a] 0.0001
15. [4.2b] 25.6 **16.** [4.3a] 39.815 **17.** [4.4a] 186.9 mi
18. [5.1a] 0.03 **19.** [5.3a] 2.$\overline{3}$ **20.** [5.2b] 2.4
21. [5.5a] $119.95 **22.** [5.4a] 160 **23.** [6.1c] 4.5
24. [6.3a] 12 **25.** [6.2b] 10.0¢/oz **26.** [7.2b] 12.5%
27. [7.2a] 0.0135 **28.** [7.5b] 40% **29.** [7.8a] $19.55
30. [8.1a, b, c] Average: 24.$\overline{3}$; median: 25; mode: 25
31. [8.1a] 18.5 **32.** [8.1a] 85 **33.** [9.1a] 72
34. [9.2a] 0.00004 **35.** [9.6a, b, c] 36π cm²; 12π cm
36. [9.3a], [9.4b] 7.5 ft²; 11 ft **37.** [9.7a] 11
38. [10.3c] 120 **39.** [10.3b] 0.005 **40.** [10.5a] 18
41. [10.1b] 2.5 **42.** [10.2b] 3052.08 cm³ **43.** [11.1e] 4.2
44. [11.5c] $-0.\overline{4}$ **45.** [11.3a] -0.1 **46.** [11.4a] $-\frac{1}{12}$
47. [12.4b] -3 **48.** [12.4b] $\frac{4}{5}$ **49.** [12.5b] $25.56
50. [12.5b] 8

CHAPTER 1

Pretest: Chapter 1, p. 2

1. [1.1c] Three million, seventy-eight thousand, fifty-nine
2. [1.1a] 6 thousands + 9 hundreds + 8 tens + 7 ones
3. [1.1d] 2,047,398,589 **4.** [1.1e] 6 ten thousands
5. [1.4a] 956,000 **6.** [1.5c] 60,000 **7.** [1.2b] 10,216
8. [1.3d] 4108 **9.** [1.5b] 22,976 **10.** [1.6c] 503 R 11
11. [1.4c] < **12.** [1.4c] > **13.** [1.7b] 5542

14. [1.7b] 22 **15.** [1.7b] 34 **16.** [1.7b] 25
17. [1.8a] 12 lb **18.** [1.8a] 126 **19.** [1.8a] 22,277,717
20. [1.8a] 2292 sq ft **21.** [1.9b] 25 **22.** [1.9b] 64
23. [1.9c] 0 **24.** [1.9d] 0

Exercise Set 1.1, p. 7

1. 5 thousands + 7 hundreds + 4 tens + 2 ones
3. 2 ten thousands + 7 thousands + 3 hundreds +
4 tens + 2 ones **5.** 5 thousands + 6 hundreds +
9 ones **7.** 2 thousands + 3 hundreds **9.** 2475
11. 68,939 **13.** 7304 **15.** 1009 **17.** Eighty-five
19. Eighty-eight thousand **21.** One hundred
twenty-three thousand, seven hundred sixty-five
23. Seven billion, seven hundred fifty-four million, two
hundred eleven thousand, five hundred seventy-seven
25. Six million, four hundred sixty-nine thousand, nine
hundred fifty-two **27.** One million, nine hundred
fifty-four thousand, one hundred sixteen **29.** 2,233,812
31. 8,000,000,000 **33.** 217,503 **35.** 1,000,187,542
37. 206,658,000 **39.** 5 thousands **41.** 5 hundreds
43. 3 **45.** 0 **47.** All 9's as digits. Answers may vary.
For an 8-digit readout, it would be 99,999,999. This
number has three periods.

Sidelights: Number Patterns: Magic Squares, p. 14

1. First row: 8, 3, 4; second row: 1, 5, 9; third row: 6, 7, 2.
2. First row: 1, 12, 14, 7; second row: 4, 15, 9, 6; third row:
13, 2, 8, 11; fourth row: 16, 5, 3, 10 **3.** 78 should be 79.
4. 18 should be 17. **5.** 3 should be 2. **6.** 50 should
be 31.

Exercise Set 1.2, p. 15

1. 3 videos + 6 videos = 9 videos   **3.** $23 + $31 = $54
5. 1300 ft   **7.** 387   **9.** 5198   **11.** 164   **13.** 100
15. 900   **17.** 1010   **19.** 1201   **21.** 847   **23.** 8310
25. 6608   **27.** 16,784   **29.** 34,432   **31.** 101,310
33. 100,111   **35.** 28   **37.** 25   **39.** 35   **41.** 87
43. 230   **45.** 130   **47.** 149   **49.** 169   **51.** 842
53. 2320   **55.** 11,679   **57.** 22,654   **59.** 12,765,097
61. 7992   **63.** 8 ten thousands   **65.** $1 + 99 = 100$,
$2 + 98 = 100, \ldots, 49 + 51 = 100$. Then $49 \cdot 100 = 4900$ and
$4900 + 50 + 100 = 5050$.

Exercise Set 1.3, p. 23

1. $2400 - 800 = $ ▨   **3.** $7 = 3 + 4$, or $7 = 4 + 3$
5. $13 = 5 + 8$, or $13 = 8 + 5$ **7.** $23 = 14 + 9$, or
$23 = 9 + 14$ **9.** $43 = 27 + 16$, or $43 = 16 + 27$
11. $6 = 15 - 9; 9 = 15 - 6$ **13.** $8 = 15 - 7; 7 = 15 - 8$
15. $17 = 23 - 6; 6 = 23 - 17$ **17.** $23 = 32 - 9$;
$9 = 32 - 23$ **19.** $190 + $ ▨ $= 220;$ ▨ $= 220 - 190$
21. 12   **23.** 44   **25.** 533   **27.** 1126   **29.** 39   **31.** 298
33. 226   **35.** 234   **37.** 5382   **39.** 1493   **41.** 2187
43. 3831   **45.** 7748   **47.** 33,794   **49.** 56   **51.** 36
53. 84   **55.** 454   **57.** 771   **59.** 2191   **61.** 3749
63. 7019   **65.** 4206   **67.** 10,305   **69.** 7 ten thousands

Exercise Set 1.4, p. 29

1. 50   **3.** 70   **5.** 730   **7.** 900   **9.** 100   **11.** 1000
13. 9100   **15.** 2900   **17.** 6000   **19.** 8000   **21.** 45,000
23. 373,000   **25.** 17,620   **27.** 5720   **29.** 220;
incorrect   **31.** 890; incorrect   **33.** 16,500   **35.** 5200
37. 1600   **39.** 1500   **41.** 31,000   **43.** 69,000   **45.** <
47. >   **49.** <   **51.** >   **53.** >   **55.** >   **57.** 86,754
59. 48,824   **61.** 30,411   **63.** 69,594

Exercise Set 1.5, p. 37

1. $32 \cdot $10 = 320 **3.** $8 \cdot 8 = 64$ **5.** $3 \cdot 6 = 18$ ft^2
7. 870   **9.** 2,340,000   **11.** 520   **13.** 564   **15.** 1527
17. 64,603   **19.** 4770   **21.** 3995   **23.** 46,080
25. 14,652   **27.** 207,672   **29.** 798,408   **31.** 166,260
33. 11,794,332   **35.** 20,723,872   **37.** 362,128
39. 20,064,048   **41.** 25,236,000   **43.** 302,220
45. 49,101,136   **47.** 30,525   **49.** 298,738   **51.** $50 \cdot 70 =$
3500   **53.** $30 \cdot 30 = 900$ **55.** $900 \cdot 300 = 270,000$
57. $400 \cdot 200 = 80,000$ **59.** $6000 \cdot 5000 = 30,000,000$
61. $8000 \cdot 6000 = 48,000,000$ **63.** 4370   **65.** 2350;
2300; 2000

Exercise Set 1.6, p. 47

1. $176 \div 4 = $ ▨   **3.** $184,000 \div $23,000 = $ ▨
5. $18 = 3 \cdot 6$, or $18 = 6 \cdot 3$ **7.** $22 = 22 \cdot 1$, or $22 = 1 \cdot 22$
9. $54 = 6 \cdot 9$, or $54 = 9 \cdot 6$ **11.** $37 = 1 \cdot 37$, or $37 = 37 \cdot 1$
13. $9 = 45 \div 5; 5 = 45 \div 9$ **15.** $37 = 37 \div 1$;
$1 = 37 \div 37$ **17.** $8 = 64 \div 8$ **19.** $11 = 66 \div 6$;
$6 = 66 \div 11$ **21.** 55 R 2   **23.** 108   **25.** 307
27. 753 R 3   **29.** 74 R 1   **31.** 92 R 2   **33.** 1703
35. 987 R 5   **37.** 52 R 52   **39.** 29 R 5   **41.** 40 R 12
43. 90 R 22   **45.** 29   **47.** 105 R 3   **49.** 1609 R 2

51. 1007 R 1   **53.** 23   **55.** 107 R 1   **57.** 370
59. 609 R 15   **61.** 304   **63.** 3508 R 219   **65.** 8070
67. 7 thousands + 8 hundreds + 8 tens + 2 ones   **69.** 30

Exercise Set 1.7, p. 53

1. 14   **3.** 0   **5.** 29   **7.** 0   **9.** 8   **11.** 14   **13.** 1035
15. 25   **17.** 450   **19.** 90,900   **21.** 32   **23.** 143
25. 79   **27.** 45   **29.** 324   **31.** 743   **33.** 37   **35.** 66
37. 15   **39.** 48   **41.** 175   **43.** 335   **45.** 104   **47.** 45
49. 4056   **51.** 17,603   **53.** 18,252   **55.** 205   **57.** 55
59. $6 = 48 \div 8; 8 = 48 \div 6$ **61.** >

Exercise Set 1.8, p. 63

1. (a) 8447, 65; **(b)** 7401, 141   **3.** 449 m   **5.** 2995 cubic
centimeters   **7.** 137 megabytes   **9.** 11,883,479   **11.** $91
13. 665 cal   **15.** 3600 sec   **17.** 2808 ft^2 **19.** 7815 mi
21. $4098   **23.** 5130 yd^2 **25.** 38   **27.** 15   **29.** $27
31. 38 bags; 11 kg left over   **33.** 16   **35.** 11 in.; 770 mi
37. 480   **39.** 525 min, or 8 hr 45 min   **41.** 235,000
43. 186,000 mi

Exercise Set 1.9, p. 71

1. 3^4 **3.** 5^2 **5.** 7^5 **7.** 10^3 **9.** 49   **11.** 729
13. 20,736   **15.** 121   **17.** 22   **19.** 20   **21.** 100   **23.** 1
25. 49   **27.** 27   **29.** 434   **31.** 41   **33.** 88   **35.** 4
37. 303   **39.** 20   **41.** 70   **43.** 295   **45.** 32   **47.** 906
49. 62   **51.** 102   **53.** $94   **55.** 110   **57.** 7   **59.** 544
61. 708   **63.** $24; 1 + 5 \cdot (4 + 3) = 36$
65. $7; 12 \div (4 + 2) \cdot 3 - 2 = 4$

Summary and Review: Chapter 1, p. 75

1. [1.1a] 2 thousands + 7 hundreds + 9 tens + 3 ones
2. [1.1c] Two million, seven hundred eighty-one thousand,
four hundred twenty-seven   **3.** [1.1e] 7 ten thousands
4. [1.1d] $5,685,800,000,000   **5.** [1.2b] 5979
6. [1.2b] 66,024   **7.** [1.2b] 22,098   **8.** [1.2b] 98,921
9. [1.3d] 1153   **10.** [1.3d] 1147   **11.** [1.3d] 2274
12. [1.3d] 17,757   **13.** [1.3d] 444   **14.** [1.3d] 4766
15. [1.5b] 420,000   **16.** [1.5b] 6,276,800   **17.** [1.5b] 684
18. [1.5b] 44,758   **19.** [1.5b] 3404   **20.** [1.5b] 506,748
21. [1.5b] 27,589   **22.** [1.5b] 3,456,000   **23.** [1.6c] 5
24. [1.6c] 12 R 3   **25.** [1.6c] 80   **26.** [1.6c] 207 R 2
27. [1.6c] 384 R 1   **28.** [1.6c] 4 R 46   **29.** [1.6c] 54
30. [1.6c] 452   **31.** [1.6c] 5008   **32.** [1.6c] 4389
33. [1.7b] 45   **34.** [1.7b] 546   **35.** [1.7b] 8
36. [1.8a] 1982   **37.** [1.8a] 2785 bu   **38.** [1.8a] $19,748
39. [1.8a] 2825 cal   **40.** [1.8a] $501
41. [1.8a] 137 beakers, 13 mL left over   **42.** [1.8a] 10
43. [1.4a] 345,800   **44.** [1.4a] 345,760
45. [1.4a] 346,000   **46.** [1.4b] $41,300 + 19,700 = 61,000$
47. [1.4b] $38,700 - 24,500 = 14,200$
48. [1.5c] $400 \cdot 700 = 280,000$ **49.** [1.4c] >
50. [1.4c] <   **51.** [1.9a] 8^3 **52.** [1.9b] 16   **53.** [1.9b] 36
54. [1.9c] 65   **55.** [1.9c] 233   **56.** [1.9c] 56
57. [1.9c] 32   **58.** [1.9d] 260   **59.** [1.9c] 165
60. [1.2a], [1.5a] 112 ft; 784 ft^2
61. [1.2a], [1.5a] 986 yd; 46,000 yd^2

Test: Chapter 1, p. 77

1. [1.1a] 8 thousands + 8 hundreds + 4 tens + 3 ones
2. [1.1c] Thirty-eight million, four hundred three thousand, two hundred seventy-seven   **3.** [1.1e] 5
4. [1.2b] 9989   **5.** [1.2b] 63,791   **6.** [1.2b] 34
7. [1.2b] 10,515   **8.** [1.3d] 3630   **9.** [1.3d] 1039
10. [1.3d] 6848   **11.** [1.3d] 5175   **12.** [1.5b] 41,112
13. [1.5b] 5,325,600   **14.** [1.5b] 2405
15. [1.5b] 534,264   **16.** [1.6c] 3 R 3   **17.** [1.6c] 70
18. [1.6c] 97   **19.** [1.6c] 805 R 8   **20.** [1.8a] 1955
21. [1.8a] 92 packages, 3 left over   **22.** [1.8a] 18
23. [1.8a] 120,000 m^2 **24.** [1.8a] 1808 lb   **25.** [1.8a] 20
26. [1.8a] 305 mi^2 **27.** [1.8a] 56   **28.** [1.8a] 66,444 mi^2
29. [1.8a] $271   **30.** [1.7b] 46   **31.** [1.7b] 13
32. [1.7b] 14   **33.** [1.4a] 35,000   **34.** [1.4a] 34,580
35. [1.4a] 34,600   **36.** [1.4b] 23,600 + 54,700 = 78,300
37. [1.4b] 54,800 − 23,600 = 31,200
38. [1.5c] 800 · 500 = 400,000   **39.** [1.4c] >
40. [1.4c] <   **41.** [1.9a] 12^4 **42.** [1.9b] 343   **43.** [1.9b] 8
44. [1.9c] 64   **45.** [1.9c] 96   **46.** [1.9c] 2   **47.** [1.9d] 216
48. [1.9c] 18   **49.** [1.9c] 92   **50.** [1.2a], [1.5a] 226 ft; 2322 sq ft

CHAPTER 2

Pretest: Chapter 2, p. 80

1. [2.1c] Prime   **2.** [2.1d] 2 · 2 · 5 · 7   **3.** [2.2a] Yes
4. [2.2a] No   **5.** [2.3c] 1   **6.** [2.3c] 68   **7.** [2.3c] 0
8. [2.5b] $\frac{1}{4}$ **9.** [2.6a] $\frac{6}{5}$ **10.** [2.6a] 20   **11.** [2.6a] $\frac{5}{4}$
12. [2.7a] $\frac{8}{7}$ **13.** [2.7a] $\frac{1}{11}$ **14.** [2.7b] 24   **15.** [2.7b] $\frac{3}{4}$
16. [2.7c] 30   **17.** [2.5c] ≠   **18.** [2.6b] $36
19. [2.7d] $\frac{1}{24}$ m

Sidelights: Factors and Sums, p. 86

First row: 48, 90, 432, 63; second row: 7, 18, 36, 14, 12, 6, 21, 11; third row: 9, 2, 2, 10, 8, 10, 21; fourth row: 29, 19, 42

Exercise Set 2.1, p. 87

1. 1, 2, 3, 6, 9, 18   **3.** 1, 2, 3, 6, 9, 18, 27, 54   **5.** 1, 2, 4
7. 1, 7   **9.** 1   **11.** 1, 2, 7, 14, 49, 98   **13.** 4, 8, 12, 16, 20, 24, 28, 32, 36, 40   **15.** 20, 40, 60, 80, 100, 120, 140, 160, 180, 200   **17.** 3, 6, 9, 12, 15, 18, 21, 24, 27, 30   **19.** 12, 24, 36, 48, 60, 72, 84, 96, 108, 120   **21.** 10, 20, 30, 40, 50, 60, 70, 80, 90, 100   **23.** 9, 18, 27, 36, 45, 54, 63, 72, 81, 90
25. No   **27.** Yes   **29.** Yes   **31.** No   **33.** No
35. Neither   **37.** Composite   **39.** Prime   **41.** Prime
43. 2 · 2 · 2   **45.** 2 · 7   **47.** 2 · 3 · 7   **49.** 5 · 5
51. 2 · 5 · 5   **53.** 13 · 13   **55.** 2 · 2 · 5 · 5   **57.** 5 · 7
59. 2 · 2 · 2 · 3 · 3   **61.** 7 · 11   **63.** 2 · 2 · 3 · 13
65. 2 · 2 · 3 · 5 · 5   **67.** 26   **69.** 425   **71.** 0   **73.** 1
75. A rectangular array of 6 rows of 9 objects each, or 9 rows of 6 objects each

Exercise Set 2.2, p. 93

1. 46; 300; 224; 36; 45,270; 4444   **3.** 300; 224; 36; 4444

5. 300; 36; 45,270   **7.** 36; 711; 45,270   **9.** 75; 324; 42; 501; 3009; 2001   **11.** 200; 75; 2345; 55,555   **13.** 324
15. 200   **17.** 138   **19.** 56   **21.** $680
23. 2 · 2 · 2 · 3 · 5 · 5 · 13   **25.** 2 · 2 · 3 · 3 · 7 · 11

Exercise Set 2.3, p. 99

1. 3 numerator; 4 denominator   **3.** 11 numerator; 20 denominator   **5.** $\frac{2}{4}$ **7.** $\frac{1}{8}$ **9.** $\frac{2}{3}$ **11.** $\frac{3}{4}$ **13.** $\frac{4}{8}$
15. $\frac{6}{12}$ **17.** $\frac{5}{8}$ **19.** $\frac{3}{5}$ **21.** 0   **23.** 8   **25.** 1   **27.** 1
29. 0   **31.** 1   **33.** 1   **35.** 1   **37.** 1   **39.** 8
41. Not defined   **43.** Not defined   **45.** 34,560
47. 35,000   **49.** 3728 lb   **51.** $\frac{1200}{2700}$; $\frac{540}{2700}$; $\frac{360}{2700}$; $\frac{600}{2700}$

Exercise Set 2.4, p. 105

1. $\frac{3}{5}$ **3.** $\frac{5}{8}$ **5.** $\frac{8}{11}$ **7.** $\frac{70}{9}$ **9.** $\frac{2}{5}$ **11.** $\frac{6}{5}$ **13.** $\frac{21}{4}$ **15.** $\frac{85}{6}$
17. $\frac{1}{6}$ **19.** $\frac{1}{40}$ **21.** $\frac{2}{15}$ **23.** $\frac{4}{15}$ **25.** $\frac{9}{16}$ **27.** $\frac{14}{39}$ **29.** $\frac{7}{100}$
31. $\frac{49}{64}$ **33.** $\frac{1}{1000}$ **35.** $\frac{182}{285}$ **37.** $\frac{12}{25}$ m^2 **39.** $\frac{1}{1521}$
41. $\frac{3}{8}$ cup   **43.** $\frac{56}{100}$ **45.** 204   **47.** 3001
49. 6 hundred thousands

Exercise Set 2.5, p. 111

1. $\frac{5}{10}$ **3.** $\frac{20}{32}$ **5.** $\frac{27}{30}$ **7.** $\frac{28}{32}$ **9.** $\frac{20}{48}$ **11.** $\frac{51}{54}$ **13.** $\frac{75}{45}$
15. $\frac{42}{132}$ **17.** $\frac{1}{2}$ **19.** $\frac{3}{4}$ **21.** $\frac{1}{5}$ **23.** 3   **25.** $\frac{3}{4}$ **27.** $\frac{7}{8}$
29. $\frac{6}{5}$ **31.** $\frac{1}{3}$ **33.** 6   **35.** $\frac{1}{3}$ **37.** =   **39.** ≠   **41.** =
43. ≠   **45.** =   **47.** ≠   **49.** =   **51.** ≠   **53.** 4992 ft^2
55. 11   **57.** 186   **59.** $\frac{2}{5}$; $\frac{3}{5}$ **61.** No. $\frac{116}{384} \neq \frac{172}{532}$ because 116 · 532 ≠ 384 · 172.

Exercise Set 2.6, p. 115

1. $\frac{1}{3}$ **3.** $\frac{1}{8}$ **5.** $\frac{1}{10}$ **7.** $\frac{1}{6}$ **9.** $\frac{27}{10}$ **11.** $\frac{14}{9}$ **13.** 1   **15.** 1
17. 1   **19.** 1   **21.** 2   **23.** 4   **25.** 9   **27.** 9   **29.** $\frac{26}{5}$
31. $\frac{98}{5}$ **33.** 60   **35.** 30   **37.** $\frac{1}{5}$ **39.** $\frac{9}{25}$ **41.** $\frac{11}{40}$ **43.** $\frac{5}{14}$
45. $27   **47.** 625   **49.** $\frac{1}{3}$ cup   **51.** $1600   **53.** 160 mi
55. Food: $6750; housing: $5400; clothing: $2700; savings: $3000; taxes: $6750; other: $2400   **57.** 35   **59.** 4989
61. 4673

Exercise Set 2.7, p. 123

1. $\frac{6}{5}$ **3.** $\frac{1}{6}$ **5.** 6   **7.** $\frac{3}{10}$ **9.** $\frac{4}{5}$ **11.** $\frac{4}{15}$ **13.** 4   **15.** 2
17. $\frac{1}{8}$ **19.** $\frac{3}{7}$ **21.** 8   **23.** 35   **25.** 1   **27.** $\frac{2}{3}$ **29.** $\frac{9}{4}$
31. 144   **33.** 75   **35.** 2   **37.** $\frac{3}{5}$ **39.** 315   **41.** $\frac{1}{10}$ m
43. 32   **45.** 24   **47.** 16 L   **49.** 288 km; 108 km
51. 67   **53.** 285 R 2   **55.** $\frac{3}{8}$

Summary and Review: Chapter 2, p. 127

1. [2.1d] 2 · 5 · 7   **2.** [2.1d] 2 · 3 · 5   **3.** [2.1d] 3 · 3 · 5
4. [2.1d] 2 · 3 · 5 · 5   **5.** [2.2a] No   **6.** [2.2a] No
7. [2.2a] Yes   **8.** [2.2a] No   **9.** [2.1c] Prime
10. [2.3a] Numerator: 2; denominator: 7   **11.** [2.3b] $\frac{3}{5}$
12. [2.5b] $\frac{3}{100}$; $\frac{8}{100} = \frac{2}{25}$; $\frac{10}{100} = \frac{1}{10}$; $\frac{15}{100} = \frac{3}{20}$; $\frac{21}{100}$; $\frac{43}{100}$
13. [2.3c] 0   **14.** [2.3c] 1   **15.** [2.3c] 48   **16.** [2.5b] 6
17. [2.5b] $\frac{2}{3}$ **18.** [2.5b] $\frac{1}{4}$ **19.** [2.3c] 1   **20.** [2.3c] 0
21. [2.5b] $\frac{2}{5}$ **22.** [2.3c] 18   **23.** [2.5b] 4   **24.** [2.5b] $\frac{1}{3}$

25. [2.3c] Not defined   **26.** [2.3c] Not defined
27. [2.5c] $\neq$ **28.** [2.5c] $=$ **29.** [2.5c] $\neq$
30. [2.5c] $=$ **31.** [2.6a] $\frac{3}{2}$ **32.** [2.6a] 56   **33.** [2.6a] $\frac{5}{2}$
34. [2.6a] 24   **35.** [2.6a] $\frac{2}{3}$ **36.** [2.6a] $\frac{1}{14}$ **37.** [2.6a] $\frac{2}{3}$
38. [2.6a] $\frac{1}{22}$ **39.** [2.7a] $\frac{5}{4}$ **40.** [2.7a] $\frac{1}{3}$ **41.** [2.7a] 9
42. [2.7a] $\frac{36}{47}$ **43.** [2.7b] 2   **44.** [2.7b] $\frac{9}{2}$ **45.** [2.7b] $\frac{11}{6}$
46. [2.7b] $\frac{1}{4}$ **47.** [2.7b] 300   **48.** [2.7b] $\frac{9}{4}$ **49.** [2.7b] 1
50. [2.7b] $\frac{4}{9}$ **51.** [2.7c] $\frac{3}{10}$ **52.** [2.7c] 240
53. [2.7d] 160 km   **54.** [2.6b] $\frac{1}{2}$ cup   **55.** [2.6b] $6
56. [2.7d] 18   **57.** [1.7b] 24   **58.** [1.7b] 469
59. [1.8a] 1118 mi   **60.** [1.8a] $512   **61.** [1.6c] 408 R 9
62. [1.3d] 3607

Test: Chapter 2, p. 129

1. [2.1d] $2 \cdot 3 \cdot 3$ **2.** [2.1d] $2 \cdot 2 \cdot 3 \cdot 5$ **3.** [2.2a] Yes
4. [2.2a] No   **5.** [2.3a] 4 numerator; 9 denominator
6. [2.3b] $\frac{3}{4}$ **7.** [2.3c] 26   **8.** [2.3c] 1   **9.** [2.3c] 0
10. [2.5b] $\frac{1}{2}$ **11.** [2.5b] 6   **12.** [2.5b] $\frac{1}{14}$ **13.** [2.3c] Not defined   **14.** [2.3c] Not defined   **15.** [2.5c] $=$
16. [2.5c] $\neq$ **17.** [2.6a] 32   **18.** [2.6a] $\frac{3}{2}$ **19.** [2.6a] $\frac{5}{2}$
20. [2.6a] $\frac{1}{10}$ **21.** [2.7a] $\frac{8}{5}$ **22.** [2.7a] 4   **23.** [2.7a] $\frac{1}{18}$
24. [2.7b] $\frac{3}{4}$ **25.** [2.7b] $\frac{8}{5}$ **26.** [2.7b] 18   **27.** [2.7c] 64
28. [2.7c] $\frac{7}{4}$ **29.** [2.6b] 28 lb   **30.** [2.7d] $\frac{3}{40}$ m
31. [1.7b] 1805   **32.** [1.7b] 101   **33.** [1.8a] 3635 mi
34. [1.6c] 380 R 7   **35.** [1.3d] 4434

Cumulative Review: Chapters 1–2, p. 131

1. [1.1d] 584,017,800   **2.** [1.1c] Five million, three hundred eighty thousand, six hundred twenty-one   **3.** [1.1e] 0
4. [1.2b] 17,797   **5.** [1.2b] 8866   **6.** [1.3d] 4946
7. [1.3d] 1425   **8.** [1.5b] 16,767   **9.** [1.5b] 8,266,500
10. [2.6a] $\frac{3}{20}$ **11.** [2.6a] $\frac{1}{6}$ **12.** [1.6c] 241 R 1
13. [1.6c] 62   **14.** [2.7b] $\frac{3}{50}$ **15.** [2.7b] $\frac{16}{45}$
16. [1.4a] 428,000   **17.** [1.4a] 5300
18. [1.4b] $749,600 + 301,400 = 1,051,000$
19. [1.5c] $700 \times 500 = 350,000$ **20.** [1.4c] $>$
21. [2.5c] $\neq$ **22.** [1.9b] 81   **23.** [1.9c] 36   **24.** [1.9d] 2
25. [2.1a] 1, 2, 4, 7, 14, 28   **26.** [2.1d] $28 = 2 \cdot 2 \cdot 7$
27. [2.1c] Composite   **28.** [2.2a] Yes   **29.** [2.2a] No
30. [2.3c] 35   **31.** [2.5b] 7   **32.** [2.5b] $\frac{2}{7}$ **33.** [2.3c] 0
34. [1.7b] 37   **35.** [2.7c] $\frac{3}{2}$ **36.** [1.7b] 3   **37.** [1.7b] 24
38. [1.8a] $474,001   **39.** [1.8a] 11,719   **40.** [1.8a] $75
41. [2.6b] 3 cups   **42.** [2.7d] 8 days
43. [1.8a] Westside   **44.** [1.8a], [2.6b] Yes; $530
45. [2.3b] $\frac{3}{6}$, or $\frac{1}{2}$

Chapter 3

Pretest: Chapter 3, p. 134

1. [3.1a] 120   **2.** [3.3b] $<$ **3.** [3.4a] $\frac{61}{8}$ **4.** [3.4b] $5\frac{1}{2}$
5. [3.4c] $399\frac{1}{12}$ **6.** [3.5a] $11\frac{31}{60}$ **7.** [3.5b] $6\frac{1}{6}$
8. [3.6a] $13\frac{3}{5}$ **9.** [3.6a] $21\frac{2}{3}$ **10.** [3.6b] 6   **11.** [3.6b] $1\frac{2}{3}$
12. [3.3c] $\frac{2}{9}$ **13.** [3.5c] $21\frac{1}{4}$ lb   **14.** [3.6c] $4\frac{1}{4}$ cu ft
15. [3.5c] $351\frac{1}{5}$ mi   **16.** [3.6c] $22\frac{1}{2}$ cups

Sidelights: Application of LCMs: Planet Orbits, p. 140

1. Every 60 yr   **2.** Every 420 yr   **3.** Every 420 yr

Exercise Set 3.1, p. 141

1. 4   **3.** 50   **5.** 40   **7.** 54   **9.** 150   **11.** 120   **13.** 72
15. 420   **17.** 144   **19.** 288   **21.** 30   **23.** 105   **25.** 72
27. 60   **29.** 36   **31.** 24   **33.** 48   **35.** 50   **37.** 143
39. 420   **41.** 378   **43.** 810   **45.** 250   **47.** $\frac{2}{3}$
49. 24 in.   **51.** 2592

Exercise Set 3.2, p. 147

1. 1   **3.** $\frac{3}{4}$ **5.** $\frac{3}{2}$ **7.** $\frac{7}{24}$ **9.** $\frac{3}{2}$ **11.** $\frac{19}{24}$ **13.** $\frac{9}{10}$ **15.** $\frac{29}{18}$
17. $\frac{31}{100}$ **19.** $\frac{41}{60}$ **21.** $\frac{189}{100}$ **23.** $\frac{7}{8}$ **25.** $\frac{13}{24}$ **27.** $\frac{17}{24}$ **29.** $\frac{3}{4}$
31. $\frac{437}{500}$ **33.** $\frac{53}{40}$ **35.** $\frac{391}{144}$ **37.** $\frac{5}{6}$ lb   **39.** $\frac{23}{12}$ mi
41. 690 kg; $\frac{14}{23}$ cement, $\frac{5}{23}$ stone, $\frac{4}{23}$ sand; 1   **43.** $\frac{173}{100}$ in.
45. 210,528   **47.** 3,387,807   **49.** $739

Exercise Set 3.3, p. 153

1. $\frac{2}{3}$ **3.** $\frac{3}{4}$ **5.** $\frac{5}{8}$ **7.** $\frac{1}{24}$ **9.** $\frac{1}{2}$ **11.** $\frac{9}{14}$ **13.** $\frac{3}{5}$ **15.** $\frac{7}{10}$
17. $\frac{17}{60}$ **19.** $\frac{53}{100}$ **21.** $\frac{26}{75}$ **23.** $\frac{9}{100}$ **25.** $\frac{13}{24}$ **27.** $\frac{1}{10}$
29. $\frac{1}{24}$ **31.** $\frac{13}{16}$ **33.** $\frac{31}{75}$ **35.** $\frac{13}{75}$ **37.** $<$ **39.** $>$ **41.** $<$
43. $<$ **45.** $>$ **47.** $>$ **49.** $<$ **51.** $\frac{1}{15}$ **53.** $\frac{2}{15}$ **55.** $\frac{1}{15}$
57. $\frac{5}{16}$ **59.** $\frac{3}{2}$ **61.** Roberts: 0.323   **63.** $\frac{19}{24}$ **65.** $\frac{145}{144}$

Exercise Set 3.4, p. 159

1. $\frac{17}{3}$ **3.** $\frac{13}{4}$ **5.** $\frac{81}{8}$ **7.** $\frac{51}{10}$ **9.** $\frac{103}{5}$ **11.** $\frac{59}{6}$ **13.** $\frac{73}{10}$
15. $\frac{13}{8}$ **17.** $\frac{51}{4}$ **19.** $\frac{43}{10}$ **21.** $\frac{203}{100}$ **23.** $\frac{200}{3}$ **25.** $\frac{279}{50}$
27. $3\frac{3}{5}$ **29.** $4\frac{2}{3}$ **31.** $4\frac{1}{2}$ **33.** $5\frac{7}{10}$ **35.** $7\frac{4}{7}$ **37.** $7\frac{1}{2}$
29. $11\frac{1}{2}$ **41.** $1\frac{1}{2}$ **43.** $7\frac{57}{100}$ **45.** $43\frac{1}{8}$ **47.** $108\frac{5}{8}$
49. $618\frac{1}{5}$ **51.** $40\frac{4}{7}$ **53.** $55\frac{1}{51}$ **55.** 18   **57.** $\frac{1}{4}$ **59.** 24
61. $\frac{2560}{3}$ **63.** $8\frac{2}{3}$ **65.** $52\frac{2}{7}$

Exercise Set 3.5, p. 165

1. $6\frac{1}{2}$ **3.** $2\frac{11}{12}$ **5.** $14\frac{7}{12}$ **7.** $12\frac{1}{10}$ **9.** $16\frac{5}{24}$ **11.** $21\frac{1}{2}$
13. $27\frac{7}{8}$ **15.** $27\frac{13}{24}$ **17.** $1\frac{3}{5}$ **19.** $4\frac{1}{10}$ **21.** $21\frac{17}{24}$
23. $12\frac{1}{4}$ **25.** $15\frac{3}{8}$ **27.** $7\frac{5}{12}$ **29.** $13\frac{3}{8}$ **31.** $11\frac{5}{18}$
33. $7\frac{5}{12}$ lb   **35.** $6\frac{5}{6}$ in.   **37.** $19\frac{1}{16}$ ft   **39.** $95\frac{1}{5}$ mi
41. $36\frac{1}{2}$ in.   **43.** $103\frac{3}{8}$ **45.** $3\frac{1}{6}$ gal   **47.** $78\frac{1}{12}$ in.
49. $3\frac{4}{5}$ hr   **51.** $28\frac{3}{4}$ yd   **53.** $7\frac{3}{8}$ ft   **55.** $1\frac{9}{16}$ in.   **57.** $\frac{1}{10}$
59. $8\frac{7}{12}$

Exercise Set 3.6, p. 173

1. $22\frac{2}{3}$ **3.** $2\frac{5}{12}$ **5.** $8\frac{1}{6}$ **7.** $9\frac{31}{40}$ **9.** $24\frac{91}{100}$ **11.** $209\frac{1}{10}$
13. $6\frac{1}{4}$ **15.** $1\frac{1}{5}$ **17.** $3\frac{9}{16}$ **19.** $1\frac{1}{8}$ **21.** $1\frac{8}{43}$ **23.** $\frac{9}{40}$
25. 938   **27.** 7 oz   **29.** $343\frac{3}{4}$ lb   **31.** 1 chicken bouillon cube, $\frac{3}{4}$ cup hot water, $1\frac{1}{2}$ tbsp margarine, $1\frac{1}{2}$ tbsp flour, $1\frac{1}{4}$ cups diced cooked chicken, $\frac{1}{2}$ cup cooked peas, 2 oz sliced mushrooms (drained), $\frac{1}{6}$ cup sliced cooked carrots, $\frac{1}{8}$ cup chopped onion, 1 tbsp chopped pimiento, $\frac{1}{2}$ tsp salt;

6 chicken bouillon cubes, $4\frac{1}{2}$ cups hot water, 9 tbsp margarine, 9 tbsp flour, $7\frac{1}{2}$ cups diced cooked chicken, 3 cups cooked peas, 3 4-oz cans (12 oz) sliced mushrooms (drained), 1 cup sliced cooked carrots, $\frac{3}{4}$ cup chopped onion, 6 tbsp chopped pimiento, 3 tsp salt **33.** 68°F **35.** 28 min **37.** 15 mpg **39.** 4 cu ft **41.** 16 **43.** $35\frac{115}{256}$ sq in. **45.** $59,538\frac{1}{8}$ sq ft **47.** 1,429,017 **49.** 588 **51.** $\frac{1}{6}$ **53.** $35\frac{57}{64}$ **55.** $\frac{4}{9}$ **57.** $\frac{9}{5}$, or $1\frac{4}{5}$

Exercise Set 3.7, p. 179

1. $\frac{1}{24}$ **3.** $\frac{2}{5}$ **5.** $\frac{4}{7}$ **7.** $\frac{59}{30}$, or $1\frac{29}{30}$ **9.** $\frac{1}{24}$ **11.** $\frac{211}{8}$, or $26\frac{3}{8}$ **13.** $\frac{7}{16}$ **15.** $\frac{1}{36}$ **17.** $\frac{3}{8}$ **19.** $\frac{37}{48}$ **21.** $\frac{25}{72}$ **23.** $\frac{103}{16}$, or $6\frac{7}{16}$ **25.** $\frac{17}{6}$, or $2\frac{5}{6}$ **27.** $\frac{8395}{84}$, or $99\frac{79}{84}$ **29.** 3402 **31.** 59 R 77 **33.** $\frac{8}{3}$, or $2\frac{2}{3}$ **35.(a)** $13 \cdot 9\frac{1}{4} + 8\frac{1}{4} \cdot 7\frac{1}{4}$; **(b)** $\frac{2881}{16}$, or $180\frac{1}{16}$ in²; **(c)** multiply before adding

Summary and Review: Chapter 3, p. 183

1. [3.1a] 36 **2.** [3.1a] 90 **3.** [3.1a] 30 **4.** [3.2b] $\frac{63}{40}$ **5.** [3.2b] $\frac{19}{48}$ **6.** [3.2b] $\frac{29}{15}$ **7.** [3.2b] $\frac{7}{16}$ **8.** [3.3a] $\frac{1}{3}$ **9.** [3.3a] $\frac{1}{8}$ **10.** [3.3a] $\frac{5}{27}$ **11.** [3.3a] $\frac{11}{18}$ **12.** [3.3b] > **13.** [3.3b] > **14.** [3.3c] $\frac{19}{40}$ **15.** [3.3c] $\frac{2}{5}$ **16.** [3.4a] $\frac{15}{2}$ **17.** [3.4a] $\frac{67}{8}$ **18.** [3.4a] $\frac{13}{3}$ **19.** [3.4a] $\frac{75}{7}$ **20.** [3.4b] $2\frac{1}{3}$ **21.** [3.4b] $6\frac{3}{4}$ **22.** [3.4b] $12\frac{3}{5}$ **23.** [3.4b] $3\frac{1}{2}$ **24.** [3.4c] $877\frac{1}{3}$ **25.** [3.4c] $456\frac{5}{23}$ **26.** [3.5a] $10\frac{2}{5}$ **27.** [3.5a] $11\frac{11}{15}$ **28.** [3.5a] $10\frac{2}{3}$ **29.** [3.5a] $8\frac{1}{4}$ **30.** [3.5b] $7\frac{7}{9}$ **31.** [3.5b] $4\frac{11}{18}$ **32.** [3.5b] $4\frac{3}{20}$ **33.** [3.5b] $13\frac{3}{8}$ **34.** [3.6a] 16 **35.** [3.6a] $3\frac{1}{2}$ **36.** [3.6a] $2\frac{21}{50}$ **37.** [3.6a] 6 **38.** [3.6b] 12 **39.** [3.6b] $1\frac{7}{17}$ **40.** [3.6b] $\frac{1}{8}$ **41.** [3.6b] $\frac{9}{10}$ **42.** [3.6c] 15 **43.** [3.5c] $\$70\frac{3}{8}$ **44.** [3.6c] 30 cups **45.** [3.5c] $8\frac{3}{8}$ cups **46.** [3.7a] $177\frac{3}{4}$ in² **47.** [3.5c] $50\frac{1}{4}$ in² **48.** [3.7a] 1 **49.** [3.7a] $\frac{77}{240}$ **50.** [2.6a] $\frac{6}{5}$ **51.** [2.7b] $\frac{3}{2}$ **52.** [1.5b] 708,048 **53.** [1.8a] 17 days

Test: Chapter 3, p. 185

1. [3.1a] 48 **2.** [3.2a] 3 **3.** [3.2b] $\frac{37}{24}$ **4.** [3.2b] $\frac{79}{100}$ **5.** [3.3a] $\frac{1}{3}$ **6.** [3.3a] $\frac{1}{12}$ **7.** [3.3a] $\frac{1}{12}$ **8.** [3.3b] > **9.** [3.3c] $\frac{1}{4}$ **10.** [3.4a] $\frac{7}{2}$ **11.** [3.4a] $\frac{79}{8}$ **12.** [3.4b] $4\frac{1}{2}$ **13.** [3.4b] $8\frac{2}{9}$ **14.** [3.4c] $162\frac{7}{11}$ **15.** [3.5a] $14\frac{1}{5}$ **16.** [3.5a] $14\frac{5}{12}$ **17.** [3.5b] $4\frac{7}{24}$ **18.** [3.5b] $6\frac{1}{6}$ **19.** [3.6a] 39 **20.** [3.6a] $4\frac{1}{2}$ **21.** [3.6a] $5\frac{5}{6}$ **22.** [3.6b] 6 **23.** [3.6b] 2 **24.** [3.6b] $\frac{1}{36}$ **25.** [3.6c] $17\frac{1}{2}$ cups **26.** [3.6c] 80 **27.** [3.5c] $360\frac{5}{12}$ lb **28.** [3.5c] $2\frac{1}{2}$ in. **29.** [3.7a] $3\frac{1}{2}$ **30.** [3.7a] $\frac{11}{20}$ **31.** [1.5b] 346,636 **32.** [2.7b] $\frac{8}{5}$ **33.** [2.6a] $\frac{10}{9}$ **34.** [1.8a] 535 bottles, 10 oz left over

Cumulative Review: Chapters 1–3, p. 187

1. [1.1e] 5 **2.** [1.1a] 6 thousands + 7 tens + 5 ones **3.** [1.1c] Twenty-nine thousand, five hundred

4. [1.2b] 899 **5.** [1.2b] 8982 **6.** [3.2b] $\frac{5}{12}$ **7.** [3.5a] $8\frac{1}{4}$ **8.** [1.3d] 5124 **9.** [1.3d] 4518 **10.** [3.3a] $\frac{5}{12}$ **11.** [3.5b] $1\frac{1}{6}$ **12.** [1.5b] 5004 **13.** [1.5b] 293,232 **14.** [2.6a] $\frac{3}{2}$ **15.** [2.6a] 15 **16.** [3.6a] $7\frac{1}{3}$ **17.** [1.6c] 715 **18.** [1.6c] 56 R 11 **19.** [3.4c] $56\frac{11}{45}$ **20.** [2.7b] $\frac{4}{7}$ **21.** [3.6b] $7\frac{1}{3}$ **22.** [1.4a] 38,500 **23.** [3.1a] 72 **24.** [2.2a] No **25.** [2.1a] 1, 2, 4, 8, 16 **26.** [2.3b] $\frac{1}{4}$ **27.** [3.3b] > **28.** [3.3b] < **29.** [2.5b] $\frac{4}{5}$ **30.** [2.5b] 32 **31.** [3.4a] $\frac{37}{8}$ **32.** [3.4b] $5\frac{2}{3}$ **33.** [1.7b] 93 **34.** [3.3c] $\frac{5}{9}$ **35.** [2.7c] $\frac{12}{7}$ **36.** [1.7b] 905 **37.** [1.8a] $235 **38.** [1.8a] $108 **39.** [1.8a] 297 sq ft **40.** [1.8a] 31 **41.** [2.6b] $\frac{2}{5}$ tsp **42.** [3.6c] 39 lb **43.** [3.6c] 16 **44.** [3.2c] $\frac{33}{20}$ mi

CHAPTER 4

Pretest: Chapter 4, p. 190

1. [4.1a] Two and three hundred forty-seven thousandths **2.** [4.1a] Three thousand, two hundred sixty-four and $\frac{78}{100}$ dollars **3.** [4.1b] $\frac{21}{100}$ **4.** [4.1b] $\frac{5408}{1000}$ **5.** [4.1c] 3.79 **6.** [4.1c] 0.0539 **7.** [4.2a] 3.2 **8.** [4.2a] 0.099 **9.** [4.2a] 0.562 **10.** [4.2b] 21.0 **11.** [4.2b] 21.04 **12.** [4.2b] 21.045 **13.** [4.3a] 607.219 **14.** [4.3a] 0.8684 **15.** [4.3a] 113.664 **16.** [4.3a] 6.7437 **17.** [4.3b] 91.732 **18.** [4.3b] 39.0901 **19.** [4.3b] 7.9951 **20.** [4.3b] 252.9937 **21.** [4.3c] 3.27 **22.** [4.3c] 6345.2117 **23.** [4.4a] $285.95 **24.** [4.4a] 1081.6

Exercise Set 4.1, p. 195

1. Twenty-three and nine tenths **3.** One hundred seventeen and sixty-five hundredths **5.** Thirty-four and eight hundred ninety-one thousandths **7.** Three hundred twenty-six and $\frac{48}{100}$ dollars **9.** Thirty-six and $\frac{72}{100}$ dollars **11.** $\frac{68}{10}$ **13.** $\frac{17}{100}$ **15.** $\frac{821}{100}$ **17.** $\frac{2046}{10}$ **19.** $\frac{1509}{1000}$ **21.** $\frac{4603}{100}$ **23.** $\frac{13}{100,000}$ **25.** $\frac{20,003}{1000}$ **27.** $\frac{10,008}{10,000}$ **29.** $\frac{5321}{10,000}$ **31.** 0.8 **33.** 0.92 **35.** 9.3 **37.** 8.89 **39.** 250.8 **41.** 3.798 **43.** 0.0078 **45.** 0.56788 **47.** 21.73 **49.** 0.66 **51.** 0.853 **53.** 0.376193 **55.** 6170 **57.** 6000 **59.** 379 **61.** 4.909

Exercise Set 4.2, p. 199

1. 0.58 **3.** 0.91 **5.** 0.001 **7.** 235.07 **9.** 0.4545 **11.** $\frac{4}{100}$ **13.** 0.78 **15.** 0.84384 **17.** 1.9 **19.** 0.1 **21.** 0.2 **23.** 0.6 **25.** 2.7 **27.** 13.4 **29.** 123.7 **31.** 0.89 **33.** 0.67 **35.** 0.42 **37.** 1.44 **39.** 3.58 **41.** 0.01 **43.** 37.70 **45.** 1.00 **47.** 0.325 **49.** 0.667 **51.** 17.002 **53.** 0.001 **55.** 10.101 **57.** 0.116 **59.** 800 **61.** 809.473 **63.** 809 **65.** 34.5439 **67.** 34.54 **69.** 35 **71.** 830 **73.** 182 **75.** 6.78346 **77.** 99.99999

Sidelights: A Number Puzzle, p. 204

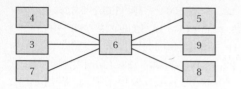

Exercise Set 4.3, p. 205

1. 334.37   **3.** 1576.215   **5.** 132.560   **7.** 84.417
9. 50.0248   **11.** 40.007   **13.** 771.967   **15.** 20.8649
17. 227.4680   **19.** 8754.8221   **21.** 1.3   **23.** 49.02
25. 45.61   **27.** 85.921   **29.** 2.4975   **31.** 3.397
33. 8.85   **35.** 3.37   **37.** 1.045   **39.** 3.703   **41.** 0.9902
43. 99.66   **45.** 4.88   **47.** 0.994   **49.** 17.802   **51.** 51.13
53. 2.491   **55.** 32.7386   **57.** 1.6666   **59.** 2344.90886
61. 11.65   **63.** 19.251   **65.** 384.68   **67.** 582.97
69. 15,335.3   **71.** 35,000   **73.** $\frac{1}{6}$ **75.** 6166   **77.** 345.8

Exercise Set 4.4, p. 211

1. $3.01   **3.** 118.5 gal   **5.** 102.8°F   **7.** 22,691.5
9. 8.9 billion   **11.** $356.89   **13.** 93.9 km   **15.** 18.09 min
17.(a) 36.85 in.; **(b)** 93.599 cm   **19.** 225.8 mi
21. $1171.74   **23.** 28.8 million   **25.** 79.6 million
27. $84.70   **29.** 78.1 cm   **31.** 2.31 cm   **33.** 1.4°F
35. No   **37.** 6335   **39.** $\frac{23}{15}$ **41.** 2803   **43.** $\frac{2}{15}$

Summary and Review: Chapter 4, p. 217

1. [4.1a] Three and forty-seven hundredths
2. [4.1a] Thirty-one thousandths   **3.** [4.1a] Five hundred
ninety-seven and $\frac{25}{100}$ dollars   **4.** [4.1a] Zero and
$\frac{98}{100}$ dollars   **5.** [4.1b] $\frac{9}{100}$ **6.** [4.1b] $\frac{4561}{1000}$ **7.** [4.1b] $\frac{89}{1000}$
8. [4.1b] $\frac{30,227}{10,000}$ **9.** [4.1c] 0.034   **10.** [4.1c] 4.2603
11. [4.1c] 27.91   **12.** [4.1c] 0.006   **13.** [4.2a] 0.034
14. [4.2a] 0.91   **15.** [4.2a] 0.741   **16.** [4.2a] 1.041
17. [4.2b] 17.4   **18.** [4.2b] 17.43   **19.** [4.2b] 17.429
20. [4.2b] 4.272   **21.** [4.2b] 4.27   **22.** [4.2b] 4.3
23. [4.3a] 574.519   **24.** [4.3a] 0.6838   **25.** [4.3a] 499.829
26. [4.3a] 62.6932   **27.** [4.3a] 229.1   **28.** [4.3a] 45.601
29. [4.3b] 29.2092   **30.** [4.3b] 790.29   **31.** [4.3b] 29.148
32. [4.3b] 70.7891   **33.** [4.3b] 685.0519   **34.** [4.3b] 7.9953
35. [4.3c] 496.2795   **36.** [4.3c] 4.9911   **37.** [4.4a] 2.9 yr
38. [4.4a] 11.16   **39.** [4.4a] $5888.74
40. [4.4a] 6365.1 bu   **41.** [1.2b] 14,605   **42.** [3.2b] $\frac{49}{30}$
43. [1.3d] 3389   **44.** [3.3a] $\frac{1}{30}$

Test: Chapter 4, p. 219

1. [4.1a] Two and thirty-four hundredths   **2.** [4.1a] One
thousand, two hundred thirty-four and $\frac{78}{100}$ dollars
3. [4.1b] $\frac{91}{100}$ **4.** [4.1b] $\frac{2769}{1000}$ **5.** [4.1c] 0.74
6. [4.1c] 3.7047   **7.** [4.2a] 0.162   **8.** [4.2a] 0.9
9. [4.2a] 0.078   **10.** [4.2b] 5.7   **11.** [4.2b] 5.68
12. [4.2b] 5.678   **13.** [4.3a] 405.219   **14.** [4.3a] 0.7902

15. [4.3a] 186.5   **16.** [4.3a] 1033.23   **17.** [4.3b] 48.357
18. [4.3b] 19.0901   **19.** [4.3b] 1.9946   **20.** [4.3b] 152.8934
21. [4.3c] 8.982   **22.** [4.3c] 3365.6597
23. [4.4a] 29.27 mph   **24.** [4.4a] $3627.65   **25.** [3.2b] $\frac{19}{18}$
26. [1.2b] 13,652   **27.** [3.3a] $\frac{11}{18}$ **28.** [1.3d] 2155

Cumulative Review: Chapters 1–4, p. 221

1. [1.1a] 1 ten thousand + 2 thousands + 7 hundreds +
5 tens + 8 ones   **2.** [4.1a] Eight hundred two and
$\frac{53}{100}$ dollars   **3.** [4.1b] $\frac{1009}{100}$ **4.** [3.4a] $\frac{27}{8}$ **5.** [4.1c] 0.035
6. [2.1a] 1, 2, 3, 6, 11, 22, 33, 66   **7.** [2.1d] $2 \cdot 3 \cdot 11$
8. [3.1a] 140   **9.** [4.2b] 7000   **10.** [4.2b] 6962.47
11. [3.5a] $6\frac{2}{9}$ **12.** [4.3a] 235.397   **13.** [1.2b] 5495
14. [3.2b] $\frac{1}{2}$ **15.** [1.3d] 826   **16.** [4.3b] 8446.53
17. [3.3a] $\frac{1}{72}$ **18.** [3.5b] $3\frac{2}{5}$ **19.** [1.5b] 182,820
20. [2.6a] $\frac{2}{7}$ **21.** [3.6a] $13\frac{7}{11}$ **22.** [2.6a] $\frac{3}{2}$
23. [3.6b] $1\frac{1}{2}$ **24.** [2.7b] $\frac{48}{35}$ **25.** [1.6c] 38
26. [3.4c] $205\frac{20}{21}$ **27.** [3.3b] <   **28.** [4.2a] >
29. [2.5c] =   **30.** [4.2a] <   **31.** [4.3c] 0.795
32. [2.7c] $\frac{25}{8}$ **33.** [1.7b] 121   **34.** [3.3c] $\frac{7}{45}$
35. [4.3c] 4.985   **36.** [2.7c] $\frac{6}{5}$ **37.** [1.8a] 807,643
38. [1.8a] 40,800   **39.** [4.4a] 7.1 billion
40. [4.4a] $364.54   **41.** [3.6c] $1\frac{1}{8}$ cups   **42.** [3.5c] $1\frac{7}{8}$ yd
43. [3.2c] $\frac{7}{24}$ **44.** [1.8a] 2560   **45.** [1.9c], [2.6a], [3.3a],
[3.5b] $\frac{9}{32}$ **46.** [4.1c], [4.3a] 17.887

Chapter 5

Pretest: Chapter 5, p. 224

1. [5.1a] 38.54   **2.** [5.1a] 0.6179   **3.** [5.1a] 46.3
4. [5.1a] 435.4724   **5.** [5.1a] 3.672   **6.** [5.1a] 0.42735
7. [5.1a] 0.32456   **8.** [5.1a] 739.62   **9.** [5.2a] 3.625
10. [5.2a] 1.32   **11.** [5.2a] 0.48   **12.** [5.2a] 30.4
13. [5.2a] 0.38   **14.** [5.2a] 0.57698   **15.** [5.2a] 75,689
16. [5.2a] 0.00004653   **17.** [5.2b] 84.26   **18.** [5.4a] 224
19. [5.4a] 3.5   **20.** [5.3a] 1.4   **21.** [5.3a] 1.4375
22. [5.3a] 13.25   **23.** [5.3a] 2.75   **24.** [5.3a] $0.\overline{7}$
25. [5.3a] $4.\overline{142857}$ **26.** [5.3b] 4.1   **27.** [5.3b] 4.14
28. [5.3b] 4.143   **29.** [5.5a] $89.70   **30.** [5.5a] 2.17 km
31. [5.5a] $3397.71   **32.** [5.1b] 7496¢   **33.** [5.1b] $135.49
34. [5.1b] 48,600,000,000,000   **35.** [5.2c] 38.66475
36. [5.2c] 1548.8836   **37.** [5.3c] 49.34375
38. [5.3c] 58.17

Sidelights: Number Patterns, p. 230

1. 24; 264; 2664; 26,664; 266,664   **2.** 81; 8811; 888,111;
88,881,111; 8,888,811,111   **3.** 6006; 60,606; 606,606;
6,066,606; 60,666,606   **4.** 999,999; 1,999,998; 2,999,997;
3,999,996; 4,999,995

Exercise Set 5.1, p. 231

1. 60.2   **3.** 6.72   **5.** 0.252   **7.** 0.522   **9.** 237.6

11. 583,686.852 **13.** 780 **15.** 8.923 **17.** 0.09768
19. 0.782 **21.** 521.6 **23.** 3.2472 **25.** 897.6
27. 322.07 **29.** 55.68 **31.** 3487.5 **33.** 50.0004
35. 114.42902 **37.** 13.284 **39.** 90.72 **41.** 0.0028728
43. 0.72523 **45.** 1.872115 **47.** 45,678 **49.** 2888¢
51. 66¢ **53.** $0.34 **55.** $34.45
57. $4,030,000,000,000 **59.** 4,700,000 **61.** $11\frac{1}{5}$
63. 342 **65.** 4566 **67.** 10^{21}

Exercise Set 5.2, p. 239

1. 2.99 **3.** 23.78 **5.** 7.48 **7.** 7.2 **9.** 1.143 **11.** 4.041
13. 0.07 **15.** 70 **17.** 20 **19.** 0.4 **21.** 0.41 **23.** 8.5
25. 9.3 **27.** 0.625 **29.** 0.26 **31.** 15.625 **33.** 2.34
35. 0.47 **37.** 0.2134567 **39.** 21.34567 **41.** 1023.7
43. 9.3 **45.** 0.0090678 **47.** 45.6 **49.** 2107
51. 303.003 **53.** 446.208 **55.** 24.14 **57.** 13.0072
59. 19.3204 **61.** 96.13 **63.** 10.49 **65.** 911.13 **67.** 205
69. $1288.355 **71.** $15\frac{1}{8}$ **73.** $\frac{6}{7}$ **75.** $2 \cdot 2 \cdot 3 \cdot 3 \cdot 19$

Exercise Set 5.3, p. 247

1. 0.6 **3.** 0.325 **5.** 0.2 **7.** 0.85 **9.** 0.475 **11.** 0.975
13. 0.52 **15.** 20.016 **17.** 0.25 **19.** 0.575 **21.** 0.72
23. 1.1875 **25.** $0.2\overline{6}$ **27.** $0.\overline{3}$ **29.** $1.\overline{3}$ **31.** $1.1\overline{6}$
33. $0.\overline{571428}$ **35.** $0.91\overline{6}$ **37.** 0.3; 0.27; 0.267 **39.** 0.3;
0.33; 0.333 **41.** 1.3; 1.33; 1.333 **43.** 1.2; 1.17; 1.167
45. 0.6; 0.57; 0.571 **47.** 0.9; 0.92; 0.917 **49.** 0.2; 0.18;
0.182 **51.** 0.3; 0.28; 0.278 **53.** 11.06 **55.** 8.4
57. $417.51\overline{6}$ **59.** 0 **61.** 2.8125 **63.** 0.20425
65. 317.14 **67.** 0.1825 **69.** 18 **71.** 2.736 **73.** 21
75. $3\frac{2}{5}$ **77.** 325

Sidelights: Finding Whole-Number Remainders in Division, p. 252

1. 28 R 2 **2.** 116 R 3 **3.** 74 R 10 **4.** 415 R 3

Exercise Set 5.4, p. 253

1. (d) **3.** (c) **5.** (a) **7.** (c) **9.** 1.6 **11.** 6 **13.** 60
15. 2.3 **17.** 180 **19.** (a) **21.** (c) **23.** (b) **25.** (b)
27. $2 \cdot 2 \cdot 3 \cdot 3 \cdot 3$ **29.** $5 \cdot 5 \cdot 13$ **31.** $\frac{5}{16}$ **33.** $\frac{8}{9}$
35. Yes

Exercise Set 5.5, p. 259

1. $39.60 **3.** $19.45 **5.** 62.5 mi **7.** $5.95
9. 250,205.04 ft^2 **11.** $150,000 **13.** $349.44
15. $465.78 **17.** 887.4 km **19.** $57.35 **21.** 20.2 mpg
23. 11.9752 cu ft **25.** $10 **27.** 1032 cal **29.** 0.420
31. $7.76 **33.** $8.70 **35.** $3.99 **37.** $19,090.86
39. 2152.56 yd^2 **41.** $316,987.20; $196,987.20 **43.** $13\frac{7}{12}$
45. $34\frac{11}{12}$ **47.** $258\frac{2}{3}$

Summary and Review: Chapter 5, p. 265

1. [5.1a] 12.96 **2.** [5.1a] 0.14442 **3.** [5.1a] 4.3
4. [5.1a] 0.2784 **5.** [5.1a] 1.073 **6.** [5.1a] 0.2184

7. [5.1a] 0.02468 **8.** [5.1a] 24,680 **9.** [5.2a] 7.5
10. [5.2a] 3.2 **11.** [5.2a] 0.45 **12.** [5.2a] 45.2
13. [5.2a] 1.6 **14.** [5.2a] 1.022 **15.** [5.2a] 2.763
16. [5.2a] 0.2763 **17.** [5.2a] 1.274 **18.** [5.2a] 1389.2
19. [5.2b] 6.95 **20.** [5.2b] 42.54 **21.** [5.4a] 272
22. [5.4a] 4 **23.** [5.4a] 216 **24.** [5.4a] $125
25. [5.3a] 2.6 **26.** [5.3a] 1.28 **27.** [5.3a] 2.75
28. [5.3a] 3.25 **29.** [5.3a] $1.1\overline{6}$ **30.** [5.3a] $1.\overline{54}$
31. [5.3b] 1.5 **32.** [5.3b] 1.55 **33.** [5.3b] 1.545
34. [5.5a] 82.67 mi **35.** [5.5a] $239.80
36. [5.5a] 24.36 cups; 104.4 cups **37.** [5.5a] $4.56
38. [5.5a] $15.98 **39.** [5.1b] $82.73 **40.** [5.1b] $4.87
41. [5.1b] 2493¢ **42.** [5.1b] 986¢
43. [5.1b] 3,400,000,000 **44.** [5.1b] 1,200,000
45. [5.2c] 1.8045 **46.** [5.2c] 57.1449 **47.** [5.3c] 15.6375
48. [5.3c] $41.537\overline{3}$ **49.** [3.6a] $43\frac{3}{4}$ **50.** [3.6b] $3\frac{3}{4}$
51. [3.5a] $19\frac{4}{5}$ **52.** [3.5b] $6\frac{3}{8}$ **53.** [2.5b] $\frac{1}{2}$
54. [2.1d] $2 \cdot 2 \cdot 2 \cdot 2 \cdot 2 \cdot 2 \cdot 3$

Test: Chapter 5, p. 267

1. [5.1a] 8 **2.** [5.1a] 0.03 **3.** [5.1a] 3.7 **4.** [5.1a] 0.2079
5. [5.1a] 1.088 **6.** [5.1a] 0.31824 **7.** [5.1a] 0.21345
8. [5.1a] 739.62 **9.** [5.2a] 4.75 **10.** [5.2a] 0.44
11. [5.2a] 0.24 **12.** [5.2a] 30.4 **13.** [5.2a] 0.19
14. [5.2a] 0.34689 **15.** [5.2a] 34,689
16. [5.2a] 0.0000123 **17.** [5.2b] 84.26 **18.** [5.4a] 198
19. [5.4a] 4 **20.** [5.3a] 1.6 **21.** [5.3a] 0.88
22. [5.3a] 5.25 **23.** [5.3a] 0.75 **24.** [5.3a] $1.\overline{2}$
25. [5.3a] $2.\overline{142857}$ **26.** [5.3b] 2.1 **27.** [5.3b] 2.14
28. [5.3b] 2.143 **29.** [5.5a] $479.70 **30.** [5.5a] 4.97 km
31. [5.5a] $1675.50 **32.** [5.1b] 8795¢ **33.** [5.1b] $9.49
34. [5.1b] 38,700,000,000,000 **35.** [5.2c] 40.0065
36. [5.2c] 384.8464 **37.** [5.3c] 302.4 **38.** [5.3c] $52.339\overline{4}$
39. [3.5b] $26\frac{1}{2}$ **40.** [3.5a] $2\frac{11}{16}$ **41.** [3.6b] $1\frac{1}{8}$
42. [3.6a] 14 **43.** [2.5b] $\frac{11}{18}$ **44.** [2.1d] $2 \cdot 2 \cdot 2 \cdot 3 \cdot 3 \cdot 5$

Cumulative Review: Chapters 1–5, p. 269

1. [3.4a] $\frac{20}{9}$ **2.** [4.1b] $\frac{3052}{1000}$ **3.** [5.3a] 1.4 **4.** [5.3a] $0.\overline{54}$
5. [2.1c] Prime **6.** [2.2a] Yes **7.** [1.9c] 1754
8. [5.2c] 4.364 **9.** [4.2b] 584.90 **10.** [5.3b] 218.56
11. [5.4a] 160 **12.** [5.4a] 4 **13.** [1.5c] 12,800,000
14. [5.4a] 6 **15.** [3.5a] $6\frac{1}{20}$ **16.** [1.2b] 139,116
17. [3.2b] $\frac{31}{18}$ **18.** [4.3a] 145.953 **19.** [1.3d] 710,137
20. [4.3b] 13.097 **21.** [3.5b] $\frac{5}{7}$ **22.** [3.3a] $\frac{1}{110}$
23. [2.6a] $\frac{1}{6}$ **24.** [1.5b] 5,317,200 **25.** [5.1a] 4.78
26. [5.1a] 0.0279431 **27.** [5.2a] 2.122 **28.** [1.6c] 1843
29. [5.2a] 13,862.1 **30.** [2.7b] $\frac{5}{6}$ **31.** [4.3c] 0.78
32. [1.7b] 28 **33.** [5.2b] 8.62 **34.** [1.7b] 367,251
35. [3.3c] $\frac{1}{18}$ **36.** [2.7c] $\frac{1}{2}$ **37.** [1.8a] 11,222
38. [2.7d] $500 **39.** [1.8a] 86,400 **40.** [2.6b] $2400
41. [4.4a] $258.77 **42.** [3.5c] $6\frac{1}{2}$ lb **43.** [3.2c] 2 lb
44. [5.5a] 467.28 **45.** [3.6c] 144 **46.** [5.5a] $0.91

CHAPTER 6

Pretest: Chapter 6, p. 272

1. [6.1a] $\frac{35}{43}$ **2.** [6.1a] $\frac{0.079}{1.043}$ **3.** [6.3b] 22.5   **4.** [6.3b] 0.75
5. [6.3b] $\frac{117}{14}$, or $8\frac{5}{14}$ **6.** [6.2a] 25.5 mi/gal
7. [6.2a] $\frac{2}{9}$ qt/min   **8.** [6.2b] 5.79 ¢/oz   **9.** [6.2b] Brand B
10. [6.4a] 1944 km   **11.** [6.4a] 22   **12.** [6.4a] 12 min
13. [6.4a] 393.75 miles   **14.** [6.5a] $x = 15$, $y = 9$

Exercise Set 6.1, p. 275

1. $\frac{4}{5}$ **3.** $\frac{178}{572}$ **5.** $\frac{0.4}{12}$ **7.** $\frac{3.8}{7.4}$ **9.** $\frac{56.78}{98.35}$ **11.** $\frac{8\frac{3}{4}}{9\frac{5}{6}}$
13. $\frac{36.1}{1000}$; $\frac{1000}{36.1}$ **15.** $\frac{2.7}{13.1}$; $\frac{13.1}{2.7}$ **17.** $\frac{478}{213}$; $\frac{213}{478}$ **19.** $\frac{2}{3}$ **21.** $\frac{3}{4}$
23. $\frac{12}{25}$ **25.** $\frac{7}{9}$ **27.** $\frac{2}{3}$ **29.** $\frac{14}{25}$ **31.** $\frac{1}{2}$ **33.** $\frac{3}{4}$ **35.** $\frac{27}{50}$
37. $\frac{32}{101}$ **39.** =   **41.** 50   **43.** 14.5
45. $\frac{13,339,000}{145,304,000} \approx 0.09$; $\frac{145,304,000}{13,339,000} \approx 10.89$

Exercise Set 6.2, p. 279

1. 40 km/h   **3.** 11 m/sec   **5.** 152 yd/day   **7.** 25 mi/hr;
0.04 hr/mi   **9.** 0.623 gal/sq ft   **11.** 57.5 ¢/min
13. 186,000 mi/sec   **15.** 124 km/h   **17.** 560 mi/hr
19. $9.50/yd   **21.** 26.84 ¢/oz   **23.** $4.34/lb   **25.** A
27. B   **29.** A   **31.** A   **33.** Four 12-oz bottles
35. Package A   **37.** 1.7 million   **39.** 67,819
41. 5833.56   **43.** **(a)** 10.83¢; 10.91¢; **(b)** 1.14¢; 1.22¢

Exercise Set 6.3, p. 287

1. No   **3.** Yes   **5.** Yes   **7.** No   **9.** 45   **11.** 12   **13.** 10
15. 20   **17.** 5   **19.** 18   **21.** 22   **23.** 28   **25.** $9\frac{1}{3}$
27. $2\frac{8}{9}$ **29.** 0.06   **31.** 5   **33.** 1   **35.** 1   **37.** 14
39. $2\frac{3}{16}$ **41.** $\frac{51}{16}$, or $3\frac{3}{16}$ **43.** 12.5725   **45.** $\frac{1748}{249}$, or $7\frac{5}{249}$
47. ≠   **49.** ≠   **51.** Approximately 2731.4

Sidelights: Applications: Quarterback Comparisons, p. 292

1. $\frac{302}{473} \approx 0.64$ **2.** $\frac{920}{1528} \approx 0.602$ **3.** $\frac{39}{59} \approx 0.66$ **4.** $\frac{60}{1528} \approx$
0.039   **5.** $\frac{54}{1528} \approx 0.035$ **6.** $\frac{269}{462} \approx 0.58$ **7.** $\frac{1824}{3024} \approx 0.603$
8. $\frac{17}{24} \approx 0.71$ **9.** $\frac{108}{3024} \approx 0.036$ **10.** $\frac{161}{3024} \approx 0.053$
11. Too close to call; Aikman   **12.** 60; 4   **13.** 60; 5

Exercise Set 6.4, p. 293

1. 702 km   **3.** $84.60   **5.** 3.11   **7.** 954   **9.** $61\frac{2}{3}$ lb
11. 322   **13.** 120 lb   **15.** 1980   **17.** 58.1 mi   **19.** $7\frac{1}{3}$ in.
21. No. The money will be gone in 30 weeks; $53.33 more.
23. 2150

Exercise Set 6.5, p. 297

1. 25   **3.** $\frac{4}{3}$, or $1\frac{1}{3}$ **5.** $x = \frac{27}{4}$, or $6\frac{3}{4}$; $y = 9$ **7.** $x = 7.5$;
$y = 7.2$ **9.** 1.25 m   **11.** 36 ft   **13.** $59.81
15. 679.4928   **17.** 27,456.8   **19.** 100 ft

Summary and Review: Chapter 6, p. 301

1. [6.1a] $\frac{47}{84}$ **2.** [6.1a] $\frac{46}{1.27}$ **3.** [6.1a] $\frac{83}{100}$ **4.** [6.1a] $\frac{0.72}{197}$
5. [6.1a] $\frac{5200}{1070}$ **6.** [6.1b] $\frac{3}{4}$ **7.** [6.1b] $\frac{9}{16}$
8. [6.2a] 23.54 mi/hr   **9.** [6.2a] 0.638 gal/sq ft
10. [6.2a] $25.36/kg   **11.** [6.2a] 0.72 serving/lb
12. [6.2b] 5.7 ¢/oz   **13.** [6.2b] 5.45 ¢/oz   **14.** [6.2b] B
15. [6.2b] B   **16.** [6.3a] No   **17.** [6.3a] No   **18.** [6.3b] 32
19. [6.3b] $\frac{1}{40}$ **20.** [6.3b] 7   **21.** [6.3b] 24
22. [6.4a] $4.45   **23.** [6.4a] 351   **24.** [6.4a] 832 mi
25. [6.4a] 27 acres   **26.** [6.4a] 2,392,000 kg
27. [6.4a] 6 in.   **28.** [6.4a] 2622   **29.** [6.5a] $x = 20$;
$y = 16$ **30.** [4.4a] $1630.40   **31.** [2.5c] =   **32.** [2.5c] ≠
33. [5.1a] 10,672.74   **34.** [5.2a] 45.5

Test: Chapter 6, p. 303

1. [6.1a] $\frac{85}{97}$ **2.** [6.1a] $\frac{0.34}{124}$ **3.** [6.1b] $\frac{9}{10}$ **4.** [6.1b] $\frac{25}{32}$
5. [6.2a] 0.625 ft/sec   **6.** [6.2a] $1\frac{1}{3}$ servings/lb
7. [6.2b] 19.39 ¢/oz   **8.** [6.2b] B   **9.** [6.3a] Yes
10. [6.3a] No   **11.** [6.3b] 12   **12.** [6.3b] 360
13. [6.4a] 1512 km   **14.** [6.4a] 44   **15.** [6.4a] 4.8 min
16. [6.4a] 525 mi   **17.** [6.5a] 66 m
18. [4.4a] 25.8 million lb   **19.** [2.5c] ≠
20. [5.1a] 17,324.14   **21.** [5.2a] 0.9944

Cumulative Review: Chapters 1–6, p. 305

1. [4.3a] 513.996   **2.** [3.5a] $6\frac{3}{4}$ **3.** [3.2b] $\frac{7}{20}$
4. [4.3b] 30.491   **5.** [4.3b] 72.912   **6.** [3.3a] $\frac{7}{60}$
7. [5.1a] 222.076   **8.** [5.1a] 567.8   **9.** [3.6a] 3
10. [5.2a] 43   **11.** [1.6c] 899   **12.** [2.7b] $\frac{3}{2}$
13. [1.1a] 3 ten thousands + 7 tens + 4 ones
14. [4.1a] One hundred twenty and seven hundredths
15. [4.2a] 0.7   **16.** [4.2a] 0.8   **17.** [2.1d] $2 \cdot 2 \cdot 2 \cdot 2 \cdot 3 \cdot 3$
18. [3.1a] 140   **19.** [2.3b] $\frac{5}{8}$ **20.** [2.5b] $\frac{5}{8}$
21. [5.3c] 5.718   **22.** [5.3c] 0.179   **23.** [6.1a] $\frac{0.3}{15}$
24. [6.3a] Yes   **25.** [6.2a] 55 m/sec
26. [6.2b] The 14-oz jar   **27.** [6.3b] 30.24
28. [5.2b] 26.4375   **29.** [2.7c] $\frac{8}{9}$ **30.** [6.3b] 128
31. [4.3c] 33.34   **32.** [3.3c] $\frac{76}{175}$ **33.** [2.6b] 390 cal
34. [1.8a] 1,154,782   **35.** [4.4a] 976.9 mi
36. [6.4a] 7 min   **37.** [3.5c] $2\frac{1}{4}$ cups   **38.** [2.7d] 12
39. [5.5a] 42.2025 mi   **40.** [5.5a] 132   **41.** [6.2a] 60 mph
42. [6.2b] The 12-oz bag

CHAPTER 7

Pretest: Chapter 7, p. 308

1. [7.1b] 0.87   **2.** [7.1c] 53.7%   **3.** [7.2a] 75%
4. [7.2b] $\frac{37}{100}$ **5.** [7.3a, b] $x = 60\% \times 75$; 45
6. [7.4a, b] $\frac{n}{100} = \frac{35}{50}$; 70%   **7.** [7.5a] 90 lb   **8.** [7.5b] 20%
9. [7.6a] $14.30; $300.30   **10.** [7.7a] $5152
11. [7.7b] $112.50 discount; $337.50 sale price
12. [7.8a] $99.60   **13.** [7.8a] $20   **14.** [7.8b] $7128.60

Exercise Set 7.1, p. 313

1. $\frac{90}{100}$; $90 \times \frac{1}{100}$; 90×0.01 **3.** $\frac{12.5}{100}$; $12.5 \times \frac{1}{100}$; 12.5×0.01
5. 0.67 **7.** 0.456 **9.** 0.5901 **11.** 0.1 **13.** 0.01 **15.** 2
17. 0.001 **19.** 0.0009 **21.** 0.0018 **23.** 0.2319 **25.** 0.9
27. 0.108 **29.** 0.458 **31.** 47% **33.** 3% **35.** 100%
37. 33.4% **39.** 75% **41.** 40% **43.** 0.6% **45.** 1.7%
47. 27.18% **49.** 2.39% **51.** 2.5% **53.** 68.4% **55.** $33\frac{1}{3}$
57. $9\frac{3}{8}$ **59.** $0.\overline{6}$ **61.** $0.8\overline{3}$ **63.** Multiply by 100.

Sidelights: Applications of Ratio and Percent: The Price–Earnings Ratio and Stock Yields, p. 318

1. 16.1; 4.3% **2.** 11.6; 4.0% **3.** 30.2; 1.3%
4. 19.9; 2.6%

Exercise Set 7.2, p. 319

1. 41% **3.** 5% **5.** 20% **7.** 30% **9.** 50% **11.** 62.5%,
or $62\frac{1}{2}$% **13.** 80% **15.** $66.\overline{6}$%, or $66\frac{2}{3}$% **17.** $16.\overline{6}$%, or
$16\frac{2}{3}$% **19.** 16% **21.** 5% **23.** 34% **25.** 36% **27.** 3%
29. 64% **31.** $\frac{17}{20}$ **33.** $\frac{5}{8}$ **35.** $\frac{1}{3}$ **37.** $\frac{1}{6}$ **39.** $\frac{29}{400}$
41. $\frac{1}{125}$ **43.** $\frac{203}{800}$ **45.** $\frac{176}{225}$ **47.** $\frac{711}{1100}$ **49.** $\frac{3}{2}$ **51.** $\frac{13}{40,000}$
53. $\frac{1}{3}$ **55.** $\frac{11}{20}$ **57.** $\frac{19}{50}$ **59.** $\frac{11}{100}$ **61.** $\frac{6}{25}$ **63.** $\frac{11}{40}$

65.

FRACTIONAL NOTATION	DECIMAL NOTATION	PERCENT NOTATION
$\frac{1}{8}$	0.125	$12\frac{1}{2}$%, or 12.5%
$\frac{1}{6}$	$0.1\overline{6}$	$16\frac{2}{3}$%, or $16.\overline{6}$%
$\frac{1}{5}$	0.2	20%
$\frac{1}{4}$	0.25	25%
$\frac{1}{3}$	$0.\overline{3}$	$33\frac{1}{3}$%, or $33.\overline{3}$%
$\frac{3}{8}$	0.375	$37\frac{1}{2}$%, or 37.5%
$\frac{2}{5}$	0.4	40%
$\frac{1}{2}$	0.5	50%

67.

FRACTIONAL NOTATION	DECIMAL NOTATION	PERCENT NOTATION
$\frac{1}{2}$	0.5	50%
$\frac{1}{3}$	$0.\overline{3}$	$33\frac{1}{3}$%, or $33.\overline{3}$%
$\frac{1}{4}$	0.25	25%
$\frac{1}{6}$	$0.1\overline{6}$	$16\frac{2}{3}$%, or $16.\overline{6}$%
$\frac{1}{8}$	0.125	$12\frac{1}{2}$%, or 12.5%
$\frac{3}{4}$	0.75	75%
$\frac{5}{6}$	$0.8\overline{3}$	$83\frac{1}{3}$%, or $83.\overline{3}$%
$\frac{3}{8}$	0.375	$37\frac{1}{2}$%, or 37.5%

69. 72.5 **71.** 400 **73.** 23.125 **75.** $11.\overline{1}$% **77.** $0.01\overline{5}$

Exercise Set 7.3, p. 327

1. $y = 32\% \times 78$ **3.** $89 = a \times 99$ **5.** $13 = 25\% \times y$
7. 234.6 **9.** 45 **11.** $18 **13.** 1.9 **15.** 78% **17.** 200%
19. 50% **21.** 125% **23.** 40 **25.** $40 **27.** 88
29. 20 **31.** 6.25 **33.** $846.60 **35.** $\frac{9}{100}$ **37.** $\frac{875}{1000}$, or $\frac{7}{8}$
39. 0.89 **41.** 0.3 **43.** $880 (can vary); $843.20

Exercise Set 7.4, p. 333

1. $\frac{37}{100} = \frac{a}{74}$ **3.** $\frac{n}{100} = \frac{4.3}{5.9}$ **5.** $\frac{25}{100} = \frac{14}{b}$ **7.** 68.4 **9.** 462
11. 40 **13.** 2.88 **15.** 25% **17.** 102% **19.** 25%
21. 93.75% **23.** $72 **25.** 90 **27.** 88 **29.** 20 **31.** 25
33. $780.20 **35.** 8 **37.** 8 **39.** 100 **41.** $1134 (can
vary); $1118.64

Exercise Set 7.5, p. 341

1. 27; 133 **3.** 536; 264 **5.** 32.5%; 67.5% **7.** 20.4 mL;
659.6 mL **9.** 25% **11.** 45%; $37\frac{1}{2}$%; $17\frac{1}{2}$% **13.** 8%
15. 20% **17.** $19,530 **19.** $12,600 **21.** 5.7 billion;
5.8 billion; 5.9 billion **23.** $16,400 **25.** 166; 156; 146;
136; 122 **27.** $39.\overline{3}$% **29.** Neither; they are the same.
31. About 5 ft, 6 in.

Exercise Set 7.6, p. 347

1. $20.46; $268.46 **3.** $11.40; $201.35 **5.** 5% **7.** 4%
9. $2000 **11.** $800 **13.** $46.55 **15.** 5.6% **17.** 37
19. 8 **21.** $1.\overline{18}$ **23.** $5214.72

Sidelights: An Application: Water Loss, p. 352

1. 120 lb **2.** 1.2 lb **3.** 9.0 lb **4.** 12 lb **5.** 24 lb

Exercise Set 7.7, p. 353

1. $2700 **3.** 5% **5.** $980 **7.** $6860 **9.** $519.80
11. $30; $270 **13.** $52.50; $297.50 **15.** $125; $112.50
17. 40%; $360 **19.** $387; $30\frac{6}{17}$% **21.** $0.\overline{5}$ **23.** $0.91\overline{6}$
25. $2.\overline{142857}$ **27.** $5924.25; $73,065.75

Exercise Set 7.8, p. 359

1. $26 **3.** $248 **5.** $150.50 **7.** $120.83 **9.** $413.60
11. $484 **13.** $236.75 **15.** $428.49 **17.** $2184.05
19. 4.5 **21.** 8 **23.** $33\frac{1}{3}$ **25.** $12\frac{2}{3}$ **27.** $1434.53
29. $28,130.05

Summary and Review: Chapter 7, p. 363

1. [7.1c] 48.3% **2.** [7.1c] 36% **3.** [7.2a] 37.5%
4. [7.2a] $33.\overline{3}$%, or $33\frac{1}{3}$% **5.** [7.1b] 0.735 **6.** [7.1b] 0.065
7. [7.2b] $\frac{6}{25}$ **8.** [7.2b] $\frac{63}{1000}$ **9.** [7.3a, b] $30.6 = x\% \times 90$;
34% **10.** [7.3a, b] $63 = 84\% \times n$; 75 **11.** [7.3a, b] $y = 38\frac{1}{2}\% \times 168$; 64.68 **12.** [7.4a, b] $\frac{24}{100} = \frac{16.8}{b}$; 70

13. [7.4a, b] $\frac{n}{100} = \frac{22.2}{30}$; 74% **14.** [7.4a, b] $\frac{38\frac{1}{2}}{100} = \frac{a}{168}$; 64.68
15. [7.5a] $182 **16.** [7.5b] 12% **17.** [7.5b] 82,400
18. [7.5b] 20% **19.** [7.5a] 168 **20.** [7.6a] $14.40
21. [7.6a] 5% **22.** [7.7a] 11% **23.** [7.7b] $42; $308
24. [7.7b] $42.70; $262.30 **25.** [7.7a] $29.40
26. [7.8a] $3.60 **27.** [7.8a] $12.60 **28.** [7.8a] $12.10
29. [7.8a] $5.25 **30.** [7.8b] $216.32 **31.** [7.8b] $178.22
32. [6.3b] $18\frac{2}{3}$ **33.** [5.2b] 64 **34.** [5.2b] 7.6123
35. [6.3b] 42 **36.** [5.3a] $3.\overline{6}$ **37.** [5.3a] $1.5\overline{71428}$
38. [3.4b] $3\frac{2}{3}$ **39.** [3:4b] $17\frac{2}{7}$

Test: Chapter 7, p. 365

1. [7.1b] 0.89 **2.** [7.1c] 67.4% **3.** [7.2a] 87.5%
4. [7.2b] $\frac{13}{20}$ **5.** [7.3a, b] $m = 40\% \times 55$; 22
6. [7.4a, b] $\frac{n}{100} = \frac{65}{80}$; 81.25% **7.** [7.5a] 50 lb
8. [7.5b] 20% **9.** [7.6a] $16.20; $340.20 **10.** [7.7a] $630
11. [7.7b] $40; $160 **12.** [7.8a] $8.52 **13.** [7.8a] $4.30
14. [7.8b] $110.25 **15.** [5.2b] 222 **16.** [6.3b] 16
17. [5.3a] $1.41\overline{6}$ **18.** [3.4b] $3\frac{21}{44}$

Cumulative Review: Chapters 1–7, p. 367

1. [4.1b] $\frac{91}{1000}$ **2.** [5.3a] $2.1\overline{6}$ **3.** [7.1b] 0.03

4. [7.2a] 112.5% **5.** [6.1a] $\frac{10}{1}$ **6.** [6.2a] $23\frac{1}{3}$ km/h
7. [3.3b] < **8.** [3.3b] < **9.** [1.4b] 296,200
10. [1.4b] 50,000 **11.** [1.9d] 13 **12.** [5.2c] 1.5
13. [3.5a] $3\frac{1}{30}$ **14.** [4.3a] 49.74 **15.** [1.2b] 515,150
16. [4.3b] 0.02 **17.** [3.5b] $\frac{2}{3}$ **18.** [3.3a] $\frac{2}{63}$ **19.** [2.6a] $\frac{1}{6}$
20. [1.5b] 853,142,400 **21.** [5.1a] 1.38036 **22.** [3.6b] $1\frac{1}{2}$
23. [5.2a] 12.25 **24.** [1.6c], [3.4c] $123\frac{1}{3}$ **25.** [1.7b] 95
26. [4.3c] 8.13 **27.** [2.7c] 9 **28.** [3.3c] $\frac{1}{12}$ **29.** [6.3b] 40
30. [6.3b] $8\frac{8}{21}$ **31.** [4.4a] $6878.84 **32.** [5.5a] $11.50
33. [6.2b] 31 ¢/oz **34.** [6.4a] 608 km **35.** [7.5b] 30%
36. [7.5a] 1,033,710 sq mi **37.** [1.8a] 1887 **38.** [3.6c] 5
39. [3.2c] $1\frac{1}{2}$ mi **40.** [6.4a] 60 mi
41. [7.8a, b] First National Bank
42. [3.6b], [7.5b] 12.5% increase

CHAPTER 8

Pretest: Chapter 8, p. 370

1. [8.1a, b, c] **(a)** 51; **(b)** 51.5; **(c)** 46, 50, 53, 55
2. [8.1a, b, c] **(a)** 3; **(b)** 3; **(c)** 5, 4, 3, 2, 1
3. [8.1a, b, c] **(a)** 12.75; **(b)** 17; **(c)** 4 **4.** [8.1a] 55 mi/hr
5. [8.1a] 76
6. [8.4b]

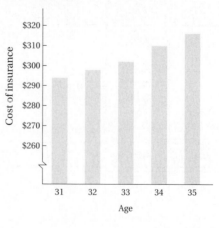

Health 51%
Relieve stress 11%
Lose weight 38%

7. [8.2a] **(a)** $298; **(b)** $172; **(c)** $134
8. [8.3b]

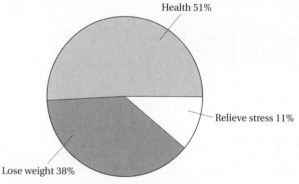

9. [8.3d]

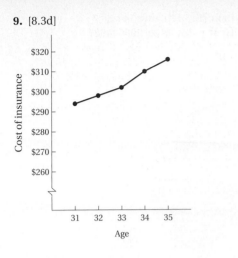

10. [8.3c] 260   **11.** [8.3c] 160

Exercise Set 8.1, p. 375

1. Average: 20; median: 18; mode: 29   **3.** Average: 20; median: 20; mode: 5, 20   **5.** Average: 5.2; median: 5.7; mode: 7.4   **7.** Average: 239.5; median: 234; mode: 234
9. Average: 256.25 lb; median: 257.5 lb; mode: 260 lb
11. Average: 40°; median: 40°; mode: 43°, 40°, 23°, 38°, 54°, 35°, 47°   **13.** 29 mpg   **15.** 2.75   **17.** Average: $9.75; median: $9.79; mode: $9.79   **19.** 90   **21.** $18,460   **23.** $\frac{4}{9}$
25. 1.999396   **27.** 171

Exercise Set 8.2, p. 383

1. 5 in.   **3.** 94   **5.** 21   **7.** 52   **9.** 2740   **11.** 294
13. Calisthenics   **15.** Aerobic dance   **17.** 660
19. Moderate walking   **21.** 121   **23.** 1995   **25.** 1991 and 1992   **27.** 5000   **29.** 1994   **31.** 3   **33.** 16   **35.** 13
37. Out
39.

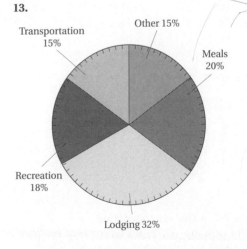

41. 27,859.5 mi²

Exercise Set 8.3, p. 393

1. Los Angeles   **3.** 22   **5.** 2   **7.** Syracuse
9. Approximately $270   **11.** San Francisco

15.

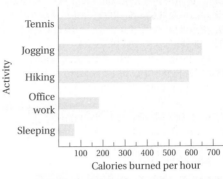

17. Hiking   **19.** 3:00 P.M.   **21.** Approximately 15°
23. Between 5:00 P.M. and 6:00 P.M.   **25.** 1995
27. Approximately $17.0 million   **29.** Approximately $1.5 million
31.

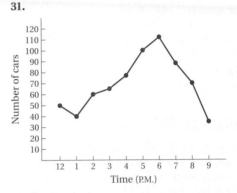

33. Between 8 P.M. and 9 P.M.   **35.** 69.7   **37.** 18%
39. 66.$\overline{6}$%, or 66$\frac{2}{3}$%

Exercise Set 8.4, p. 401

1. 3.7%   **3.** 270   **5.** 6.8%   **7.** Gas purchased   **9.** 4¢
11. 16¢
13.

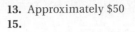

15. 25%   **17.** $\frac{3}{4}$

Summary and Review: Chapter 8, p. 405

1. [8.1a] 38.5 **2.** [8.1a] 13.4 **3.** [8.1a] 1.55
4. [8.1a] 1840 **5.** [8.1a] $16.$\overline{6}$ **6.** [8.1a] 321.$\overline{6}$
7. [8.1b] 38.5 **8.** [8.1b] 14 **9.** [8.1b] 1.8 **10.** [8.1b] 1900
11. [8.1b] $17 **12.** [8.1b] 375 **13.** [8.1c] 26
14. [8.1c] 11; 17 **15.** [8.1c] 0.2 **16.** [8.1c] 700; 800
17. [8.1c] $17 **18.** [8.1c] 20 **19.** [8.1a, b] $110.5; $107
20. [8.1a] 66.1$\overline{6}$° **21.** [8.1a] 96 **22.** [8.2a] 166 lb
23. [8.2a] 124 lb **24.** [8.2a] 5 ft, 7 in., large frame
25. [8.2a] 5 ft, 11 in., large frame **26.** [8.2a] 5 lb
27. [8.2a] 128.$\overline{6}$ lb **28.** [8.2b] 5500 **29.** [8.2b] 1993
30. [8.2b] 1995 **31.** [8.2b] Approximately 4075
32. [8.3a] $6.25 **33.** [8.3a] Hong Kong
34. [8.3a] New York City **35.** [8.3a] $2.90 in Dallas
36. [8.3a] $4.50 **37.** [8.3a] Sydney **38.** [8.3c] Under 20
39. [8.3c] Approximately 12 **40.** [8.3c] Approximately 13
per 100 drivers **41.** [8.3c] Between 45 and 74
42. [8.3c] Approximately 11 per 100 drivers
43. [8.3c] Under 20 **44.** [8.4a] 30% **45.** [8.4a] Outlying
suburbs **46.** [8.4a] 64% **47.** [8.4a] 7.5
48. [6.4a] 12,600 mi **49.** [7.5a] 4.968 billion
50. [7.3b], [7.4b] 222.$\overline{2}$%, or 222$\frac{2}{9}$% **51.** [7.3b], [7.4b] 50%
52. [2.7b] $\frac{9}{10}$ **53.** [2.7b] $\frac{5}{12}$

23. [8.3c] $1.9 billion; answers may vary.
24. [8.3c] $1.6 billion **25.** [8.3c] 1986–1987
26. [8.3c] 1990
27. [8.4b]

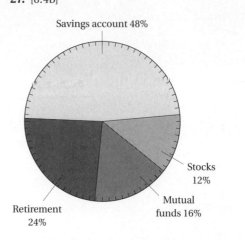

28. [2.7b] $\frac{25}{4}$ **29.** [7.3b], [7.4b] 68 **30.** [7.5a] 15,600
31. [6.4a] 340

Test: Chapter 8, p. 409

1. [8.1a] 50 **2.** [8.1a] 3 **3.** [8.1a] 15.5
4. [8.1b, c] Median: 50.5; mode: 45, 49, 52, 54
5. [8.1b, c] Median: 3; mode: 1, 2, 3, 4, 5
6. [8.1b, c] Median: 17.5; mode: 17, 18 **7.** [8.1a] 58 km/h
8. [8.1a] 76 **9.** [8.2a] Hiking with 20-lb load
10. [8.2a] Hiking with 10-lb load **11.** [8.2a] Walking
3.5 mph **12.** [8.2a] Fitness walking
13. [8.3b]

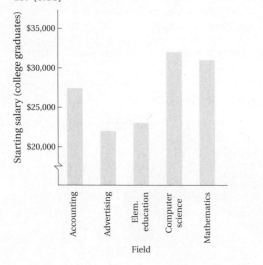

14. [8.2b] 125 **15.** [8.2b] Terry Pendleton
16. [8.2b] Mark Carreon **17.** [8.2b] Terry Pendleton
18. [8.2b] 10 **19.** [8.3c] Revenue is increasing.
20. [8.3c] 1987–1988 and 1990–1991
21. [8.3c] $3.0 billion **22.** [8.3c] $2.4 billion

Cumulative Review: Chapters 1–8, p. 413

1. [1.1e] 5 hundreds **2.** [1.9c] 128 **3.** [2.1a] 1, 2, 3, 4, 5,
6, 10, 12, 15, 20, 30, 60 **4.** [4.2b] 52.0 **5.** [3.4a] $\frac{33}{10}$
6. [5.1b] $2.10 **7.** [5.1b] $3,250,000,000 **8.** [6.3a] No
9. [3.5a] 6$\frac{7}{10}$ **10.** [4.3a] 44.6351 **11.** [3.3a] $\frac{1}{3}$
12. [4.3b] 325.43 **13.** [3.6a] 15 **14.** [1.5b] 2,740,320
15. [2.7b] $\frac{9}{10}$ **16.** [1.6c], [3.4c] 4361$\frac{1}{2}$ **17.** [6.3b] 9$\frac{3}{5}$
18. [2.7c] $\frac{3}{4}$ **19.** [5.2b] 6.8 **20.** [1.7b] 15,312
21. [6.2b] 20.6 ¢/oz **22.** [7.5a] 3324 **23.** [4.4a] 6.2 lb
24. [3.6c] $\frac{1}{4}$ yd **25.** [1.8a] 2572 billion
26. [2.4c] $\frac{3}{8}$ cup **27.** [3.3d] $\frac{1}{4}$ **28.** [6.4a] 1122
29. [5.5a] $9.55 **30.** [7.7a] 7% **31.** [8.4a] 56%
32. [8.4a] 94% **33.** [8.4a] 15
34. [8.3b]

35. [8.1c] 5 **36.** [8.1b] 62
37. [7.5b], [8.1a] 12% decrease

CHAPTER 9

Pretest: Chapter 9, p. 418

1. [9.1a] 96　**2.** [9.1a] $\frac{5}{12}$　**3.** [9.2a] 8460　**4.** [9.2a] 0.92
5. [9.3a] 131 mm　**6.** [9.4a] 100 ft^2　**7.** [9.5a] 22 cm^2
8. [9.5a] $32\frac{1}{2}$ ft^2　**9.** [9.5a] 4 m^2　**10.** [9.6a] 9.6 m
11. [9.6b] 30.144 m　**12.** [9.6c] 72.3456 m^2
13. [9.4b] 92 in^2　**14.** [9.7a] 9　**15.** [9.7b] 9.849
16. [9.7c] $c = 20$　**17.** [9.7c] $b = \sqrt{45}$; $b \approx 6.708$

Sidelights: Applications to Baseball, p. 422

1. (a) 0.367; (b) 185; (c) 1.000　**2.** (a) 0.756; (b) 0.557;
(c) 4.000　**3.** (a) 333; (b) 4; (c) 2129　**4.** (a) 12,364;
(b) 1051; (c) Yes　**5.** About 96,281; about 138,551

Exercise Set 9.1, p. 423

1. 12　**3.** $\frac{1}{12}$　**5.** 5280　**7.** 108　**9.** 7　**11.** $1\frac{1}{2}$
13. 26,400　**15.** 3　**17.** $3\frac{1}{3}$　**19.** 52,800　**21.** $1\frac{1}{2}$　**23.** 1
25. 110　**27.** 2　**29.** 300　**31.** 30　**33.** $\frac{1}{36}$　**35.** 126,720
37. $\frac{37}{400}$　**39.** 137.5%　**41.** 25%　**43.** 0.0041 in.

Exercise Set 9.2, p. 431

1. (a) 1000; (b) 0.001　**3.** (a) 10; (b) 0.1　**5.** (a) 0.01;
(b) 100　**7.** 6700　**9.** 0.98　**11.** 8.921　**13.** 0.05666
15. 566,600　**17.** 4.77　**19.** 688　**21.** 0.1　**23.** 100,000
25. 142　**27.** 0.82　**29.** 450　**31.** 0.000024　**33.** 0.688
35. 230　**37.** 3.92　**39.** 100　**41.** 727.409592
43. 104.585　**45.** 1.75　**47.** 0.234　**49.** 13.85　**51.** 4

Exercise Set 9.3, p. 435

1. 17 mm　**3.** 15.25 in.　**5.** 13 m　**7.** 30 ft
9. 79.14 cm　**11.** 88 ft　**13.** 182 mm　**15.** 826 m;
$1197.70　**17.** 122 cm　**19.** (a) 228 ft; (b) $1046.52
21. 0.561　**23.** 112.5%, or $112\frac{1}{2}$%　**25.** 100

Exercise Set 9.4, p. 439

1. 15 km^2　**3.** 1.4 in^2　**5.** $6\frac{1}{4}$ yd^2　**7.** 8100 ft^2　**9.** 50 ft^2
11. 169.883 cm^2　**13.** $41\frac{2}{9}$ in^2　**15.** 484 ft^2
17. 3237.61 km^2　**19.** $28\frac{57}{64}$ yd^2　**21.** 1197 m^2
23. 630.36 m^2　**25.** (a) 819.75 ft^2; (b) 10 gal; (c) $179.50
27. 80 cm^2　**29.** 45.2%　**31.** 55%

Exercise Set 9.5, p. 445

1. 32 cm^2　**3.** 60 in^2　**5.** 104 ft^2　**7.** 64 m^2
9. 45.5 in^2　**11.** 8.05 cm^2　**13.** 297 cm^2　**15.** 7 m^2
17. $55\frac{1}{8}$ ft^2　**19.** 675 cm^2　**21.** 8944 in^2　**23.** 852.04 ft^2
25. 24 cm; 25 cm^2

Exercise Set 9.6, p. 451

1. 14 cm　**3.** $1\frac{1}{2}$ in.　**5.** 16 ft　**7.** 0.7 cm　**9.** 44 cm
11. $4\frac{5}{7}$ in.　**13.** 100.48 ft　**15.** 4.396 cm　**17.** 154 cm^2

19. $1\frac{43}{56}$ in^2　**21.** 803.84 ft^2　**23.** 1.5386 cm^2　**25.** 3 cm;
18.84 cm; 28.26 cm^2　**27.** 151,976 mi^2　**29.** 3.454 ft
31. 2.5 cm; 1.25 cm; 4.90625 cm^2　**33.** 65.94 yd^2
35. 45.68 ft　**37.** 26.84 yd　**39.** 45.7 yd　**41.** 100.48 m^2
43. 6.9972 cm^2　**45.** 48.8886 in^2　**47.** 87.5%
49. 66.$\overline{6}$%　**51.** 37.5%　**53.** 66.$\overline{6}$%　**55.** 3.142　**57.** $3d$;
πd; circumference of one ball, since $\pi > 3$

Exercise Set 9.7, p. 459

1. 10　**3.** 21　**5.** 25　**7.** 19　**9.** 23　**11.** 100
13. 6.928　**15.** 2.828　**17.** 4.243　**19.** 2.449　**21.** 3.162
23. 8.660　**25.** 14　**27.** 13.528　**29.** $c = \sqrt{34}$; $c \approx 5.831$
31. $c = \sqrt{98}$; $c \approx 9.899$　**33.** $a = 5$　**35.** $b = 8$
37. $c = 13$　**39.** $b = 24$　**41.** $a = \sqrt{399}$; $a \approx 19.975$
43. $b = \sqrt{224}$; $b \approx 14.967$　**45.** $\sqrt{250}$ m ≈ 15.811 m
47. $\sqrt{8450}$ ft ≈ 91.924 ft　**49.** $h = \sqrt{500}$ ft; $h \approx 22.361$ ft
51. 14,532.7 ft　**53.** 0.456　**55.** 1.23　**57.** 0.0041
59. The areas are the same.

Summary and Review: Chapter 9, p. 465

1. [9.1a] $2\frac{2}{3}$　**2.** [9.1a] 30　**3.** [9.2a] 0.03　**4.** [9.2a] 0.004
5. [9.1a] 72　**6.** [9.2a] 400,000　**7.** [9.1a] $1\frac{1}{6}$
8. [9.2a] 0.15　**9.** [9.2b] 220　**10.** [9.2b] 32.18
11. [9.3a] 23 m　**12.** [9.3a] 4.4 m
13. [9.3b], [9.4b] 228 ft; 2808 ft^2　**14.** [9.7c] 85.9 ft
15. [9.3a], [9,4a] 36 ft; 81 ft^2　**16.** [9.3a], [9.4a] 17.6 cm;
12.6 cm^2　**17.** [9.5a] 60 cm^2　**18.** [9.5a] 35 mm^2
19. [9.5a] 22.5 m^2　**20.** [9.5a] 27.5 cm^2　**21.** [9.5a] 88 m^2
22. [9.5a] 126 in^2　**23.** [9.4b] 840 ft^2　**24.** [9.6a] 8 m
25. [9.6a] $\frac{14}{11}$ in., or $1\frac{3}{11}$ in.　**26.** [9.6a] 14 ft
27. [9.6a] 20 cm　**28.** [9.6b] 50.24 m　**29.** [9.6b] 8 in.
30. [9.6c] 200.96 m^2　**31.** [9.6c] $5\frac{1}{11}$ in^2
32. [9.6d] 1038.555 ft^2　**33.** [9.7a] 8　**34.** [9.7b] 9.110
35. [9.7c] $c = \sqrt{850}$; $c \approx 29.155$　**36.** [9.7c] $b = \sqrt{51}$;
$b \approx 7.141$　**37.** [9.7c] $c = \sqrt{89}$ ft; $c \approx 9.434$ ft
38. [9.7c] $a = \sqrt{76}$ cm; $a \approx 8.718$ cm　**39.** [9.7c] 28.8 ft
40. [9.7c] 44.7 ft　**41.** [7.1c] 47%　**42.** [7.2a] 92%
43. [7.1b] 0.567　**44.** [7.2b] $\frac{73}{100}$

Test: Chapter 9, p. 469

1. [9.1a] 48　**2.** [9.1a] $\frac{1}{3}$　**3.** [9.2a] 6000　**4.** [9.2a] 0.87
5. [9.2b] 181.$\overline{81}$　**6.** [9.2b] 1490.4
7. [9.3a], [9.4a] 32.82 cm; 65.894 cm^2
8. [9.3a], [9.4a] 100 m; 625 m^2　**9.** [9.5a] 25 cm^2
10. [9.5a] 12 m^2　**11.** [9.5a] 18 ft^2　**12.** [9.6a] $\frac{1}{4}$ in.
13. [9.6a] 9 cm　**14.** [9.6b] $\frac{11}{14}$ in.　**15.** [9.6c] 254.34 cm^2
16. [9.6d] 103.815 km^2　**17.** [9.7a] 15　**18.** [9.7b] 9.327
19. [9.7c] $c = 40$　**20.** [9.7c] $b = \sqrt{60}$; $b \approx 7.746$
21. [9.7c] $c = \sqrt{2}$; $c \approx 1.414$　**22.** [9.7c] $b = \sqrt{51}$;
$b \approx 7.141$　**23.** [9.7c] 15.8 m　**24.** [7.1c] 93%
25. [7.2a] 81.25%　**26.** [7.1b] 0.932　**27.** [7.2b] $\frac{1}{3}$

Cumulative Review: Chapters 1–9, p. 471

1. [1.5b] 50,854,100　**2.** [2.4b] $\frac{7}{12}$　**3.** [5.3c] 15.2
4. [3.6b] $\frac{2}{3}$　**5.** [5.2a] 35.6　**6.** [3.5b] $\frac{5}{6}$　**7.** [2.2a] Yes

8. [2.2a] No **9.** [2.1d] $3 \cdot 3 \cdot 11$ **10.** [3.1a] 245
11. [5.3b] 35.8 **12.** [4.1a] One hundred three and
sixty-four thousandths **13.** [8.1a, b] $17.8\overline{3}$, 17.5
14. [7.1c] 8% **15.** [7.2a] 60% **16.** [9.7a] 11
17. [9.7b] 5.39 **18.** [9.1a] 6 **19.** [9.3a] 14.3 m
20. [9.5a] 297.5 yd^2 **21.** [5.2b] 150.5 **22.** [1.7b] 19,248
23. [2.7c] $\frac{15}{2}$ **24.** [3.3c] $\frac{2}{35}$ **25.** [8.2a] 5.5 hr
26. [8.2a] 13.8 hr **27.** [8.2a] 2.5 hr **28.** [8.2a] 9 hr
29. [7.7b] $11\frac{9}{11}$% **30.** [1.8a] 165,000,000,000 lb
31. [9.4b], [9.6d] 248.64 in^2 **32.** [6.4a] $8\frac{3}{4}$ lb
33. [7.5a] Yes **34.** [3.2c] $1\frac{1}{4}$ hr **35.** [4.4a] 255.8 mi
36. [2.4c] $6600 **37.** [7.5b], [9.4b] $41\frac{2}{3}$% increase

CHAPTER 10

Pretest: Chapter 10, p. 474

1. [10.1b] 2.304 **2.** [10.1b] 2400 **3.** [10.3a] 80
4. [10.3a] 8800 **5.** [10.3b] 4800 **6.** [10.3b] 0.62
7. [10.3b] 3.4 **8.** [10.3c] 420 **9.** [10.3c] 384
10. [10.1b] 64 **11.** [10.1b] 2560 **12.** [10.1b] 24
13. [10.4b] 25°C **14.** [10.4b] 98.6°F **15.** [10.5a] 144
16. [10.5b] 2,000,000 **17.** [10.3b] 1000 g
18. [10.1a] 160 cm^3 **19.** [10.2a] 1256 ft^3
20. [10.2b] $33,493.\overline{3}$ yd^3 **21.** [10.2c] 150.72 cm^3

Exercise Set 10.1, p. 479

1. 768 cm^3 **3.** 45 in^3 **5.** 75 m^3 **7.** $357\frac{1}{2}$ yd^3
9. 1000; 1000 **11.** 87,000 **13.** 0.049 **15.** 0.000401
17. 78,100 **19.** 320 **21.** 10 **23.** 32 **25.** 500 mL
27. 125 mL **29.** 1.75 gal/week; 7.5 gal/month;
91.25 gal/year; 65,250,000 gal/day; 23,816,250,000 gal/year
31. $19.20 **33.** 1000 **35.** 49 **37.** 57,480 in^3; 33.3 ft^3

Exercise Set 10.2, p. 483

1. 803.84 in^3 **3.** 353.25 cm^3 **5.** 41,580,000 yd^3
7. $4,186,666\frac{2}{3}$ in^3 **9.** 124.72 m^3 **11.** $1437\frac{1}{3}$ km^3
13. 113,982 ft^3 **15.** 24.64 cm^3 **17.** 33,880 yd^3
19. 367.38 m^3 **21.** 143.72 cm^3 **23.** 32,993,440,000 mi^3
25. $6.9\overline{7}$ ft^3; $52.\overline{3}$ gal

Exercise Set 10.3, p. 489

1. 2000 **3.** 3 **5.** 64 **7.** 12,640 **9.** 0.1 **11.** 5
13. 1000 **15.** 10 **17.** $\frac{1}{100}$, or 0.01 **19.** 1000 **21.** 10
23. 234,000 **25.** 5.2 **27.** 6.7 **29.** 0.0502 **31.** 8.492
33. 58.5 **35.** 800,000 **37.** 1000 **39.** 0.0034 **41.** 24
43. 60 **45.** $365\frac{1}{4}$ **47.** 0.05 **49.** 8.2 **51.** 6.5
53. 10.75 **55.** 336 **57.** 4.5 **59.** 56 **61.** 16 **63.** 125
65. 543 **67.** 0.4535 **69.** 2 oz **71.** About 31.7 yr
73. 1000 **75.** 0.125 mg **77.** 4 **79.** 8 mL

Exercise Set 10.4, p. 495

1. 80°C **3.** 60°C **5.** 20°C **7.** −10°C **9.** 190°F
11. 140°F **13.** 10°F **15.** 40°F **17.** 77°F **19.** 104°F

21. 5432°F **23.** 30°C **25.** 55°C **27.** 37°C **29.** 234
31. 336 **33.** 7240 **35.** 260.6°F

Exercise Set 10.5, p. 499

1. 144 **3.** 640 **5.** $\frac{1}{144}$ **7.** 198 **9.** 396 **11.** 12,800
13. 27,878,400 **15.** 5 **17.** 1 **19.** $\frac{1}{640}$, or 0.0015625
21. 5,210,000 **23.** 140 **25.** 23.456 **27.** 0.085214
29. 2500 **31.** 0.4728 **33.** $210 **35.** 4.2 **37.** 10.89
39. 1.65 **41.** 1852.6 m^2

Summary and Review: Chapter 10, p. 503

1. [10.1a] 93.6 m^3 **2.** [10.1a] 193.2 cm^3 **3.** [10.3a] 112
4. [10.3b] 0.004 **5.** [10.3c] $\frac{4}{15}$ **6.** [10.1b] 0.464
7. [10.3c] 180 **8.** [10.3b] 4700 **9.** [10.3a] 16,140
10. [10.1b] 830 **11.** [10.3c] $\frac{1}{4}$ **12.** [10.3b] 0.04
13. [10.3b] 200 **14.** [10.3b] 30 **15.** [10.1b] 0.06
16. [10.3a] 1600 **17.** [10.1b] 400 **18.** [10.3a] $1\frac{1}{4}$
19. [10.3c] 50 **20.** [10.1b] 160 **21.** [10.1b] 7.5
22. [10.1b] 13.5 **23.** [10.4b] 80.6°F **24.** [10.4b] 20°C
25. [10.2a] 31,400 ft^3 **26.** [10.2b] $33.49\overline{3}$ cm^3
27. [10.2c] 4.71 in^3 **28.** [10.5a] 36 **29.** [10.5b] 300,000
30. [10.5a] 14.375 **31.** [10.5b] 0.06 **32.** [10.1c] 18.72 L
33. [7.8a] $39.58 **34.** [1.9b] 27 **35.** [1.9b] 22.09
36. [1.9b] 103.823 **37.** [1.9b] $\frac{1}{16}$ **38.** [9.1a] 13,200
39. [9.1a] 4 **40.** [9.2a] 45.68 **41.** [9.2a] 45,680

Test: Chapter 10, p. 505

1. [10.1a] 84 cm^3 **2.** [10.1b] 3.08 **3.** [10.1b] 240
4. [10.3a] 64 **5.** [10.3a] 8220 **6.** [10.3b] 3800
7. [10.3b] 0.4325 **8.** [10.3b] 2.2 **9.** [10.3c] 300
10. [10.3c] 360 **11.** [10.1b] 32 **12.** [10.1b] 1280
13. [10.1b] 40 **14.** [10.4b] 35°C **15.** [10.4b] 138.2°F
16. [10.5a] 1728 **17.** [10.5b] 0.0003 **18.** [10.1a] 420 in^3
19. [10.2a] 1177.5 ft^3 **20.** [10.2b] $4186.\overline{6}$ yd^3
21. [10.2c] 113.04 cm^3 **22.** [7.8a] $680 **23.** [1.9b] 1000
24. [1.9b] $\frac{1}{16}$ **25.** [1.9b] 9.8596 **26.** [1.9b] 0.00001
27. [9.1a] 42 **28.** [9.1a] 250 **29.** [9.2a] 2300
30. [9.2a] 3400

Cumulative Review: Chapters 1–10, p. 507

1. [3.5a] $4\frac{1}{6}$ **2.** [5.3c] 49.2 **3.** [4.3b] 87.52
4. [1.6c] 1234 **5.** [1.9d] 2 **6.** [1.9c] 1565 **7.** [4.1b] $\frac{1209}{1000}$
8. [7.2b] $\frac{17}{100}$ **9.** [3.3b] < **10.** [2.5c] = **11.** [10.3a] $\frac{3}{8}$
12. [10.4b] 59 **13.** [10.1b] 87 **14.** [10.3c] $\frac{3}{4}$
15. [10.5a] 27 **16.** [9.2a] 0.17 **17.** [9.3a], [9.5a] 380 cm;
5500 cm^2 **18.** [9.3a], [9.5a] 32.3 ft; 56.55 ft^2
19. [8.3c] 22 **20.** [8.3c] February **21.** [6.3b] $14\frac{2}{5}$
22. [6.3b] $4\frac{2}{7}$ **23.** [5.2b] 113.4 **24.** [3.3c] $\frac{1}{8}$
25. [8.1a] 99 **26.** [1.8a] 528 million
27. [10.2b] 4187 cm^3 **28.** [7.8a] $24 **29.** [9.7c] 17 m
30. [7.6a] 6% **31.** [3.5c] $2\frac{1}{8}$ yd **32.** [5.5a] $21.82
33. [6.2b] The 8-qt box **34.** [2.6b] $\frac{7}{20}$ km **35.** [9.1a],
[10.1a] 278 ft^3

CHAPTER 11

Pretest: Chapter 11, p. 510

1. [11.1d] >   **2.** [11.1d] >   **3.** [11.1d] >   **4.** [11.1d] <
5. [11.1c] -0.625 **6.** [11.1c] $-0.\overline{6}$ **7.** [11.1c] $-0.\overline{90}$
8. [11.1e] 12   **9.** [11.1e] 2.3   **10.** [11.1e] 0
11. [11.2b] -5.4 **12.** [11.2b] $\frac{2}{3}$ **13.** [11.2a] -17
14. [11.3a] 38.6   **15.** [11.3a] $-\frac{17}{15}$ **16.** [11.2a] -5
17. [11.4a] 63   **18.** [11.4a] $-\frac{5}{12}$ **19.** [11.5c] -98
20. [11.5a] 8   **21.** [11.3a] 24   **22.** [11.5d] 26

Exercise Set 11.1, p. 517

1. -200; 600   **3.** -34; 15   **5.** 750; -125
7. 20; -150; 300
9.

11.

13. -0.625 **15.** $-1.\overline{6}$ **17.** $-1.1\overline{6}$ **19.** -0.875
21. -0.35 **23.** >   **25.** >   **27.** >   **29.** <   **31.** >
33. >   **35.** <   **37.** >   **39.** >   **41.** <   **43.** 7   **45.** 0
47. 4   **49.** 325   **51.** $\frac{10}{7}$ **53.** 14.8
55. $2 \cdot 2 \cdot 2 \cdot 2 \cdot 2 \cdot 3$ **57.** $2 \cdot 2 \cdot 5 \cdot 13$ **59.** 72
61. 72   **63.** <   **65.** $-\frac{5}{6}, -\frac{3}{4}, -\frac{2}{3}, \frac{1}{6}, \frac{3}{8}, \frac{1}{2}$

Exercise Set 11.2, p. 523

1. -7 **3.** -4 **5.** 0   **7.** -8 **9.** -7 **11.** -27 **13.** 0
15. -42 **17.** 0   **19.** 0   **21.** 3   **23.** -9 **25.** 7   **27.** 0
29. 45   **31.** -1.8 **33.** -8.1 **35.** $-\frac{1}{5}$ **37.** $-\frac{8}{7}$
39. $-\frac{3}{8}$ **41.** $-\frac{29}{35}$ **43.** $-\frac{11}{15}$ **45.** -6.3 **47.** $\frac{7}{16}$ **49.** 39
51. 50   **53.** -1093 **55.** -24 **57.** 26.9   **59.** -9
61. $\frac{14}{3}$ **63.** -65 **65.** $\frac{5}{3}$ **67.** 14   **69.** -10 **71.** All
positive   **73.** Positive   **75.** Negative

Exercise Set 11.3, p. 527

1. -4 **3.** -7 **5.** -6 **7.** 0   **9.** -4 **11.** -7 **13.** -6
15. 0   **17.** 11   **19.** -14 **21.** 5   **23.** -7 **25.** -5
27. -3 **29.** -23 **31.** -68 **33.** -73 **35.** 116
37. -2.8 **39.** $-\frac{1}{4}$ **41.** $\frac{1}{12}$ **43.** $-\frac{17}{12}$ **45.** $\frac{1}{8}$ **47.** 19.9
49. -9 **51.** -0.01 **53.** -2.7 **55.** -3.53 **57.** $-\frac{1}{2}$
59. $\frac{6}{7}$ **61.** $-\frac{41}{30}$ **63.** $-\frac{1}{156}$ **65.** 37   **67.** -62 **69.** 6
71. 107   **73.** 219   **75.** 96.6 cm^2 **77.** 108   **79.** 1767 m

Exercise Set 11.4, p. 531

1. -16 **3.** -24 **5.** -72 **7.** 16   **9.** 42   **11.** -120
13. -238 **15.** 1200   **17.** 98   **19.** -12.4 **21.** 24
23. 21.7   **25.** $-\frac{2}{5}$ **27.** $\frac{1}{12}$ **29.** -17.01 **31.** $-\frac{5}{12}$
33. 420   **35.** $\frac{2}{7}$ **37.** -60 **39.** 150   **41.** $-\frac{2}{45}$
43. 1911   **45.** 50.4   **47.** $\frac{10}{189}$ **49.** -960 **51.** 17.64
53. $-\frac{5}{784}$ **55.** -756 **57.** -720 **59.** $-30,240$

61. $2 \cdot 2 \cdot 2 \cdot 2 \cdot 2 \cdot 2 \cdot 2 \cdot 2 \cdot 2 \cdot 3 \cdot 3$ **63.** $33\frac{1}{3}$%, or
$33.\overline{3}$%   **65.** 72   **67.** 72   **69.** 1944   **71.** -17
73. **(a)** One must be negative and one must be positive.
(b) Either or both must be zero. **(c)** Both must be negative
or both must be positive.

Exercise Set 11.5, p. 537

1. -6 **3.** -13 **5.** -2 **7.** 4   **9.** -8 **11.** 2   **13.** -12
15. -8 **17.** Undefined   **19.** -9 **21.** $-\frac{7}{15}$ **23.** $\frac{1}{13}$
25. $-\frac{9}{8}$ **27.** $\frac{5}{3}$ **29.** $\frac{9}{14}$ **31.** $\frac{9}{64}$ **33.** -2 **35.** $\frac{11}{13}$
37. -16.2 **39.** Undefined   **41.** -7 **43.** -7
45. -334 **47.** 14   **49.** 1880   **51.** 12   **53.** 8
55. -86 **57.** 37   **59.** -1 **61.** -10 **63.** 25
65. -7988 **67.** -3000 **69.** 60   **71.** 1   **73.** 10
75. $-\frac{13}{45}$ **77.** $-\frac{4}{3}$

Summary and Review: Chapter 11, p. 541

1. [11.1e] 38   **2.** [11.1e] 7.3   **3.** [11.1e] $\frac{5}{2}$ **4.** [11.1e] -0.2
5. [11.1c] -1.25 **6.** [11.1c] $-0.8\overline{3}$ **7.** [11.1c] $-0.41\overline{6}$
8. [11.1c] $-0.\overline{27}$
9. [11.1b]

10. [11.1b]

11. [11.1d] <   **12.** [11.1d] >   **13.** [11.1d] >
14. [11.1d] <   **15.** [11.2b] -3.8 **16.** [11.2b] $\frac{3}{4}$
17. [11.2b] 34   **18.** [11.2b] 5   **19.** [11.5b] $\frac{8}{3}$
20. [11.5b] $-\frac{1}{7}$ **21.** [11.5b] -10 **22.** [11.2a] -3
23. [11.2a] $-\frac{7}{12}$ **24.** [11.2a] -4 **25.** [11.2a] -5
26. [11.3a] 4   **27.** [11.3a] $-\frac{7}{5}$ **28.** [11.3a] -7.9
29. [11.4a] 54   **30.** [11.4a] -9.18 **31.** [11.4a] $-\frac{2}{7}$
32. [11.4a] -210 **33.** [11.5a] -7 **34.** [11.5c] -3
35. [11.5c] $\frac{3}{4}$ **36.** [11.5d] 40.4   **37.** [11.5d] -62
38. [11.5d] -5 **39.** [9.4a] 210 cm^2 **40.** [3.1a] 270
41. [2.1d] $2 \cdot 2 \cdot 2 \cdot 3 \cdot 3 \cdot 3 \cdot 3$ **42.** [7.3b] 36%
43. [11.1e], [11.3a] $-\frac{5}{8}$ **44.** [11.1e], [11.5d] -2.1

Test: Chapter 11, p. 543

1. [11.1d] <   **2.** [11.1d] >   **3.** [11.1d] >   **4.** [11.1d] <
5. [11.1c] -0.125 **6.** [11.1c] $-0.\overline{4}$ **7.** [11.1c] $-0.\overline{18}$
8. [11.1e] 7   **9.** [11.1e] $\frac{9}{4}$ **10.** [11.1e] -2.7
11. [11.2b] $-\frac{2}{3}$ **12.** [11.2b] 1.4   **13.** [11.2b] 8
14. [11.5b] $-\frac{1}{2}$ **15.** [11.5b] $\frac{7}{4}$ **16.** [11.3a] 7.8
17. [11.2a] -8 **18.** [11.2a] $\frac{7}{40}$ **19.** [11.3a] 10
20. [11.3a] -2.5 **21.** [11.3a] $\frac{7}{8}$ **22.** [11.4a] -48
23. [11.4a] $\frac{3}{16}$ **24.** [11.5a] -9 **25.** [11.5c] $\frac{3}{4}$
26. [11.5c] -9.728 **27.** [11.5d] 109   **28.** [9.4a] 55.8 ft^2
29. [7.3b] 48%   **30.** [2.1d] $2 \cdot 2 \cdot 2 \cdot 5 \cdot 7$ **31.** [3.1a] 240
32. [11.1e], [11.3a] 15

Cumulative Review: Chapters 1–11, p. 545

1. [7.1b] 0.263 **2.** [11.1c] $-0.\overline{45}$ **3.** [10.3b] 834
4. [10.5b] 0.0275 **5.** [11.1e] 4.5 **6.** [11.3a] -11
7. [6.2a] 12.5 m/sec **8.** [9.6a, b, c] 35 mi; 220 mi;
3850 mi^2 **9.** [9.7a] 15 **10.** [9.7b] 8.31 **11.** [11.4a] -10
12. [11.5a] 3 **13.** [11.2a] 8
14. [8.2c]

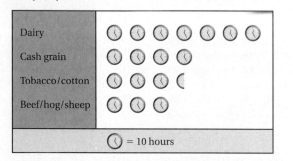

Dairy	$\bigcirc$ $\bigcirc$ $\bigcirc$ $\bigcirc$ $\bigcirc$ $\bigcirc$ $\bigcirc$
Cash grain	$\bigcirc$ $\bigcirc$ $\bigcirc$ $\bigcirc$
Tobacco/cotton	$\bigcirc$ $\bigcirc$ $\bigcirc$ $\bigcirc$
Beef/hog/sheep	$\bigcirc$ $\bigcirc$ $\bigcirc$

$\bigcirc$ = 10 hours

15. [5.1a] 0.01485 **16.** [11.3a] -3 **17.** [11.3a] $-\frac{3}{22}$
18. [11.4a] $-\frac{1}{24}$ **19.** [3.5b] $1\frac{5}{6}$ **20.** [11.5c] $-\frac{1}{4}$
21. [11.5d] -57.5 **22.** [1.5b] 9,640,500,000
23. [11.2a] 32.49 **24.** [11.5d] Undefined
25. [7.3b] 87.5% **26.** [7.3b] 32 **27.** [4.4a] $133.05
28. [10.2a] 307.72 cm^3 **29.** [8.1a] $44.8\overline{3}°$ **30.** [7.5a] 78
31. [9.4b] 6902.5 m^2 **32.** [3.5c] $2\frac{11}{12}$ cups **33.** [1.8a],
[8.1a] $239,441; $59,860.25 **34.** [5.5a] 4.55 km

CHAPTER 12

Pretest: Chapter 12, p. 548

1. [12.1a] $\frac{5}{16}$ **2.** [12.5a] 78%x, or 0.78x **3.** [12.1b] $9z - 18$
4. [12.1b] $-4a - 2b + 10c$ **5.** [12.1c] $4(x - 3)$
6. [12.1c] $3(2y - 3z - 6)$ **7.** [12.1d] $-3x$
8. [12.1d] $2x + 2y + 18$ **9.** [12.3a] -7 **10.** [12.4b] -1
11. [12.4a] 2 **12.** [12.4c] 3 **13.** [12.5b] Width: 34 m;
length: 39 m **14.** [12.5b] $780

Exercise Set 12.1, p. 555

1. 42 **3.** 3 **5.** 1 **7.** 6 **9.** 240; 240 **11.** 160; 160
13. $2b + 10$ **15.** $7 - 7t$ **17.** $30x + 12$
19. $7x + 28 + 42y$ **21.** $-7y + 14$ **23.** $45x + 54y - 72$
25. $-4x + 12y + 8z$ **27.** $-3.72x + 9.92y - 3.41$
29. $2(x + 2)$ **31.** $5(6 + y)$ **33.** $7(2x + 3y)$
35. $5(x + 2 + 3y)$ **37.** $8(x - 3)$ **39.** $4(8 - y)$
41. $2(4x + 5y - 11)$ **43.** $6(-3x - 2y + 1)$,
or $-6(3x + 2y - 1)$ **45.** $19a$ **47.** $9a$ **49.** $8x + 9z$
51. $-19a + 88$ **53.** $4t + 6y - 4$ **55.** $8x$ **57.** $5n$
59. $-16y$ **61.** $17a - 12b - 1$ **63.** $4x + 2y$
65. $7x + y$ **67.** $0.8x + 0.5y$

Exercise Set 12.2, p. 559

1. 7 **3.** -20 **5.** -14 **7.** -18 **9.** 15 **11.** -14
13. 2 **15.** 20 **17.** -6 **19.** $\frac{7}{3}$ **21.** $-\frac{7}{4}$ **23.** $\frac{41}{24}$
25. $-\frac{1}{20}$ **27.** 5.1 **29.** 12.4 **31.** -5 **33.** $1\frac{5}{6}$ **35.** $-\frac{10}{21}$
37. -11 **39.** $-\frac{5}{12}$ **41.** 13.12 **43.** 342.246 **45.** $-\frac{26}{15}$
47. -10 **49.** All real numbers **51.** $-\frac{5}{17}$

Exercise Set 12.3, p. 563

1. 6 **3.** 9 **5.** 12 **7.** -40 **9.** 5 **11.** -7 **13.** -6
15. 6 **17.** -63 **19.** 36 **21.** -21 **23.** $-\frac{3}{5}$ **25.** $-\frac{3}{2}$
27. $\frac{9}{2}$ **29.** 7 **31.** -7 **33.** 8 **35.** 15.9 **37.** 62.8 ft;
20 ft; 314 ft^2 **39.** 8000 ft^3 **41.** -8655 **43.** No solution
45. No solution

Exercise Set 12.4, p. 571

1. 5 **3.** 8 **5.** 10 **7.** 14 **9.** -8 **11.** -8 **13.** -7
15. 15 **17.** 6 **19.** 4 **21.** 6 **23.** -3 **25.** 1
27. -20 **29.** 6 **31.** 7 **33.** 2 **35.** 5 **37.** 2 **39.** 10
41. 4 **43.** 0 **45.** -1 **47.** $-\frac{4}{3}$ **49.** $\frac{2}{5}$ **51.** -2
53. -4 **55.** $\frac{4}{5}$ **57.** $-\frac{28}{27}$ **59.** 6 **61.** 2 **63.** 6 **65.** 8
67. 1 **69.** 17 **71.** $-\frac{5}{3}$ **73.** -3 **75.** 2 **77.** $\frac{4}{7}$
79. $-\frac{51}{31}$ **81.** 2 **83.** -6.5 **85.** $<$ **87.** 2 **89.** -2
91. 8

Exercise Set 12.5, p. 581

1. $b + 6$, or $6 + b$ **3.** $c - 9$ **5.** $6 + q$, or $q + 6$
7. $b + a$, or $a + b$ **9.** $y - x$ **11.** $x + w$, or $w + x$
13. $r + s$, or $s + r$ **15.** $2x$ **17.** $5t$ **19.** 97%x, or
$0.97x$ **21.** 32 **23.** 667.5 hr **25.** $2.39 **27.** 57
29. -10 **31.** 20 m, 40 m, 120 m **33.** Width: 100 ft;
length: 160 ft; area: 16,000 ft^2 **35.** Length: 27.9 cm;
width: 21.6 cm **37.** 450.5 mi **39.** 1863 =
$1776 + (4s + 7)$ **41.** Length: 12 cm; width: 9 cm
43. 5 half dollars, 10 quarters, 20 dimes, 60 nickels

Summary and Review: Chapter 12, p. 585

1. [12.2a] -22 **2.** [12.3a] 7 **3.** [12.3a] -192
4. [12.2a] 1 **5.** [12.3a] $-\frac{7}{3}$ **6.** [12.2a] 25 **7.** [12.2a] $\frac{1}{2}$
8. [12.3a] $-\frac{15}{64}$ **9.** [12.2a] 9.99 **10.** [12.4a] -8
11. [12.4b] -5 **12.** [12.4b] $-\frac{1}{3}$ **13.** [12.4a] 4
14. [12.4b] 3 **15.** [12.4b] 4 **16.** [12.4b] 16 **17.** [12.4c] 6
18. [12.4c] -3 **19.** [12.4c] 12 **20.** [12.4c] 4
21. [12.5a] 19%x, or 0.19x **22.** [12.5b] $591
23. [12.5b] 27 **24.** [12.5b] 3 m, 5 m **25.** [12.5b] 9
26. [12.5b] Width: 11 cm; length: 17 cm
27. [12.5b] 1800.4 mi **28.** [9.6a, b, c] 40 ft; 125.6 ft;
1256 ft^2 **29.** [10.1a] 1702 cm^3 **30.** [11.2a] -45
31. [11.4a] -60 **32.** [12.5b] Amazon: 6437 km;
Nile: 6671 km **33.** [11.1e], [12.4a] 23, -23 **34.** [11.1e],
[12.3a] 20, -20

Test: Chapter 12, p. 587

1. [12.2a] 8 **2.** [12.2a] 26 **3.** [12.3a] -6 **4.** [12.3a] 49
5. [12.4b] -12 **6.** [12.4a] 2 **7.** [12.4a] -8
8. [12.2a] $-\frac{7}{20}$ **9.** [12.4b] 2.5 **10.** [12.4c] 7
11. [12.4c] $\frac{5}{3}$ **12.** [12.5b] Width: 7 cm; length: 11 cm
13. [12.5b] 6 **14.** [12.5a] $x - 9$ **15.** [11.4a] 180
16. [11.2a] $-\frac{2}{9}$ **17.** [9.6a, b, c] 140 yd; 439.6 yd; 15,386 yd^2
18. [10.1a] 1320 ft^3 **19.** [11.1e], [12.4a] 15, -15
20. [12.5b] 60

Cumulative Review: Chapters 1–12, p. 589

1. [1.1e] 7   2. [1.1a] 7 thousands + 4 hundreds + 5 ones
3. [1.2b] 1012   4. [1.2b] 21,085   5. [3.2b] $\frac{5}{26}$
6. [3.5a] $5\frac{7}{9}$ 7. [4.3a] 493.971   8. [4.3a] 802.876
9. [1.3d] 152   10. [1.3d] 674   11. [3.3a] $\frac{5}{24}$
12. [3.5b] $2\frac{17}{24}$ 13. [4.3b] 19.9973   14. [4.3b] 34.241
15. [1.5b] 4752   16. [1.5b] 266,287   17. [3.6a] $4\frac{1}{12}$
18. [2.6a] $\frac{6}{5}$ 19. [2.6a] 10   20. [5.1a] 259.084
21. [1.6c] 573   22. [1.6c] 56 R 10   23. [3.4c] $56\frac{5}{17}$
24. [2.7b] $\frac{3}{2}$ 25. [3.6b] $\frac{7}{90}$ 26. [5.2a] 39
27. [1.4a] 68,000   28. [4.2b] 0.428   29. [5.3b] 21.84
30. [2.2a] Yes   31. [2.1a] 1, 3, 5, 15   32. [3.1a] 800
33. [2.5b] $\frac{7}{10}$ 34. [2.5b] 55   35. [3.4b] $3\frac{3}{5}$ 36. [2.5c] $\neq$
37. [3.3b] <   38. [4.2a] 1.001   39. [1.4c] >   40. [2.3b] $\frac{3}{5}$
41. [4.1c] 0.037   42. [5.3a] 0.52   43. [5.3a] $0.\overline{8}$
44. [7.1b] 0.07   45. [4.1b] $\frac{463}{100}$ 46. [3.4a] $\frac{29}{4}$ 47. [7.2b] $\frac{2}{5}$
48. [7.2a] 85%   49. [7.1c] 150%   50. [1.7b] 555
51. [5.2b] 64   52. [2.7c] $\frac{5}{4}$ 53. [6.3b] $76\frac{1}{2}$, or 76.5
54. [1.8a] $675   55. [1.8a] 65 min   56. [3.5c] $25\frac{3}{4}$
57. [4.4a] 485.9 mi   58. [1.8a] $8100   59. [1.8a] $423
60. [6.4a] $\frac{3}{10}$ km   61. [5.5a] $84.96   62. [6.4a] 13 gal
63. [6.2b] 17 cents/oz   64. [7.8a] $150   65. [7.7a] 7%
66. [7.5b] 30, 160   67. [8.1a, b, c] 28; 26; 18
68. [1.9b] 324   69. [1.9b] 400   70. [9.7a] 3   71. [9.7a] 11
72. [9.7b] 4.472   73. [9.7c] $c = \sqrt{50}$ ft; $c \approx 7.071$ ft
74. [9.1a] 12   75. [9.2a] 428   76. [10.3c] 72
77. [10.3b] 20   78. [10.3a] 80   79. [10.3b] 0.08
80. [10.1b] 8.19   81. [10.1b] 5   82. [9.3a], [9.4a] 25.6 m; 25.75 m²   83. [9.5a] 25 in²   84. [9.5a] 128.65 yd²
85. [9.5a] 61.6 cm²   86. [9.6a, b, c] 20.8 in.; 65.312 in.; 339.6224 in²   87. [10.1a] 52.9 m³   88. [10.2a] 803.84 ft³
89. [10.2c] $267.94\overline{6}$ cm³   90. [10.2b] $267.94\overline{6}$ mi³
91. [1.9c] 238   92. [1.9c] 172   93. [11.1e], [11.4a] 3
94. [11.2a] 14   95. [11.3a] $\frac{1}{3}$ 96. [11.4a] 30
97. [11.4a] $-\frac{2}{7}$ 98. [11.5a] −8   99. [12.4a] −5
100. [12.3a] 4   101. [12.4b] −8   102. [12.4c] $\frac{2}{3}$
103. [12.5a] $y + 17$ 104. [12.5a] 38%x, or 0.38x
105. [12.5b] 39   106. [12.5b] 36

FINAL EXAMINATION, P. 593

1. [1.1a] 8 thousands + 3 hundreds + 4 tens + 5 ones
2. [1.1e] 7   3. [4.3a] 215.6177   4. [3.5a] $8\frac{3}{4}$ 5. [3.2b] $\frac{5}{3}$
6. [1.2b] 759   7. [1.2b] 21,562   8. [4.3a] 158.8837
9. [1.3d] 5561   10. [1.3d] 5937   11. [4.3b] 30.392
12. [3.5b] $\frac{5}{8}$ 13. [3.3a] $\frac{1}{12}$ 14. [4.3b] 99.16
15. [3.6a] $24\frac{3}{8}$ 16. [2.6a] $\frac{2}{3}$ 17. [2.6a] 6   18. [1.5b] 5814
19. [1.5b] 234,906   20. [5.1a] 226.327   21. [1.6c] 517 R 1
22. [3.4c] $517\frac{1}{8}$ 23. [1.6c] 197   24. [2.7b] $\frac{2}{3}$
25. [3.6b] $1\frac{19}{66}$ 26. [5.2a] 48   27. [1.4a] 43,000
28. [4.2b] 6.79   29. [5.3b] 7.384   30. [2.2a] Yes
31. [2.1a] 1, 2, 4, 8   32. [3.1a] 230   33. [2.5b] $\frac{3}{2}$
34. [2.5b] 10   35. [3.4b] $7\frac{2}{3}$ 36. [2.5c] =   37. [3.3b] <
38. [4.2a] 0.9   39. [1.4c] <   40. [2.3b] $\frac{2}{6}$, or $\frac{1}{3}$
41. [7.1b] 0.499   42. [5.3a] 0.24   43. [5.3a] $0.\overline{27}$

44. [4.1c] 7.86   45. [3.4a] $\frac{23}{4}$ 46. [7.2b] $\frac{37}{100}$
47. [4.1b] $\frac{897}{1000}$ 48. [7.1c] 77%   49. [7.2a] 96%
50. [6.3b] 3.84, or $3\frac{21}{25}$ 51. [3.3c] $\frac{1}{25}$ 52. [1.7b] 25
53. [4.3c] 245.7   54. [7.5b] 5%   55. [4.4a] $312.60
56. [2.7d] 80   57. [1.8a] $348   58. [3.5c] $3\frac{3}{4}$ m
59. [1.8a] $1311   60. [1.8a] $39   61. [1.8a] 13 mpg
62. [5.5a] $24.98   63. [6.4a] 540 km   64. [6.2b] $3.98
65. [7.5a] 39   66. [7.8a] $60   67. [8.1a, b, c] $15.17; $12; $12   68. [8.4a] (a) Blue; (b) 1050   69. [1.9b] 625
70. [1.9b] 256   71. [9.7a] 7   72. [9.7a] 25
73. [9.7b] 4.899   74. [9.7c] $a = \sqrt{85}$ ft; $a \approx 9.220$ ft
75. [9.1a] 5   76. [9.2a] 2.371   77. [10.1b] 5000
78. [10.3a] 14,000   79. [10.3c] 1440   80. [10.3b] 5340
81. [10.3b] 7.54   82. [10.1b] 5   83. [9.3a], [9.4a] 24.8 m; 26.88 m²   84. [9.5a] 38.775 ft²   85. [9.5a] 153 m²
86. [9.5a] 216 cm²   87. [9.6a, b, c] 4.3 yd; 27.004 yd; 58.0586 yd²   88. [10.1a] 68.921 ft³
89. [10.2a] 314,000 mi³   90. [10.2b] $4186.\overline{6}$ m³
91. [10.2c] $104,666.\overline{6}$ in³   92. [1.9c] 383   93. [1.9c] 99
94. [11.1e] 32   95. [11.3a] −22   96. [11.2a] −22
97. [11.4a] 30   98. [11.4a] $-\frac{5}{9}$ 99. [11.5a] −6
100. [12.2a] −76.4   101. [12.3a] 27   102. [12.4b] −15
103. [12.5b] 532.5   104. [12.5b] 40

DEVELOPMENTAL UNITS

Exercise Set A, p. 603

1. 11   2. 17   3. 15   4. 13   5. 15   6. 14   7. 12
8. 12   9. 13   10. 11   11. 16   12. 17   13. 16   14. 12
15. 10   16. 10   17. 11   18. 7   19. 7   20. 11   21. 8
22. 0   23. 3   24. 18   25. 14   26. 9   27. 10
28. 14   29. 4   30. 14   31. 11   32. 2   33. 15
34. 16   35. 16   36. 9   37. 11   38. 13   39. 14
40. 5   41. 11   42. 7   43. 13   44. 14   45. 12
46. 6   47. 10   48. 12   49. 10   50. 8   51. 2   52. 9
53. 13   54. 8   55. 10   56. 9   57. 10   58. 6
59. 13   60. 9   61. 8   62. 11   63. 16   64. 7   65. 12
66. 12   67. 8   68. 4   69. 7   70. 10   71. 15   72. 13
73. 4   74. 6   75. 8   76. 8   77. 15   78. 12   79. 13
80. 10   81. 12   82. 17   83. 13   84. 10   85. 16
86. 16   87. 19   88. 39   89. 95   90. 89   91. 87
92. 79   93. 849   94. 999   95. 900   96. 868
97. 875   98. 999   99. 848   100. 877   101. 1689
102. 17,680   103. 10,873   104. 4699   105. 9747
106. 10,867   107. 9895   108. 3998   109. 9998
110. 18,222   111. 7777   112. 16,889   113. 18,999
114. 64,489   115. 99,999   116. 77,777   117. 46
118. 26   119. 55   120. 101   121. 1643   122. 1010
123. 846   124. 628   125. 1204   126. 607
127. 10,000   128. 1010   129. 1110   130. 1227
131. 1111   132. 1717   133. 10,138   134. 6554
135. 6111   136. 8427   137. 9890   138. 11,612
139. 11,125   140. 15,543   141. 16,774   142. 68,675
143. 34,437   144. 166,444   145. 101,315   146. 49,449
147. 49, 61, 73, 85, 97   148. 1 6 15 20 15 6 1
1 7 21 35 35 21 7 1

Exercise Set S, p. 613

1. 4 2. 1 3. 3 4. 2 5. 0 6. 0 7. 5 8. 3
9. 1 10. 1 11. 7 12. 5 13. 7 14. 9 15. 0
16. 0 17. 5 18. 3 19. 8 20. 8 21. 6 22. 5
23. 7 24. 3 25. 7 26. 9 27. 6 28. 6 29. 2
30. 2 31. 0 32. 0 33. 4 34. 4 35. 5 36. 4
37. 4 38. 5 39. 9 40. 9 41. 7 42. 9 43. 5
44. 6 45. 6 46. 2 47. 6 48. 7 49. 1 50. 8
51. 11 52. 13 53. 33 54. 21 55. 47 56. 67
57. 247 58. 193 59. 500 60. 766 61. 654
62. 333 63. 202 64. 617 65. 288 66. 1220
67. 2231 68. 2261 69. 5003 70. 4126 71. 3764
72. 12,522 73. 1691 74. 12,100 75. 1226
76. 99,998 77. 10 78. 10,101 79. 11,043 80. 11,111
81. 18 82. 39 83. 65 84. 29 85. 8 86. 9
87. 308 88. 317 89. 126 90. 617 91. 214 92. 89
93. 4402 94. 3555 95. 1503 96. 5889 97. 2387
98. 3649 99. 3832 100. 8144 101. 7750
102. 10,445 103. 33,793 104. 26,869 105. 16
106. 13 107. 16 108. 17 109. 86 110. 281
111. 455 112. 389 113. 571 114. 6148 115. 2200
116. 2113 117. 3748 118. 4313 119. 1063 120. 2418
121. 5206 122. 1459 123. 305 124. 4455 125. Fill
the 3-gal bucket and pour it into the 5-gal bucket. Fill the
3-gal bucket again, and pour it into the 5-gal bucket until
full. One gallon remains.

Exercise Set M, p. 623

1. 12 2. 3 3. 24 4. 0 5. 7 6. 0 7. 24 8. 4
9. 10 10. 30 11. 10 12. 10 13. 10 14. 63
15. 21 16. 54 17. 12 18. 0 19. 72 20. 18
21. 72 22. 8 23. 0 24. 0 25. 28 26. 24 27. 7
28. 45 29. 0 30. 18 31. 42 32. 7 33. 0
34. 35 35. 9 36. 45 37. 40 38. 0 39. 40
40. 16 41. 16 42. 25 43. 64 44. 81 45. 1
46. 0 47. 9 48. 4 49. 36 50. 8 51. 48 52. 0
53. 27 54. 18 55. 45 56. 0 57. 42 58. 10
59. 0 60. 48 61. 54 62. 0 63. 25 64. 72
65. 49 66. 15 67. 27 68. 0 69. 8 70. 9 71. 2
72. 5 73. 32 74. 6 75. 15 76. 6 77. 8 78. 8
79. 20 80. 20 81. 64 82. 16 83. 10 84. 0
85. 80 86. 60 87. 90 88. 90 89. 70 90. 160
91. 210 92. 450 93. 780 94. 450 95. 560
96. 360 97. 800 98. 300 99. 900 100. 1000
101. 6700 102. 32,100 103. 345,700 104. 1200
105. 4900 106. 4000 107. 4200 108. 10,000
109. 7000 110. 9000 111. 2000 112. 457,000
113. 7,888,000 114. 6,769,000 115. 18,000
116. 20,000 117. 48,000 118. 16,000 119. 6000

120. 1,000,000 121. 1200 122. 4900 123. 200
124. 4000 125. 2500 126. 12,000 127. 49,000
128. 6000 129. 63,000 130. 120,000 131. 150,000
132. 800,000 133. 120,000 134. 24,000,000
135. 16,000,000 136. 80,000 137. 147 138. 444
139. 2965 140. 4872 141. 6293 142. 3460
143. 16,236 144. 29,376 145. 30,924 146. 13,508
147. 87,554 148. 195,384 149. 3480 150. 2790
151. 3360 152. 7020 153. 20,760 154. 36,150
155. 6840 156. 10,600 157. 358,800 158. 109,800
159. 583,800 160. 299,700 161. 11,346,000
162. 23,390,000 163. 61,092,000 164. 73,032,000

Exercise Set D, p. 635

1. 3 2. 8 3. 4 4. 6 5. 1 6. 32 7. 9 8. 5
9. 7 10. 5 11. 37 12. 29 13. 5 14. 9 15. 4
16. 5 17. 6 18. 9 19. 5 20. 9 21. 8 22. 9
23. 6 24. 7 25. 3 26. 2 27. 9 28. 6 29. 2
30. 3 31. 7 32. 2 33. 8 34. 4 35. 7 36. 3
37. 2 38. 3 39. 6 40. 6 41. 1 42. 4 43. 6
44. 3 45. 7 46. 9 47. 1 48. 0 49. Not defined
50. Not defined 51. 7 52. 9 53. 8 54. 7 55. 1
56. 5 57. 0 58. 1 59. 0 60. 3 61. 1 62. 4
63. 7 64. 1 65. 6 66. 0 67. 1 68. 5 69. 2
70. 8 71. 0 72. 4 73. 0 74. 8 75. 3 76. 8
77. Not defined 78. Not defined 79. 4 80. 1
81. 69 R 1 82. 199 R 1 83. 92 R 1 84. 138 R 3
85. 1723 R 4 86. 2925 87. 864 R 1 88. 522 R 3
89. 527 R 2 90. 2897 91. 43 R 15 92. 32 R 27
93. 47 R 12 94. 42 R 9 95. 91 R 22 96. 52 R 13
97. 29 98. 55 R 2 99. 228 R 22 100. 20 R 379
101. 21 102. 118 103. 91 R 1 104. 321 R 2
105. 964 R 3 106. 679 R 5 107. 1328 R 2 108. 27 R 8
109. 23 110. 321 R 18 111. 224 R 10 112. 27 R 60
113. 87 R 1 114. 75 115. 22 R 85 116. 49 R 46
117. 383 R 91 118. 74 R 440

APPENDIX: USING A SCIENTIFIC CALCULATOR, P. C-1

1. 6561 2. 3375 3. 529 4. 1024 5. 2401
6. 10.4976 7. 0.00243 8. 81 9. 118 10. 39 11. 71
12. 450 13. 100 14. 3 15. $\frac{13}{24}$ 16. $\frac{31}{60}$ 17. $13\frac{1}{12}$, or
$\frac{157}{12}$ 18. $\frac{1}{110}$ 19. 2 20. 2.775 21. 1.625 22. 0.3125
23. 0.01 24. 13 25. 250 26. 0.042 27. 3.317
28. 16.432 29. 2.530 30. 0.095 31. 8 32. 16
33. 36 34. 153 35. -11 36. 9 37. 114 38. Yes
39. No 40. Yes 41. No 42. Yes

Index

Common denominators, 109, 135, 143, 150
Common multiple, least, 135, 568
Commutative law
  of addition, 12, 599
  of multiplication, 35
Composite numbers, 83
Compound interest, 357, 358
Cone, circular, 482
Consecutive numbers, 125, 539
Constant, 549
Converting
  American measures to metric measures, 430
  area units, 497, 498
  capacity units, 476, 478
  cents to dollars, 229
  decimal notation to fractional notation, 193
  decimal notation to percent notation, 311
  decimal notation to a word name, 192
  dollars to cents, 229
  expanded notation to standard notation, 4
  fractional notation to decimal notation, 194, 243, 513
  fractional notation to percent notation, 315
  length units, 420, 427
  mass units, 487
  mentally, in metric system, 429, 487, 498
  metric to American measures, 430
  percent notation to decimal notation, 310
  percent notation to fractional notation, 317
  standard notation to expanded notation, 3
  standard notation to word names, 4
  temperature units, 494
  time units, 488
  weight units, 485
  word names to standard notation, 5
Cross products, 110, 283
Cube, unit, 475
Cubic centimeter, 477
Cup, 476
Cylinder, circular, 481

D
Day, 488

Decigram, 486
Decimal equivalent, 317
Decimal notation, 191
  addition with, 201
  converting
    from/to fractional notation, 193, 194, 243
    from/to percent notation, 310, 311
  division with, 233
  and fractional notation together, 246
  for irrational numbers, 514
  and money, 192, 229
  multiplication with, 225, 228
  and order, 197
  order of operations with, 237
  place-value, 191
  for rational numbers, 513, 514
  repeating, 244, 514
    rounding, 245
  rounding, 197
  subtraction with, 202
  terminating, 244, 513
  word names for, 191
Decimals, clearing, 568, 569
Decimeter, 425
Decrease, percent of, 337
Dekagram, 486
Dekameter, 425
Denominators, 95
  common, 109
  least common, 135, 143, 568
  of zero, 98
Depreciation, 343
Diameter, 447
Difference, 19, 20
  estimating, 28, 249
Digits, 5
Discount, 351
Distance, and absolute value, 516
Distributive laws, 33, 552
  and factoring, 553
  and multiplying, 552
Dividend, 41
Divisibility, 82
  by 2, 89
  by 3, 90
  by 4, 92
  by 5, 91
  by 6, 90
  by 7, 92
  by 8, 92
  by 9, 91
  by 10, 91
Division
  basic, 627–634

checking, 42
  with decimal notation, 233
  definition, 533
  dividend in, 41
  divisor in, 41
  by estimating multiples, 632
  estimating quotients, 250, 251
  with fractional notation, 119
  by guess, multiply, and subtract, 631
  and LCMs, 139
  with mixed numerals, 170
  of a number by itself, 630
  by one, 629
  partial quotient, 631
  by a power of 10, 236
  quotient in, 41
  of real numbers, 533, 535
  and reciprocals, 120, 535
  and rectangular arrays, 39, 627
  related multiplication sentence, 40, 628
  with remainders, 42, 252, 631
  as repeated subtraction, 41
  of whole numbers, 39–45
  by zero, 533, 629, 630
  of zero by a nonzero number, 533, 630
  zeros in quotients, 45
Divisor, 41
Dollars, converting from/to cents, 229

E
Earned run average, 290
Eight, divisibility by, 92
English system of measures, *see* American system of measures
Equality of fractions, test for, 109, 110
Equation, 28, 49
  equivalent, 557
  Pythagorean, 456
  solution of, 49
  solving, 49–52, 121, 151, 203, 237, 557, 561, 565–570. *See also* Solving equations.
Equivalent equations, 557
Equivalent expressions, 550
Equivalent numbers, 107
Estimating. *See also* Rounding.
  differences, 28, 249
  in division process, 43, 632
  with fractions, 182
  products, 36, 250
  quotients, 250, 251

Substituting for a variable, 549
Subtraction
 and additive inverses, 525
 basic, 607
 and borrowing, 22, 609–611
 checking, 22, 611
 with decimal notation, 202
 definition, 525
 difference, 19, 20
  estimating, 28
 using fractional notation, 149
 as "how much more?", 20
 minuend, 19
 using mixed numerals, 162
 no borrowing, 609
 and opposites, 525
 of real numbers, 525
 related addition sentence, 20, 607
 subtrahend, 19
 as "take away," 19
 of whole numbers, 19–22,
  607–612
 zeros in, 22, 611
Subtrahend, 19
Subunits, 419
Sum, 9, 598
 estimating, 27, 249
 and factors, 86
Symbol
 approximately equal to ($\approx$), 28
 greater than ($>$), 28
 less than ($<$), 28
 for multiplication ($\times$ and $\cdot$), 31
 radical ($\sqrt{}$), 455
Systems of measure, *see* American
  system of measures; Metric
  system of measures

T

Table of data, 377
Table of primes, 84
"Take away" situation, 19
Tax, sales, 345
Temperature, 493, 494, 496

Ten, divisibility by, 91
Ten-thousandths, 191
Tenths, 191
Terminating decimal, 244, 513
Terms, 552
 like, 554
Thousands period, 4
Thousandths, 191
Three, divisibility by, 90
Time, 488
Tip, estimating, 362
Ton
 American, 485
 metric, 486
Total price, 345
Translating
 to algebraic expressions, 575
 to equations, *see* Solving
  problems
Trapezoid, 443
Treadmill test, 343
Tree, factor, 85
Triangle
 area, 442
 right, 456
 similar, 295
Triangular number, 125
Trillions period, 4
Truncating, 26, 200
Twin primes, 125
Two, divisibility by, 89

U

Unit cube, 475
Unit price, 278
Unit segment, 419

V

Value of an expression, 549
Variable, 49, 549
 substituting for, 549
Volume
 of a circular cone, 482
 of a circular cylinder, 481

of a rectangular solid, 475
of a sphere, 482

W

Water loss, 352
Week, 488
Weight, 485. *See also* Mass.
Whole numbers, 3
 addition, 9–13, 590–602
 division, 39–45, 627–634
 expanded notation, 3
 fractional notation for, 98
 multiplication, 31–36, 617–622
  by fractions, 101
 order, 28
 rounding, 25
 standard notation, 3
 subtraction, 19–22, 607–612
 word names for, 4
Wildlife population, estimating,
  291
Word names
 for decimal notation, 191
 for whole numbers, 4

Y

Yard, 419
Year, 488
Yield, 318

Z

Zero
 addition of, 598
 denominator of, 98
 divided by a nonzero number,
  533, 630
 division by, 533, 629, 630
 fractional notation for, 98
 identity property, 520
 in multiplication, 35, 617
 multiplication by, 617
 in quotients, 45
 reciprocal, 119, 534
 in subtraction, 22, 611